International Conference on Advanced Materials for Better Future 2016

IOP Conference Series: Materials Science and Engineering
Volume 176

Surakarta, Indonesia
3 – 4 October 2016

ISBN: 978-1-5108-3744-7
ISSN: 1757-8981

Printed from e-media with permission by:

Curran Associates, Inc.
57 Morehouse Lane
Red Hook, NY 12571

CURRAN ASSOCIATES INC. proceedings.com

Some format issues inherent in the e-media version may also appear in this print version.

Printed by Curran Associates, Inc. (2017)

Institute of Physics
Dirac House, Temple Back
Bristol BS1 6BE UK

Phone: 44 1 17 929 7481
Fax: 44 1 17 920 0979

techtracking@iop.org

Additional copies of this publication are available from:

Curran Associates, Inc.
57 Morehouse Lane
Red Hook, NY 12571 USA
Phone: 845-758-0400
Fax: 845-758-2633
Email: curran@proceedings.com
Web: www.proceedings.com

Table of contents

Volume 176

International Conference on Advanced Materials for Better Future 2016

3–4 October 2016, Surakarta, Indonesia

Accepted papers received: 3 February 2017Published online: 8 March 2017

Preface

Drug and Biomaterial Technology (DB)

Electronic Materials Sensing (EM)

Fuctional Materials (FM)

Metallurgy (MT)

Nanotechnology Inorganic Composite (NI)

Polymer Composite (PC)

Preface

The first International Conference on Advanced Material for Better Future (ICAMBF 2016) was held October 3-4th, 2016 at Solo Paragon Hotel, Surakarta, Central Java, Indonesia. This conference is organized by Faculty of Mathematics & Natural Science, Sebelas Maret University.

ICAMBF 2016 aims to present research in a wide variety of scientific areas with a common interest in Advanced Material. The conferences gathered over 90 participants that get a platform to exchange idea and build new colleague over two days. It included plenary and parallel session. The scope of advanced material including Electronic Materials/Sensing, Nanotechnology/Inorganic Composite, Polymer Composite, Mechanical Engineering & Information Technology, Mettalurgy, Functional Materials, Drug & Biomaterial Technology.

We are pleased to our outstanding keynote speakers Prof. Santiago Gomez, Prof. Hanafi Ismail, Prof. Roger Narayan, Prof. Neng Sri Suharty, and other invited speakers for sharing the best knowledge in Advanced Material.

We also thanks to all researchers and student who participate in conference by submitting their papers to convey the current science discussed at ICAMBF 2016, all of the organizing committee wish you get awesome conference experience, and reviewers for their valuable revision of these papers.

Prof. Ari Handono Ramelan

Chairman of ICAMBF 2016
Dean of Faculty Mathematics & Natural Sciences

IOP Conf. Series: Materials Science and Engineering **176** (2017) 011001 doi:10.1088/1757-899X/176/1/011001

CONFERENCE COMMITTEE/ORGANIZER

Chairman	Prof. Ir. Ari Handono Ramelan, M.Sc., (Hons), Ph.D.	Indonesia
Co-Chairman	Dr. Sayekti Wahyuningsih, MSi	Indonesia
Secretary	Ahmad Marzuki, SSi, PhD	Indonesia
Scientific Committee	Prof. Ir. Ari Handono Ramelan, M.Sc., (Hons), Ph.D.	Indonesia
	Prof. Dra Neng Sri Suharty MS, Ph.D	Indonesia
	Prof. Sulistyo Saputro, M.Si., Ph.D.	Indonesia
	Prof. Roger Narayan	USA
	Prof. Dr. Santiago Gómez-Ruiz	Spain
	Prof. Dr. Hanafi Ismail	Malaysia
	Ahmad Marzuki, SSi, PhD	Indonesia
	Nuryani, MSi, PhD	Indonesia
	Venty Suryanti. M.Phil. PhD	Indonesia
	Dr. Eng Budi Purnama, M.Si	Indonesia
	Dr. Sayekti Wahyuningsih, MSi	Indonesia

Peer review statement

All papers published in this volume of *IOP Conference Series: Materials Science and Engineering* have been peer reviewed through processes administered by the proceedings Editors. Reviews were conducted by expert referees to the professional and scientific standards expected of a proceedings journal published by IOP Publishing.

IOP Conf. Series: Materials Science and Engineering **176** (2017) 012001 doi:10.1088/1757-899X/176/1/012001

Preparation and Mechanical Properties of Chitosan-graft Maleic Anhydride Reinforced with Montmorillonite

A Fajrin[1], L A Sari[1], N Rahmawati[1], O A Saputra[1] and V Suryanti[1,*]
[1]Department of Chemistry, Faculty of Mathematics and Natural Sciences, Sebelas Maret University, Jl. Ir. Sutami 36A Surakarta 57126 Indonesia

*E-mail: venty@mipa.uns.ac.id

Abstract. The research aims to develop biodegradable composites as bio-based plastics from chitosan. The composites were prepared *via* solution casting method by introducing the maleic anhydride (MAH) as grafting agent and montmorillonite (MMt) as reinforcement. The grafting process of chitosan was conducted by varying concentrations of MAH which were 10, 20, and 30% w/w. It was observed that the chitosan-graft-maleic anhydride (Cs-g-MAH) containing 10% w/w of MAH increased its tensile strength by 70%. Reinforcement material was added to the Cs-g-MAH by varying MMt concentrations, e.g. 3, 6, 9 and 12% w/w. It was noted that the presence of 9% w/w of MMt in the Cs-g-MAH gave the best mechanical properties of the Cs-g-MAH/MMt composite.

1. Introduction

The use of synthetic polymer materials which are usually derivate products of petroleum have been increased enormously. Petroleum-derived polymer are not biodegradable, worsening the problem of the waste disposal. Therefore, the use of biodegradable materials bio-based polymers from renewable resources is a growing demand for limited availability of petroleum resources and environmentally-friendly [1-4].

Chitosan is a natural polymer which is a de-acetylated product of the chitin. It is non-toxic, biocompatible and biodegradable with antimicrobial activity [5]. Chitosan has been applied for bioplastic materials which have notable mechanical properties. However, their water sensitivity, mechanical strength and barrier properties are not as good as synthetic polymer plastics. Therefore, these disadvantages need to be solved so that these bioplastics can compete with the synthetic polymers [6]. Graft copolymerization is one of the technique which can be used for chemical modification of natural and synthetic polymers [5]. It has been a growing intertest in grafting of chitosan which allows the formation of functional derivatives onto the chitosan backbone by covalent binding of a molecule with amino and hydroxyl groups of chitosan. The carboxylation of chitosan can be achieved easily by the reaction between its amine groups and anhydride groups from various anhydride derivatives such as succinic anhydride, phtalic anhydride and maleic anhydride (MAH) [7-13]. In this work, MAH was used to increase the mechanical properties of chitosan polymer matrix.

The inclusion of a small amount of clay particles such as bentonite, halloysite, kaolin, or montmorillonite (MMt) into a polymeric matrix has received considerable research attention due to the improvement in mechanical and thermal properties [14-15]. When clay is blended with a polymer, clay particles are dispersed within the polymer matrix and these interactions increase the mechanical properties [16]. MMt is commonly used as cationic clay of the smectite group for preparation of

IOP Conf. Series: Materials Science and Engineering **176** (2017) 012001 doi:10.1088/1757-899X/176/1/012001

polymer nanocomposites. In this work, MMt was used as reinforcement to enhance the mechanical properties of the biocomposites. The aim of this research was to fabricate novel bioplastic composites based on chitosan to improve their properties by MAH grafting and adding of MMt.

2. Experimental

2.1. Materials

Chitosan (Cs) was purchased from PT Biotech Surindo, Indonesia. All chemicals were used are analytical grade from e-Merck, otherwise is stated. Montmorillonite K10 (Aldrich) was prepared by calcination at 400°C for 12 h before used.

2.2. Synthesis of chitosan-g-poly(maleic anhydride) (Cs-g-MAH)

Chitosan (2 g) was dissolved in 70 mL of acetic acid at room temperature at constant stirring condition. Maleic anhydride (MAH) as much as 10, 20 and 30% w/w, 5% w/w benzoyl peroxide (BPO) and 10 mL acetic acid were dispersed using sonicator for 30 mins. This mixtures were then added into chitosan solution (Table 1). Co-polymerization was conducted at 80°C for 1 h. The mixtures were poured into stainless steel templates and dried for 12 h at 50°C. The composites were then airing for acetic acid removal for 12 h.

Table 1. Chitosan-g-poly(maleic anhydride) (Cs-g-MAH) composite compositions.

Composites	Composition (g)		
	Chitosan	MAH	BPO
Cs	2	0	0
Cs-g-MAH-10	2	0.2	0.01
Cs-g-MAH-20	2	0.4	0.02
Cs-g-MAH-30	2	0.6	0.03

2.3. Synthesis of chitosan/montmorillonite g-poly(maleic anhydride) (Cs-g-MAH/MMt)

The fabrication of chitosan/montmorillonite co-poly(maleic anhydrides) (Cs-g-MAH/MMt) composites were conducted using the same procedures for the synthesis of Cs-g-MAH. The MMt as much as 3, 6, 9, and 12% w/w was added after 1 h of blending the chitosan, MAH and BPO (Table 2). The mixtures were blended for 1 h and then poured into stainless steel templates.

Table 2. *Chitosan/montmorillonite g-poly(maleic anhydride)* (Cs/MAH/MMt) composite compositions.

Composites	Composition (g)			
	Chitosan	MAH	BPO	MMt
Cs-g-MAH/MMt -3	1.940	0.194	0.009	0.060
Cs-g-MAH/MMt -6	1.880	0.188	0.009	0.120
Cs-g-MAH/MMt -9	1.820	0.182	0.009	0.180
Cs-g-MAH/MMt -12	1.760	0.176	0.009	0.240

2.4. The characterization of composites

The composites were characterized for their functional groups by FT-IR spectrophotometer. The mechanical testing of Cs-g-MAH and Cs-g-MAH/MMt were performed according to ASTM D-638. The specimen tests were prepared with 6 mm of section width, 33 mm of section length and under 4 mm of thickness. Mechanical tests were conducted using Mechanical Testing Machine Ray-Run M500 50CT, with 5 mm/min of applying speed.

The water absorption tests were performed for analysing the swelling phenomena of Cs-g-MAH and Cs-g-MAH/MMt. The initial weights of samples were recorded as *Wo*. The sample with 1 x 1 cm^2 (Ao) of dimension were placed into chamber containing distilled water for 1 h. The final sample weight were recorded as *Wt*. The samples were also measured for swelling area by calculating the area before (*Ao*) and after 1 h of soaking in water (*At*). Water absorption (*WA*) and swelling area (*SA*) of samples were then calculated using equation (1) and (2).

$$WA = \frac{(Wt-Wo)}{Wo} \times 100\% \text{---} \quad (1)$$

$$SA = \frac{(At-Ao)}{Ao} \times 100\% \text{---} \quad (2)$$

3. Result and discussion

3.1. Composites of chitosan-g-poly(maleic anhydride) (Cs-g-MAH)

The chitosan-g-poly(maleic anhydride (Cs-g-MAH) composites were prepared and their mechanical properties such as tensile strength, strain and young's modules were investigated. Increasing of the tensile strength (TS) after copolymerization chitosan matrixs with MAH were noted (Table 3). The highest TS was achieved when the amount of 10% w/w of MAH was applied. In this composite, compatibilized matrixs with MAH had a significant effect on both tensile strength and strain at break. TS increased from 21.11 to 35.90 Mpa and strain decreased from 13.46 to 9.18%. An increased of young's modulus was also observed from 156.78 to 892.28 MPa. It indicated that the addition of MAH reduced the brittle of chitosan plastics. This is consistent with several studies involving polymer matrixs compatibilized MAH system [17, 18].

Table 3. The mechanical properties of Cs-g-MAH composites.

Composites	Tensile stength (MPa)	Strain (%)	Young's modulus (MPa)
Cs	21.11	13.47	156.78
Cs-g-MAH-10	35.90	9.18	494.61
Cs-g-MAH-20	30.97	3.42	892.28
Cs-g-MAH-30	33.94	4.29	781.95

Water absorption is the key to indicate the bioplastics moisture sensitivity which is an important aspect for packaging applications. The swelling behaviors at 100% relative humidity were evaluated for 1 h. Figure 1 shows the swelling properties of Cs-g-MAH composites. In respect to the chitosan, the swelling degrees decreased upon the addition of MAH. The addition of 10% w/w of MAH insignificantly decreased the swelling degree by 11.76%, nevertheless the addition of 20 and 30% w/w of MAH significantly decreased the swelling degree by 45.38 and 44.45%, respectively. In was noted that the addition of 10% of MAH did not affect the swelling area, whereas the addition of 20 and 30% w/w of MAH decreased the swelling area significantly by 77.27 and 52.27%, respectively. It was reported that compatibilization of natural polymers decreased the tendency of the composites to absorb water [19, 20]. Increasing cross-linking decreased water absorption due to the reduction of hydroxyl group available to absorb water [19, 21]. In this work, the Cs-g-MAH-10 composite was observed to have best water resistance, where the swelling degree and swelling area remained almost unaffected by water.

Based on the mechanical properties and the swelling behaviour data, the Cs-g-MAH-10 composite has been chosen for further studies. Montmorillonite as reinforcement was then added to the composite giving Cs-g-MAH/MMt composites in order to improve their properties.

3.2. Composites of chitosan/montmorillonite g-poly(maleic anhydride) (Cs-g-MAH/MMt)

The mechanical properties of the Cs/MAH/MMt composites are shown on Table 4. In respect to composite without MMt, which was Cs-g-MAH-10 composite, the increased of TS was observed upon addition of MMt and the highest TS of 42.14 Mpa was reached when the amount of 9% w/w of MMt was used. The decreased of strain from 9.18 to 7.14% were observed. The brittle of biocomposite diminished which was indicated by increasing of young's modulus from 494.61 to 707.68 MPa. Improving the mechanical properties upon addition of MMt indicated that the composite of Cs-g-MAH-10 was compatible with MMt due to the interaction among the polar hydroxyl group of chitosan, polar carbonyl group of MAH and polar hidroxyl group of montmorillonite [17].

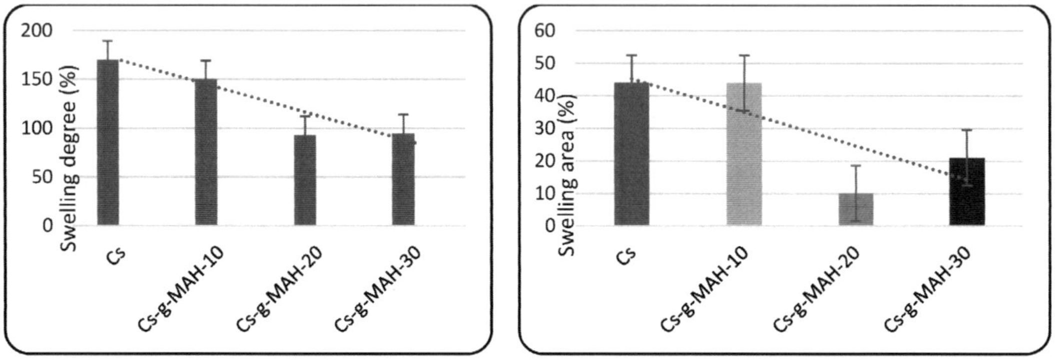

Figure 1. Swelling degree (a) and swelling area (b) of Cs-g-MAH composites.

Table 4. The mechanical properties of Cs-g-MAH/MMt composites.

Sample	Tensile stength (MPa)	Strain (%)	Young's modulus (MPa)
Cs-g-MAH-10	35.90	9.18	494.61
Cs-g-MAH/MMt-3	26.97	9.49	333.95
Cs-g-MAH/MMt-6	39.14	7.14	552.12
Cs-g-MAH/MMt-9	42.14	6.04	707.68
Cs-g-MAH/MMt-12	41.92	5.49	766.96

The swelling properties of Cs-g-MAH/MMt composites are shown in Figure 2. The addition of MMt into Cs-g-MAH-10 composite increased the swelling degree and swelling area of Cs-g-MAH/MMt composites. Upon addition of MMt into matrix polymer increased the water absorbtion due to the hydrophilic activity of clay [22]. The Cs-g-MAH/MMt-9 composite was observed to have best water resistance since it had the lowest swelling degree respect to the Cs-g-MAH-10 composite. The swelling degree of Cs-g-MAH/MMt-9 composite increased from 170 to 233 %.

3.3. FTIR analysis chitosan, Cs-g-MAH-10 and Cs-g-MAH/MMt-9

Figure 3 shows the FT-IR spectra of chitosan, Cs-g-MAH-10 and Cs-g-MAH/MMt-9 composites. The FT-IR spectrum shows characteristics peaks for chitosan [23]. The spectrum shows a peak at 3080 cm^{-1} due to the overlapped of -NH symmetric stretching vibration and -OH stretching vibration. Peaks at 1664, 1588 and 1324 cm^{-1} were observed due to vibration of C=O secondary amide, protonated amide and stretching vibration of $-CH_3$ tertiary amide of chitosan, respectively. A peak at 2886 cm^{-1} was obtained due to the asymmetric stretching vibration of aliphatic ($-CH_2$), peaks at 1422 and 1388 cm^{-1}

appeared for the characteristic peaks of δ(C-H) and a peak at 894 cm^{-1} was characteristic for wagging of C-H. The peaks at 1150 and 1033 cm^{-1} were characteristics for the symmetric and asymmetric stretching vibration of C-O-C. The characteristic peaks of C-O and C-N stretching vibration appeared at 1033 and 1422 cm^{-1}, respectively.

The spectrum of Cs-g-MAH-10 showed that a peak at 3080 cm^{-1} was disappeared and a peak at 3435 cm-1 was appeared. A strong band at 1566 cm^{-1} was observed due to C=C bond. In addition, the Cs-g-MAH-10 FT-IR spectrum showed a new absorption peak at 1704 cm^{-1} corresponding to the CO stretching vibration of carboxyl groups [14]. These results provided evidences that the MAH was interacted with chitosan matrix [23].

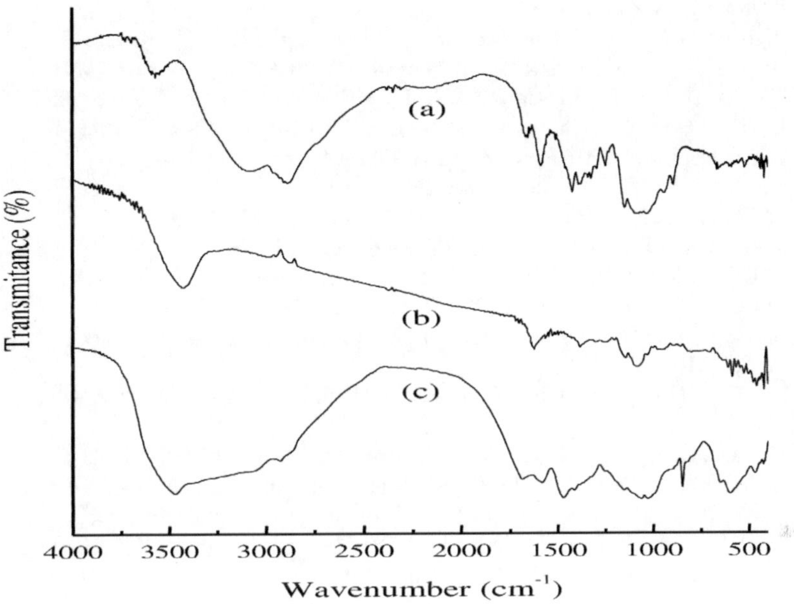

Figure 3. FT-IR spectra of: (a). chitosan, (b). Cs-g-MAH-10 and (c). Cs-g-MAH/MMt-9.

The peaks at 814, 835 and 946 cm^{-1} shown in Cs-g-MAH-10 FT-IR spectrum were shifted to the 823, 848 and 903 cm^{-1} in Cs-g-MAH/MMt-9 spectrum, respectively. Those peaks was due to the the bending vibration of Al-Mg-OH, Al-Fe-OH and Al-Al-OH. The characteristic absorption bands corresponding to the water molecules associated with the samples and with the protonated amine group in the chitosan chain were also shifted. The peaks at 1588 and 1644 cm^{-1} in chitosan spectrum was shifted into 1566 and 1625 cm^{-1} in Cs-g-MAH-10 FT-IR spectrum and 1583 and 1688 cm^{-1} in Cs-g-MAH/MMt-9 spectrum. These results confirmed that there were interactions between MMt and the chitosan-MAH composite.

4. Conclusions

The graft co-polymer of chitosan-maleic anhydride reinforced with montmorillonite composites have been successfully produced. FTIR analysis revealed that compatibilizing matrix among chitosan, MAH and MMt was fully achieved. A chitosan composite containing 10% w/w MAH and 9% w/w MMt was observed to have the best mechanical properties. However, the water resistance decreased upon addition of MMt into a chitosan-maleic anhydride composite. Further studies by varying amount of the each component and the thickness of composites still need to be evaluated to produce the bioplastics with better water resistance.

Acknowledgments
Authors would like to acknowledge their thanks to Ministry of Research, Technology and Higher Education, Indonesia for financial support through Student Creativity Program (Program Kreativitas Mahasiswa) with contract number: 117/B3.1/KM/2016 dated 19 February 2016.

References

[1] Avella M, Vlieger J J D, Errico M E, Fischer S, Vacca P and Volpe M G 2005 *Food Chem.* **93** 467-474.
[2] Deng J, Bowen Y, Chao C and Junya L 2015 *ACS Sustainable Chem. Eng.* **3** 509-605.
[3] Hudson R, Samuel G, Alexandra B and Jaffery L K 2015 *J. Chem. Educ.* Doi: 10.*1021/acs.jchemed.5b00108*
[4] Fernandez J G and Donald E I 2014 *Macromol. Mater. Eng.* **299** 932-938.
[5] Jayakumar R, Prabaharan M, Reis R L and Mano J F 2005 *J. Carb. Pol.* **62** 142-158.
[6] Ghani S W A, Aznizam A B and Sani A S 2015 *J. of Plastic Film & Sheeting.* **32(2)** 140-162.
[7] Kim S J, Shin B S, Hong J L, Cho W J, and Ha C S 2001 *Polym.* **42** 4073-4080.
[8] Pagnoulle C and Jerome R 2001 *Polymer* **42** 1893-1906.
[9] Zhang J F and Xiuzhi S 2004 *Biomacromol.* **5** 1446-1451.
[10] Rocco A M, Pareira R P and Felisberti M I 2001 *Polym.* **42** 5199-5205.
[11] Don T M and Chen H R 2005 *Carbohyd. Polym* **61** 334-347.
[12] Huang M, Shen X, Sheng Y and Fang Y 2005 *Int. J. Biol.Macromol.* **36** 98-102.
[13] Zhang W, Li G, Fang Y and Wang X 2007 *J. Mem. Sci.* **295** 130-138.
[14] Hong S I, Lee J H, Bae H J, Koo H S, Choi J H, Kim D H, Park S H and Park H J 2011 *J. Appl. Polym. Sci.* **119** 2742-2749.
[15] Lin K F, Hsu C Y, Huang T S, Chiu W Y, Lee Y H, and Young T H 2005 *J. Appl. Polym. Sci.* **98** 2042-2047.
[16] Xu Y, Ren X and Hanna M A 2006 *J. Appl. Polym. Sci.* **99** 1684-1691
[17] Troltzsch J, Jonas S, Katja H, Isabelle R, Frank H and Lothar K 2016 *J. Mater. Sci. Res.* **5** 111-121.
[18] Lee T K, Jae B L, Dong H P and Woo N K 2012 *J. Appl. Polym. Sci.* DOI: 10.1002/app.37504
[19] Diao C, Xia H and Parnas R S 2015 *ACS. Appl. Mater. Interfaces.* DOI: 10.1021/acsami.5b07136
[20] Gunning M A, Geever L M, Killion J A, Lyons J G and Higginbotham C L 2014 *Polym. Plast. Technol. Eng.* **53** 1223-1235.
[21] Tserki V, Matzinos P and Zafeiropoulos N E P 2006 *J. Appl. Polym. Sci.* **100** 4703-4710.
[22] Saputra O A, Saputra D A, Rini K S and Pramono E 2016 *J. Pure. Appl. Chem. Res.* **5** 85-94.
[23] Gopal Reddi M R, Gomathi T and Sudha P N 2014 *Der Pharmacia Lettre* **6 4** 296-305

IOP Conf. Series: Materials Science and Engineering **176** (2017) 012002 doi:10.1088/1757-899X/176/1/012002

Polyketide Derivatives from *Annona muricata* Linn Leaves as Potencial Anticancer Material by Combination Treatment With Doxorubicin on Hela Cell Line

A N Artanti[1], O P Astirin[2], A Prayito[3], R F Widiyaningsih[2], and F Prihapsara[1]

[1]Department of Pharmacy, Sebelas Maret University, Surakarta, Indonesia

[2]Department of Biology, Sebelas Maret University, Surakarta, Indonesia

[3]Department of Pathobiology Sebelas Maret University, Surakarta, Indonesia

E-mail: antantinuranif@gmail.com

Abstract. One of the compounds found effication as an anticancer agent on cervical cancer is acetogenin, a polyketide compound that is abundant in *Annona muricata* L. leaves. This study has been done to examine polyketide derivatives was isolated from *Annona muricata* L. which has potency to induce apoptosis by p53 expression on hela cell line. An approach recently develop to overcome side effect of chemoterapeutic agent is used of combined chemoterapeutic agent, i.e doxorubicin. The determination of cytotoxic combination activity from polyketide derivative and doxorubicin was evaluated using MTT assay to obtain the value of CI (*combination index*). The expression of p53 profile was evaluated by immunohistochemistry on hela cell line. Data analysis showed that combination of polyketide derivative from *Annona muricata* L. (38,5 µg/ml) and doxorubicin with all of concentration performed synergistic effect on hela cell line with CI value from 0,33 – 0,65. The analysis on immucytochemistry showed that polyketide derivative from *Annona muricata* L. leaves could enhance p53 pathway significantly on hela cell line.

1. Introduction

Monotetrahydrofuran annonaceous acetogenins is one of the active compound were isolated from *Annona muricata* L leaves. They exhibit a broad range of biological properties such as cytotoxic, immunosuppressive, antiparasitic and antimicrobial [1]. Especially, their potential to inhibit tumor cells and the activity as anticancer agent. Annonaceous acetogenin often referred to inhibit the growth of cancer cell. Acetogenin is a polyketide compound with structure C-34 or C-37 unbranched carbon chain attached to the 2-propanol group at C-2 to from a lactone [2]. The main focus of this study was to observe the citotoxic activity of polyketide derivative from *Annona muricata* L by combination treatment with doxorubicin on Hela cell line. Cervical cancer is the second most common found in women all over the world. There are 532.232 cases of cervical cancer in 100 million women [3]. The viral factor associated with oncogenic include the persisten infection, elevated viral load, virus genotype chemotherapy, and combination of these. To date, the use of chemotherapeutic agent is the mostly used treatment [4] but it cause many side effects, such as severe nausea and vomiting, acute and chronic nephrotoxicity, anemia [5] [6], acute cochlear toxicity (ototoxicity) [7] and also developed resistance [8]. Many studies was conducted on natural compound as an anticancer agent [9] that has shown potentially of anti-cancer that have toxicity selectively without damaging in normal cells [10]. One of the natural compound which has potential effect as anticancer agent is *Annona muricata* L..

There are four compound of *Annona muricata* L.. have a potential anticancer namely are monotetrahydofuran acetogenins, muricin H, muricin I and cis-annomontacin [11]. Therefore, the aim of this research was to develop a new approach of combination treatment using polyketide derivatives (Me (CH_2) 7O C(O) (CH_2) 4C(O) O (CH_2) 7 Me) from *Annona muricata* L. [12] and doxorubicin as co-chemotherapeutic agent. Hela cells are known to express 2 kinds of oncogenes, E6 and E7. E6 bounds with p53 phosphorilated protein meanwhile the E7 bounds with pRb protein [13]. The bonding of p53 protein by the E6 causing the increase of protein's degradation rate. Poliketide derivatives from *Annona muricata* L reported to have ability in increasing the p53 expression thereby activates apoptosis process on Raji cell line [12].

2. Experimental

2.1 Sampel preparation

Annona muricata L. leaves was obtained from Karanganyar, Central Java, Indonesia. It was extracted with ethanol, evaporated and partitioned with chloroform, *n*-hexane and ethyl acetate by vaccum coloumb chromatography (VLC). The fractions was evaporated by using rotary evaporator and identified with KLT. Polyketide was isolated by chloroform fraction using vaccum coloumb chromatography (VLC). Each sample from TLC process which have one spot is joined and evaporated again to FTIR.

2.2 Cell culture

Hela cells were obtained from LPPT, Gadjah Mada University and were grown in RPMI (Gibco), supplemented with 10% Fetal Bovine Serum (FBS Qualified, Gibco, Invitrogen USA), 1.5% (w/w) penisilin-streptomisin (Gibco, Invitrogen USA), and 0.5 % fungizone (Gibco, Invitrogen USA). Cells were incubated at 37°C and 5% CO_2.

2.3 Cytotoxic assay

The cytotoxic activity were done by MTT assay to obtain the value of Combination Index (CI). Hela cells were distributed in 96-well plate before being exposured to sample treatment with 12000 in cells each well. For cell viability assay, cells were treated for 24 hours with polyketide derivetive from *Annona muricata* L both for the single and combinational cytotoxic assay. As negative control, only growth medium was added. In the end of the incubation time, each well were added by 100 μg/ml of MTT solution (0.5 mg/mlin PBS) continued with incubation for 3 hours at 37°C. There action was stopped by dilution with 10% (w/v) Sodium Dodecyl Sulphate in 0.01 N HCl, and cells were incubated overnight. The absorbance was determined by using ELISA reader at λ 595 nm. [14].

2.4 Immunohistochemistry

Hela cells were grown with the density of 50000 cells/cover slip in 24-well plate for 24 hours prior to treatment of.Cells were treated for 15 hours and then fixed with cold methanol p.a for 10 minutes and washed twice using PBS and sterile water. Cells were blocked with H_2O_2 and then prediluted blocking serum were added, continued with incubation for 10 minutes. Cells were washed using PBS and incubated with primary monoclonal antibody antip53 (mouse monoclonal anti-p53 antibody, BioGenex) overnight. Cells were washed using PBS, then universal detection kit was added (Star Trek Universal HRP Detection Kit, Ref STUHRP 700L10-KIT, Biocare Medical). Cells were then incubated with biotinylated universal secondary antibody for 20 minutes, followed by streptavidin-enzyme horse radish peroxidase for 10 minutes.Substrate solution DAB used as a chromogenic agent was exposed for 10 minutes and cells were counterstained with Mayer's Haematoxylin (Dako) for 1 minute and fixed with ethanol and xylol. Between each immunostaining step, cells were washed briefly in PBS pH 7.4. Negative controls were prepared by replacing the primary antibody with PBS.

2.5 Data analysis

Single citotoxic assay was determined by linier regression between concentration and % cell viability to calculate IC_{50} value so that can be determined the cytotoxic potency. It was analyzed statistically by using Microsoft Excell and statistical significance was estimated by using ANOVA test. Statistical significance was placed at $p<0.05$. Combinational treatment was evaluated by calculating the Combinational Index (CI) value which has the formula as follows : CI = D1/Dx1 + D2/Dx2

D1 and D2 represent concentrations used in combinational treatment, while Dx1 and Dx2 are single treatment concentration giving same response as D1 and D2, respectively. The interpretation of CI value was done based on classification listed in Table 1.

Table 1. Interpretation of CI value representing potency of combination application

CI value	Interpretation	CI value	Interpretation
< 0.1	Very strong synergist	**0.9 – 1.1**	Closely additive
0.1 – 0.3	Strongly synergist	**1.1 – 1.45**	Middle antagonist
0.3 - 0.7	Synergist	**1.45 – 3.3**	Antagonist
0.7 – 0.9	Middle synergist	**>3.3**	Strongly antagonist

3. Result and Discussions

3.1 Fourier Transform - InfraRed (FTIR)

There are 16 points uptake representing 16 functional groups contain in isolates of *Annona muricata* L. The absorption interpretation was done up to maximum wavelenght 500 cm^{-1} and 15 points uptake that appropriate wuth standard table. The absorbance at 3.397,76 cm^{-1} indicates alcohol O-H groups. Other typical uptake absorbance 2.928,07 cm-1 indicates the chain C-H asymmetric and 2.851,88 cm^{-1} indicates CH symmetric, both groups showed adjacent C-H vibration. The absorbance at 1.465.96 cm-1 indicates vibration C-H as the form of vibration cut off [15]. The absorbance at 1.743,72 cm^{-1} indicates the absorpsion of the lactone groups (Fig.1). It was an special absorbance that comes from the group C=O in γ-butyrolactone which estimated as lacton of acetogenis [16]. The existence of these different groups with ester lactones which have a longer absorption. This is reinforced by the absorption at 1.095,61cm-1 as the O-C-C that indicates the presence of a lactone group. According Blessing *et al.* (2015), alkena has absorbance in 1.680-1.600 cm^{-1}[17]. In Table 1, the group C = C bonds (alkenes) are shown in absorbance 1.614,49cm^{-1}. This existence by the absorption at wave numbers 960.59 nm, 799.53nm dan 720.44nm which indicate the presence of aromatic C-H bending. Molecular masses of polyketide isolate from *Annona muricata* L are consisted of 20 or 22 carbon atoms apart for few exception such as fatty acid esters.

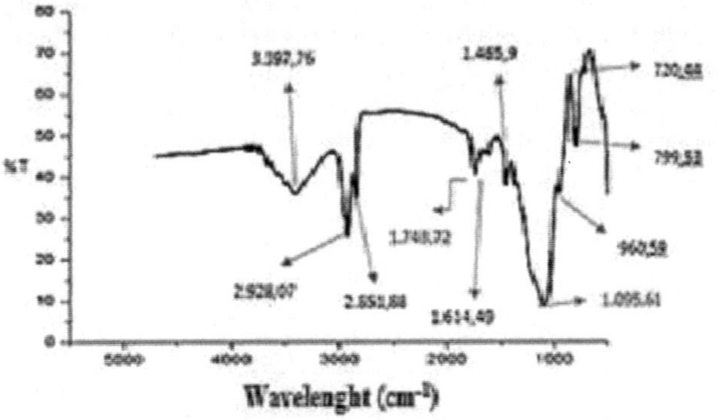

Figure 1. FTIR interpretation of polyketide isolate from *Annona muricata* L.

3.2 Effect of polyketide derivatives from Annona muricata L. on Hela Cells growth
Polyketide derivatives from *Annona muricata* L. leaves shows potent cytotoxicity on Hela cells. Furthermore, IC_{50} value acquired as the parameter of concentration that needed to yield 50% cells' growth inhibition of Polyketide derivatives from *Annona muricata* L. leaves on Hela cells. The effect of polyketode derivatives to cell viability of 24-hours treatment. Treated cells showed cytotoxic effect with change of cell morphology and decreased cell viability. Viable cells had epithelial shape, but after being treated with certain concentration of samples they gave spherical-shape and shrunk cell wall. Polyketide derivatives from *Annona muricata* L. caused inhibition of cell growth in a concentration-dependent manner (Fig.1). Cell viability after treatment with polyketide derivatives from *Annona muricata* L. was significantly decreased by IC_{50} value is 77.09 µg/ml on Hela cells (Fig. 2). According to Prayong *et al.* (2008) this active compound exhibited potent cytotoxicity ($IC_{50} \leq$ 100µg/ml) [18]. These results above suggested that polyketide derivatives from *Annona muricata* L. leaves indicate as the most potential concentration which has potential cytotoxic effect.

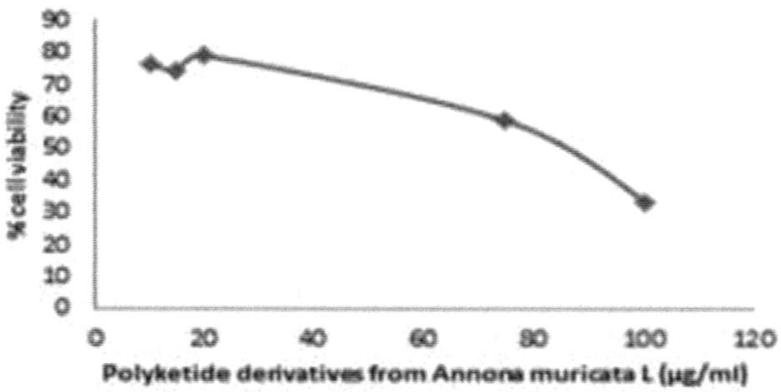

Figure 2. The effect of polyketide derivative from Annona muricata L. on the growth Hela cells. The cells were incubated in the 96 well plates and given the treatment the sample in the range 15 µg/ml - 100 µg/ml.

Based on the study, the IC_{50} of doxorubicin on Hela cells is 8.88 µM. Hela cells has potency resistence of doxorubicin by the antiapoptotic protein in Hela cells and efflux mechanism of doxorubicin from Hela cells [19]. Therefore, the observation of combinational cytotoxicity of polyketide derivatives from Annona muricata L and doxorubicin is needed to be done.

3.3 Effect of polyketide derivatives from Annnona muricata L combined with doxorubicin on Hela cell growth
Combination assay was done by combining ½ IC_{50}, ¼ IC_{50}, $^1/_8$ IC_{50}, $^1/_{10}$ IC_{50} of polyketide derivatives from Annona muricata L with ½ IC_{50}, ¼ IC_{50}, $^1/_8$ IC_{50}, $^1/_{10}$ IC_{50} of doxorubicin. The cell viability were quantified by *Combination Index* (CI). The optimum of CI value were obtained were combining ½ IC_{50} polyketide derivatives from *Annnona muricata* L with ¼ IC_{50}, $^1/_8$ IC_{50}, $^1/_{10}$ IC_{50} respectively and the combination of $^1/_8$ IC_{50} polyketide derivatives from Annnona muricata L with ¼ IC_{50} of doxorubicin with CI value in range 0.3 – 0.7 (Table 1). the synergistic effect of doxorubicin in the combination with polyketide derivatives from Annona muricata L. probably through the apoptosis induction mechanis so that the observation of these should be done.

Table 2. Combination Index (CI) value of polyketide derivatives from *Annona muricata* L. and doxorubicin

Polyketide derivatives (µg/mL)	**Doxorubicin (µM)**			
	0.9	**1.125**	**2.25**	**4.5**
7.7	0.40	0.87	1.34	0.78
9.625	0.85	1.08	1.76	0.35
19.25	1.22	0.60	2.61	2.97
38.5	0.71	0.34	0.57	4.06

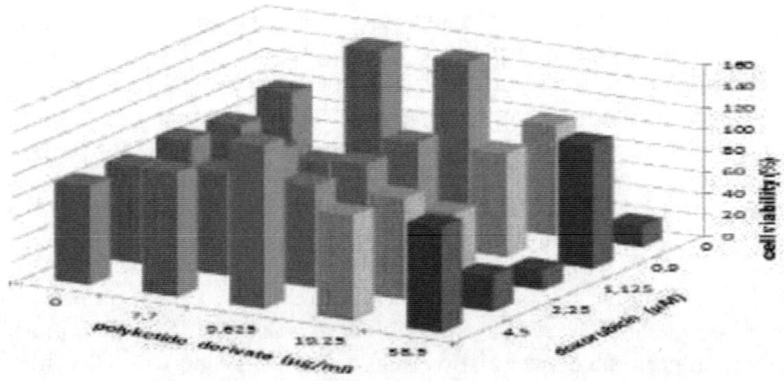

Figure 3. The effects of combination treatment of polyketide derivatives from Annona muricata L. with doxorubicin on Hela cells viability. Cells were incubated overnight and given the treatment of these two compound by combining ½ IC_{50}, ¼ IC_{50}, $^1/_8$ IC_{50}, $^1/_{10}$ IC_{50}. The graph shows Hela cells viability after the treatment ($p<0.05$)

3.4 Effect of polyketide derivatives from Annona muricata L. and Doxorubicin on p53 expression

To confirm the mechanism of polyketide derivatives from *Annona muricata* L. in Hela cells, therefore molecular target protein p53 was investigated. Hela cells have the characteristic of degrade p53 and loss of function as tumor suppresor protein causing uncontrooled proliverative cells [20]. As shown in Fig 4. immunocytochemistry evaluation indicated that p53 level increased significantly ($p<0.05$) with intensive brown colour in nucleus after being treated with 77,09 µg/ml Annona's polyketide derivatives. The result suggest that Annona's polyketide derivatives has cytotoxicity efficacy on Hela cells through p53 stabilization. Besides, Annona's polyketide derivative addition to doxorubicin enhance cytotoxic efficacy on Hela cell line.

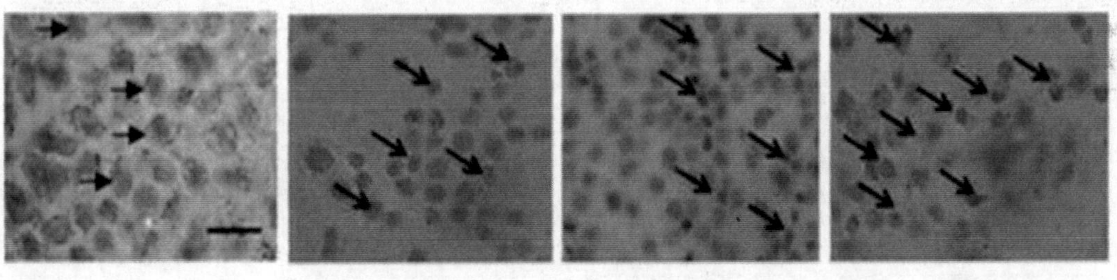

A B C D

Figure 4. Doxorubicin in combination with Annona's polyketide derivative increased p53 level in nucleus. Immunocytochemistry of (A) cell control overnight treatment with (B) 77 µg/ml Annona's polyketide derivative (C) 8 µM doxorubicin (D) 8 µM doxorubicin in combination with 77 µg/ml Annona's polyketide derivative

The study for anticancer agent from natural source has been succesfull worldwide. Active constituens have been purified and nowadays used to treat in human tumours. The etnopharmacological knowledge is helpful to lead the study for plant which has potential cytotoxic activity. *Annona muricata* L. reported to be utilises as remedies againts cancer, however is not especially for cancer caused virus. Our result demonstrated that polyketide derivatives from *Annona muricata* L. have growth inhibitory and cytotoxic effect on cervical cancer cell line. Polyketide derivatives from *Annona muricata* L. leaves performed potent cytotoxic effect on Hela cells with IC_{50} value of 77,09 µg/ml.

Decreasing cell viability may be because of either cell death or cell cycle arrest. The mechanism of cell cycle distribution is also associated with some of celuller protein especially p53 protein. p53 is a tumor suppresor protein. In this study, we observed that polyketide derivative from *Annona muricata* L. treatment increased p53 level in nucleus. Therefore, poliketide isolation may be amandable as viral inhibitor agent and as competitor of vaccine to prevent the development of cervical cancer. However, this speculation still needs further investigation by in silico study. In conclusion, polyketide derivatives from *Annona muricata* L. leaves indicate has potential to be developed as a co-chemotherapeutic agent on Hela cell lines, it can exhibit potential abbility with p53 stabilization. Further molecular target detection to investigate its cellular pathway needs to be conducted.

4. Conclusions

The combination of polyketide derivative from *Annona muricata* L. (38,5 μg/ml) and doxorubicin with all of concentration performed synergistic effect on hela cell line with CI value from 0,33 – 0,65. The analysis on immucytochemistry showed that polyketide derivative from *Annona muricata* L. leaves could enhance p53 pathway significantly on hela cell line.

Acknowledgment
We thank to Universitas Sebelas Maret for Grand Featured Research Universities 2016 which has funded this research and Natural Maedicine Research Group, Pharmacy Department of Universitas Sebelas Maret.

References

[1] Agatonovic K, Morton D W and Yusof A P 2013 *Mod Chem appl* **1** 4
[2] Zuhud E A 2011 *Bukti Kedahsyatan Sirsak Menumpas Kanker* (Jakarta:Agromedia Pustaka)
[3] Blessing L D T, Budeguer F, Ramos J, Bardon A, Diaz S, Brovetto M, Seoane G and Neske A 2015 *Journal of Agricultural Chemistry and Environment* **4** 56-61
[4] Leaver D and Labonte G 2010 HPV and Cervical Cancer. *Radiation* **19** 27-44
[5] Miller R P, Tadagavadi R K, Ramesh G, and Reeves W B 2010 Mechanisms of Cisplatin Nephrotoxicity *Toxins* **2** 2490-2518
[6] Rabik C A and Dolan M E 2007 *Cancer Treatment Reviews* **33** 9– 23
[7] Mukherjea D and Rybak L P 2011 *Pharmacogenomics* **12** 1039–1050
[8] Sirichanchuen B, Pengsuparp T and Chanvorachote P 2012 *Mol Cell Biochem* **364** 11–18
[9] Newman D J and Cragg G M 2007 *J. Nat. Prod* **70** 461-477
[10] Lee Y M, Jeong G S, Lim H D, An R B, Kim Y C and Kim E C 2010 *J. Nat. Prod.* **67** 470–475
[11] Liaw C C, F R Chang, C Y Lin, C J Chou, H F Chiu, M J Wu and Y C Wu 2002 *Journal of Natural Products* **65** 470-475
[12] Astirin O P, Prayitno A, Artanti N A, Fitria M S, Witianingsih D A, Pranatami D A and Putra S T 2014 Int.J. Cancer Ther Oncol **2** 20-28
[13] Minaguchi T, Kanamori Y and Matsushima M 1998 *Cancer Res* **58** 4585-4586
[14] Cheng H, Lshu J S, Huang L W, Hsieh B S, Hu Y C, Hung T C and Chang K L 2010 *Molecular Cancer* **9** 126
[15] Silverstein R M, Bassler G C and Morrill T C 2005 *Spectrometric Identification of Organic Compounds* (New York:John Wiley) pp 109-130
[16] Zeng L, Ye Q, Oberlies N H, Shi G, Gu Z M, He K and McLaughlin J L 1996 Recent advances in *Organic Compounds* (New York: John Wiley) pp 109-130
[17] Blessing L D T, Budeguer F, Ramos J, Bardon A, Diaz S, Brovetto M, Seoane G and Neske A 2015 *Journal of Agricultural Chemistry and Environment* **4** 56-61
[18] Prayong P, Barusrux S and Weerapreeyakul N 2008 *Fitoterapia* **79** 598-601
[19] Bravo C A, Ortiz L P C, Lerma D J M, Dominguez R J R, Jave S L F, Aguilar L A, Toro A S, Celis C R, Sahagun F J E, Alba G J E and Hernandez F G 2010 *Molecular Cancer* **9** 114
[20] Minaguchi T, Kanamori Y and Matsushima M 1998 *Cancer Res* **58** 4585-4586

Physicochemical Properties of Biosurfactant Produced by *Pseudomonas fluorescens* Grown on Whey Tofu

V Suryanti[1,*], D S Handayani[1], S D Marliyana[1] and S Suratmi[1]
[1]Department of Chemistry, Faculty of Mathematics and Natural Sciences, Sebelas Maret University, Jl. Ir. Sutami 36A Surakarta 57126 Indonesia

[*]E-mail: venty@mipa.uns.ac.id

Abstract. The research aims to examine the physicochemical properties of biosurfactant produced by *Pseudomonas fluorescens*. Biosurfactant was produced in whey tofu media containing 8 g/L nutrient broth and 5 g/L NaCl which was fermented for 2 days at room temperature. Biosurfactant was identified as rhamnolipids which had critical micelle concentration (CMC) value of 638 mg/L and surface tension of 54 mN/m. The biosurfactant had water in oil (w/o) emulsion type. The biosurfactant was able to decrease the interfacial tension more than 40% for emulsion of water with hexane, pentane, benzene, lubricants or kerosene. The stable emulsions were reached up to 30 days with the E24 value of about 50% when paraffin, toluene, lubricants or palm oil was used as an immiscible compound. Commercial surfactants, such as Triton X-100 and Tween-80 were investigated to compare their emulsification activities and emulsion stabilities with the produced biosurfactant.

1. Introduction

Due to the limitation of surfactants derived from petroleum such as non-biodegradability, ability to accumulate and toxicity, there has been increasing interest in finding replacement of chemically surfactants with biological products. Biosurfactants are amphiphilic surface active compounds produced by microorganisms [1-3]. Biosurfactants have potential in the various application fields such as in bioremediation, biodegradation, oil recovery, food and pharmaceutics. However, they have not being widely used yet due to high production cost [4].

Optimizing growth factors in the fermentation technology for biosurfactant production is of paramount importance. One possible strategy for reducing production costs is the use of cheaper and renewable substrates from agricultural waste, or that produced by the food industries, and oil waste industries. Several studies have reported the biosurfactants production using agricultural industrial waste, such as molasses, starchy substances, lactic whey, distillery wastes which are known for containing high levels of carbohydrates and of lipids which are necessary for substrates in the production of biosurfactants [5-9].

We have reported the production of biosurfactants using vegetable oils (such as soybean oil and corn oil) as well as agroindustrial wastes (such as cassava flour wastewater and molasses). We also have studied the application of biosurfactants for environmental remediation, such as removal of heavy metals and enhancement degradation of phenol [10-14]. The production of biosurfactants by *Pseudomonas fluorescens* has been studies using various substrates. However, whey tofu has not been used yet. Therefore, as extension of our work, we are interested in the production of biosurfactant by

P. fluorescens using whey tofu as a substrate. The physicochemical properties of the biosurfactant are discussed in this work.

2. Experimental

2.1. Materials

All chemical were used are analytical grade from e-Merck. The whey tofu was obtained from a local tofu industry. The strain used in this work, *P. fluorescens* FNCC 0063, was purchased from IUC Food and Nutrition, Universitas Gadjah Mada, Indonesia.

2.2. Media Used and Growth Condition

Cultures of bacteria were maintained on nutrient agar. The media were used for biosurfactants production composed of whey tofu containing 8 g/L nutrient broth and 5 g/L NaCl. Whey tofu was heated to boil and then filtered. Nutrient broth and NaCl were then added to the filtered whey tofu and the mixture was sterilized prior used. Fermentation was carried out at 150 rpm for 2 days at room temperature.

2.3. Biosurfactant recovery

Culture liquid of *P. fluorescens* was centrifuged at 10,000*g* for 10 mins. The supernatant was acidified to pH 2.0 with HCl 6N and leaved overnight at 4°C. The supernatant then was extracted using chloroform:methanol = 2:1. Sodium sulphate anhydrous was then added to the organic layer. The organic layer which was free form water was then evaporated.

2.4. Biosurfactant Characterization

The capillary rise method was used for measuring the surface tension and interfacial tensions. The CMC value was obtained by dissolving of biosurfactant in distilled water and the surface tension was measured with various concentrations of the biosurfactant. The CMC was indicated by sudden changes in the surface tension, which was determined by plotting the surface tension as a function of biosurfactant concentration.

The emulsion type of biosurfactant was investigated by conductivity test. Sodium chloride (1, 2, 3 and 4% w/w) as an electrolyte was added to the emulsion and the conductivity was measured. If the conductance increases with increasing of concentration of sodium chloride, the emulsion is oil in water type. On the other hand, it is water in oil type if there is no significant changes in its conductance. The dilution test was also applied for the emulsion type investigation [15].

The E24 of the biosurfactants were investigated by addition of 0.1 mg of biosurfactant to a crew-capped tube containing 1 mL of distilled water and 1 mL of immicible compounds. The mixtures were then vortexed for 2 mins and leaved to stand for 24 h. The emulsification index (E24) is given as percentage of height of emulsified layer (mm) divided by total height of the liquid column (mm). The E24 of the formed emulsions were monitored for 30 days.

Biosurfactant was dissolved at the CMC value in 4 mL distilled water and the immiscible compounds (1 mL) were added. The mixtures were shaken vigorously for 2 mins at room temperature. The resulting emulsion was allowed to stand for 10 min, and its absorbance was measured at 540 nm every 10 mins for 1 h. One unit of emulsification activity was defined as the amount of biosurfactant that affected an emulsion with an absorbance at 540 nm of 1.0. The log of the absorbance was plotted versus time, and its slope was taken as the decay constant, K_d [16-20].

3. Results and Discussion

The FT-IR spectrum of the produced biosurfactant is shown in Fig. 1. A strong and broad band at 3410 cm^{-1} due to the O-H stretching vibration has been observed. The –CH stretch of methylene was confirmed by the band at 2931 cm^{-1}. The asymmetric stretching vibration of C=O was characterised by the band at 1720 cm^{-1} for the carboxylate and 1651 cm^{-1} for the ketone. The band at 1404 were observed as a result of deformation and bending vibrations of aliphatic chains (-CH_3- and CH_2-), suggesting the presence of alkyl chains in the compounds. The stretching of C-O ester was found by the presence of the intense and strong band at 1126 cm^{-1}. The obtained peaks of biosurfactant suggested that rhamnolipids, glycolipids type biosurfactants, were produced by *P. fluorescens* as reported by other researchers that *Pseudomonas* sp produces rhamnolipids [21-23].

The produced biosurfactant presented the CMC value of 638 mg/L and the surface tension value of 54 mN/m. The higher value of CMC and the presence of carboxylate and hydroxyl groups suggested that the produced biosurfactant was ionic surfactant types.

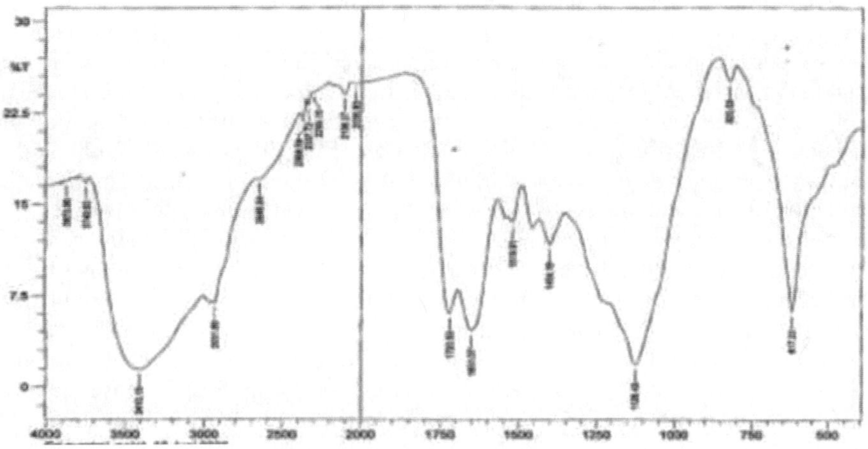

Figure 1. FT-IR spectrum of the produced biosurfactant.

Table 1 shows the conductivity of the solution for investigation of biosurfactant emulsion type. The conductivity of water as a positive control increased upon increasing addition of sodium chloride concentration. For the negative controls, the conductivity of the mixture of palm oil and water as well as water itself were measured. The conductivity of these negative controls were remained zero upon increasing addition of sodium chloride concentration. The addition of sodium chloride did not affect the conductance of the emulsion of water and palm oil with addition of biosurfactant, indicating that the emulsion type of the biosurfactant was water in oil (w/o).

Table 1. Biosurfactant emulsion type investigation by addition of NaCl.

Samples	Conductivity (μs)				
	Before addition of NaCl	After addition of NaCl (w/w)			
		1 %	2 %	3 %	4 %
Water	3.60	5.50	7.30	13.10	24.80
Palm Oil	0	0	0	0	0
Water and Palm Oil	0	0	0	0	0
Biosurfactant in water (650 ppm) and palm oil	0	0	0	0	0
Biosurfactant in water (1000 ppm) and palm oil	0	0	0	0	0

The emulsion type of biosurfactant was also determined by dilution test. The dilution test is based on the principle that an emulsion can be diluted with its continuous phase. The emulsion was not diluted upon addition of palm oil. On the other hand, separation of the emulsion was observed upon addition of water. These results confirmed that the emulsion type was w/o emulsion. As controls, the dilution test was also performed for commercially available surfactants, such as Triton X-100 and Span 60 which have w/o and o/w emulsion type, respectively.

The produced biosurfactant was characterized for main surfactant properties, such as their ability to lower the surface tension of water and immiscible compounds mixtures (Table 2). The immiscible compounds which were used in this study were palm oil, soybean oil, olive oil, lubricant oil, kerosene, solar, paraffin, pentane, hexane, benzene, benzaldehyde, benzyl chloride, aniline, anisaldehyde, and toluene. The emulsion stabilities were observed for 30 days by monitoring the E24 (Table 3).

The biosurfactant could decreased the surface tension of water and benzene emulsion about 60.6 %. The decreased surface tension of emulsion water and lubricant oil, pentane or hexane were about 40-50%, whereas the decreased surface tension of emulsion water and soybean oil, kerosene or toluene were about 30-40%. These results suggested that the produced biosurfactant was a good emulsifier. The biosurfactant could emulsified the mixture of water and palm oil, soybean oil, olive oil, lubricant oil, solar, paraffin, benzaldehyde or toluene with the E24 value more than 50%. The stable emulsions were reached up to 30 days with the E24 value of about 50% when palm oil, lubricant oil, paraffin or toluene was used as an immiscible compound.

The emulsification activities and decay constant of the biosurfactant have been investigated (Table 4). The immiscible compounds used in these investigations were palm oil, paraffin and toluene because these compounds had the highest emulsion stabilities (Table 3). The emulsion activities and decay constants of biosurfactant have been compared with commercially available surfactants, such as Triton X-100 and Tween 80. These two surfactants were used as comparison because these surfactants have the same emulsion type with the produced biosurfactant which is water in oil emulsion (w/o).

Table 2. The surface tension of emulsion between water and immiscible compounds.

No.	Samples	Surface Tension		Decreased surface tension (%)
		Without Biosurfactant	With Biosurfactant	
1.	Palm oil	64.3	59.1	8.0
2.	Soybean oil	32.8	22.1	32.5
3.	Olive oil	54.1	44.0	18.8
4.	Lubricant oil	54.1	27.3	49.6
5.	Kerosene	50.7	30.5	39.9
6.	Solar	26.5	21.4	19.2
7.	Paraffin	58.3	43.2	25.9
8.	Pentane	54.0	30.6	43.2
9.	Hexane	52.5	27.3	48.6
10.	Benzene	42.4	16.7	60.6
11.	Benzaldehyde	56.0	45.6	18.6
12.	Benzyl chloride	58.6	46.9	20.0
13.	Aniline	56.7	39.8	29.9
14.	Anisaldehyde	65.0	53.3	18.1
15.	Toluene	64.1	43.0	32.8

The emulsification activities of biosurfactant for palm oil, paraffin and toluene was higher than that of Tween 80 but they are lower than that of Triton X-100. The best emulsification activities of biosurfactant was achieved at 0.60 for toluene. Decay constants reflect the stability of the emulsion, the higher the decay constants, the better the stability of the emulsion [18-20]. Biosurfactant had the highest decay constant for palm oil among other surfactants studied. However, biosurfactant had higher decay constants than that of Tween 80 and lower decay constant than that of Triton X-100 for paraffin and toluene. These results indicated that the biosurfactant could be useful as an emulsifying and emulsion-stabilizing agent.

Table 3. The emulsification index of emulsion between water and immiscible compounds.

No.	**Samples**	**Emulsification Index (E24) (%)**															
		Day															
		1	**2**	**4**	**6**	**8**	**10**	**12**	**14**	**16**	**18**	**20**	**22**	**24**	**26**	**28**	**30**
1.	Palm oil	63	60	60	60	60	58	58	56	56	56	56	56	56	56	56	56
2.	Soybean oil	50	48	44	44	44	0	0	0	0	0	0	0	0	0	0	0
3.	Olive oil	52	52	48	46	44	44	0	0	0	0	0	0	0	0	0	0
4.	Lubricant oil	61	57	56	56	56	52	52	48	48	48	48	48	48	48	48	48
5.	Kerosene	19	14	14	14	14	14	14	11	0	0	0	0	0	0	0	0
6.	Solar	52	52	42	42	0	0	0	0	0	0	0	0	0	0	0	0
7.	Paraffin	65	65	65	54	54	54	54	54	54	54	54	54	54	54	54	54
8.	Pentane	0	0	0	0	0	0	0	0	0	0	0	0	0	0	0	0
9.	Hexane	64	63	55	50	44	35	29	15	8	0	0	0	0	0	0	0
10.	Benzene	0	0	0	0	0	0	0	0	0	0	0	0	0	0	0	0
11.	Benzaldehyde	52	48	44	41	41	0	0	0	0	0	0	0	0	0	0	0
12.	Benzyl chloride	48	41	41	41	41	0	0	0	0	0	0	0	0	0	0	0
13.	Aniline	8	8	8	8	8	8	0	0	0	0	0	0	0	0	0	0
14.	Anisaldehide	48	48	39	37	37	37	37	0	0	0	0	0	0	0	0	0
15.	Toluene	62	62	61	61	61	59	59	59	59	59	59	59	59	59	59	59

Table 4. The emulsification activities and decay constants of biosurfactant and commercially available surfactants.

No.	**Surfactant**	**Immiscible Compound**	**Emulsification Activities (U)**	**Decay Constant K_d (10^{-3})**
1.	Biosurfactants	Palm oil	0.43	-1.22
2.	Triton X-100	Palm oil	0.81	-12.44
3.	Tween-80	Palm oil	0.05	-5.69
4.	Biosurfactants	Paraffin	0.39	-2.45
5.	Triton X-100	Paraffin	3.41	-1.22
6.	Tween-80	Paraffin	0.07	-12.08
7.	Biosurfactants	Toluene	0.63	-2.94
8.	Triton X-100	Toluene	2.21	-1.53
9.	Tween-80	Toluene	0.06	-7.29

4. Conclusions

The rhamnolipids biosurfactant has been produced by *P. fluorescens* with whey tofu as a substrate. The physicochemical properties of the produced biosurfactant were evaluated and the results showed that it has effective emulsifier properties. The emulsion type of the biosurfactant was water in oil emulsion type (w/o). The biosurfactant could lower the surface tension of emulsion with various immiscible compounds and the obtained emulsification index was relatively high. The emulsification activities and the decay constants are comparable to the commercially available surfactants, such as Triton X-100 and Tween 80.

Acknowledgments
The authors acknowledge their thanks to the Directorate General of Higher Education, Department of National Education, Indonesia for the financial support to this research in the form of Hibah Bersaing Program.

References
[1] Pacwa-Płociniczak M, Płaza G A, Piotrowska-Seget Z and Cameotra S S 2011 *Int. J. Mol. Sci.* **12** 633-65.
[2] Nguyen T T and Sabatini D A 2011 *Int. J. Mol. Sci.* **12** 1232-124.
[3] Makkar R S and Cameotra S S 2002 *Appl. Microbiol. Biotechnol.* **58** 428-434.
[4] Banat I M, Makkar R S and Cameotra S S 2000 *Appl. Microbiol. Biotechnol.* **53 5** 495-508.
[5] Banat I M, Satpute S K Cameotra S S, Patil R and Nyayanit N V 2014 *Front. Microbiol.* **5 697** 1-18.
[6] Maneerat S 2005 *Songklanakarin J. Sci. Technol.* **27 3** 675-683.
[7] Chooklin C S, Phertmean S, Cheirsilp B, Maneerat S and Saimmai A 2013 *Songklanakarin J. Sci. Technol.* **35 2** 167-176.
[8] Rodrigues L R, Teixeira J A and Oliveira R 2006 *Biochem. Eng. J.* **32** 135-142.
[9] Nitschke M 2004, *Brazilian J. Microbiol.* **35** 81-85.
[10] Suryanti V, Hastuti S, Wahyuningsih T W, Mudasir and Muliawati D I 2009 *Indo. J. Chem.* **9 1** 107-112.
[11] Suryanti V, Marliyana S D, Handayani D S and Ratnaningrum D 2013 *Indo. J. Chem.* **13 3** 229-235.
[12] Suryanti V, Marliyana S D and Wulandari A 2015 *AIP Conf. Proc.* **1699** 040003 doi: 10.1063/1.4938318.
[13] Suryanti V, Marliyana S D and Andriani D 2016 *IOP Conf. Ser. Mater. Sci. Eng.* **107** 012018 doi:10.1088/1757-899X/107/1/012018.
[14]Suryanti V, Hastuti S and Pujiastuti D 2016 *AIP Conf. Proc.* **1710** 030016 doi: 10.1063/1.4941482.
[15] Khan B A, Akhtar N, Khan H M S, Waseem K, Mahmood T, Rasul A, Iqbal M and Khan H 2011 *Afr. J. Pharm. Pharmacol.* **5 25** 2715-2725.
[16] Cirigliano M C and Carman G M 1984 *Appl. Environ. Microbiol.* **48 4** 747-750.
[17] Lima A S, Sampaio R M and Alegre R M 1997 *Tecnologia química* **17 2** 18-22.
[18] Lee S-C, Yoo J-S, Kim S-H, Chung S-Y, Hwang C-W, Joo W-H and Choi Y-L 2006 *J. Microbiol. Biotechnol.* **16 5** 716-723.
[19] Amaral P F F, da Silva J M, Lehocky M, Barros-Timmons A M V, Coelho M A Z, Marrucho I M and Coutinho J A P 2006 *Process Biochem.* **41** 1894-1898.
[20] Lima A S and Alegre R M 2009 *Braz. Arch. Biol. Technol.* **52 2** 285-290.
[21] Onbasli D and Aslim B 2009 *J. Environ Biol.* **30 1** 161-163.
[22] Santa Anna L M, Sebastian G V, Menezes E P, Alves T L M, Santos A S, Pereira Jr N and Freire D M G 2002 *Braz. J. Chem. Eng.* **19 2** 159-166.
[23] Lee Y, Lee S Y and Yang J-W 1999 *Biosci. Biotechnol. Biochem.* **63 5** 946-947.

IOP Conf. Series: Materials Science and Engineering **176** (2017) 012004 doi:10.1088/1757-899X/176/1/012004

Antidiabetic Activity of Self Nanoemulsifying Drug Delivery System from Bay Leaves (*Eugenia polyantha* Wight) Ethyl Acetate Fraction

F Prihapsara[1], M Harini[2], T Widiyani[2], A N Artanti[1], and I L Ani[2]
[1]Department of Pharmacy, Faculty of Mathematics and Natural Science, Sebelas Maret University
[2]Department of Biology, Faculty of Mathematics and Natural Science, Sebelas Maret University

E-mail: feapri87@gmail.com

Abstract. Insulin resistance is caused by inability of target tissues to insulin response. Bay leaves (*Eugenia polyantha* Wight) fraction or extract have been used for the treatment of antidibetic mellitus type-2 resistance insulin (ADMRI) but it has low solubility and bioavailability. To overcome these problems, ethyl acetate fraction of bay leaves was formulated into self nanoemulsifying drug delivery system (SNEDDS) using Virgin Coconut Oil (VCO) as a carrier oil. This study aims to produce nanoherbal medicine, determine effect of nanoherbal preparation derived from bay leaves as an anti-ADMRI. The results showed that the optimum SNEDDS formula was tween 80 : PEG 400 : Virgin Coconut Oil (30% : 60% : 10%) in 5 mL. It has emulsification time 13.00 seconds with the average of droplet size value 84.5 nanometer and zeta potential value ± 0.2 mV. Morphological observation showed the nanoemulsion particles has spherical shaped and stable in different pH media. Hypoglycaemic effect of single dose metformin, SNEDDS, combination a-half dose of SNEEDS with metformin value is 28.3 % ; 15.6% ; 34.6% respectively.

1. Introduction

Bay leaves as one of herb plants has some therapeutic effects which are decreasing blood glucose degree so it can be used to be anti-diabetes agent [1]. Formula in the form of nanoemulsion is a choice that is expected to increase solubility and oral bioavailability of Bay leaves ethyl acetate fraction. Nanoemulsion can be formulated through Self-Nanoemulsifying Drug Delivery System (SNEDDS) which is a nanoemulsion pre-concentrate formed from the mixture of isotropic drug, oil, and surfactant. When they are combined with water phase in agitation condition, slowly, it will form nanoemulsion of oil phase in water (M/A) spontaneously [2].

The suitable oil in SNEDDS formulation as drug carrier is oil containing high medium-chain fat acid so it will be easier to be emulsified than long-chain fat acid. It is appropriate with Virgin Coconut Oil (VCO) that contains high lauric acid while lauric acid is medium-chain saturated fat acid [3].

Tween 20 and Tween 80 are surfactant having alcil-chain structure that has oil penetration effect to surfactant layer that enables nanoemulsion formation. Thus, the both are appropriate surfactant to be used in SNEDDS formulation [4]. Besides, Tween 20 and Tween 80 have high HLB values which are

[1]Coressponding autor : feapri87@gmail.com

16.7 for tween 20 and 15 for tween 80. The higher HLB value, the easier it will decrease the surface tension between oil and water [5].

Meanwhile, co-surfactants that are choosen are *propylene glycol* and *polyethylene glycol* 400, they are commonly used in formulating SNEDDS. It is because they are short-chain alcohol so it is more effective in decreasing surface tension [6]. They are ampifilic compound having affinity towards water phase and oil phase [7].

Ethyl acetate fraction from bay leaves is used as a sample research because it has low solubility in the body. The successfulness of this research is expected to be an innovation for nanoemulsion availability of ethyl acetate fraction from bay leaves and be a product that can be used as alternative of herb medical antidiabetic treatment with high bioavailability.

Diabetes mellitus (DM) is a disease which can be identified from the over blood glucose level (hyperglycaemic) as the result of the lack of insulin [8]. According to International Diabetes Federation (IDF), DM is the top seven death causes in the world [9]. Diabetes mellitus is divided into two categories. They are DM type 1 and DM type 2. Diabetes mellitus type 1 is caused by autoimmune damage that is mediated by pancreas cell causing insulin deficiency [10]. Meanwhile, hyperinsulinemia occurs in DM 2. However, insulin cannot bring glucose to get into tissue because of the occurrence of insulin resistance which is the reduction of insulin ability to stimulate glucose removal by peripheral tissue [11]. The use of Oral Hypoglycaemic Drug (OHM), for example metformin, still have side effect [12]. Recently, human tend to return to the nature, including the use of antidiabetes herb drugs that is Bay leaves (*Eugenia polyantha Wight*). The use of plant active extract and fraction still has weaknesses because of the lack of oral bioavailability [13]. Therefore, in order to make herb drug bioavailability becomes maximum, the size of the particle must be changed into nano-size (nanoemulsion).

Based on the background above, a test was conducted to examine the availability of ethyl acetate fraction from bay leaves in SNEEDS formula as antidiabetes 2 which is resistance to insulin.

2. Experimental

2.1 Tool and Materials

Materials which were used in this research are Bay leaves were obtained from Sukoharjo, Indonesia; component of SNEDDS: aquades, chloroform, ethyl acetate, Virgin Coconut Oil (VCO), PEG 400, propylene glycol, tween 20, tween 80 (Brataco); materials of artificial gastric fluid (AGF) and artificial intestinal fluid (AIF) consist of aquades (general), NaCl, sulfuric acid, MgCl2, CaCl2, NaHCO3, and KCl (E. Merck); male white mice (*Rattus norvegicus*) on Wistar strain around 3 months old with weight about 150-200 gram. It was diet administration in composition of woof (BR comfeed) (80%), lard (15%), duck yolk (5%), fructose; Na-CMC liquid 0,5%; reagent Glucose GOD (FS dyasis), metformin and glibenclamide (Generic).

Equipment which were used in this research are oven, blender, flannel, stopwatch, rotary evaporator ((RVO 400 SD Boeco Germany), waterbath (Grant), vortex (Maxi Mix II Thermolyne), sonicator (Branson 1510), magnetic stirrer (Cimarec), yellow tip (Kan Jian), flacon, spectrophotometer UV/Vis (Perkin Elmer Lambda 25), micro pipettes 50-200μL, particle size analyzer (Horiba SZ-100), Transmission Electron Microscope, analytical scale (Metler Teledo), and glass made tools (Pyrex).

2.2. Procedure

2.2.1. Ethyl acetate fraction from bay leaves

Bay leaves simplisia powder was extracted by using maceration method with chloroform solvent for 5 days. Macerate was filtered through glass funnel to separate macerate from simplisia powder. The simplisia powder then remaceration treatment with ethyl acetate. Thus, evaporate with rotary evaporator in 55^{0}C temperature and 6 rotary speeds until it became concentrated.

2.2.2 *Optimization of surfactant, consurfactant, and oil composition*
Ethyl acetate fraction from bay leaves was added to 5 ml carrier component (carrier oil: VCO, surfactant: tween 20 and/or tween 80, and consurfactant: PEG 400 or propyleneglycol) with optimal composition of surfactant and cosurfactant which was appropriate with optimization results that had been done. Then, the comparison of surfactant and cosurfactant with oil is 9:1. The mixture was made to be homogeneous, then sonicated for 15 minutes, and was placed in 45^0C temperature for 10 minutes. The result of the mixture was keep for 24 hours in the room temperature to find out the homogeneity. The homogeneous formula is the chosen formula for SNEDDS formula [14].

2.2.3. *SNEDDS formula selection*
100 μL of SNEDDS formula candidate was added with aquades until the last volume becomes 50 ml . Mixture homogenization was done by using vortex for 30 minutes. The mixture result which is in the form of homogeneous emulsion and give clear visual look are the beginning marks of the successfulness of nanoemulsion making. SNEDDS which had been emulsified was measured on its transmittance by using spectrophotometer on the length wave 650nm with aquades blank to know the clarity level.
The calculation of emulsification time was done towards herb extract nanoemulsion in three medium which were aquades, artificial gastric fluid without pepsin, and artificial intestinal fluid without pancreatic. 500 ml of each medium were placed in 37^0c temperature on the magnetic stirrer with speed about 120rpm. SNEDDS containing 1 ml of herb extract was dropped into medium quickly. The observation was done towards the time needed from the first drop until it forms emulsion. Nanoemulsion that was formed was marked by herb extract SNEDDS which was perfectly dissolved in the medium [15].
Nanoparticle size, size distribution, and zeta potential was measured using Particle Size Analyzer (PSA). Two drops of SNEDDS sample were emulsified into 5 ml of aquades, taken 3 ml and poured into covet to be analyzed. Then, to know the morphology of nanoemulsion particle visually by doing observation using Transmission Electron Microscope (TEM).

2.2.4. Hypoglicemic Effect Test
To make mice resistance to insulin, that it can be done by feeding the mice with high fat woof which consists of woof (80%), lard (15%) and duck yolk (5%). Meanwhile, diet on fructose was also given 3,3 gram/kg BB of perioral mice for 80 days. To know whether the mice are resistance to insulin or not, a measurement was conducted on blood glucose level which comes from glibenclamide hypoglycaemic activity.
The test of pharmacology compound activity was conducted for 15 days. the sample were divided into 5 groups, which are, group I (normal), group II (negative), group III (metformin 91.75 mg/kg of weight), group IV (SNEDDS of Bay leaves ethyl acetate fraction 183,5 mg/kg of weight), and group V (the combination of both metformin 22.5 gram/kg of weight and SNEDDS of Bay leaves ethyl acetate fraction 91.75mg/kg of weight). Blood measurement was done by using reagent Glucose GOD FS. Blood measurement data was obtained from the 0 day, 80th day, and 95th day. The results of blood glucose level measurement were analysed by using paired t-test, independent t-test, and Anova test which is continued by LSD test.

3. Result and Discussions
3.1. Optimization or surfactant and cosurfactant, and VCO

This optimization aims to determine the composition ratio of surfactant and co-surfactant with oil (Virgin Coconut Oil) which is capable of producing a homogeneous phase after mixing. Composition optimization results surfactant and co-surfactant with Virgin Coconut Oil (VCO) (Table 1) showed that combination of tween 80 : propylene glycol : VCO was not able to produce a homogeneous mixture in various proportions, while combination of tween 80 : PEG 400 : VCO capable of producing a homogeneous mixture in the ratio of surfactant: co surfactant is 1:2 and 1:3. Combination of tween 20 : PEG 400 : VCO was not able to produce a homogeneous mixture in various proportions, while combination of tween 20 : propylene glycol : VCO capable of producing a homogeneous mixture in the ratio of surfactant: co-surfactant is 1:1 and 2:3. Combination of surfactant (Tween 80 and Tween 20), co-surfactant (propylene glycol or PEG 400) and VCO was not able to produce a homogeneous mixture in various proportions.

Table 1. Optimization of surfactant (Tween 80, Tween 20 and combination Tween 80 : Tween 20), co-surfactant (Propylene Glycol and PEG 400), and oil phase (Virgin Coconut Oil)

Surfactant	**Composition of surfactant : co-surfactant**	**VCO : surfactant-cosurfactant (1 : 9)**	
		PG	**PEG**
T80	1 : 1	x	X
	2 : 1	x	X
	3 : 1	x	X
	3 : 2	x	X
	2 : 3	x	X
	1 : 3	x	v*(F1)
	1 : 2	x	v*(F2)
T20	1 : 1	v*(F3)	X
	2 : 1	x	X
	3 : 1	x	X
	3 : 2	x	X
	2 : 3	v*(F4)	X
	1 : 3	x	X
	1 : 2	x	X
T80 : T20	**Composition of surfactant**	**T80/T20 : PG (3 : 1)**	**T80/T20 : PEG (3 : 1)**
	1 : 1	x	X
	2 : 1	x	X
	3 : 1	x	X
	3 : 2	x	X
	2 : 3	x	X
	1 : 3	x	X
	1 : 2	x	X

x : not homogeneous (split); v: homogeneous (not split)

3.2. SNEDDS Formula Selection

This selection was done on SNEDDS formula which is chosen from optimization phase of surfactant, cosurfactant, and oil composition which are formula F1, F2, F3, and F4. Transmittance observation result (%) at 650 nm wave length shows that formula with the highest transmittance is formula F2. The average transmittance value F2 is 90,95% while emulsification time average in aquades is 13.00 seconds, in AGF is 13.09 seconds, and in AIF is 16.05 seconds. Nanoemulsion particle size was measured by using Dynamic light scattering method using PSA. From the measurement, the result of nanoemulsion size is 84,5 Nm. Nanoemulsion has size about 1-100nm [16].

An emulsion can be said as nano-sized emulsion if it is passed by light having wave length about 650 nm [17[. The light can be continued until the solution looks transparent. The high transmittance value which is closer to aquades is 100%. Emulsification tima is time that is calculated from the beginning of sprinkling until the formation of nanoemulsion in three medium, those are aquades, Artificial Gastric Fluid (AGF) without pepsin, and Artificial Intestinal Fluid (AIF) without pancreatin. The faster process will fasten the nanoemulsion formation.

From the visual observation of nanoemulsion morphology of formula 2 SNEDDS availability using Transmission Electron Microscope (TEM), the obtained results are as follows:

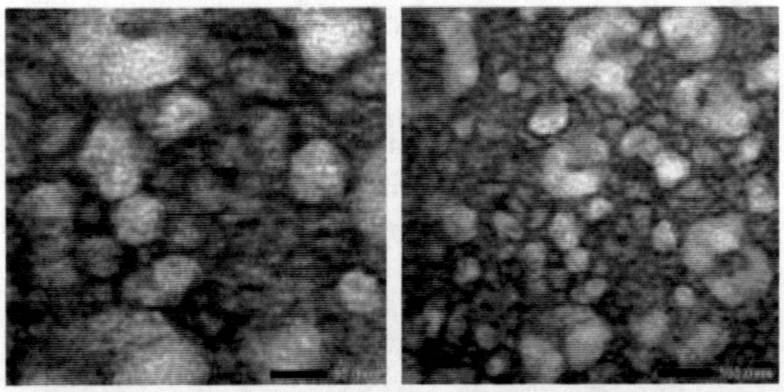

Figure 1. Nanoemulsion morphology of formula 2 SNEDDS using transmission Electron Microscope (TEM) with $2x10^5$ and 10^5 x magnification

The result (figure 1) observation above shows *sferis* particle though there is a minor part of the particle that is not perfectly *sferis*. Particle morphology is important because *un-sferis* particle form will enable inter-particle contact which will lead to aggregation (Couvreur et al., 2002) [18].

3.3. Hypoglycaemic Effect

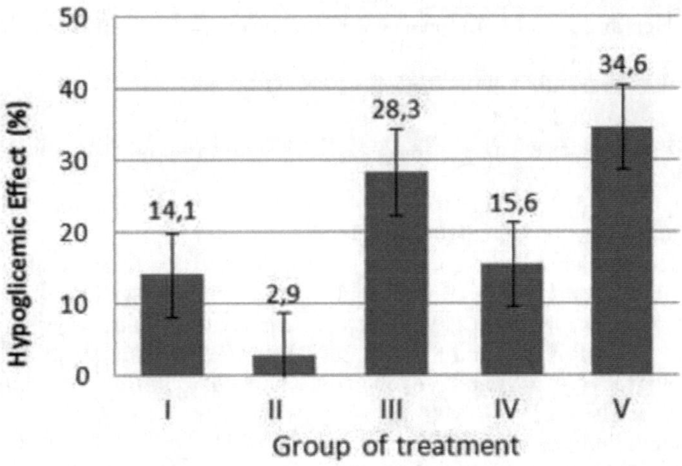

Figure 2. Hypoglycaemic effect of bay leaves treatment. The figure showed CMC Na treatment on normal rat (I), CMC Na treatment on insulin resistance rat (II), single dose of metformin treatment on insulin resistance rat (III), SNEDDS bay leaves fraction on insulin resistance rat (IV) and combination treatment of a half dose SNEDDS bay leaves fraction with metformin on insulin resistance rat (V)

According to figure 2 showed that group III and group V have higher hypoglycaemic effect with value 28.3% and 34.6%% after compared with control group.The result of LSD statistical analysis on the 80th day and 95 day between treatment on group III (single dose of metformin), group IV (SNEDDS), and group V (the combination a half dose of metformin and SNEDDS) showed that there is a significant difference ($P>0.05$). It means that combination treatment of SNEEDS bay leaves fraction and metformin has potensial to develop as co-therapy of ADMRI. Further molecular target detection to investigate its mechanism need to be conducted.

4. Conclusions

The results showed that the optimum SNEDDS formula was tween 80 : PEG 400 : Virgin Coconut Oil (30% : 60% : 10%) in 5 mL. SNEDDS fraction of bay leaves had emulsification time with value 13.00 seconds, the average of droplet size value 84.5 nanometer and zeta potential value ± 0.2 mV. Morphological observation showed that the nanoemulsion particles had spherical shaped and stable in different pH media. SNEDDS from bay leaves ethyl acetate fraction can decrease blood glucose level with hypoglycaemic effect was 15.6% but combination a- half dose of SNEDDS and metformin from ethyl acetate fraction has higher hypoglycaemic effect (34.6%) compared to a-single dose of metformin (hypoglycaemic effect was 28.3%) on the animal testing with diabetes mellitus type 2 resistance insulin.

References

[1] Taufiqurrohman 2015 *Journal Majority* **4** 101-108
[2] Date A A, Desai N, Dixit R and Nagarsenker M 2010 *Nanomedicine* **5** 1595–1616
[3] Enig G M 2004 *Health and Nutritional Benefits from Coconut Oil and Its Advantages Over Competing Oils* (Maryland: Enig Associates Inc.)
[4] Debnath S, Satayanarayana and Kumar G V 2011 *Int. J. Adv. Pharm. Sci.* **2** 72–83
[5] Wakerly M G, Pouton C W, Meakin B J and Morton F S 1986 ACS *Symp. Series* **311** 242-255
[6] Rowe R C, Sheskey P and Quinn M E 2009 *Handbook of Pharmaceutical Excipients 6th Edition* (London: Pharmaceutical Press)
[7] Makadia H A, Bhatt A Y, Parmar R B, Paun J S and Tank H M 2013 *Asian J. Pharm. Res.* **3** 21–27
[8] Geurin J C and Reveillere H P 2011 *J. Nat.Prod.* **52** 171-173
[9] Department of Health Republic Indonesia 2007 *Pharmaceutical Care for Diabetes Mellitus (*Jakarta: Kemenkes RI Press)
[10] American Diabetes Association 2008 *Diabetes Care* **31** 62-66
[11] Ndraha S 2014 *Medicinus* **27** 9-16
[12] Heather R G, Jennie C B, Anne W T, Sharon E, Patty C and George A W 2001 *Diabetes Care* **24** 1137-1143
[13] Bansal A, Munjal B and Patel S 2010 Self-Nano-Emulsifying Curcuminoids Composition with Enhanced Bioavailability WO/2010/010431
[14] Meirista I 2014 Formulation and Nanoherbal Anti-LDL Activity from Combination Temulawak Extract (*Curcuma xanthorrhiza* Roxb.) and Sambung Nyawa (*Gynura procumbens* (Lour.) Merr.) Using Myritol 318 as Oil Phase *Thesis* Pharmacy UGM Yogyakarta
[15] Patel J Kevin G Patel A Raval M and Sheth N 2011 *Int. J. Pharm. Investigation* **1** 112–118
[16] Hosokawa M, Nogi K, Naito M and Yokoyama T 2007 *Nanoparticle Technology Handbook* 1st *Edition* (Oxford: Elsevier)
[17] Wahyuningsih I and Putranti W 2015 *Pharmacy* **12** 224-241
[18] Couvreur P Barrat G Fattal E Legrand P and Vauthier C 2002 *Crit. Rev. Ther. Drug Carrier Sys.* **19** 99-134

IOP Conf. Series: Materials Science and Engineering **176** (2017) 012005 doi:10.1088/1757-899X/176/1/012005

Fabrication and Characterization of Immobilized Biosurfactant Produced by *Pseudomonas aeruginosa* Grown on Cassava Industrial Wastewater into Activated Allophane as an Adsorbent

V Suryanti[1,*], D M Widjonarko[1], Windrawati[1] and V Widyaningsih[1]
[1]Department of Chemistry, Faculty of Mathematics and Natural Sciences, Sebelas Maret University, Jl. Ir. Sutami 36A Surakarta 57126 Indonesia

*E-mail: venty@mipa.uns.ac.id

Abstract. The immobilization of biosurfactant into activated allophane has been conducted with mass ratio of biosurfactant:allophane of 1:5; 1:7 and 1:10 and contact time of 24 and 48 h. The optimum condition for immobilization was reached when the mass ratio of biosurfactant: allophane of 1:10 with the contact time of 24 h was applied. The result yielded the immobilization product having the specific surface area of 82.42 m^2/g and the surface acidity of 9.12 mmol/g. A better adsorbent has been produced. In respect to the activated allophane, there was a decreasing of specific surface area about 20% and increasing of surface acidity value about 120%.

1. Introduction

Allophane is an amorphous clay mineral which is naturally available in volcanic soils. It is found in the mountain areas with an altitude of 0-3000 m above sea level [1-2]. Allophane is found in many Indonesian volcanic mountains. Natural allophane has been reported as an adsorbent for heavy metal ions, such as Cr, Fe, Cd, Cu, Pb and Mn [3]. Natural allophane which has been activated by H_2SO_4 and NaOH giving higher specific surface area and surface acidity value [4].

On the other hand, biosurfactants have also been applied for remediation of both organic and inorganic pollutants [5-6]. The application of biosurfactants as adsorbents for heavy metals has been reported [7-8]. Biosurfactants are surface active compounds which are produced by a variety of bacteria, fungi and yeast which have various structures having many functional groups for ion binding. They possess better environmental compatibility than that of the chemical surfactants, e.g biodegradable and lower toxicity [9-11]. Therefore, research in developing method of biosurfactants application as adsorbents is still need to be conducted.

The immobilization methods have been reported to increase the properties of adsorbents [12]. Our previous work demonstrated the production of biosurfactant by *Pseudomonas aeruginosa* using cassava industrial wastewater (*manipuera*) as media [13]. The abilities of allophane and biosurfactants as adsorbents may be increased by surface modification. The purpose of this research was to develop a better adsorbent by immobilization of biosurfactant into activated allophane.

2. Experimental

2.1. Materials and Instruments

Chemicals and solvents were used are analytical grade from e-Merck. The *manipuera* used was obtained a local cassava industry. The strain *P. aeruginosa* FNCC 0063 was purchased from IUC Food and Nutrition, Universitas Gadjah Mada, Indonesia. Allophane was collected from Lawu Mount, Central Java, Indonesia. The instruments were used are Shimadzhu FTIR-8201 PC Spectrometer and X-Ray Diffractometer Shimadzu 6000.

2.2 Media Used and Growth Condition

Cultures of bacteria were maintained on nutrient agar. Biosurfactant was produced using media composed of nutrient broth (8 g/l), NaCl (5 g/l) and *manipuera*. The media were sterilized prior used. Fermentation was carried out for 4 days at room temperature on a rotary shaker (100 rpm).

2.3. Biosurfactant recovery

Culture liquid of *P. aeruginosa* was centrifuged at 12,500 *g* for 15 mins. The supernatant was acidified to pH 2.0 with HCl 6N and leaved overnight at 4 °C. The supernatant then was extracted using chloroform:methanol = 2:1. The organic layer was evaporated to give the biosurfactant.

2.4. Preparation of activated allophane

The natural allophane was washed, dried and ground. The powder obtained was sieved on sieves of 150 mesh. Activation of allophane was conducted by addition of 1 L H_2SO_4 5 M to the 200 g of allophane and then stirred for 3 h. The allophane was then washed for several times until the pH was neutral. Finally, the activated allophane was then dried in oven at 100°C for 1 h.

2.5. Immobilization of biosurfactant into allophane

Mass ratio of biosurfactant : allophane used in this immobilization process were 1:5; 1:7 and 1:10 with contact time of 24 and 48 h. Aquadest (5 mL) was added to the biosurfactant (0.4 g) and stirred for 5 mins. The activated allophane was added as much as 2, 2.8 and 4 g. The water (25 mL) was then added and the mixtures were stirred for 24 or 48 h at room temperature. The solid was filtered using Whatman filtration paper No. 42 and washed with aquadest. The resulting solid was dried in oven at 40°C.

2.6. Characterization of biosurfactant, allophane and immobilization products (IP)

Characterization of biosurfactant, allophane and immobilization products (IP) were performed for functional groups determination using FT-IR spectrophotometer, specific surface area using methylene blue absorption test and surface acidity value using ammonia adsorption method [14-15]. Additional characterizations of allophane and immobilization products were conducted by sodium-flouride (NaF) test and X-Ray diffraction [16].

3. Results and Discussion

3.1. Biosurfactants characterization

The obtained FT-IR spectra (Figure 1a) revealed a strong and broad band at 3402 cm^{-1} for O-H stretching vibration. The band at 2927and 2854 cm^{-1} was related to the –CH stretch. The asymmetric stretching vibration of C=O carboxylate was identified by the band at 1654 and 1253 cm^{-1} and the peak at 1056 cm^{-1} was confirmed the presence of ester. The weak band at 1458 and 1377 cm^{-1} were obtained for -CH_3- and CH_2-, respectively. The resulted peaks of biosurfactants confirmed that rhamnolipids biosurfactants were produced by *P. aeruginosa* [17]. The specific surface area and surface acidity value of biosurfactant were 55 m^2/g and 88.24 mmol/g, respectively.

3.2. Allophane characterizations

The presence of allophane in the soil samples can be validated by NaF test which pH sample is greater than 9.4 [3]. The pH sample was obtained for 9.5 confirming the presence of allophane in the soil samples. The resulted FT-IR spectrum also verified the presence of allophane in the sample because it showed similar characteristic peaks with the standard allophane, such as 3447 cm^{-1} for -OH (Al-OH/Si-OH); 1035 cm^{-1} for Si-O/Al-O and 550, 432 and 471 cm^{-1} for O-Al-O/O-Si-O [3].

X-Ray powder diffraction of soil sample was recorded (Figure 1) and its *d-spacings* were confirmed with JCPDS for allophane (JCPDS 38-0449) as shown in Table 1. The XRD analysis revealed the peaks at 2θ = 27.854; 27.300 and 66.470 with intensity of 100, 7 and 9, respectively. The mineral compositions of sampel soil were allophan (A) and Klo-Ver-Mont (K-V-M) with the precentage of 34.81 and 9.07%, respectively. The specific surface area and surface acidity value of allophane were 104 m^2/g dan 4.12 mmol/g, respectively.

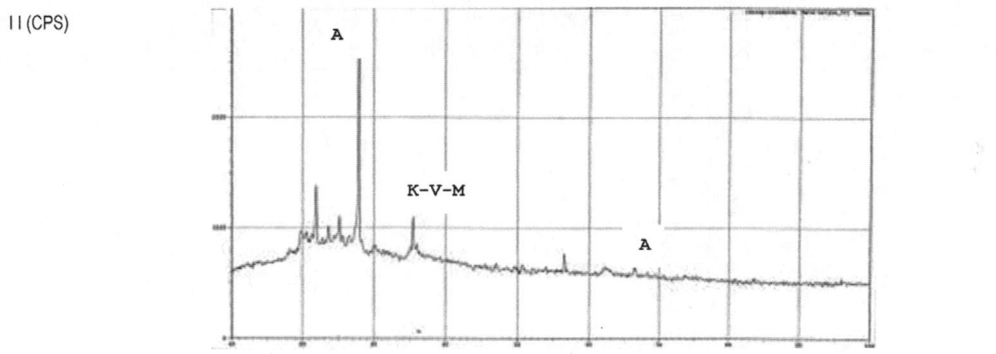

Figure 1. XRD Spectra of soil sample.

Table 1. Typical *d-spacings* data of the soil sample compared to the JCPDS 38-0449.

	d-spacings (Å)				
Standar (JCPDS 38-0449)	3.30	2.25	1.86	1.40	1.23
Sample soil	3.17	2.23	1.78	1.41	1.23

3.3. Immobilization Products (IP) characterization

Figure 2a and 2b shows the FT-IR spectra of the biosurfactant and immobilization product of sample IP-110-24, respectively. Table 2 presents the comparison peaks obtained for biosurfactant, activated allophane and immobilization products. Some peaks of activated allophane were found to be shifted in the immobilization product. However, the peaks for all immobilization product were similar to the peaks of activated allophane possibly due to the mass ratio of activated allophane was higher than that of biosurfactants.

Table 3 presents the specific surface area and surface acidity value of biosurfactant, activated allophane and immobilization products. The specific surface are of all immobilization products were smaller than that of activated allophane. These results were unexpected. Possibly, the porous of activated allophane were occupied by biosurfactants leading to the decreased of the specific surface area. The specific surface area of immobilization product with 48 h immobilization process were lower than that of with 24 h immobilization process. These findings indicated that the longer immobilization process, the more surface area were occupied by biosurfactant leading to the decreased of the specific surface area. The immobilization product with the mass ratio of 1:10 and the immobilization process of 24 h (IP-110-24) had the highest of specific surface area. In respect to the specific surface area of activated allophane, the specific surface area of IP-110-24 decreased 20.4 %.

The surface acidity value of all the immobilization products were higher than that of the activated allophane (Table 3). The surface acidity value reflects the number of ionic active sites of the samples. The higher of surface acidity value, the better adsorbent was produced. The porous of activated allophane were occupied by biosurfactant in the immobilization products resulting the higher surface acidity values were obtained These results possibly due to the additional functional groups from biosurfactants The surface acidity value of immobilization product with 24 h immobilization process were higher than that of with 48 h of immobilization process. These results suggested that the longer immobilization process, the lower number of active sites of the immobilization product was obtained resulting to the decrease of the surface acidity value. In respect to the surface acidity value of activated allophane, the highest of surface acidity value was obtained for IP-110-24 with 121.5 % increased.

(a)

(b)

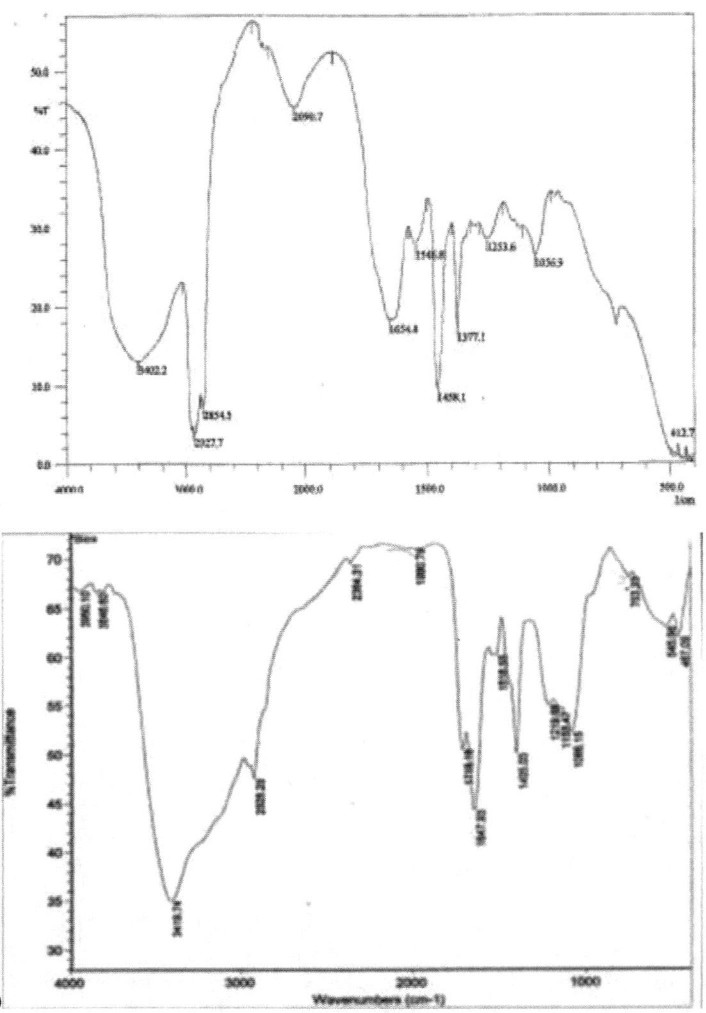

Figure 2. FT-IR spectra of: (a) biosurfactant and (b) immobilization product of sample IP-110-24.

Table 2. The peaks obtained from FT-IR spectra analysis for biosurfactant, activated allophane and immobilization products.

Samples	Wavenumber (cm^{-1}) for Fuctional Groups			
	-OH	C-O	Si-O/ Al-O	O-Al-O/ O-Si-O
Biosurfactants	3419	1086	-	-
Activated allophane	3449	1039	1039	542;471
IP-15-24	3449	1039	1039	542;471
IP-15-48	3449	1039	1039	543;471
IP-17-24	3451	1039	1039	542;471
IP-17-48	3453	1040	1040	542;471
IP-110-24	3449	1037	1037	540;471
IP-110-48	3451	1040	1040	542;471

Table 3. The specific surface area and surface acidity value of biosurfactant, activated allophane and immobilization products.

Samples	Specific Surface Area (m^2/g)	Surface Acidity Value (mmol/g)
Biosurfactants	54.86	88.24
Activated allophane	103.56	4.12
IP-15-24	73.85	9.12
IP-15-48	73.07	7.94
IP-17-24	78.44	9.12
IP-17-48	73.24	7.35
IP-110-24	82.44	9.12
IP-110-48	72.56	7.06

Based on the specific surface area and surface acidity value data, the optimum condition of immobilization process was obtained for sample IP-110-24 which had mass ratio of biosurfactant and allophane of 1:10 with 24 h of immobilization process. Although the specific surface area decreased in this immobilization products in respect to the activated allophane, its surface acidity value increased significantly by 121.5% indicating the higher number of active sites were obtained.

4. Conclusion

The novel adsorbent could be obtained by immobilization of biosurfactant produced by *P. aeruginosa* grown on *manipuera* into allophane. The optimum condition of immobilization process was achieved for sample IP-110-24 which had mass ratio of biosurfactant and allophane of 1:10 with 24 h immobilization process. The specific surface area and surface acidity value of this sample were 82.44 m^2/g and 9.12 mmol/g, respectively. In respect to activated allophane, there was a decreasing of specific surface area about 20% and increasing of surface acidity value about 120%. The significantly increasing of surface acidity value indicated that the immobilization product was better adsorbent than that of activated allophane. This adsorbent has potential applications in environmental remediation.

Acknowledgments
The authors acknowledge their thanks to Education and Culture Office, Jawa Tengah, Indonesia for the financial support to this research in the form of Applied Research Grant 2006.

References

[1].Wada K 1989 *Allophane and Imogolite* In Dixon J B and Weed S B (ed), Minerals in Soil Environments, 2nd Edition 1051-1087.

[2] Iyoda F, Hayashi S, Arakawa S and Okamoto M 2011 PPS-27 *27th World Congress of the Polymer Processing Society*, May 10-14, Marrakech, Morocco.

[3] Pranoto, Suranto K H, Sugiyarto and Ashadi, 2013 *J. Environ. Earth Sci.* **3(5)** 48-56.

[4] Widjonarko D M, Pranoto and Christina Y 2003 *Alchemy* **2(2)** 19-19.

[5] Cameotra S S and Makkar R S 2010 *Pure Appl. Chem.* **82** 1 97-116.

[6] Suryanti V, Marliyana S D and Wulandari A 2015 *AIP Conf. Proc.* **1699** 040003 doi: 10.1063/1.4938318.

[7] Suryanti V, Marliyana S D and Andriani D 2016 *IOP Conf. Ser. Mater. Sci. Eng.* **107** 012018 doi:10.1088/1757-899X/107/1/012018.

[8] Suryanti V, Hastuti S and Pujiastuti D 2016 *AIP Conf. Proc.* 1710 030016 doi: 10.1063/1.4941482.

[9] Amara P F, Coelho M A, Marrucho I M and Coutinho J A 2010 *Adv. Exp. Med. Biol.* **672** 236-244.

[10] Muthusamy K, Gopalakrishnan S, Ravi T K and Sivachidambaram P 2008 *Curr. Sci.* **94** 736-747.

[11] Mukherjee S, Das P and Sen R 2006 *Trends Biotechnol.* **24(11)** 509-515.

[12] Volesky B 1990 *Biosorption of Heavy Metals* CRC Press Florida.

[13] Suryanti V, Hastuti S, Handayani, D S and Windrawati 2014 *Alchemy* **10(1)** 22-30.

[14] Hang P T and Brindley G W 1970 *Clays and Clay Minerals* **18** 203-212 Pergamon Press Great Britain.

[15] Brown D R and Rhodes C N 1997 *Thermochim. Acta* **294 (1)** 33-37.

[16] Brydon J E and Day J H 1970 *Can. J. Soil Sci.* **50** 35-41.

[17] Makkar R S and Cameotra S S 2002 *Appl. Microbiol. Biotechnol.* **58** 428-434.

The interaction of a binary/ternary interactive mixture of hydrophobic-hydrophilic materials on the drug distribution and drug release performance in the tablet formulation

A Ainurofiq[1,3] **and S Choiri**[1,2]

[1]Department of Pharmacy, Sebelas Maret University, Ir. Sutami 36A Surakarta, Indonesia 57126

[2]Faculty of Pharmacy, Gadjah Mada University, Sekip Utara, Yogyakarta, Indonesia 55281

E-mail: rofiq@mipa.uns.ac.id

Abstract. The aim of this research was to optimize and determine an interaction of a binary/ternary mixture of hydrophobic-hydrophilic materials (H-HM) on the drug distribution, tablet characteristics, and drug release performance. The interactive mixture (IM) between carrier and H-HM was conducted using a carrier, Pruv and Cab-O-Sil as hydrophilic materials, magnesium stearate as a hydrophobic material, and a micronized nifedipine as a drug model. These interactions between binary and ternary mixtures of H-HM were assessed by a simplex centroid design (SCD) approach. The homogeneity of IM between drug and carrier was achieved at more time of mixing time. Unique effects and interactions of H-HM were observed on the drug distribution and drug release. Furthermore, the SCD had successfully determined the optimum design space of IM in the ternary mixture of H-HM.

1. Introduction

In the solid oral dosage form e.g. a tablet formulation, a mixing or blending is a fundamental process to ensure the homogeneity of the drug and achieve a high product quality. The objective of mixing is a less variation of the drug content in each unit of dosage form i.e. a homogenous mixture [1–3]. Furthermore, the mixing process of a very small particle involve the interactive mixture (IM) to obtain a good homogeneity. It is a mixture usually using an interactive constituent (a fine particle of drug) and a carrier. The mixing mechanism based on adhesive and interparticles forces, that is a fine or ultrafine drug adhere in to the surface of carrier [4–6].

A lubricant is used to reduce the friction or enhance the powder flow by reducing the interparticle friction [7,8]. The difference of lubricant properties affects the final product of a dosage form [4]. Magnesium stearate (MgSt) had negative effects on reducing the drug release and prolong the disintegration time due to its hydrophobicity [8]. The addition of MgSt in the final mixture retarded the drug release [9,10]. Enhancement of MgSt concentration enhance those negative effects although easy in the compression process due to a proper tablet ejection [10,11]. MgSt at concentrations 0,1 and 5% not significantly affected on the tensile strength and tablet crushing strength [10]. MgSt increased the drug distribution depending on the MgSt concentration and particle size [12].

[3] To whom any correspondence should be addressed.

Sodium stearyl fumarate (SSF) as a hydrophilic material has been used as a pharmaceutical lubricant [11,13]. It has some advantages include a less interaction or compatible with other excipients, no reduce bioavailability effect, and enhance a tablet appearance [13]. Colloidal silica is a hydrophilic material that is used as a flowing agent to enhance flow of the powder or granule. This material is very small and has tendency to form an agglomerate [14]. A combination between colloidal silica and MgSt did not has significant effect on the drug release (13).

A design of experiment (DoE) offers several advantages, those are to reduce number of runs, to elucidate and assess the effect and interaction of factors, and to obtain the design space [15–17]. The aim of this study was to assess and elucidate effects and interactions of single, binary or ternary mixtures of MgSt, Pruv, Cab-O-Sil (COS) on the drug distribution, tablet characteristics, and drug release performance. The most optimum combination can be determined by a quality by design (QbD) approach using a simplex centroid design (SCD).

2. Experimental

2.1. *Material*

Nifedipin (Nif) micronized form (Lot No. 400196067; d_{10} = 1.5 μm, d_{50} = 6,7 μm, and d_{90} = 9,8 μm) was obtained from Dexa Medica (Palembang, Indonesia), Pruv (d_{10} = 1 μm, d_{50} = 9 μm, and d_{90} = 29 μm) was gift from JRS Pharma (Rosenberg, Germany), (COS) d_{50} = 16 nm (Cabot; Jiujiang, China), MgSt d_{50} = 36 μm (Peter Greven; Germany), lactose (DFE Pharma; Foxhol, the Netherlands), Avicel PH101 (FMC Biopolymer, NJ, USA), Plasdone K30 (Shin-Etsu, Tokyo, Japan). All of the ingredients were used as receipt and pharmaceutical grade.

2.2. *Preparation of carrier*

Lactose 70% and Avicel 30% were mixed for 15 min at 25 revolution per minute (rpm). Plasdone K30 3% (w/v) in ethanol 50% (v/v) was used as a binder and added until the elastic mass was achieved. The mass was passed through a 20 mesh sieve (850 μm) and dried in the oven at 60°C for 4 h or the loss on drying below 2%. The used granule as carrier was passed through a 30 mesh sieve (600 μm) and retained on 50 mesh sieve (300 μm).

2.3. *Homogeneity test of binary mixture between drug and carrier*

The drug of 5% and the carrier 95% was mixed at 25 rpm. The sample was collected in the 3 different places at 100, 200, 300, 500, 800, 1200, and 1600 revolution (rev). The samples were determined on the Nif content spectrophotometrically at 238 nm and calculated the relative standard deviation (RSD) of the Nif content. The homogeneity was interpreted by a RSD.

2.4. *Tablet formulation*

A 13 runs was constructed depending on a quadratic equation using a SCD. The lubricant mixture was used at total level of 1%, 5% drug and 94% carrier. The Nif and carrier were mixed at 1600 rev, then lubricants were added to the mixture and mixed at 100 rev. The flow characteristic of the final mixture was assessed by a flow time through the funnel. The mixture were compressed using a single punch tablet (a flat-face punch diameter 8 mm) at a similar compression force.

2.5. *Tablet characterization*

The Nif was individually determined using spectrophotometer UV-Vis. The drug distribution was described by the RSD of 10 tablets. Tablet crushing strength (CS) was conducted using a Gouming YD-1 hardness tester (Tianjin, China) (20 tablets). The tablet disintegration test (DT) was determined using a Gouming DT-1 disintegration tester (Tianjin, China).

2.6. *Drug release*

The drug release was determined using an Electrolab TDT 08L (Electrolab, India) dissolution tester type apparatus II with 900 ml of HCl 0,1N as a dissolution medium, the temperature maintained at 37±0.5°C and a stirring rate of 50 rpm. Aliquots of 5 mL were withdrawn at a determined time by a replacement of same medium. All the samples were analyzed spectrophotometrically. The model dependent approach was applied to compare the drug release profiles. The dissolution efficiency (DE) was used and calculated using following equation.

$$DE = \frac{\int_0^t y\delta t}{100T} \text{ x } 100\% \quad (1)$$

Here *y* is an amount of the drug released in time "*t*" and T is a total time. The DE was calculated depending on the drug released over 60 min (DE_{60min}).

2.7. Data analysis

The obtained data were analyzed using a Design Expert® software version 7.1.5 (Stat-Ease Inc., Minneapolis, MN, USA). Several statistical parameters were used to evaluate the equation fitting i.e the determination coefficient (R^2), adjusted determination coefficient (Adj. R^2), predicted determination coefficient (Pred. R^2), adequate precision (adeq. prec) and predicted residual error sum square (PRESS). The optimized formulae was determined by critical quality attributes (CQAs) depending on quality target product profile (QTPP) that are presented in Table 1.

Table 1. CQAs and QTPP of IM between hydrophobic and hydrophilic material

Parameters	QTPP	Priority level
Flow Rate	more than 10 g/s	+++
RSD	less than 5%	+++++
CS	more than 40N	+++
DT	less than 300 s	+++.
DE_{60min}	more than 30%	+++++

3. Results and Discussion

3.1. Preparation of carrier

In the IM, a granule size plays a fundamental role in the drug distribution and homogeneity [4,5]. A sufficient granule size promote a single layer of drug thus avoid the segregation of drug and carrier [18,19]. In the carrier preparation, the granule size was controlled. Hence, the particle size diameter was d_{10} = 377.5 μm, d_{50} = 468.6 μm, and d_{90} = 553.5 μm, thus indicated the particle size normally distributed and had a narrow distribution of particle size. The granule had a flow rate of 14.50±1.05 g/s, bulk and tapped densities of 0.441±0.006 and 0.472±0.004 g/mL, respectively, an angle of repose of 21.86±0.84° and a compressibility index of 10.66±1.48%. According to these results, the granule had a good flow ability, thus can ensure the constant filling in the compression room.

3.2. Homogeneity mixture between drug and carrier

The homogeneity of mixture between drug and carrier was described by a RSD of Nif content. Figure 1 shows that the RSD pattern under different revolutions of mixing decreased exponentially. In the IM the drug as a guest substance adhere in the carrier strongly, thus avoid the segregation, an enhancement of mixing time enhanced the homogeneity [4,5,19]. In the initial mixing at a low revolution (100 – 500 rev), the RSD higher than that of revolution more than 800 rev. The desirable RSD was less than 5% that was achieved at 1600 rev.

3.3. DoE approach for assessing the effect of lubricant materials

A DoE approach using a SCD was helpful to elucidate the effect and interaction between hydrophilic and hydrophobic materials (HH-M). These effects and interactions could be knew by comparing the regression coefficient value and significant effects and interactions were showed by p-value (<0.05). Depending on the flow rate, an equation using a special cubic model, the linear mixture (MgSt, Pruv, and COS proportions) not significantly affected on the flow rate (p>0.05). Furthermore, these materials did not show a different effect due to small amount addition and we suggested that the flow characteristic was dominantly affected by the carrier. There was no a significant interaction in each material (p>0.05), meanwhile an interaction in the ternary mixture of lubricant materials showed an enhancement of flow rate significantly ($p<0.05$). A contour plot (CP) of powder flow (Fig. 2(1)a) shows a unique effect on the mixture of lubricant materials. Fitting parameters of flow rate equation showed that the difference between adjusted and predicted R^2 was more than 0.2. Furthermore, the model could not be used to predict the flow rate adequately. In the mixing process, H-HM as lubricants coat the carrier and drug. Furthermore, those are be in place in the inter particles and affect the particle bonding consequently [20]. The particle bonding (CS) described by a MLRA of CS (Table 2). The linear mixture of equation had a significant effect on CS ($p<0.05$). The MgSt had more dominant effect on an increase the CS than Pruv and COS due to a reduction in ejection force. MgSt had a different mechanism in reduce the friction [9]. The interactions of MgSt – Pruv and MgSt-COS were not significant affecting on an increase the tablet CS ($p>0.05$). Although an interaction between Pruv and COS was statistically significant on an increase the tablet CS. The CP of tablet CS (Fig 2(1b)) was used to assess and elucidate the effect of H-HM on the interparticle bonding. Generally, the addition of H-HM promotes a reduce the tablet CS due to a reduction of the interparticle bonding that is caused by coating of H-HM on the surface of particle [4].

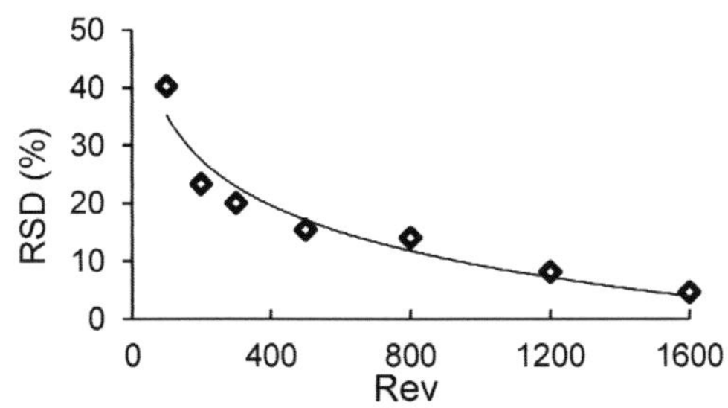

Figure 1. RSD of drug distribution in the IM between drug and carrier as a function of revolution. This point was calculated using quintuplicate replications.

The drug homogeneity is ensure the drug distribution be spread evenly in all samples. The lower RSD the more homogeneous mixture, in other word the RSD is less than 5%. The homogeneity of a binary mixture between Nif and carrier was achieved at 1600 rev of mixing process. The addition of lubricant materials could promote an alteration of homogeneity i.e. the RSD depending on their interactions between Nif, carrier, and lubricants. The RSD of all formulations ranged from 1.27 to 7.85%. This indicated that the addition of H-HM altered the drug homogeneity or induced a segregation. Depending on a quadratic equation of RSD, the linear mixture was significant affecting on the RSD. The Pruv had a better effect on drug distribution than MgSt and COS. Those were affected by particle size and interaction i.e. a competition between drug and lubricant that caused a redistribution of the adhered drug on the carrier. Interactions of MgSt-Pruv and COS – Pruv enhanced the RSD not significantly ($p>0.05$), although an interaction between MgSt and COS reduced the RSD significantly ($p<0.05$). The CP of RSD (Fig. 2(1c)) showed that Pruv had the lowest RSD and MgSt had the highest RSD, and the RSD changed as function of proportion of the mixture.

The quadratic model of DT was constructed using a logaritmic transformation to obtain the best fitting and a normally distributed data. The linear mixture was significant ($p<0.05$). Among the others materials, MgSt was 10 fold longer DT than Pruv and 35 fold than COS. All interactions between H-HM increased the DT significantly ($p<0.05$). Fig 2(2)a describes the DT pattern in a change of lubricant proportion in the mixture. The highest DT was obtained at high proportion of MgSt, and decreased logaritmically with decreasing the MgSt proportion. The lowest DT was achieved at the highest proportion of COS due to its hydrophilicity.

The drug release for all formulation are presented in Figure 3. Nif was chosen as a drug model due to practically insoluble in water, therefore it was very sensitive if a small amount of H-HM was added. The formulation with a hydrophobic material i.e MgSt (runs 1) showed that had lower drug released than a hydrophilic material about 20 and 80% during 60 min, respectively. This phenomenon was caused by coating the carrier and drug thus inhibit a water penetration and avoid or retard the disintegration process thus decreased the drug release. A binary blend reduced the drug released when compared to a single component of lubricant. The effect of material hydrophobicity was more dominant in these combinations so it reduced the drug release. A tertiary mixture of lubricant was a preferable combination except the high proportion of MgSt. An over blended or lubrication cause a problem that is associated by disintegration process and drug release [3,8].

Tabel 2. Statistical parameter of simplex centroid design

Statistics	Flow rate (g/sec)		RSD (%)		CS (N)		Log(DT) (sec)		DE_{60min} (%)	
	Coef.	p-value	Coef.	p-value	Coef.	p-value	Coef.	p-value	Coef.	p-value
A	16.86		7.11		80.90		3.10		16.74	
B	18.37	0.902*	1.99	0.002	49.95	0.000	2.10	<0.0001	53.19	0.0003
C	16.57		5.60		30.54		1.56		57.42	
AB	2.85	0.879*	5.61	0.102*	39.46	0.157*	2.43	0.008	-68.69	0.021
AC	5.62	0.765*	-10.66	0.009	40.48	0.148*	3.00	0.003	-99.09	0.004
BC	0.58	0.975*	5.73	0.097*	63.51	0.038	2.45	0.008	-118.06	0.002
ABC	535.71	0.006	-	-	-	-	-	-	687.13	0.005
Model	-	0.0289	-	0.003	-	0.0007	-	0.0001	-	0.0007
Lack of fit	-	0.0211	-	0.490*	-	0.1667*	-	0.1107*	-	0.1433*
R^2	0.8455		0.8875		0.9286		0.9593		0.9576	
Adj. R^2	0.6910		0.8071		0.8775		0.9302		0.9152	
Pred. R^2	-0.8149		0.6344		0.7019		0.8781		0.4451	
PRESS	1177.48		12.81		1141.75		0.59		1990.37	
Adeq. Prec	7.184		10.042		11.858		15.659		12.213	

* : not significant term

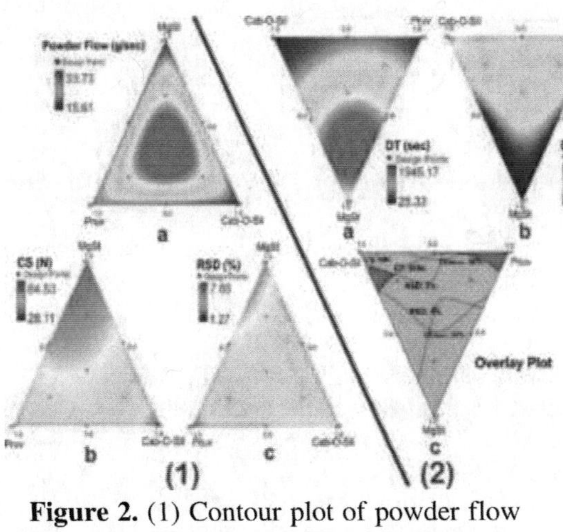

Figure 2. (1) Contour plot of powder flow (a), tablet CS (b), and RSD of drug content (c); and (2) contour plot of DT (a), DE_{60min} (b), and superimposed contour plot (c)

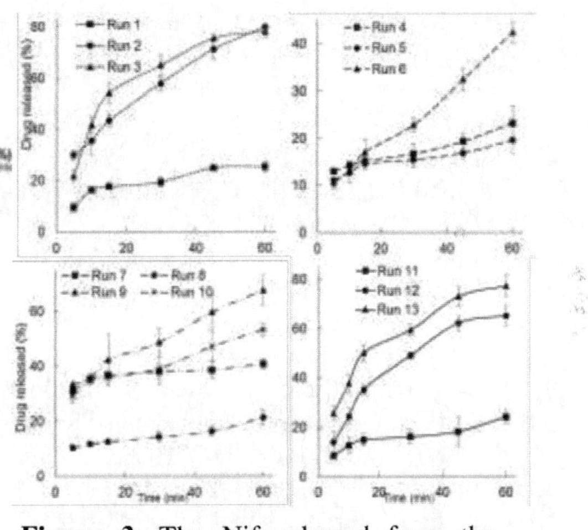

Figure 3. The Nif released from the tablet formulation using the mixture of hydrophilic and hydrophobic materials.

The drug released of all formulations (Figure 3) was compared by DE, comparing all points and possibility depending on the area under curve. A special cubic equation was constructed to describe the DE. The linear mixture had significant effect on DE_{60min} ($p < 0.05$), thus a hydrophilic material had higher in an increasing the DE than a hydrophobic material. It was discussed above due to a hydrophobicity characteristic. The interaction of two material lubricants and a combination of three materials lubricant had a significant effect on reducing and increasing DE_{60min} ($p<0.05$), respectively. The interaction of two lubricant in order from high to low in reducing the DE was Pruv-COS>MgSt-COS>MgSt-Pruv. An interaction of three lubricants increased the DE_{60min} value. Although some problems could occur due to inadequate prediction. The adequate precision describe the ratio of signal to noise, it more than 4 is the desirable model. The pattern of CP of DE_{60min} was similar to CP of DT, but the value had an inversely correlation with DT. The control strategy was obtained by combining the CP of each CQA depending on QTPP. A superimposed CP was achieved and presented in Fig 2(2)c. The key determining in the control strategy is presented in Table 1. The control strategy was a non-linear combination, therefore it could be minimize to obtain a linear combination. The desired formulation based on QTTP was achieved at a dominant Pruv region and combination of COS and MgSt.

4. Conclusion

The homogeneity of IM between drug and carrier needed more time to ensure the homogeneity. A H-HM has unique effects and interactions on the drug distribution, tablet properties and drug release. All of the interactions between H-HM reduced the dissolution performance. Furthermore, the optimized formulae was achieve at Pruv 100% or a combination between COS 84% and MgSt 16%.

Acknowledgment

The authors would like to thank JRS Pharma (Rosenberg, Germany) and Dexa medica (Palembang, Indonesia) for providing Pruv and nifedipine, respectively. This work was supported by Sebelas Maret University with Grant Mandatory Research of PNBP Sebelas Maret University 2015, batch number 624/UN27.11/PL/2015.

References

[1] Barling D, Morton DAV, Hapgood K 2015 *Powder Technol.* **270**:461–469.
[2] Fan LT, Chen Y, Lai FS 1990 *Powder Technol.* **61**(3):255–287.
[3] Koller DM, Posch A, Hörl G, Voura C, Radl S, Urbanetz N, et al. 2011 *Powder Technol.* **205**(1–3):87–96.
[4] Kuncahyo I, Choiri S 2015 *Int J Pharm Pharm Sci.* **7**(1):397–402.
[5] Mangal S, Gengenbach T, Millington-Smith D, Armstrong B, Morton DAV, Larson I. 2016 *Eur J Pharm Biopharm.* **102**:168–177.
[6] Mendez ASL, de Carli G, Garcia CV 2010 *Powder Technol.* **198** (2):310–313.
[7] Morin G, Briens L 2013 *AAPS PharmSciTech.* **14**(3):1158–1168.
[8] Pingali K, Mendez R, Lewis D, Michniak-Kohn B, Cuitino A, Muzzio F 2011 *Int J Pharm.* **409**(1–2):269–277.
[9] Nakagawa H, Kano M, Hasebe S, Suzuki T, Wakiyama N 2013 *Int J Pharm.* **441**(1–2):402–13.
[10] Tay T, Morton DAV, Gengenbach TR, Stewart PJ 2012 *Eur J Pharm Biopharm.* **80**(2):443–52.
[11] Wang J, Wen H, Desai D 2010 *Eur J Pharm Biopharm.* **75**(1):1–15.
[12] Tay T, Das S, Stewart P 2010 *Int J Pharm.* **383**(1–2):62–69.
[13] Patadia R, Vora C, Mittal K, Mashru R 2016 *Pharm Dev Technol.* **21**(3):302–310.
[14] Majerová D, Kulaviak L, Růžička M, Štěpánek F, Zámostný P 2016 *Eur J Pharm Biopharm.* **106**:2–8.
[15] Badawy SIF, Narang AS, LaMarche KR, Subramanian GA, Varia SA, Lin J, et al. 2016 *J Pharm Sci.* **105**(1):168–181.
[16] Choiri S, Sulaiman TNS, Kuncahyo I 2014 *Indonesian J Pharm.* **25**(4):255–264.
[17] Yu LX, Amidon G, Khan MA, Hoag SW, Polli J, Raju GK, et al 2014 *AAPS J.* **16**(4):771–783.
[18] Allahham A, Stewart PJ, Das SC 2013 *Int J Pharm.* **457**(1):101–109.
[19] Sulihtyowati Soebagyo S, Stewart PJ 1985 *Int J Pharm.* **25**(2):225–236.
[20] Zhou Q (Tony), Denman JA, Gengenbach T, Das S, Qu L, Zhang H, et al 2011 *J Pharm Sci.* **100**(8):3421–3430.

ANTIBACTERIAL ACTIVITY OF THE ISOLATION ETHYL ACETATE-SOLUBLE EXTRACT NONI FRUIT (*Morindra citrifolia* L.) AGAINST MEAT BACTERIAL DECAY

E R Nugraheni[1], M B E Nurrakhman[2], H Munawaroh[3], L Saputri[3]

[1]Department of Pharmacy, Faculty of Mathematic and Natural Sciences, Sebelas Maret University, Jl. Ir. Sutami 36A Surakarta 57126 INDONESIA
[2]Departement of Biology, Faculty of Mathematic and Natural Sciences, Sebelas Maret University, Jl. Ir. Sutami 36A Surakarta 57126 INDONESIA
[3]Department of Chemistry, Faculty of Mathematic and Natural Sciences, Sebelas Maret University, Jl. Ir. Sutami 36A Surakarta 57126 INDONESIA

E-mail: retnaningtyas_n@yahoo.co.id

Abstract. Noni (*Morindra citrifolia L.*) is native to Indonesia which have medicinal properties. One of them as an antibacterial. This study aims to determine the antibacterial activity of isolates from the ethanol extract noni fruit to bacterial decay meat is *Bacillus licheniformis, Klebsiella pneumonia, Bacillus alvei, Acinetobacter calcoaceticus,* and *Staphylococcus saprophyticus*. The extraction process using the maceration method, and then made a partition by centrifugation ethyl acetate. Soluble part partition showed bacterial growth inhibition activity of the strong to very strong. Furthermore, the ethyl acetate soluble partition on preparative thin layer chromatography produced 5 isolates. Isolates obtained antibacterial activity test performed with a concentration of 20% and 30%. The results of antibacterial test against bacteria test isolates, showing isolates A can not inhibit the growth of bacteria, isolates B and C have medium activity and strong, isolates D and E isolates have activity against bacteria that were tested. MIC and MBC test results showed that the isolates B gives an inhibitory effect (bacteriostatic) against all bacteria. Content analysis of compounds by TLC using the reagents cerium (IV) sulfate indicates a phenol group. Isolates B contains a major compound which can be used as an antibacterial candidate in food preservation replace chemical preservatives.

1. Introduction

Meat is a food of high nutritional value because it is rich in protein, fat, minerals and other substances that are needed the body [1]. Foodstuffs are often decay and rancidity [2]. The process caused by the biochemical reaction that comes from within and outside of foodstuffs such as microbial activity [3]. The microbial contaminant is one cause of reduced quality of the meat so as to be unsafe for consumption. Contamination of microbial spoilage leading to degradation of the protein-protein breakdown process into simple molecules such as amino acids that cause cells to become damaged flesh [4].

According to Purwani et al. (2008) [5], there are some bacteria that live in fresh meat, namely *Bacillus licheniformis, Klebsiella pneumoniae, Bacillus alvei, Acinetobacter calcoaceticus,* and *Staphylococcus saprophyticus.* The bacteria can potentially as spoilage bacteria capable of degrading protein for meat used as a nutrient for the growth and development of the bacteria.

Noni (*Morinda citrifolia L.*) is one of the plants that produce active compounds that can be used as an alternative to natural preservatives. This plant is native to Indonesia being very abundant. Part of these plants contains many active compounds found in stems, roots, leaves or fruit [6]. According to Peter (2005) [7], in part noni plant there are several compounds that can be used as antibacterials, such as flavonoids, glycosides and scopoletin.

According to [8], the ethanol extract of noni is able to inhibit the growth of spoilage bacteria on meat and fresh fish. The minimum inhibitory ethanol extract of Noni fruit by 70%. To optimize its inhibitory ability is necessary to do advanced processes such as partitioning and preparative thin layer chromatography to obtain pure isolates are active and have greater power resistor. This study uses the antibacterial activity test method to determine the potential of the results of partitioning and preparative thin layer chromatography ethanol extract of noni. The final analysis of the study by measuring Minimum Inhibitory Concentration (MIC) and Minimum Bactericidal Concentration (MBC) to determine the inhibition of the growth of the active isolates, against spoilage bacteria on meat.

2. Experimental

2.1 Isolation of Section Ethyl Acetate with Preparative Thin Layer Chromatography (TLC)

Isolation was done by Thin Layer Chromatography (TLC). Silica gel plate of GF254 was made by comparison of 7 grams of powdered silica and 14 mL of distilled water (1: 2), then put on glass plates and do activation in oven with temperature of 50 °C for ± 1 hour. Then, samples were spotted on TLC glass plate of like made ribbon until approximately 3 times. The next step was elution of TLC glass plate with mobile phase (chloroform: ethyl acetate 3: 1). Spot formed scraped to obtain isolates. Isolates of silica gel separated with methanol: chloroform in the ratio 1: 1, then filtered the isolates to obtain filtrate and residue. The filtrate was dried to be powder, then was weighed.

2.2 Thin Layer Chromatography and Determination Compound of Isolation Results

Isolates of the isolation results was detected by TLC. The stationary phase used silica gel plate of GF254 while the mobile phase using chloroform: ethyl acetate = 3: 1. The results of separation identified under UV light (254 nm and 366 nm). Furthermore, the TLC plate was sprayed with cerium (IV) sulphate to detect the general presence of compounds and with specific sprayers $FeCl_3$ and $AlCl_3$ then calculated value of Rf (retention factor).

2.3 Test of Antibacterial of Isolates Noni Fruit (M. citrifolia L.)

2.3.1 Well Diffusion Method

Sterile media of MHA 10 mL and inoculum poured in petri dish where inoculated with the bacteria of *Bacillus licheniformis, Klebsiella pneumoniae, Bacillus alvei, Acinetobacter calcoaceticus, and Staphylococcus saprophyticus.* Then made sinks with perforator. 25 microliters sample with concentration of 20 and 30% put on pitting, each concentration of 2 replicates. Furthermore, incubated at 37 °C for 24 hours.

2.3.2 Test of MIC and MBC

Isolates with optimum concentration made stock 40 mg / mL (2x of the optimum concentration). 50 mL of isolates and 50 mL inoculum were mixed in eppendorf. Then, incubated for 24 hours at 37 °C. After 24 hours, result of incubation displacement in a microplate which contains media of MHA, 25 mL for each repetition, then leveled with glass spreader. Then, incubated for 24 hours at 37 °C. If showed bacterial growth were expressed as the Minimum Inhibitory Concentration (MIC). While the absence of bacterial growth were expressed as the Minimum Bactericidal Concentration (MBC) [9].

3. Result and Discussion

3.1 Separation and Isolation Compound

Some parts of dissolved ethyl acetate separation and isolation with TLCp. The result of the isolation of soluble ethyl acetate 70% by UV_{254} showed 5 isolates could be separated and further isolates marked as A, B, C, D, and E. Isolate B has the greatest weight to the HRF 77 that allegedly contains major compound in noni fruit (*M. citrifolia L.*).

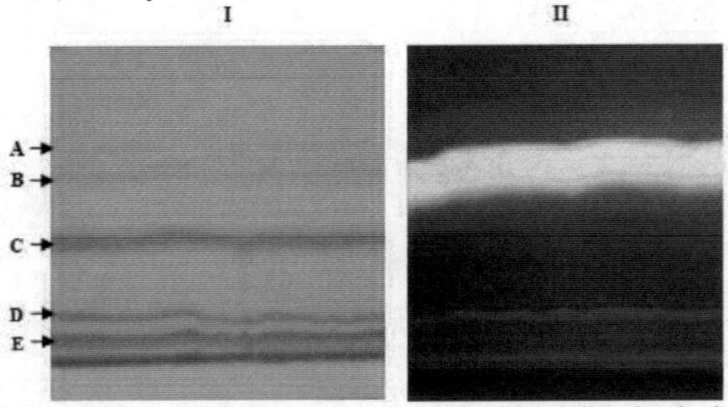

Figure 1. Profile chromatogram of the isolated section of noni dissolved ethyl acetate with 254 detection (I), UV_{366} (II). Isolates (A), (B), (C), (D) and (E).

Isolates have then tested the antibacterial against five species of meat spoilage bacteria such as *Bacillus licheniformis, Klebsiella pneumoniae, Bacillus alvei, Acinetobacter calcoaceticus,* and *Staphylococcus saprophyticus.*

3.2 Antibacterial Test

3.2.1 Well Diffusion Method

Isolates that have made the series based on the concentration of the effective concentration of the test partition that is 20 and 30% tested in five *Bacillus licheniformis bacterium, Klebsiella pneumoniae, Bacillus alvei, Acinetobacter calcoaceticus,* and *Staphylococcus saprophyticus.* Test plate was incubated at 37°C for 24 hours using amoxicillin controls 0.25%, CMC 1% and 70% ethyl acetate.
Table 1 as a whole shows that the isolated B has a diameter greater inhibition than the other isolates. This is because there is a major B isolate in a compound which can inhibit the growth of bacteria. These isolates can be used as a broad-spectrum antibacterial candidate because it can inhibit the growth of gram-positive and negative bacteria.

Table 1. The diameter of inhibition (mm) test isolate bacterial growth results TLCp

Species	Control			Isolate A		Isolate B		Isolate C	
				Konsentrasi					
	X	Y	Z	20 %	30 %	20 %	30 %	20 %	30 %
B. alvei	36	-	-	-	-	17	8	6	7
S. saprophyticus	18	-	-	-	-	7	5	5	10
A. calcoaceticus	19	-	-	-	-	9	10	7	11
K. pneumonia	20	-	-	-	-	7	13	7	10
B. licheniformis	19	-	-	-	-	10	10	5	8

Species	Control			Isolate D		Isolate E	
				Concentration			
	X	Y	Z	20 %	30 %	20 %	30 %
B. alvei	36	-	-	7	8	6	9
S. saprophyticus	18	-	-	5	9	-	-
A. calcoaceticus	19	-	-	7	9	5	8
K. pneumonia	20	-	-	8	6	-	-
B. licheniformis	19	-	-	6	8	9	6

Note:
X = Control Amoxicillin 0:25%, Y = Control Ethyl Acetate 70%, Z = Control CMC 1%

Each isolate has the potential inhibition of growth of bacteria that is different. The greater concentration of isolates greater inhibitory power. This is consistent with the statement [11], that the higher the concentration of an antibacterial material than the stronger antibacterial activity. The results of measuring the diameter of the inhibition isolate soluble partition results have inhibitory weak to strong to bacteria *B. licheniformis, K.pneumonia, B. alvei, A. calcoaceticus* and *S. Saphropyticus*. Species for gram-positive bacteria have an average zone of inhibition greater than negative bacteria, it is influenced by the structure of the bacterial cell wall. Differences in cell wall structure determine the penetration, binding and antibacterial activity of the compounds [12]. Gram-positive bacteria have a cell wall structure with more peptidoglycan, a little lipid and cell wall polysaccharides (teikoat acid). Teikoat acid is a water-soluble polymer, which serves as a positive ion transport to leave or enter. The nature of this water soluble indicating that the cell wall of gram-positive bacteria is more polar. While the gram-negative bacteria more lipid, a little peptigoglycan, formed the outer membrane bilayer (functioning as selective defense compounds exit or enter the cell and cause toxic effects). The outer membrane is composed of phospholipids (inner lining), and lipopolysaccharide (outer layer) composed of lipid A, which is nonpolar. This causes the antibacterial compounds contained in the noni fruit isolates it more difficult to enter the cell so that the antibacterial activity is weaker than the gram-positive bacteria.

3.2.2 MIC dan MBC Test

In the current study used a concentration of 20%, where the selection is based on the concentration of effective concentration in the inhibition test. From the results obtained that isolates B, C, D, and E show the growth of bacteria that is different.

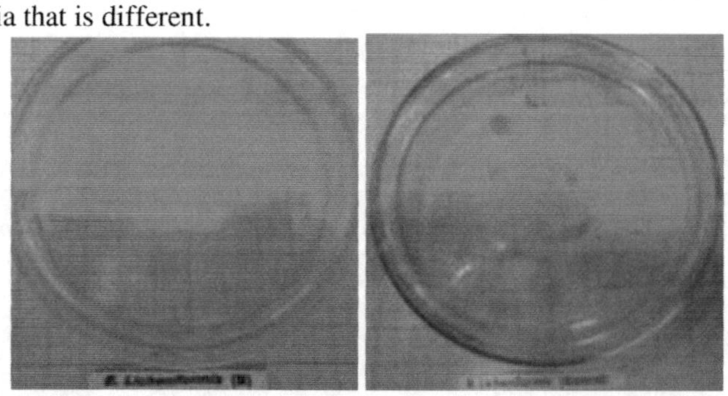

a b

Figure 2. (a) Growth *B.licheniformis* to isolate B concentration of 20% in media MHA. (b) Controls

Table 2. The test results MIC/MBC isolates results TLCp incubation at 37°C

Species	Control	Mass (g)	Area (cm^2)	Isolate B	Mass (g)	Area (cm^2)
B. alvei	Grow	0.0010	13.63	Grow	0.0002	2.83
S. saprophyticus	Grow	0.0014	2.27	Grow	0.0005	2.54
A. calcoaceticus	Grow	0.0027	4.65	Grow	0.0009	1.77
K. pneumonia	Grow	0.0028	3.25	Grow	0.0008	0.84
B. licheniformis	Grow	0.0015	6.01	Grow	0.0006	2.54

Based on Table 2. It can be seen that the isolated B concentration of 20% showed bacterial growth by various of percentage of growth. Isolates B bacteriostatis by inhibiting the growth of test bacteria. *B. alvei* growth inhibition of 20%, 35% *S.saprophyticus, A. calcoaceticus* by 30%, *K. pneumoniae* at 28%, *B. licheniformis* by 40%

3.3 Detection of compounds in Isolates with Reagent Spray and UV rays

Detection of compounds with spray reagent $FeCl_3$, $AlCl_3$ and UV rays performed on all isolates to suspect compounds in each isolate. Results of detection isolate A, B, C, D, E as shown in Table 3.

Table 3. Results of detection of isolates A, B, C, D, E with a reagent spray

No	Isolate	Reagent Spray			
		$FeCl_3$	$AlCl_3$	$AlCl_3$ dan UV_{366}	Hypothesis
1	A	Light yellow	-	Reddish yellow	Fenol
2	B	Black brown	Yellow	blue phosphorescent	Flavonoid or coumarin
3	C	-	-	-	-
4	D	-	-	Blue	Flavonoid or coumarin
5	E	-	-	Blue	Flavonoid or coumarin

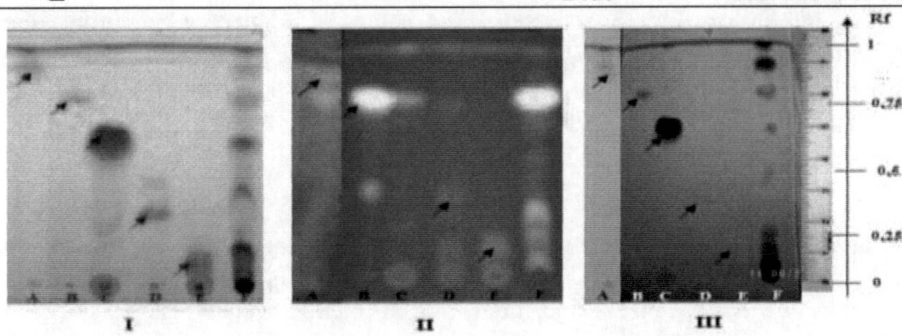

I II III

1 2 3

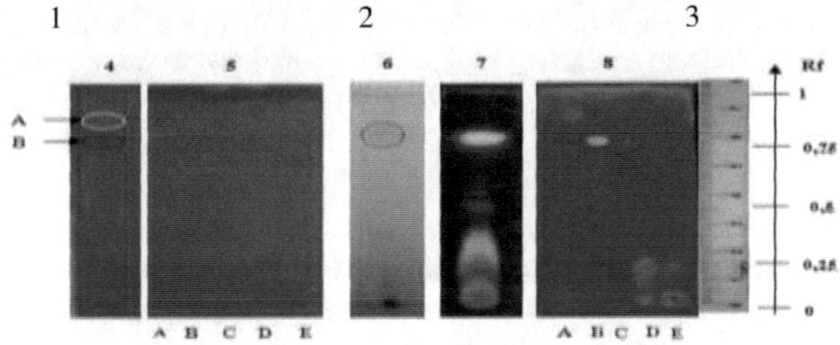

Figure 3. Profile chromatogram compound detection results.

Information: 1. Detection of isolates A, B, C, D, E with UV_{254}
2.Detection of isolates A, B, C, D, E with UV_{254}

3. Detection of isolates A, B, C, D, E with Cerium (IV) sulfate
4. Section dissolved ethyl acetate to 70% with the detection of $FeCl_3$
5. Isolate A, B, C, D, E with detection $FeCl_3$
6. Section dissolved ethyl acetate to 70% of visible light detection $AlCl_3$
7. Section dissolved ethyl acetate 70% with $AlCl_3$ detection and UV_{366}
8. Isolate A, B, C, D, E, and UV_{366} detection $AlCl_3$

Stationary phase: Silica gel 60 GF_{254}
Mobile phase: chloroform: ethyl acetate 3: 1 (v / v)

The results of the detection reagent compound with $FeCl_3$ spray against isolates of A, B, showed positive results (+) phenol compounds, but to isolate the C, D, and E showed a negative result (-). Detection of compounds with the spray reagent $AlCl_3$ followed UV_{365} showed positive results (+) to isolate B (Table 2), so that suspected isolates B contains flavonoids. Wagner (1983) [13] mentions that the compounds detected by UV_{365} blue fluoresence or blue-green indicate that these compounds including flavonoids or coumarin also depend on the chemical structure of the compound is still in a class of phenolic. E isolates also showed blue luminescence under UV365 rays that can be suspected to contain coumarin compounds or flavonoids as well, but the allegations need to be further analyzed to determine the type of compounds.

Flavonoids and coumarin synthesized from two main pathways for biosynthesis of an aromatic ring, namely the shikimic pathway and acetate malonate. Coumarin is a phenol compound derived from the shikimic pathway while flavonoid derived from a combination of both [14]. Coumarin showed to have a broad spectrum of biological activities such as antibacterial activity, antifungal, anticoagulant, anti-inflammatory, antitumor and anti-HIV [15]. The survey results revealed that a compound coumarin and xanthones, has the ability to inhibit the growth of bacteria Staphylococus aureus but not for a number of bacteria *Escherichia coli, Acinobacter spp,* and *Crypto Cocus neofumas* [16]. While the flavonoids reported to having activity inhibiting pathogenic agents in humans and are antibiotic that can inhibit gram-positive bacteria and negative [17]. these conditions indicate that isolates thought to have the content of flavonoid compounds or compound coumarin which are candidates antibacterial, especially isolates B. Detection of compounds must be strengthened with the identification using either UV-Vis spectrophotometry, IR, or NMR in order to obtain a clear chemical structure of the compound antibacterial candidate.

4. Conclusion

The antibacterial activity of isolates (B, C, D, E) of the 70% ethyl acetate soluble portion had inhibitory activity against 5 bacteria test. The inhibitory activity showed isolates B in *B. alvei* and isolates C in *A. calcoaceticus* bacteria. Concentration of 20% ethyl acetate isolates soluble part based antibacterial activity test was enough optimum inhibit growth bacterial of *B. licheniformis, K.pneumonia, B. alvei, A. calcoaceticus* and *S. saphropyticus*. Inhibitory activity against bacteria of this test is bacteriostatic. Detection of active compounds showed that isolates contains flavonoids and coumarin which both included in the class of phenol compounds.

Reference

[1] Yanti H, Hidayati and Elfawati 2008 *Jurnal Peternakan* **5**(1) 22-27.
[2] Amalia P R 2008 *Pengaruh Penambahan Bakteriosin dari Lactobacillus sp. Galur SCG 1223 Asal Susu Sapi terhadap Karakteristik Mikrobiologis Daging Dada Ayam Segar* IPB Bogor
[3] Barus P 2009 *Pemanfaatan bahan pengawet dan antioksidan alami pada industri bahan makanan* Medan: Universitas Sumatera Utara.
[4] Usmiati S and Richana N 2011 *Buletin Teknologi Pascapanen Pertanian* **7**(2).
[5] Purwani E, Retnaningtyas, Dyah Widowati 2008 *Laporan penelitian Fakultas Ilmu Kedokteran* Universitas Muhammadiyah Surakarta.
[6] Kumala S and Endro B S 2007 *Microbiology Indonesia* **1**(3) 145-148

[7] Peter 2005 *Chemical Constituents and Noni's Function Noni* News Indian Magazine Edisi Oktober (2) X.

[8] Retnaningtyas E, E Purwani, and T Purwoko 2009 *Pemanfaatan Ekstrak Buah Mengkudu (Morinda citrifolia L) dan daun Pandan (Pandanus amaryllifolius Roxb.) sebagai Pengawet Alami Daging dan Ikan Segar* LPPM UNS: Surakarta.

[9] Zakaria Z A, A M Desa, K Ramasamy, N Ahmat, A S Mohamad, D A Israf and M R Sulaiman 2010 *African Journal of Microbiology Research* **4**(1) 071-075.

[10] Pelczar M J and E C S Chan 1988 *Dasar-dasar Mikrobiologi 2* Jakarta: UI Press.

[11] Jawetz E, J L Melnick, and E A Adelberg 2001 *Mikrobiologi untuk Profesi Kesehatan (Review of Medical Mikrobiology) Translate by H. Tomang* Jakarta: EGC.

[12] Wagner H 1983 *Plant Drug Analysis* Translation of the German Edition

[13] Kristanti A N, N S Aminah, M Tanjung, and B Kurniadi 2008 *Fitokimia. Department Kimia* FMIPA UNS: Surakarta.

[14] Dekic B R, N S Radulovic, V S Dekic, R D Vukicevic, and R M Palic 2010 *Molecules* **15** 2246-2256.

[15] Pattalung P N, Thongtheeraparp W, Wiriyachitra P, and Taylor W C 1994 *Planta Med.* **60** 365-368.

[16] Verotta L, Lovaglio E, Vidari G, Finzi P V, Neri M G, Raimondi A, Parapini S, Tara melli D, Riva A, and Bombardelli E 2004 *Phytochemistry* 2867-2879.

[17] Bylka W, I Matlawska, and N N Pilewski 2004 *Review Article JANA* **7**(2) 24-31.

ZnO wide bandgap semiconductors preparation for optoelectronic devices

A H Ramelan[1], S Wahyuningsih[2], H Munawaroh[2], R Narayan[3]
[1]Electronic Materials and Energy Research Group, Faculty of Mathematics and Natural Sciences, Sebelas Maret University, Jl. Ir. Sutami No. 36A Surakarta, INDONESIA
[2]Inorganic Materials Research Group, Faculty of Mathematics and Natural Sciences, Sebelas Maret University, Jl. Ir. Sutami No. 36A Surakarta, INDONESIA
[3]Biomedical Engineering Joint Department of University of North Carolina/North Carolina State University, USA

E-mail: aramelan@mipa.uns.ac.id

Abstract. ZnO nanoparticles were successfully synthesized by sol-gel method. According to unique structural and optical properties of ZnO semiconductor material, there are many potential important applications based on that material, including as an anti-reflection coating (ARC) in solar cells. Antireflective coatings (ARC) made of ZnO on top to improve the optical properties of the coating. TiO_2 layer have been coated on a ZnO nanoparticle layer. ZnO nanoparticle was characterized by X-ray diffraction (XRD), Scanning electron Microscopy (SEM) and UV-Vis spectroscopy. ZnO annealed at a temperature of 600 °C have the greatest crystalinity and crystal size than that at a temperature of 400 °C and 500 °C. SEM images of ZnO shown agglomeration and grain size increases with increasing annealed temperature. While, the optical properties of ZnO increase with increasing annealed temperature. The optical transmittance spectra of the ZnO are shown that the increasing annealing temperature had effectively improved the optical transmittance of the films. While, reflectance (%R) properties shows that, the higher annealing temperature of ZnO preparations can decrease of %R value of ZnO thin layer. The difference properties of ZnO are due to differences of light scattering resulting from the crystal size effect. The ZnO prepared by annealed at 600 °C gain a good performance of the lowest reflectance value and highest size crystal. By the addition of ARC ZnO 600 °C we have been capable improve cell performance so that that cells achieve an efficiency of 0.27%.

1. INTRODUCTION

Zinc oxide, ZnO, is a direct wide bandgap semiconductor. Many of its properties promise for many optoelectronic device applications. One of these applications is dye sensitized solar cells (DSSC). DSSC is an attractive alternative to photovoltaic development as compared to conventional photovoltaic device (silicon solar cells) due to low prices and simple fabrication [1]. However, the efficiency of DSSC is still low than silicon solar cells. One way to increase the efficiency of DSSC is use a material that acts as an anti-reflection coating (ARC). ZnO is attractive as an ARC material because of its good transparency in the visible range and high absorption in the UV range because the wide band gap energy of ZnO monocrystal which is 3.1-3.3 eV at room temperature, and 3.44 eV at a temperature of 4 K. Meanwhile, as polycrystalline ZnO films is between 3.28 to 3.30 eV [2]. It's assisted in electron transfer from valence band to conduction band and increase light harvesting of DSSC [3].

ARC made of ZnO on top of layer will capable to improve the optical properties of the coating. A thickness of 30 nm has been considered for the ITO layer in the case of a hybrid coating. Reducing the ITO thickness down to 30 nm, compared to a more standard value around 50-70 nm, yields a lower useless absorption in the TCO layer. This reduction is made possible in the case of our hybrid design, since the top ZnO layer will also contribute to the lateral conduction of electrical charges. Great crystal size has a great internal trajectory, where the photon absorption takes place following the Lambert-Beer law. At the relatively smaller size, there are proportionately more than the surface reflection of internal photon path length, so it will be higher reflectance [4].

Synthesis of ZnO by deposition techniques such as electrodeposition [5],co-precipitation [6], hydrothermal [7], and sol–gel process [8]. Among the different deposition techniques, sol–gel process possess several distinct process advantages such as large area capability, glass compatible process, low-cost set-up and precise control of the film composition. Physical properties of sol–gel deposited thin films are strongly influence by pH of the initial precursor [9], dopant concentration [10], calcination temperature [11], solvent selection [12].

This study reports on the synthesis and characterization of ZnO by sol–gel technique. The aim of this work is to evaluate the microstructure properties of ZnO as a function of calcination temperature and evaluation of optoelectronic DSSC.

2. EXPERIMENTAL

2.1 Synthesis of ZnO nanoparticle

ZnO samples were prepared by a simple sol-gel method. 0.5 M of zinc acetate dehydrate ($Zn(CH_3COO)_2.2H_2O$, Aldrich) was dissolved in 30 mL of ethanol. Then, 30 mL of Polyethylene Glycol 1000 (PEG) 0.5 M was add into these solution. The solution was stirred at 60 °C for 2 hour till gel formation. After that, the solutions were placed in an oven at 60 °C until the solvent evaporated. Finally, ZnO were annealed at a temperature of 400 °C, 500 °C, and 600 °C. Final product are nanoparticles of ZnO. Structural and optical properties of synthesized ZnO was conducted by XRD, SEM and spectroscopy Uv-Visible.

2.2 Fabrication of DSSC

DSSC fabrication used ITO glass as a substrate. ZnO nanoparticle about 0.1 grams was dissolved in 2 mL of ethanol to obtain a paste. The paste was then coated on ITO glass with slip coating method. ITO was coated of ZnO heated at a temperature of 300 °C for 30 minutes. Then, ITO was coated TiO_2 paste. Pt paste was also superimposed on another ITO glass to built counter electrode thin film. Both ITO based electrode was heated at a temperature of 300 °C for 30 minutes. Photo anode TiO_2 with anti-reflecting ZnO was soaked in a dye solution of N3 ruthenium complexes for 24 hours. Then, both electrodes were sealed and given a distance to immerse the electrolyte solution. The electrolyte was prepared from potassium iodide (KI) of 0.4 grams and iodine (I_2) of 0.0635 grams dissolved in 5 mL acetonitrile. After the electrolyte was immersed, the DSSC were characterized by an I–V Keithley 2602A measurement.

3. RESULT AND DISCUSSION

3.1 Structure of ZnO

ZnO was synthesized using precursor $Zn(CH_3COO)_2.2H_2O$ by the simple sol-gel method. Preparation of ZnO were using polyethylene glycol as a surfactant. The solution was stirred at 60 °C for 2 hour till gel formation. The solution were dried at 60 °C following by annealing process at a temperature of 400 °C, 500 °C and 600 °C, and then it was characterized by XRD. The characterization results of XRD were shown in Fig. 1, which shows that the crystalline structure in accordance with the wurtzite hexagonal structure (JCPDS 80-0075) with dominant peak at $2\theta = 31.6; 34.4; 36.2$. In Fig. 1, it can be seen that the particles formed is ZnO and crystallinity of particle increase with increase annealing

IOP Conf. Series: Materials Science and Engineering **176** (2017) 012008 doi:10.1088/1757-899X/176/1/012008

temperature. At high temperature, 600 °C, peak of ZnO seem sharp and clearly, the particles have a high crystallinity was explained by sharp peak and the highest intensity than crystallinity at lower temperature. This is due to the higher temperature affect evaporation process and inter particle reaction more perfect, so can increase of crystalline formation. In addition, increase of annealing temperature can increases the reaction rate of decomposition of precursors [13].

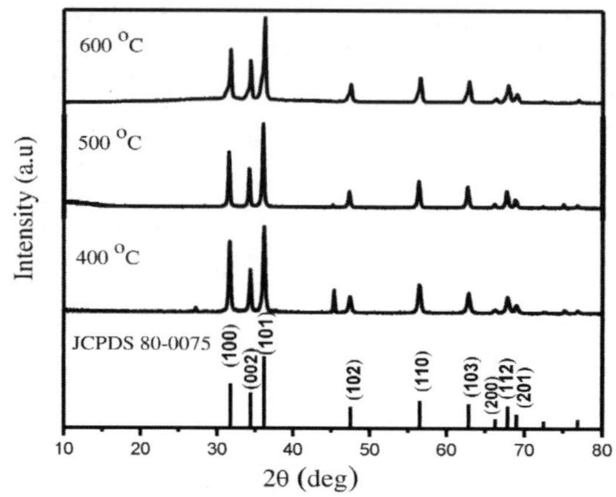

Figure 1. X-ray diffraction pattern of ZnO annealed at 400 °C, 500 °C, and 600 °C

Temperature dependence of ZnO nanoparticle crystal growth was shown in Table 1. ZnO annealed at a temperature of 600 °C have greatest crystal size than that at temperature of 400 °C and 500 °C. This is because of the higher temperature, the faster particle nucleation rate. It was also been produced by Ivanova et al. (2010) [14] and Gondal et al. (2009) [15], meanwhile increase of annealing temperature can increases crystal size. It may because the higher the calcination temperature the rate of nucleation and growth of material more quickly so that more perfect crystal growth.

Table 1. Temperature dependence of ZnO nanoparticle crystal growth

Annealed Temperature (°C)	λ (nm)	B		2θ	Cos θ	D
		(deg)	(rad)	(deg)		(nm)
400	0.154	0.403	0.007	36.407	0.949	20.750
500	0.154	0.317	0.005	36.147	0.950	26.374
600	0.154	0.290	0.005	34.660	0.954	28.651

The study morphology of ZnO were investigated using Scanning Electron Microscopy and shown in Fig. 2. It was seem that agglomeration and grain size increases with increasing annealed temperature. This is because annealed temperature affect of grain growth ZnO [16].

(a) (b) (c)

Figure 2. SEM image of ZnO (a) 400 °C,(b) 500 °C, and (c) 600 °C

3.2 Optical Properties of ZnO

Optical properties of ZnO material is characterized by a UV-VIS spectrophotometer. The analysis was performed in the wavelength range from 200 nm to 800 nm, the data obtained from this analysis is the absorbance and wavelengths are shown in Fig. 3. It can be seen that ZnO synthesized at higher annealing temperature capable increase the light absorption. The maximum wavelength absorption of ZnO is at 342 nm.

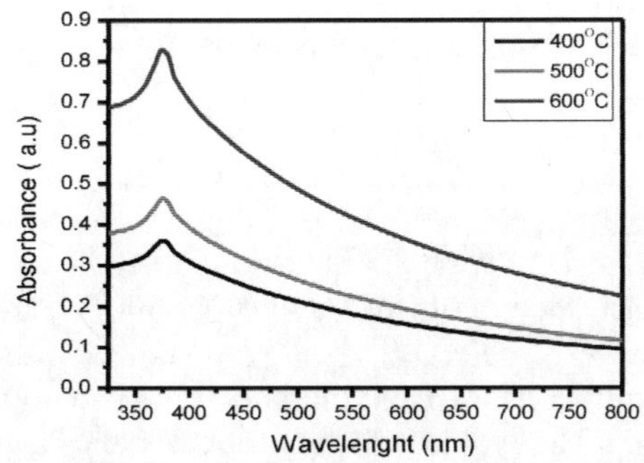

Figure 3. Absorption spectra's of ZnO

The optical transmittance spectra of the ZnO are shown in Fig. 4. In Fig. 4 shown that the increasing annealing temperature have effectively improved the optical transmittance of the films. The annealing temperature of 400 °C, 500 °C, and 600 °C have produced ZnO with average optical transmittance of 58%, 61%, and 68%, respectively. The improved optical transmittance is due to the better formed crystallite with the increase of annealing temperature.

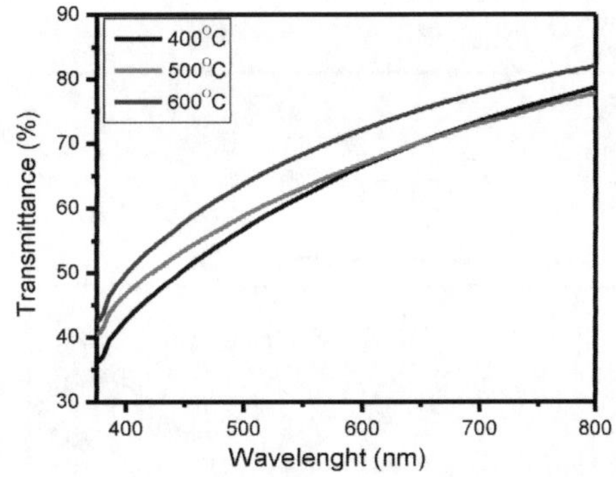

Figure 4. Transmittance spectra's of ZnO annealed at 400 °C, 500 °C, and 600 °C.

Reflectance (%R) properties show that the higher annealing temperature of ZnO preparations cause decrease of %R value of ZnO thin layer (Fig. 5). ZnO annealed at 400 °C was gained %R = 60.54%, at 500 °C was gained %R = 49.56% and at 600 °C was gained %R = 30.12%. The difference properties are due to differences of light scattering resulting from the crystal size effect.

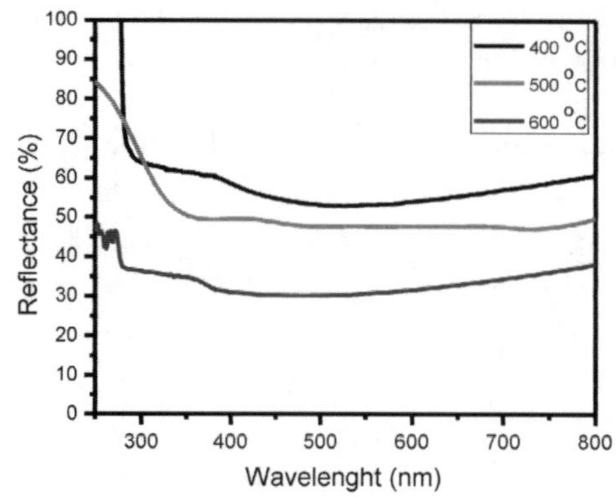

Figure 5. Reflectance spectra of ZnO at 400 °C, 500 °C, and 600 °C.

The optical band gap was determined using Tauc method [17]. The optical band gap was measure as an indirect band semiconductor. The value of the optical band gap is obtained by linearly extrapolating the linear portion to y = 0. Fig. 6 shows the function of $(\alpha h\nu)^2$ vs photon energy for ZnO with different annealing temperatures. The estimated optical band gap of ZnO at temperature 400 °C, 500 °C, and 600 °C was 3.36 eV, 3.35 eV and 3.38 eV, respectively. These values were very close to the band gap of intrinsic ZnO which is 3.37 eV [18]. The reduction of the optical band gap with increasing annealing temperature can be due to the improved crystallinity and decreased defects of the films with higher annealing temperature. Similar results have been reported in the work of Jianguo et al. [19] and Mudjat et al. [20].

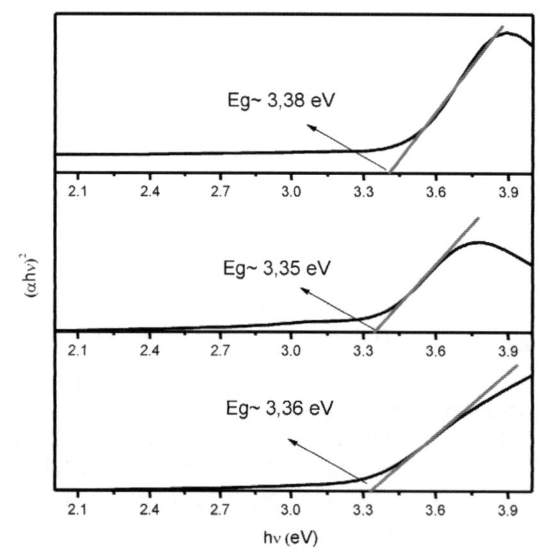

Figure 6. The optical band gap of ZnO

3.3 Fabrication of DSSC

DSSC performance conducted on TiO_2 that modified with anti-reflection ZnO layer. At this DSSC experiment, we use N3 dye as a standard dye. DSSC performance testing has been carried out with equipment Keithley IV meter with P = 1000 W/m^2. It was concluded that the ZnO prepared by annealed at 600 °C gain a good performance of the lowest reflectance value and high transparency. The maximum performance of DSSC if we use the photoanode TiO_2 at coated anti-reflection ZnO with annealing temperature of 600 °C. Fig. 5 shows the I-V curve result of DSSC performance by the addition of anti-reflection ZnO we have been capable improve cell performance so that cells achieve an efficiency of 0.27%.

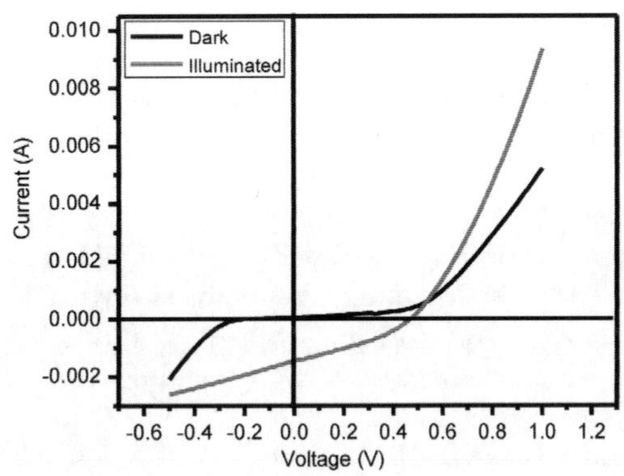

Figure 7. I-V characteristic of DSSC build by ZnO-coated TiO_2 at dark condition (black line) and under illumination condition (red line).

4. CONCLUSION

ZnO nanoparticle were synthesized by sol-gel method. The structural and optical properties of ZnO nanoparticle have been investigated. Annealed temperature influence the crystalline quality of ZnO, where increasing of annealing temperature was followed by increase of crystallinity and crystal size of ZnO. SEM images of ZnO shown agglomeration and grain size increases with increasing annealed temperature. And optical properties of ZnO increase with increasing annealed temperature. Increased optical properties of ZnO can improve efficiency of DSSC. The increase of annealed temperature can increase of optical transmittance spectra and decrease of %R value of ZnO thin layer. The ZnO prepared by annealed at 600 °C gain a good performance of the lowest reflectance value and highest size crystal. By the addition of ARC ZnO 600 °C we have been capable improve cell performance so that that cells achieve an efficiency of 0.27%.

ACKNOWLEDGMENT

The authors would like to acknowledge the International Research Collaboration and Scientific Publications Program, Ministry of Research, Technology and Higher Education, Republic of Indonesia, for supporting this research and to the Integrated Mathematics and Natural Science Laboratory of Sebelas Maret University for supporting and providing the facilities for this research.

REFERENCE

[1] Gratzel M 2003 *Photochemistry Reviews* **4** 145–153.
[2] Dengyuan Song 2005 *Tesis Doktor at University of New South Wales*.
[3] Lee, S Chang, Y L, Li-Ying H and Jung-Fu 2011 *Journal Engineering Technology and Education* **5** 545-552.
[4] Boden S A and Bagnall D M 2006 *Proc.of 4th world conference on photovoltaic energy conversion IEEE* (1-424400016-3) 1358-136.
[5] Salman KA, Omar K, and Hassan Z 2012 *Solar Energy* **86** 541–547.
[6] P K Sharma, R K Dutta, and A C Pandey 2007 *J. Colloid Interface Sci* **345** 149–153.
[7] Y Zhang, S Li, G Goh, and S Tripathy 2008 *Appl. Phys. Lett.* **93** 10251.

[8] C Knies, M T Elm, P J Klar, J Stehr, D M Hofmannn and N Romanov 2009 *J.Appl. Phys.* **105** 073918-7.
[9] L Spanhel and M A Anderson 1991 *J. Am. Chem. Soc.* **113** 2826–2833.
[10] W Tang and D Cameron 1994 *Thin Solid Films* **238** 83–87.
[11] K Kanade, B Kale, R Aiyer and B Das 2006 *Mater. Res. Bull* **41** 590–600.
[12] M Ohyama, H Kozuka and T Yoko 1998 *J. Am. Ceram. Soc.* **81** 1622–1632.
[13] Hidayat D, Ogi T, Iskandar F and Okuyama K 2008 *Materials Sciences and Engineering B.* **151** 231-237.
[14] Ivanova T, Harizanova A, Koutzarova T and Vertruyen B 2010 *Materials Letters* **64** 1147-1149.
[15] Gondal M A, Drmosh Q A, Yamani, Z H and Saleh, T A 2009 *Applied Surface Science* **256** 298-304.
[16] Zi-Neng Ng and Kah-Yoong Chan 2012 *Applied Surface Science* **258(24)** 9604–9609.
[17] J Tauc, R Grigorovici and A Vancu 1966 *Physics Status Solid B* **15 (2)** 627–637.
[18] X Zhao, J Y Lee, C R Kim, J H Heo, C M Shin, J Y Leem, H Ryu, J H Chang, H C Lee, W G Jung, C S Son, B C Shin, W J Lee, S T Tan, J Zhao, X Sun 2009 *Physica E* **41** 1423.
[19] J Lv, W Gong, K Huang, J Zhu, F Meng, X Song and Z Sun 2011 *Superlattices and Microstructures* **50** 98–106.
[20] M Caglar, Y Caglar, S Aksoy and S Ilican 2010 *Applied Surface Science* **256** 4966–4971.

Influences of mass Chlorophyll-a blends using P3HT:PCBM for efficiency of organic solar cells

E Lestari[1], A Supriyanto[1], Y Iriani[1], A H Ramelan[1], and F Nurosyid[1]
[1]Department of Physics, Faculty of Mathematics and Natural Science, Universitas Sebelas Maret, Surakarta, Indonesia

E-mail: agusf22@staff.uns.ac.id

Abstract. Organic solar cells have been made using the material poly (3-hexylthiophene)(P3HT), [6, 6]-phenyl-C61-butyric acid methyl ester (PCBM), and Chlorophyll-a with blend metods. Active layer of P3HT:PCBM:Chlorophyll-a are deposited using spin coating with rotary speed of 2500 rpm for 10 seconds and subsequently heated at 100^0C for 10 min. Mass of chlorophyll-a are 0.1 mg, 0.2 mg, and 0.3 mg. Thin layers are characterized by UV-Visible Spectrometer Lamda 25 for optical properties and Keithley 2602 for electrical properties. From the UV-Vis showed that absorbance of P3HT:PCBM:Chlorophyll-a are 400-614nm and 620-700 nm. Efficiency of P3HT:PCBM:Chlorophyll-a for mass chlorophyll 0.1 mg, 0.2 mg, and 0.3 mg are 2.68×10^{-2} %, 3.93×10^{-2} %, and 8.79×10^{-2} % respectively.

1. Introduction

Solar cells are devices whose can convert solar energy into electrical energy. Manufacture of semiconductor device using inorganic materials such as silicon, germanium. Efficiency of inorganic devices are very high, but the manufacture of the devices are very expensive. Wastes from the manufacture of inorganic devices can damage the environment. The researchers tried developing the device environmentally friendly and the cost of manufacture is also cheap. One of the materials developed are organic materials[1]. Organic solar cells (SSO) continue to be developed by mixing two types of semiconductor polymers that are known the bulk heterojunctions [2]. Researchers continue to improve efficiency of polymer solar cell based P3HT:PCBM with increasing electron and hole. Electron and hole formed from the exciton, of the process of excitation of active material when exposed to sunlight. Improved efficiency can be done to maximize the active material consisting of P3HT and PCBM with the widening of the active area, heating treatment, the use of solvents, and the addition of other materials[3]. Material P3HT and PCBM has a high absorption in the visible region in wavelenght of 400-650nm[4]. Polymer solar cells based on P3HT: PCBM with the addition of polymethine dye has an efficiency of 0.08% - 0.11%[5]. Dye is an organic dye which is often used in the SSO as an absorber of sunlight. Chlorophyll is used as a dye has a high absorption in the wavelenght of 600-700nm[6]. This study will be made of organic solar cells based on P3HT: PCBM: Chlorophyll with the structure of ITO / PEDOT: PSS / P3HT: PCBM: Chlorophyll /Al.

2. Experimental

Materials used are glass Indium Thin Oxide (ITO), HCl, aluminum, PEDOT: PSS, PCBM, P3HT, and chlorophyll. The tools used are glass beaker, Mettler Toledo AL204, Multimeters, Hot plate, and Evaporator. UV-Visible spectrometer Lamda 25 is used to determine the absorbance spectrum and

IOP Conf. Series: Materials Science and Engineering **176** (2017) 012009 doi:10.1088/1757-899X/176/1/012009

Keithley 2602A used to the I-V characteristics. P3HT, PCBM and chlorophyll are weighed using a digital balance with a variety mass of chlorohyll. PEDOT: PSS growth on the substrate ITO using a spin coating with rotary speed of 2500 rpm for 20 seconds and then heated at 100° C for 10 minutes. P3HT: PCBM: chlorophyll grown on the PEDOT : PSS layer using spin couter with a speed of 2,500 rpm for 20 seconds and heated at 100 °C for 10 minutes. Aluminum layer growth with evaporation on the layer of P3HT: PCBM: Chlorophyll.

Table 1. Mass of P3HT, PCBM, and Chlorophyll

chlorohyll	P3HT	PCBM
0.1 mg	13.33 mg	2.67 mg
0.2 mg	13.33 mg	2.67 mg
0.3 mg	13.33 mg	2.67 mg

3. Results and Discussion

3.1. Optical properties of Chlorophyll-a, P3HT, PCBM layer.

Absorption spectra of the Chlorophyll-a, P3HT, PCBM, P3HT:PCBM, and P3HT:PCBM:Chlorophyll-a are shows in Fig. 1. It is shown in this figure that absorption of Chlorophyll are 600-700 nm. The P3HT spectrum shows an absorption are 400-600 nm. The PCBM shows an absorption are 300-400nm. The P3HT blends PCBM shows absorption are 400-650 nm. Absorption of P3HT:PCBM:Chlorophyll-a are 400-700 nm. Absorption of P3HT:PCBM:Chlorophyll with variety of mass chlorophyll shows in Fig. 2. It is shows that the addition of mass chlorophyll can increase absorption and widening the range area of absorption.

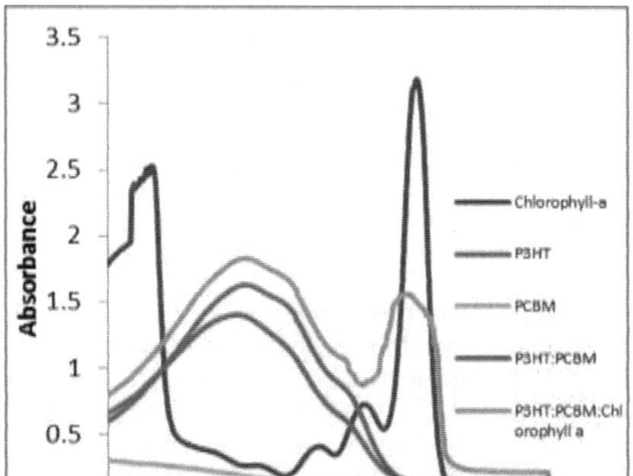

Figure 1. Spectrum Absorbance of Chlorophyll, P3HT, PCBM, P3HT:PCBM, P3HT:PCBM: Chlorophyll

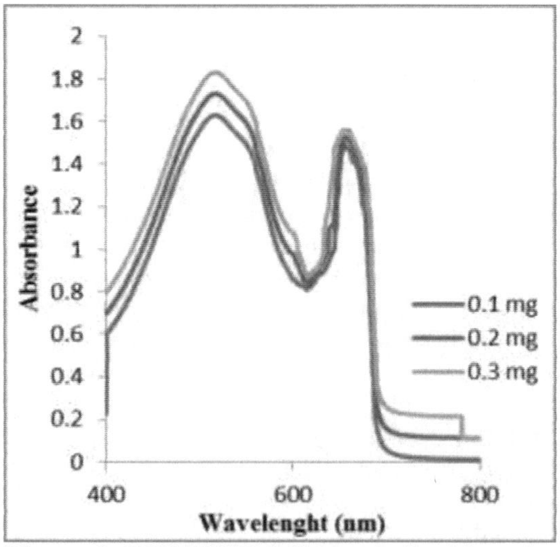

Figure 2. P3HT:PCBM:Chlorophyll with variety of mass chlorophyll-a are 0.1 mg, 0.2 mg, 0.3 mg

3.2. Electrical properties of the ITO/PEDOT:PSS/P3HT:PCBM:Chlorphyll-a/Al.

Current-voltage (I-V) characterization shows in Fig. 3. It is shows that efficiency of mass chlorophyll 0.1 mg are 2.68 x 10^{-2} %, 0.2mg are 3.93 x 10^{-2} %, and 0.3mg are 8.79 x 10^{-2} %. The addition of mass chlorophyll-a can increase efficiency, is due to the increase an electron and hole from excitons with the widening of area absorption. The results I-V characterization from Fig. 2 shows in table 1.

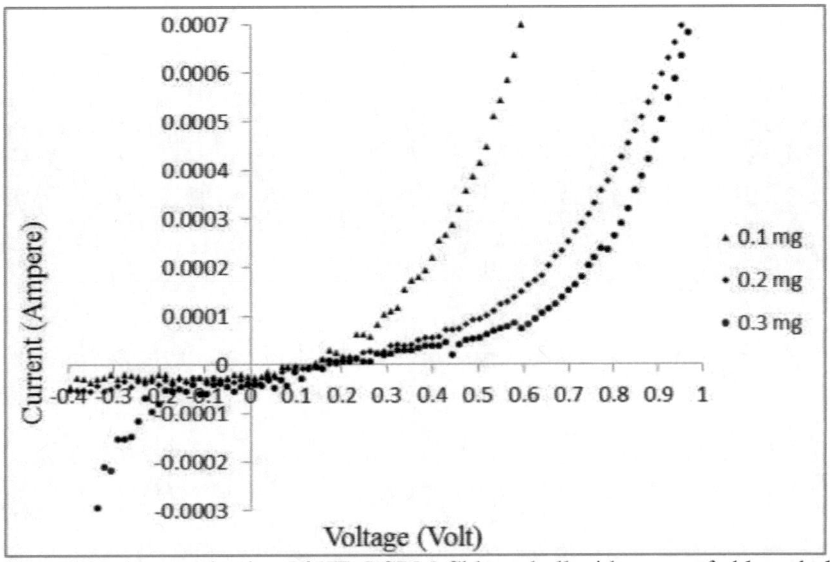

Figure 3. Curve I-V characterization P3HT: PCBM:Chlorophyll with mass of chlorophyll-a are 0.1 mg, 0.2 mg, 0.3 mg.

Table 1. Results of characterization P3HT:PCBM:Chlorphyll-a with variety of mass chlorophyll

Mass Chlorophyll	I_{sc} **(ampere)**	V_{oc} ***(volt)***	I_{max} **(ampere)**	V_{max} **(volt)**	P_{max} **(watt)**	***A (area)*** **(m^2)**	η **(x10^{-2} %)**
0.1 mg	2.76 x 10^{-5}	0.13	1.94 x 10^{-5}	0.06	1.07 x 10^{-6}	4 x 10^{-6}	2.68
0.2 mg	3.15 x 10^{-5}	0.14	2.83 x 10^{-5}	0.06	1,57 x 10^{-6}	4 x 10^{-6}	3.93
0.3 mg	4.41 x 10^{-5}	0.16	4.34 x 10^{-5}	0.08	3.52 x 10^{-6}	4 x 10^{-6}	8.79

4. Conclusion

The addition of mass chlorophyll-a can increase efficiency, is due to the increase an electron and hole from excitons with the widening of area absorption. Chlorophyll-a can increase absorption spectra from 400-650 nm to 400-700 nm. Efficiency of mass chlorophyll 0.1 mg are 2.68 x 10^{-2} %, 0.2 mg are 3.93 x 10^{-2} %, and 0.3 mg are 8.79 x 10^{-2} %.

References

[1] Freisinger B 2013 Investigation of P3HT/PCBM particle-based solar cells *Dissertation.* Universitat in Mainz.
[2] Dayal S, Zhong H, Kopidakis N, Scholes GD and Rumbles G 2015 *J. of Photonics for Energy.* Vol. **5**.
[3] Movla H, Rafi A M and Rafi N M 2015 *Optic.* **126** 1429-1432.

[4] Manceau M, Chambon S, Rivaton A, Gardette J L, Guillerez S and Lemaitre N 2010 *Solar Energy Material & Solar Cells.* **94** 1572-1577.
[5] Bliznyuk V N, Gasiorowski J, Ishchenko A A, Bulavko G V, Derevyanko N A and Sariciftci N S 2014 *Organic Electronics.* **15** 1105–1112.
[6] Eijckelhoff C and Dekker JP 1997 *Photosynthesis Research.* **52** 69-73.

Development and Testing of a Plastic Optical Fiber Grating Biosensor for Detection of Glucose in the Blood

M Yunianto[1], D Eka[1], A N Permata[1], D Ariningrum[2], S Wahyuningsih[3] and A Marzuki[1]
[1] Physics Department, Sebelas Maret University
[2] Medicines Department, Sebelas Maret University
[3] Chemistry Department, Sebelas Maret University

E-mail : mohtaryunianto@staff.uns.ac.id

Abstract. The objective of this study is to detect glucose content in human blood serum using optical fiber grating with LED wavelength corresponding to the absorption of glucose content in blood serum. The testing used a UV-Vis spectrometer and Rays spectrometers, in which in the ray spectrometer it was used optical fiber biosensor using optical fiber grating. The result obtained is the typical peak of glucose absorption in UV-Vis at 581 nm wavelength and rays spectrometer on green LED at 514.2 nm wavelength with linear regression result by 0.97 and 0.94, respectively.

1. Introduction

Diabetes Mellitus (DM) is a disease that affects many Indonesian. It is indicated by the rank of Indonesia in the fourth position in the world with the most DM disease sufferers. DM is caused by a deficiency of insulin with an increase in blood glucose levels above 130 mg/dl [1]. There have been many studies conducted on biosensor based glucose detection with cladding modification on optical fiber and the results state that it experiences rapid development. Many advantages of the use of this plastic optical fiber, are, it is small in size, high sensitivity, fast response, lasting impact on the electromagnetic and there is no electrical contact with the sample used [2,3,4].

This study uses the principles of light propagation in optical fibers with grooves treatment and the presence of light absorption when hitting the sample so the intensity of light captured by the detector will be weakened so the energy or power of the light source captured by the detector will be reduced [5]. The principle of light propagation based on Snell's law is that the light that comes into the Plastic Optical Fiber (POF) will experience a perfect reflection process due to the refractive index of the core. It is larger than the refractive index of the cladding [6]. This study is about establishing optical fiber sensor by grating to detect the content of glucose in the blood serum due to the occurrence of absorption of light intensity when the optical fiber grating biosensor hits the sample.

2. Experimental

Light spectrometer design consists of two main parts namely the hardware and software establishment. The hardware consists of a light sensor, propulsion systems, control systems and grids/slits. The software used LabView software Firmata 2012. The light sensor used was a photodiode that served to catch more or less of light that hit. The drive system used motors servo HS 311 with the movement angle that could be set. For control system, arduino UNO was used to adjust the angle shifting in data collection which was equal to 0.1

and crannies used was 600 channels/mm type which was prepared using two convex lens with diameter of 1.8 cm and 5 cm. Convex lens was used to focus the light as to be accurately received by the light sensor. Light spectrometer was designed to measure the spectrum of light produced by using fiber optics. This optical fiber was aimed to reduce the occurrence of light luminescence from the light source until it was caught by light sensor so micro positioner was used as the place for the optical fiber. Spectrometer light design using optical fibers was shown in Fig. 1. In this study, the optical fiber was treated with grating as much as 5 gratings and bending/curvature with a diameter of 5 cm.

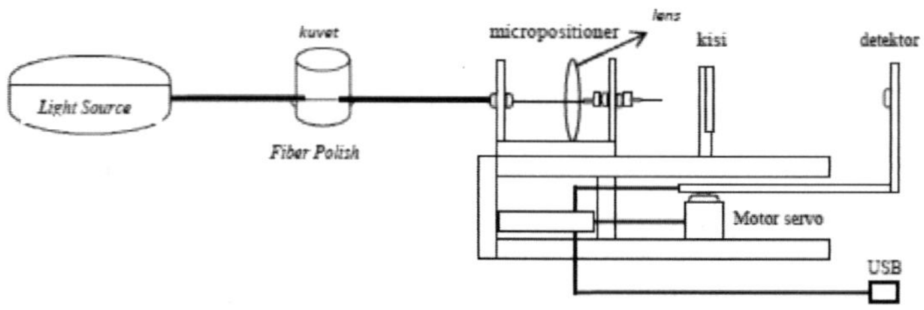

Figure1. Design of Light spectrometer

3. Result and Discussion

Data of glucose concentration value in blood serum of Klinik Budi Sehat Solo is shown in table 1. The data is used as the initial information of the glucose concentration in the blood serum that tested using a UV-Vis spectrometer and the light spectrometer.

Table 1. Glucose Concentration Data in Blood Serum

Serum samples	Glucose Concentration (mg/dl)
A	81.2
B	88.2
C	125.1
D	202.1
E	235.1

Prior to testing, device calibration was performed on light spectrometer using a HeNe laser with wavelength information at 632.80 nm. The calibration results are shown in Fig. 2.

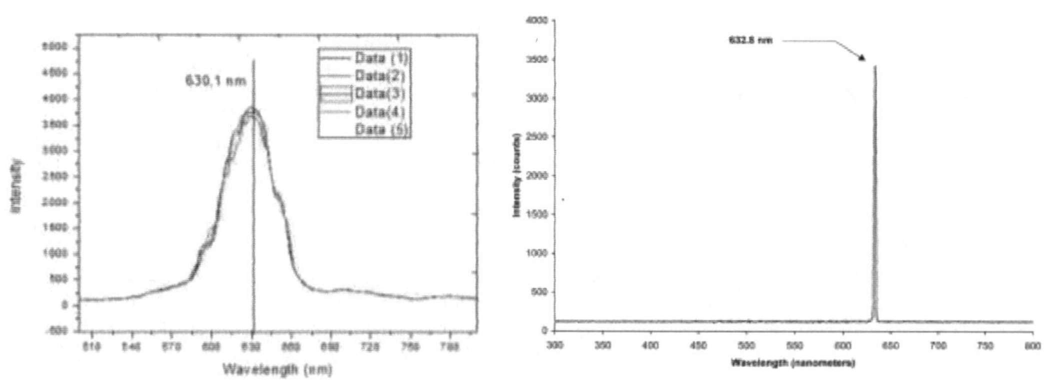

(a) (b)

IOP Conf. Series: Materials Science and Engineering **176** (2017) 012010 doi:10.1088/1757-899X/176/1/012010

Figure 2. (a) HeNe Laser spectrum measured with light spectrometer (b) Literature Graph of HeNe Laser.

Calibration results show that there is a single peak with a wavelength of 630.1 nm and a difference of 2.70 nm from the existing literature. Testing with light spectrometer uses five variations of LEDs as light sources with different wavelength ranges. Spectrum chart of each LED is shown in Fig. 3.

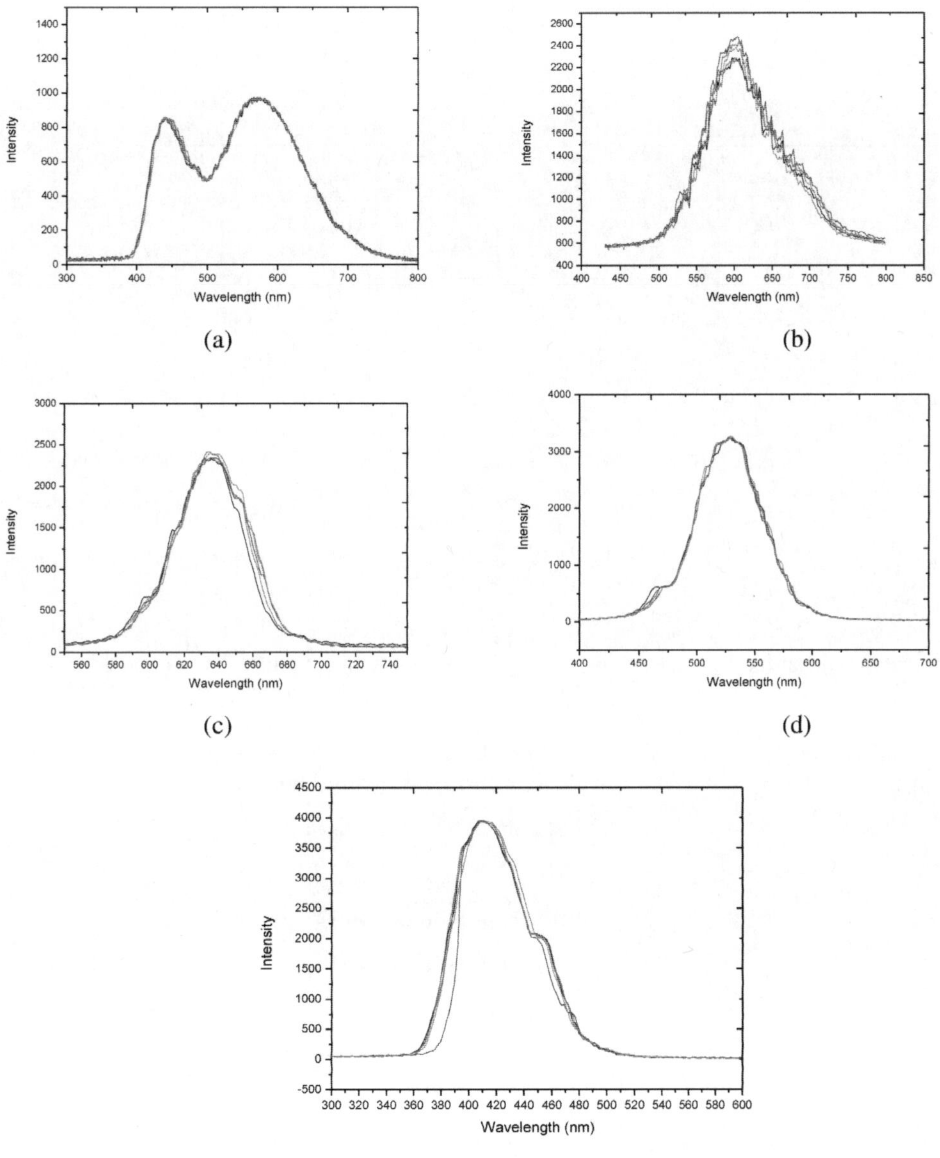

Figure 3. Graph of Spectrum (a) white (b) red-orange (c) green (d) red and (e) Blue LED

IOP Conf. Series: Materials Science and Engineering **176** (2017) 012010 doi:10.1088/1757-899X/176/1/012010

Linearity result of the light absorbance in optical fibers when hitting the blood serum with three variations of the LED is shown in Fig. 4. Linearity results of Green, Red and Blue LEDs are respectively 0.94 and 0.72.

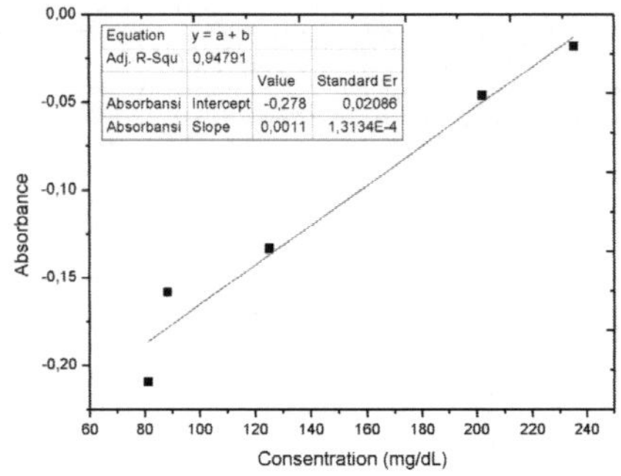

(a)

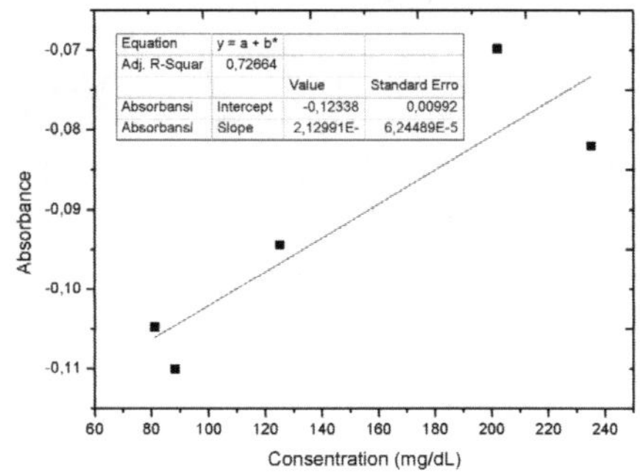

(b)

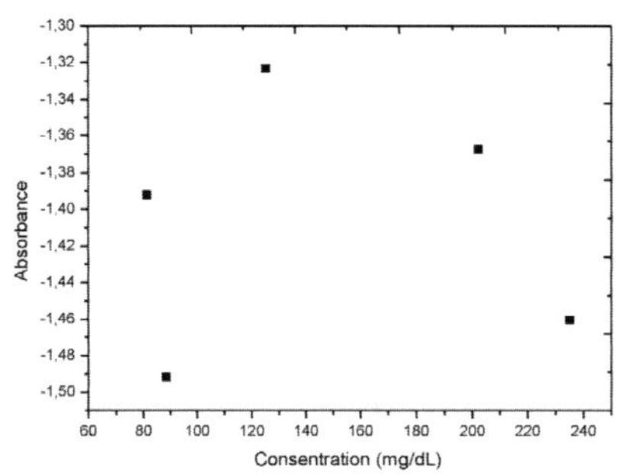

(c)

Figure 4. Linearity result of the light absorbance in optical fibers with (a) green (b) red and (c) Blue LED.

From the test results using the UV-Vis, it is obtained absorption value as wavelength function. Typical peak on UV-Vis absorption at the wavelength of 581 nm with a linearity is 0.97. These test results are shown in Fig. 5.

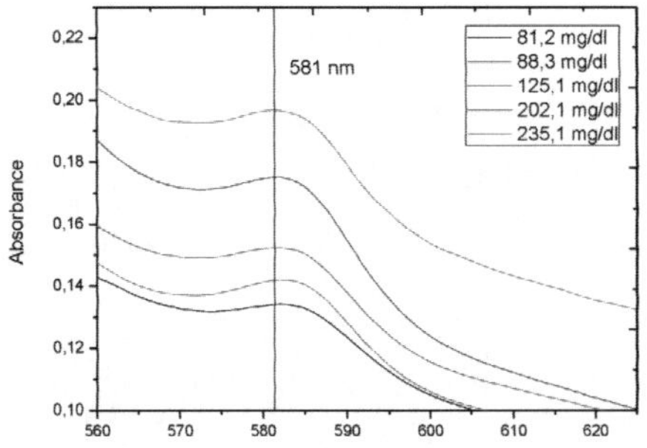

(a)

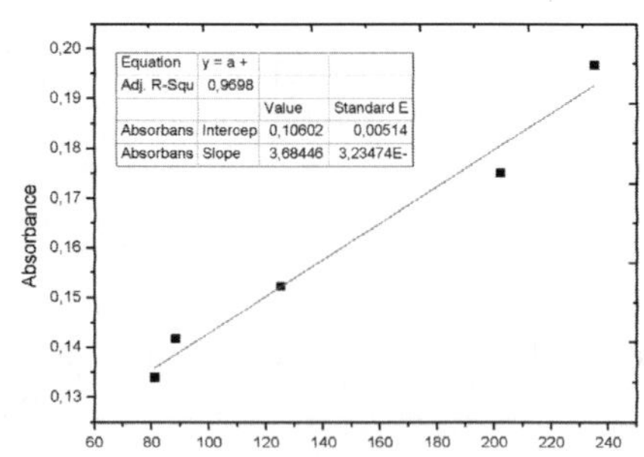

(b)

Figure 5. Relation graph of (a) absorbance as a function of wavelength and (b) absorbance as a function of glucose concentration value in blood serum

The comparison results of relation graph of absorbance as a function of the concentration of glucose content in the spectrometer light and UV-Vis show that the spectrometer light with Green LED is a graph trend generated by each point in accordance with the graphs trend generated on UV-Vis, so it can be defined that wavelength in the range of Green LED spectrum which is 514.2 nm is wavelength for the content of glucose in human blood serum using light spectrometer. The result of a combined graph on UV-Vis spectrometer and light spectrometer is shown in Fig. 6. This results have the same results with the research conducted by Pahurkar [7], the higher of glucose concentration, the absorbance is also getting bigger.

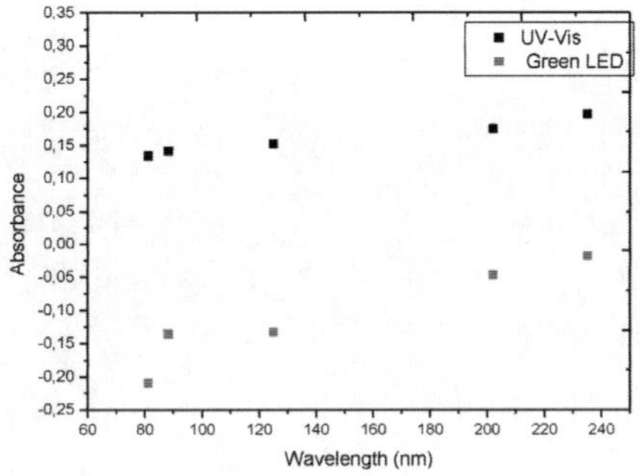

(a)

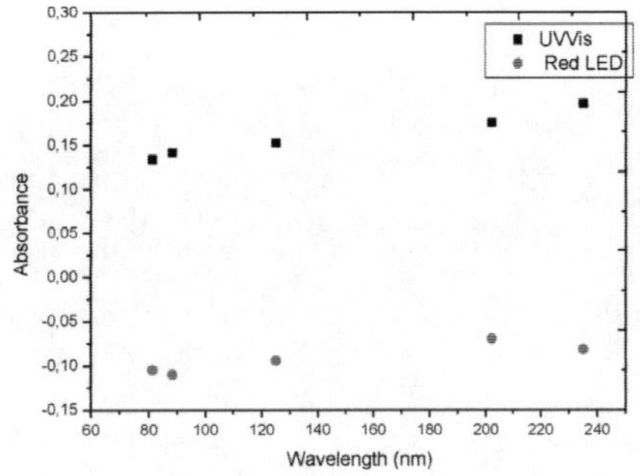

(b)

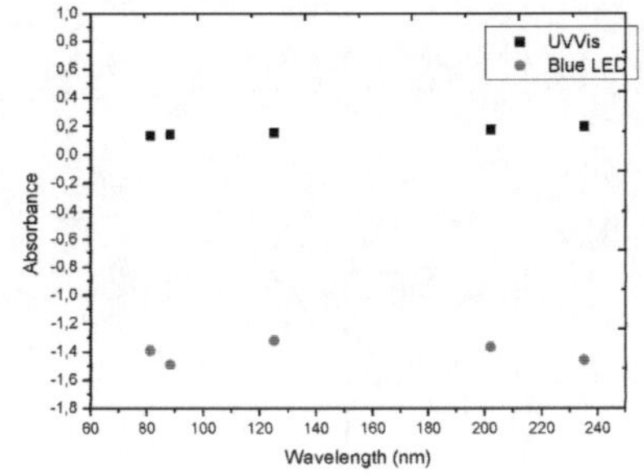

(c)

Figure 6. Combined graph on UV-Vis spectrometer and light spectrometer with (a) Green (b) red and (c) Bule LED

4. Conclusion

The conclusion of this analysis, the content of glucose in the blood serum can be detected using plastic optical fiber grated with the principles of light propagation that is capturing the intensity of light decreased due to absorption of light passing through the blood serum. Wavelength values obtained for the detection of glucose that is in the range of green LED at a wavelength of 514, 2 nm for light spectrometer and 581 nm for UV-Vis with linearity level are 0.94 and 0.97, respectively.

References

[1] Crisp J and Elliott B 2006 *Serat Optik: Sebuah Pengantar.* (Jakarta: Erlangga)

[2] Fidanboylu and Efendioglu 2009 Fiber Optic Sensors and Their Application. 5th *International Advanced Technologies Symposium (IATS'09),* Karabuk, Turkey.

[3] Lieberman R A 1993 *Sensor Actuators. B* **11** 43–55.

[4] Messica A, Greenstein A and Katzir A 1996. Appl. *Opt.* **35** *(*13) 2274–2284.
[5] Poedjiadi A 1994 *Dasar-Dasar Biokimia* (Jakarta: Universitas Indonesia Press)
[6] Sai V V R, Kundu T, Deshmukh C, Titus S, Kumar P and Mukherji S 2010. *Sensor Actuators. B* **143** 724–730.
[7] Pahurkara V G, Tamgadgeb Y S, Gambhirec A B and Muley GG 2015 *Sensors and Actuators B* **210** 362–368

Characterization of Micro Structure, Morphology and Electrical Properties Barium Titanate (BT) Doping Sr by Chemical Solution Deposition (CSD)

H D Mahmudah [1], Y Iriani[2], A H Ramelan[2]
[1]Graduate Physics Departement, Sebelas Maret University, Surakarta, Indonesia
[2]Physics Department, Faculty of Mathematics and Natural Sciences, Sebelas Maret University, Surakarta, Indonesia

E-mail: yopen_2005@yahoo.com

Abstract –Thin films successfully grown on the substrate Pt / Si is thin films BT doping Strontium (BST) by Chemical Solution Deposition (CSD) method. The percentage of doping Sr material are 1%, 3%, and 5%. The BST thin layer was characterized using X-Ray Diffraction (XRD), Scanning Electron Microscopy (SEM) and the hysteresis curve testing by Sawyer tower methods. The diffraction angle (2θ) in the XRD curve shifts to the right (bigger) along with increasing doping mol % Sr. Lattice parameters BST using the refinement GSAS program showed that a thin layer of BST has a tetragonal structure. From SEM analysis, along with increasing variation mol % Sr doping does not affect the thickness, but affects the grain size of a thin layer. BST thin layer of a ferroelectric material which has been formed mark a hysteresis curve. The characterization results to show that a thin layer of BST has been deposited on the substrate Pt/Si.

1. Introduction

Barium Titanate (BT) is a type perovskite ferroelectric and piezoelektric material first developed and studied since the 1940's. It goes through a paraelectric-ferroelectric phase transition from cubic to tetragonal at 130°C [1]. Barium Strontium Titanate (BST) is a material that has response ferroelectric and dielectric [2]. Research on BST originated from studies of barium titanate (BT), which showed that the addition of Sr ion in BT will lower the Curie temperature and change the ferroelectric properties of the BT [3]. So that the material can be used to manufacture BST thin layer which is applied as a FRAM [4].

Barium strontium titanate (BST) is a ferroelectric material. This material is the result of a mixture of Barium Titanate (BaTiO3) with strontium acetate. BST is usually obtained by substituting ions in position A and the perovskite structure ABO3 or Ba in BaTiO3 compound with Sr. The Curie temperature of barium titanate is 130°C and in the presence of doping strontium (Sr). Curie temperature decreased to room temperature and can be used in these devices that require room temperature [5]. Among thin film deposition techniques, such as wet chemical solution deposition, rf sputtering, metal-organic chemical vapor deposition and pulsed laser deposition [6]. CSD method is a process of making a thin layer with a chemical solution deposition on the substrate, and then prepared by spin coating with certain rotational speed [7]. Excellence CSD method with another method is by this method stokiometry easily controlled properly, easily prepared and carried out at low temperatures [8]. BST Samples were characterized by X-ray diffraction (XRD) to determine the thin film's microstructure, Scanning Electron Microscopy (SEM) to determine the morphology of thin film, and Sawyer Tower method to determine the hysteresis curve of a thin film.

2. Experimental

BT thin layer deposited by the method of Chemical Solution Deposition (CSD) prepared by spin coating method. BT has created a layer of doped Sr 0.5 M on the substrate Pt/Si. Barium acetate, titanium isopropoxide, used as precursor and acetic acid as a solvent. Ethylene glycol is also used in the manufacture of the BT solution which serves to prevent the occurrence of a crack. Growing thin layer of BT doping strontium (Sr) carried on a substrate Pt/Si with doping Sr mole percent variation is 1%, 3%, and 5%.

The steps in this research are the substrate preparation, preparation of the solution, solution deposition, and characterization. There are three sections in the thin layer deposition method Chemical Solution Deposition (CSD), it is the preparation of the solution, deposition on the substrate using a spin coater a rotation speed of 4000 rpm for 30 seconds. After the spin coating process, the substrate that has been coated with the solution is then heated using a hot plate. Heat treatment is hydrolysis and annealing, annealing performed at a temperature of 800°C; the heating rate is 3°C per minute with a holding time is 3 hours. Characterization is carried out in the study include testing the microstructure of thin layers BT doped Sr using X- Ray Diffraction (XRD), test morphology using SEM and electrical properties using Sawyer tower method.

3. Result and Discussion

The use of XRD equipment produces characterization in the form of a graph of diffraction angle and intensity. Figure 1 showed the peaks assumed that the belong to BT and BST at an angle (110). This indicates that the sample BST made is crystal. From Figure 1 it is showed that there is the shift of the peaks with increasing Sr.

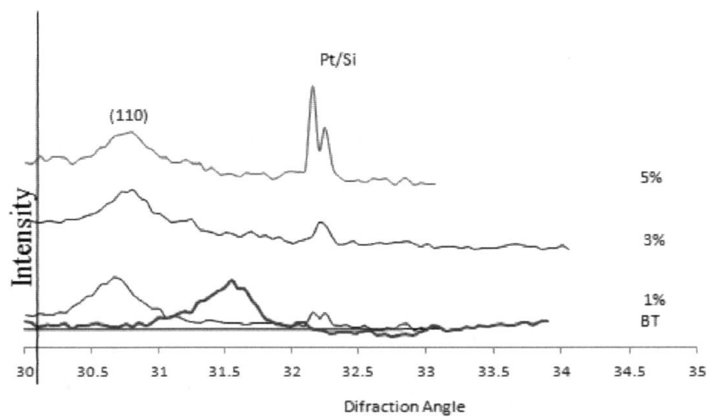

Figure 1. Difraction Pattern XRD (a) $BaTiO_3$ (b) $Ba_{0,99}Sr_{0,01}TiO_3$, (c) $Ba_{0,97}Sr_{0,03}TiO_3$, (d) $Ba_{0,95}Sr_{0,05}TiO_3$

The data taken from XRD is analyzed by working GSAS, this analyze produce the lattice parameters with each values are $a=b\neq c$ and $\alpha=\beta=\gamma=90^0$ on 101 orientation, which means the crystal structure is tetragonal. Field (110) is used to calculate the value of FWHM. The value of FWHM affected by a high intensity value for each sample. The higher the intensity, the FWHM values obtained will be smaller, because the smaller the peak width. FWHM value can be used to determine the size of the crystals. Based Scherrer equation which states that the value of FWHM is inversely proportional to the size of the crystals, in order to get a large crystal size for a small FWHM values. Material Sr replace Ba has a radius larger, in order to obtain the greater the number of doping resulted in changes in crystal size greater.

Assumption of the peak that belongs to sample wanted is done by refinement using Rietveld analysis with GSAS program. This method is matching result pattern of raw data XRD with refinement GSAS program. Figure 2 is the result of refinement thin layer of BT and BT doped 1%, 3%, and 5% mol Sr. Figure 2 showed that the doping Sr, BT have entered the compound, as in data processing used occupancy GSAS data that entered in the atom that makes the thin layer BT is appropriate that is Ba, Sr, Ti, and O3.

The processing of GSAS is done by refinement background so obtained approach between theoretical curve (calculated) which the curve experiment (observed) to obtain the chi square (λ^2) value is small to form convergenitas with indicators of success chi square small.

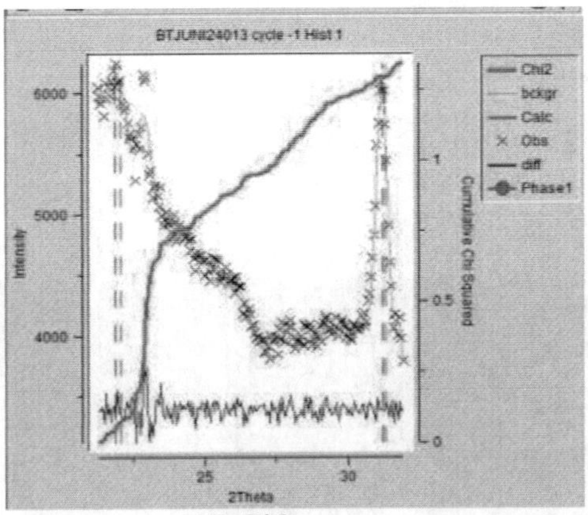

(a)BaTiO$_3$

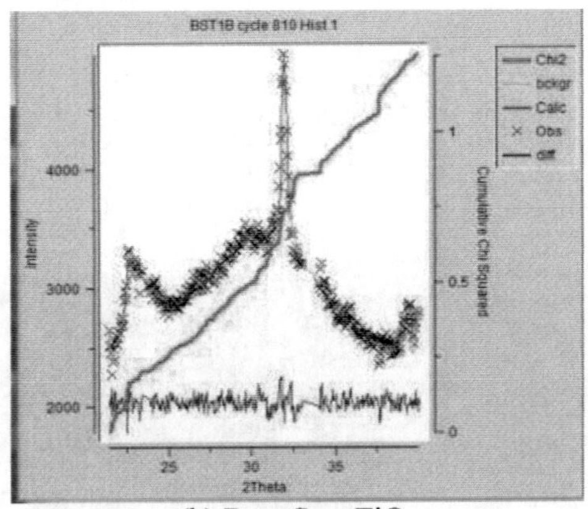

(b) Ba$_{0,99}$Sr$_{0,01}$TiO$_3$

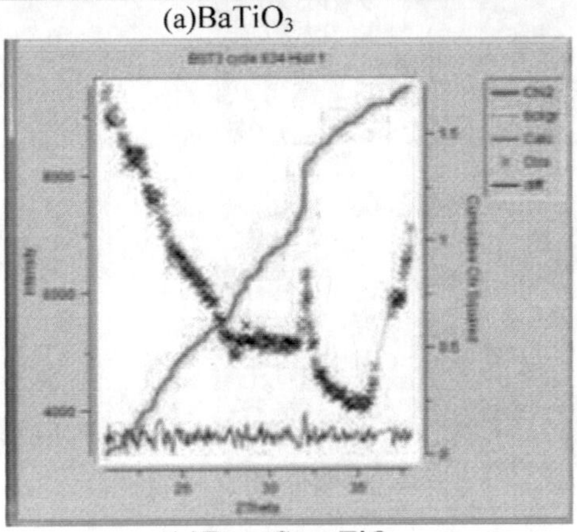

(c)Ba$_{0,97}$Sr$_{0,03}$TiO$_3$

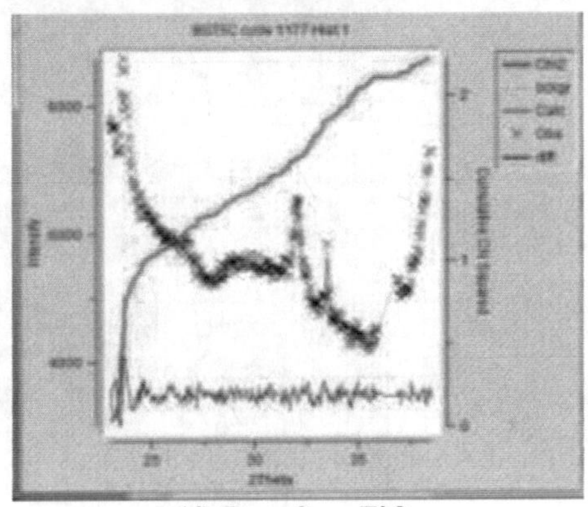

(d) Ba$_{0,95}$Sr$_{0,05}$TiO$_3$

Figure 2. Refinement Result with GSAS

Based on Table 1 it can be seen that there is the shift of angle, theta angle is smaller with the increasing of Sr doping mole percent on the orientation of the field of (110), shifted to the left. Without the addition of Sr on BT that have at peak, diffraction angles 31.21 with the addition of Sr in the (110) the angle shifted to 30.66. From Table 2 it appears that the addition of the composition of the doping Sr caused changes in lattice parameter values. The more doping Sr are added, then the value of the lattice parameter increases. This occurs because the atoms Sr and Barium (Ba) have a radius size of atomic are almost identical so that the atoms Sr^{2+} and Ba^{2+} will occupy the same position in the side of the unit cell diagonal.

Table 1. The diffractionangle doping Sr

Sample	Diffraction Angle
BaTiO$_3$	31.21
Ba$_{0,99}$Sr$_{0,01}$TiO$_3$	30.66
Ba$_{0,97}$Sr$_{0,03}$TiO$_3$	30.76
Ba$_{0,95}$Sr$_{0,05}$TiO$_3$	30.81

Tabel 2. Lattice Parameter BST

Sample	Lattice	Parameter		λ^2
	a(Å)	c(Å)	c/a	
$BaTiO_3$	3.995	4.035	1.010	1,683
$Ba_{0,99}Sr_{0,01}TiO_3$	3.958	4.006	1.011	1,632
$Ba_{0,97}Sr_{0,03}TiO_3$	3.858	4.114	1.066	1,732
$Ba_{0,95}Sr_{0,05}TiO_3$	3.806	4.168	1.096	2,230

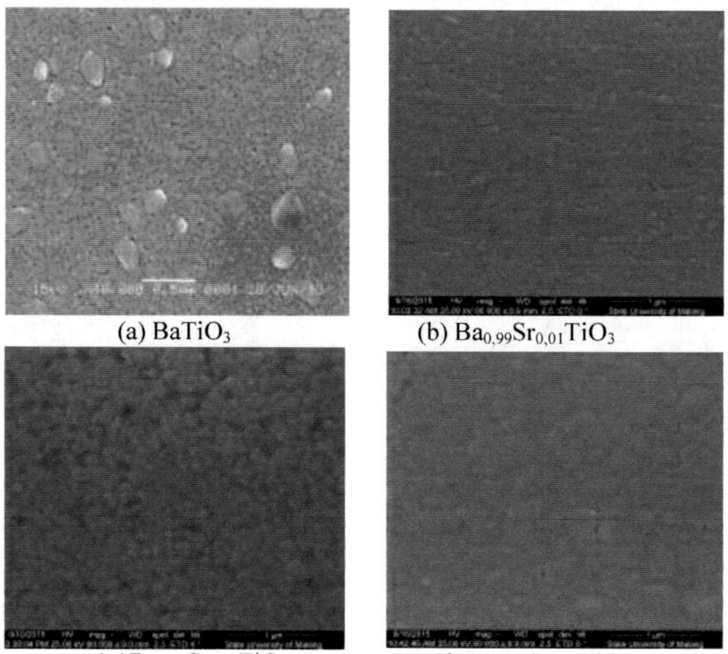

(a) $BaTiO_3$ (b) $Ba_{0,99}Sr_{0,01}TiO_3$

(c)$Ba_{0,97}Sr_{0,03}TiO_{33}$ (d) $Ba_{0,95}Sr_{0,05}TiO_3$

Figure 3. Photo SEM BT and BST Layer with doping Mol Variation

Photo SEM results in Figure 3 for the total sample it doesn't see happen to crack on the surface morphology of thin layers of BT and BST. It showed a thin layer of BT and BT with a variety of moles Sr doping was deposited well on the substrate Pt/Si with 3 number of layer, the annealing temperature 800°C for 3 hours and heating rate of 3°C/min. Based on the picture, it appears that the morphology of the thin film look homogeneous and the grain size is not visible.

Figure 4 is a cross-sectional SEM images samples of BT and BT with a variety of moles doping Sr. The results of the image are used to determine the thickness of thin layer of BT and BT with a variety of moles doping Sr. It appears that BT and BST thin layer of light-colored and dark beside a substrate Pt/Si. The result of the calculation of film thickness for all samples it appears that the overall thickness of the thin layer of BT and BT with Sr doping mole variations in size tend to have similar thickness is about 350 nanometers. Figure 5 shows the hysteresis curve of a thin layer of BT and Ba1-ySr yTiO3 (BST) with y (mole percent doping) is 1%, 3%, and 5%. Of the four samples, it appears that the hysteresis curve has been fully formed. This indicates that the thin layer formed BT and BST are ferroelectric.

(a) $BaTiO_3$ (b) $Ba_{0,99}Sr_{0,01}TiO_3$

(c)$Ba_{0,97}Sr_{0,03}TiO_{33}$ (d) $Ba_{0,95}Sr_{0,05}TiO_3$

Figure 4. The Thickness Thin Films

Figure 5 showed variations in the number of moles doping 1% , 3%, and 5% respectively show the change on the Y-axis, means that any changes in remanent polarization. The addition of a small amount of doping which has remanent polarization bigger than the number of larger doping. This means that the value of remanent polarization is getting smaller with an increasing number of doping Sr on the base material that is BT. Samples some of doping material Sr 1% had the larger electric field (E) than the material in doping Sr 3% and 5%, which means the addition of the percent number doping caused smaller of the electric field.

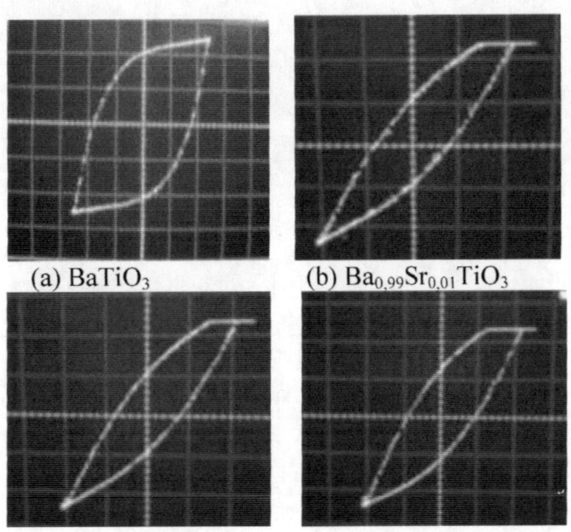

(a) $BaTiO_3$ (b) $Ba_{0,99}Sr_{0,01}TiO_3$

(c)$Ba_{0,97}Sr_{0,03}TiO_{33}$ (d) $Ba_{0,95}Sr_{0,05}TiO_3$

Figure 5. Hysteresis Curve

IOP Conf. Series: Materials Science and Engineering **176** (2017) 012011 doi:10.1088/1757-899X/176/1/012011

4. Conclusion

The addition of Sr doping on the crystal structure of Barium Titanate caused diffraction angle shift to the right, and the size of the crystal lattice parameter increases, the thickness of the thin layer is almost the same, the smaller caused, the smaller coercive field and remanent polarization.

Reference

[1] Jiangying W, Xiaoying Z, Jingji Z., Huiling L., and Zhengfa L 2012 *Science Direct* 957-960

[2] Pontes F, Leite E, Varela J, and Longo E 2001 *Science Direct* 91-98

[3] Dus B W, Czekaj A L, Ortiz T, Adamczyk M, Osinska K, Kozielski L, and Czekaj 2007 *Science-Poland* **25** 3

[4] Seo J Y and Park S W 2004 *Journal of Korean Physical Society* **45** 3 769-772

[5] Remmel Thomas 1999 *Characterization of Barium Strontium Titanate Films Using XRD* (Mesa, Arizona)

[6] Kim J, Kim S, Kim W, Ha T, Kim I -S and Song J S 2006 *Science Direct* 2322-2325

[7] Muhammad Hikam, Edy Sarwono dan Irzaman 2004 *Makara Sains* **8** 108-115

[8] Schwartz Robert W 1997 *Chemical Solution Deposition of Perovskite Thin Films. Department of Ceramic and Materials Engineering (*Clemson: University Clemson South Carolina) **9** 11

IOP Conf. Series: Materials Science and Engineering **176** (2017) 012012 doi:10.1088/1757-899X/176/1/012012

Efficiency enhancement of dye-sensitized solar cells (DSSC) by addition of synthetic dye into natural dye (anthocyanin)

D D Pratiwi[2], F Nurosyid[1], A Supriyanto[1] and R Suryana[1]
[1]Department of Physics, Faculty of Mathematics and Natural Sciences, Universitas Sebelas Maret, Jl. Ir. Sutami 36A Kentingan Jebres Surakarta 57126, Indonesia
[2]Department of Physics, Graduate Program, Universitas Sebelas Maret, Jl. Ir. Sutami 36A Kentingan Jebres Surakarta 57126, Indonesia

E-mail: nurosyid@yahoo.com

Abstract. This article reported combination of anthocyanin and synthetic dyes in dye-sensitized solar cells (DSSC) applications. This study aims was to improve the performance of DSSC by addition of synthetic dye into anthocyanin dye. Anthocyanin dye was extracted from red cabbage and synthetic dye was obtained from N719. We prepared anthocyanin and synthetic dyes at 2 different volume, anthocyanin dye at volume of 10 ml and combination dyes with anthocyanin and synthetic dyes at volume of 8 mL : 2 mL. The DSSCs were designed into sandwich structure on the fluorine-doped tin oxide (FTO) substrates using TiO_2 electrode, carbon electrode, anthocyanin and synthetic dyes, and redox electrolyte. The absorption wavelength of anthocyanin dye of red cabbage was 450 nm – 580 nm, the combination of anthocyanin and synthetic dyes can increase the absorbance peak only. The IPCE characteristic with anthocyanin dye of red cabbage and combination dyes resulted quantum efficiency of 0.081% and 0.092% at wavelength maximum about 430 nm. The DSSC by anthocyanin dye of red cabbage achieved a conversion efficiency of 0.024%, while the DSSC by combination dyes achieved a conversion efficiency of 0.054%, combination dyes by addition synthetic dye into anthocyanin dye enhanced the conversion efficiency up to 125%.

1. Introduction

Dye-sensitized solar cells (DSSCs) are solar cells based on dye moleculs that can absorb the light and convert light energy into electricity [1,2]. The advantages of DSSCs are low cost materials, simple to fabricate, easily obtained the materials, good conversion efficiencies, and many colour design [3]. The DSSC device consists of five components: conducting glass made of transparent conducting oxide as substrate (e.g. FTO or ITO), semiconductor electrode (TiO_2, SnO_2 or ZnO), counter electrode (Pt or C), dye as photosensitizer, and electrolyte as redox mediator [4]. One of the most important components of a DSSC is photosensitizer.

The research have been developed the synthetic and natural dyes. The synthetic dyes based on Ru with metal complexes such as N3, N749, and N719 which later be achieved efficiency up to 8.2%, 10.4%, and 10.1%, respectively [5-7]. However, synthetic dyes have expensive production and basic materials cost. They also contain heavy metals which make environmental pollution. There is alternatively towards the used of natural dyes as environment friendliness photosensitizer. The natural dyes are organic materials, they can be widely available, low cost, environment friendliness, various colors modified, and studied [8]. The pigments of natural dyes are present in the various part of the plant, vegetable, fruits, flowers, and leaves. Natural pigments containing betalains, carotenoids, chlorophyll, and flavonoids have been successfully as photosensitizer in DSSCs [9]. Many researchers have been developed in the various natural dyes. Teoli et al. in the year 2016 have been successfully extracted anthocyanin from blueberry and reported highest efficiency of 1.13% [10]. Chang et al., Taya et al., Syafinar et al., and Hasoon et al. developed the chlorophyll dye from spinach and it has given 0.016%-0.29% [11-14]. The study took extracts from Bougainvillea flowers and carotenoid pigments from Sponge gourd reporting conversion efficiency of 0.325% and 0.13% [5,16]. Chang et al. used chlorophyll dye from wormwood and anthocyanin dye from purple cabbage as natural dyes, the conversion efficiency of DSSCs were 0.54% and 0.75%, respectively and then after mixing the extracted wormwood chlorophyll and purple cabbage anthocyanin at volume ratio of 1:1, the conversion efficiency of DSSC was increased to 1.47% [17]. This result can be explained that it has widening of visible light absorption range, thus increasing the excitation of electrons were transmitted to TiO_2 semiconductor to enhance the photocurrent of DSSC. The novelty of this research is to improve the performance of natural dye with addition of synthetic dye, so that the good performance and low-cost DSSCs were obtained. The natural dye used red cabbage anthocyanin and synthetic dye was obtained from ruthenium complex N719. The addition of synthetic dye aims was to increase the efficiency of natural dye. The function of dye absorbed the light energy to produce excitation of electrons [18]. Selection of dye in DSSC plays vital role in improving the absorption of photons, so that the great efficiency was obtained.

2. Experimental

The dyes used by this study was extracted from red cabbage anthocyanin, while synthetic dye was obtained from ruthenium complex N719, fluorine-doped tin oxide (FTO) from Dyesol as substrate, the semiconductor layer from TiO_2 powder 21 nm as working electrode, carbon layer as counter electrode, the solution of PEG 400, potassium iodide, and iodine as redox electrolyte.

The working electrodes with TiO_2 paste were coated by spin coating. This study used UV-Visible Spectrophotometer Lambda 25 to measure the absorption spectra of dyes solution, incident photon-to-current efficiency (IPCE) Measurement System QEX7 Serial #150 to measure the quantum efficiency versus wavelength of DSSCs, and Keithley I-V Meter 2602A to measure the photoelectric conversion efficiencies.

Red cabbage was cut into small pieces and added them in solvents of distilled water, methanol, and acetic acid. It was kept for 24 hours. The extract of red cabbage anthocyanin dye solution was filtered using filter paper. The absorption spectra of anthocyanin dye solution was characterized by using UV-Vis spectrophotometer. An active area of 2 cm^2 FTO conductive glass was deposited with TiO_2 paste using spin coating technique. The TiO_2 films were annealed at 400ºC for 10 minutes. The TiO_2 films were soaked for 24 hours in solution containing the anthocyanin dye of red cabbage at volume of 10 ml and combination dyes with anthocyanin and synthetic dyes (N719) at volume of 8 ml : 2 ml.

The counter electrodes were prepared from carbon catalyst. Fabrication of DSSCs were arranged into a sandwich structure. Electrolyte solution was injected into the cell. The DSSCs were measured using incident photon-to-current efficiency (IPCE) and *I-V* Meter.

3. Results and Discussion

3.1. Absorption spectra of dyes solutions

The absorption spectra in Figure 1 shows the UV-Visible result of synthetic dye (N719) solution at wavelength in the range of 500 nm – 600 nm. The absorption wavelength anthocyanin dye of red cabbage was 450 nm – 580 nm at wavelength maximum about 530 nm. The curve by Figure 1 known that red cabbage evidently has absorbance in the visible light zone, so it can be used as photosensitizer for DSSC, but it has absorbance lower than the absorbance of synthetic dye (N719).

The UV-Visible result of anthocyanin dye of red cabbage with addition of 2 ml synthetic dye (N719) was not show the expansion visible light absorption range, but it can increase the absorbance peak only at wavelength maximum about 530 nm. This result showed that synthetic dye contributed to enhancement the absorbance peak of anthocyanin dye of red cabbage.

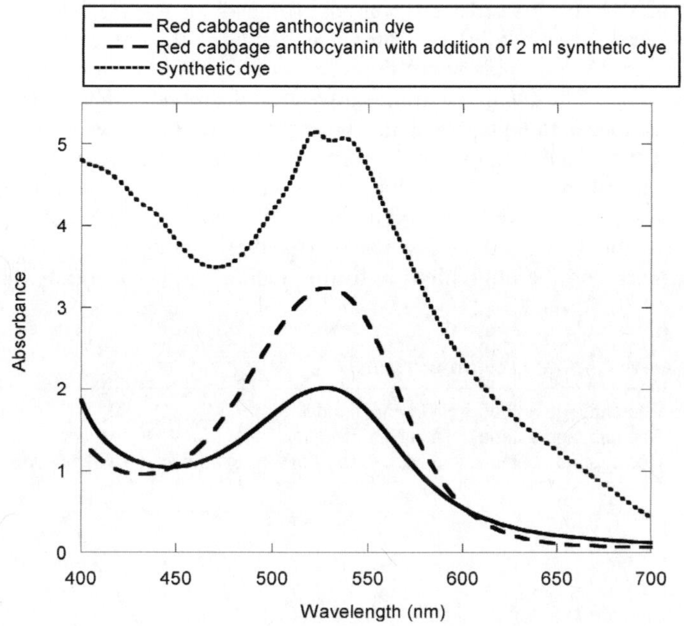

Figure 1. Absorption spectra of anthocyanin dye, combination of anthocyanin and synthetic dyes, and synthetic dye solutions

3.2. IPCE Characterization Results

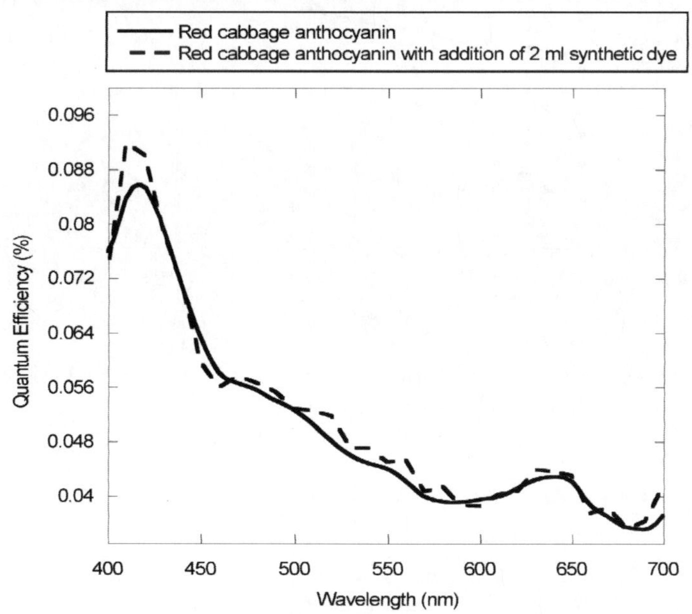

Figure 2. IPCE spectrum of DSSCs with anthocyanin dye and combination of anthocyanin and synthetic dyes

The IPCE spectrum by Figure 2 showed that quantum efficiency value of anthocyanin dye with addition of 2 ml synthetic dye (N719) increase quantum efficiency at wavelength about 430 nm. The absorption of photons occured optimally of absorption wavelength in the range of 400 nm – 450 nm at wavelength máximum about 430 nm. This result shows the contribution of photons to the dye yielded high energy and photocurrent. The IPCE spectrum showed the decrease in the visible light zone, thus the absorption of photons appear to be decreasing. The reason is that the adsorptions of dyes on the TiO2 surface were relatively small because the absorption abilities of dyes were not too dominant [19]. The decrease of absorbance can also happen because great enhancement in the exciton recombination, so that the electrons excitation would be slightly and electrons transfer from dye into the conduction band of semiconductor was not work effectively. The dye can be considered as a negative contributor (contributed the only slightly) to the overall of photocurrent when irradiated in the visible light zone. [20].

3.3. Current – Voltage (I-V) Characterization Results

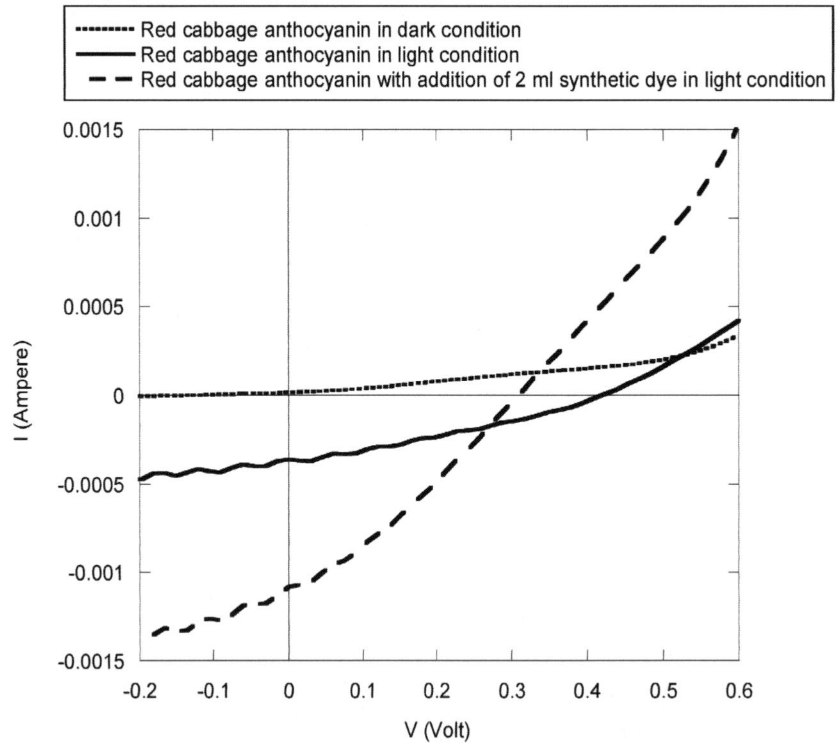

Figure 3. Current-voltage curves for the DSSCs by anthocyanin dye and combination of anthocyanin and synthetic dyes

The *I-V* curves of the DSSCs based on anthocyanin and combination dyes are shown in Fig.3 and their photoelectrical parameters are listed in Table 1 :

Table 1. Photoelectrical parameters of DSSCs by anthocyanin dye and combination of anthocyanin and N719 dyes

Red cabbage : N719	I_{max} (mA)	V_{max} (V)	I_{sc} (mA)	V_{oc} (V)	FF	η (%)
10 ml : 0 ml	0.17	0.28	0.23	0.34	0.63	0.024
8 ml : 2 ml	0.35	0.31	0.49	0.43	0.51	0.054

The conversion efficiency of DSSC by combination dyes at volume of 8 ml : 2 ml can achieve better conversion efficiency. This result was comparable with quantum efficiency by IPCE curve in Fig.2. The quantum efficiency was higher, then the conversion efficiency of DSSC was also higher.

As known from the photoelectrical parameters results shown in Table 1, open-circuit voltage (V_{oc}), short-circuit current (I_{sc}), and conversion efficiency by the combination dyes at volume of 8 ml : 2 ml has the better performance than the volume of 10 ml : 0 ml. The reason for this result was that the combination dyes with red cabbage anthocyanin and synthetic (N719) dyes at volume of 8 ml : 2 ml can increase light absorption. The higher light absorption can absorb the greater photons, thus the electrons in excited state were also increased to enhance the measured photocurrent. The DSSC with combination of synthetic (N719) and natural dyes was obtained the photocurrent increased. The photocurrent was increased, so that the conversion efficiency of DSSC by addition of 2 ml synthetic dye into red cabbage anthocyanin can increase the efficiency up to 125 %. Composition variations of dyes can increase the efficiency of natural dye, as a result the low cost DSSCs were obtained.

4. Conclusion

In conclusion, the results showed that the absorption wavelength of anthocyanin dye of red cabbage was 450 nm – 580 nm, while the combination of red cabbage anthocyanin and synthetic dyes can increase the absorbance peak only. The IPCE characteristic with anthocyanin dye of red cabbage and combination of anthocyanin and synthetic dyes resulted quantum efficiency of 0.081% and 0.092% at wavelength maximum about 430 nm and they contributed in the highest current formation of absorption wavelength in the range of 400 nm – 450 nm. The *I-V* characteristic with anthocyanin dye of red cabbage anthocyanin and combination of anthocyanin and synthetic dyes resulted efficiency of 0.024% and 0.054%, the DSSC efficiency by addition synthetic dye into anthocyanin dye significantly enhanced the conversion efficiency up to 125%.

Acknowledgment

This work is supported by Directorate of Research and Community services, Ministry of Research, Technology, and Higher Education and Institute for Research and Community services, Sebelas Maret University by PUPT Grant Program. (Contracts Number : 353/UN27.21/PN/2016)

References

[1] Grätzel M 2003 *J.Photochems.Photobio* **4** 145-153
[2] Grätzel M 2005 *Inorg.Chem.* 44 6841-6851
[3] Dodoo-Arhin D, Howe RCT, Hu G, Zhang Y, Hiralal P, Bello A, Amaratunga G and Hasan T 2016 *Elsevier.Ltd.* **105** 2016 33-41
[4] Sharma S, Jain KK, and Sharma A 2015 *Mater.Sci.Appl.* **6** 1145-1155
[5] Nazeeruddin MK, Zakeeruddin SM, Humpry-Baker R, Jirousek M, Liska P, Vlachopoulos N, Shklover V, Fischer C-H, and Grätzel M 1999 *Inorg.Chem.* **38** 6298-6305
[6] Nazeeruddin MK, Splivallo R, Liska P, Comte P, Grätzel M 2003. *Chem. Commun.* **9** 12 1456
[7] Robertson N 2006 *Angew.Chem.Int.Ed.* **45** 15 2338–2345
[8] Bose S, Soni V, and Genwa KR 2015 *Int.J.Scientific and Research Publications.* **5** 4 2250-3153
[9] Bhogaita M, Shukla AD, and Nalini RP 2016 *Sol.Energy.* **137** 2016 212-224
[10] Teoli F, Lucioli S, Nota P, Frattarelli A, Matteocci F, Di Carlo A, Caboni E, and Forni C 2016

J.Photochem.Photobiol.A Chem. **316** 24–30
[11] Chang H, Lo Y-J 2010 *Sol.Energy.* **84** 1833–1837
[12] Taya SA, El-Agez TM, El-Ghamri HS, and Abdel-Latif MS 2013 *Int.J.Mater.Sci.Appl.* **2** 37–42
[13] Syafinar R, Gomesh N, Irwanto M, Fareq M, and Irwan YM 2015 *Energy.Procedia.Elsevier B.V.* **11** 584
[14] Hasoon S, Al-haddad RMS, Shakir OT, and Ibrahim IM 2015 *Int.J.Sci.Eng.Res.* **6** 2229–5518
[15] Godibo DJ, Anshebo ST, and Anshebo TY 2015 *J.Braz.Chem.Soc.* **26** 92–101
[16] Maurya IC, Srivastava P, and Bahadur L 2016 *Opt.Mater.(Amst.).* **52** 150–156
[17] Chang H, Kao M, Chen T, Chen C, Cho K, and Lai X 2013 *J.Photoenergy.* **2013** 8
[18] Chiba Y, Islam A, Watanab Y, Komiya R, Koide N, and Han L 2006 *Japanese.J.Appl.Phys.* **45** 1638–1640
[19] Hoffmann MR, Martin ST, Choi W, and Bahnemann DW 1995 *Chem.Rev.* **95** 69-96
[20] Brennan LJ, Milton FP, Salmeron AS, Zhang H, Govorov AO, Fedorov AV, and Gun'ko YK 2015 *Nanoscale Research Letters.* DOI 10.1186/s11671-014-0710-5

TiO_2 Nanostructure Synthesized by Sol-Gel for Dye Sensitized Solar Cells as Renewable Energy Source

A H Ramelan[1], S Wahyuningsih[1], S Saputro[1], E Supriyanto[2], and Q A Hanif[1]
[1]Sebelas Maret University (UNS), Jl. Ir. Sutami No. 36A Surakarta 57126, Indonesia.
[2]Jember University (UNEJ), Jember 68188 Indonesia.

E-mail: aramelan@mipa.uns.ac.id

Abstract. The use of renewable materials as a constituent of a smart alternative energy such as the use of natural dyes for light harvesting needs to be developed. Synthesis of anatase titanium dioxide (TiO_2) and fabrication Dye-Sensitized Solar Cell (DSSC) using dye-based of anthocyanin from purple sweet potato (*Ipomoea batatas* L.) as a photosensitizer had been done. Synthesis TiO_2 through sol-gel process with the addition of triblock copolymer Pluronic F127 template was controlled at pH 3 whereas calcination was carried out at a temperature of 500 °C, 550 °C and 600 °C. The obtained TiO_2 were analyzed by XRD, SAA, and SEM. The conclusion is anatase TiO_2 obtained until annealing up to 600 °C. Self-assembly Pluronic F127 triblock copolymer capable of restraining the growth of TiO_2 crystals. Retention growth of TiO_2 mesoporous produces material character that can be used as builders photoanode DSSC with natural sensitizer anthocyanin from purple sweet potatoes. Based on the analysis of X-ray diffraction patterns and surface area analyser, the higher the calcination temperature the greater the size of the anatase crystals is obtained, however, the smaller its surface area. Purple sweet potato anthocyanin's dyed on to TiO_2 was obtained a good enough performance for DSSC's and gain the optimum performance from DSSC's system built with mesoporous TiO_2 annealed 550 °C using flavylium form anthocyanin.

1. Introduction

Environmental pollution has increasingly become a worldwide concern in the past few decades. Thus, how to enhance the efficiency of natural energy use and to recycle regenerated energy has become an important research field.

Photochemical reactions have been extensively studied for water decomposition, degradation of toxic organic pollutants, organic synthesis, dye-sensitized solar cells, and energy conversion [1,2]. Titanium dioxide (TiO_2) is one of the most widely used in photocatalytic application due to strong oxidizing power of its holes, its redox selectivity, high photostability, and easy preparation. An important requirement for high TiO_2 photocatalytic efficiency is a large surface area which increases both amount of photon generated electron-hole pairs and surface adsorbates. A number of methods such as chemical precipitation [3], microemulsion [4], hydrothermal crystallization [5], and sol-gel [6] have been used to enlarge specific surface area of TiO_2, mostly, by reducing the particle size down to nanoscale. Sol-gel is one of the most successful techniques for preparing nanosized metallic oxide materials [7]. Previously, we have successfully prepared the TiO_2 nanoparticles by a sol-gel procedure [2].

A DSSC that pioneered by Grätzel and O'Regan typically consists of a monolayer of photoactive dye molecules anchored onto the nanoparticles of a wide band gap semiconductor, an electrolyte and a counter electrode [8]. Early DSSC use transition metal coordinated compounds (e.g ruthenium polypyridyl complexes) as a photoactive dye [8,9]. Although the costly synthesis and undesired environmental impact of those prototypes call for cheaper, simpler, and safer dyes as alternatives [9,10,11]. Natural pigments, including chlorophyll, carotene, and cyanin, are available in plant leaves, flowers, and fruits and fulfill these requirements. Experimentally, natural-dye sensitized TiO_2 solar cells have reached an efficiency of 7.1% while over 8.0% using synthetic organic dyes [9,12].

Purple sweet potatoes (*Ipomoea batatas L*) that widely available in Indonesia has affordable prices and purple flesh skin that contains a lot of pigment anthocyanin. Anthocyanin can bind to the surface of TiO_2 via hydroxyl groups then form cyanine-Ti2 complex [13,14]. One of the important ways to improve adsorption capacity of anthocyanin to TiO_2 surface is enhanced surface area. Triblock copolymer in the sol-gel synthesis of TiO_2 was introduced to build self-assembly arrange mesoporous TiO_2 structure.

In this paper, we examine the effect of annealing temperature on the structural and electronic properties of TiO_2 nanostructures based films from 500 to 600 °C. It is found that the higher the calcination temperature the greater the size of the anatase crystals. Self-assembly of Pluronic F127 triblock copolymers have capable to restrain the growth of TiO_2 crystals. Purple sweet potato anthocyanin's dyed on to TiO_2 was obtained a good enough performance for DSSC's. The optimum performance from DSSC system was built with TiO_2 annealed 550 °C using flavylium form anthocyanin at initial pH 3.

2. Experimental

2.1. Synthesize Titanium Dioxide

Chemicals used for the synthesis of TiO_2 is $TiCl_4$ (E.Merck) Pluronic F127 i.e. $(PEO)_{100}(PPO)_{65}(PEO)_{100}$ (Aldrich), and ethanol (E.Merck). The titania nanoparticles were synthesized by drop wise addition of titanium tetrachloride: $TiCl_4$ in ethanol and Pluronic F127. 6 g Pluronic F127 dissolved in 76 mL of ethanol, then added 3.5 mL of $TiCl_4$ in order to obtain the mole ratio of Pluronic F127: ethanol: $TiCl_4$ = 1.0: 0.41: 21.7 in order to obtain pH 3. The solution was heated at 40 °C for 7 days for the aging process to obtain a gel. The reaction was performed at room temperature while stirring under a fume hood due to the large amount of Cl_2 and HCl gases evolved in this reaction [5]. The gel subsequently was calcinated at a temperature of 500 °C (A1), 550 °C (A2) and 600 °C (A3) for 4 hours with a heating rate 5 °C/min.

2.2. Characterization of TiO_2 materials

X-ray Diffractometer (XRD) patterns were recorded using a Bruker D8 Advance diffractometer equipped with a graphite crystal monochromator, operating with a Cu anode and a sealed X-ray tube. The 2θ scans were recorded using Cu Kα radiation of wavelength 1.54 Å in the range 20-80° with 0.05° step size while the crystallite size was calculated using Scherrer equation:

$$d_c = \frac{k\lambda}{\beta\cos\theta} \quad (1)$$

While the Surface Area Analysis (SAA) performed by the Brunauer-Emmett-Teller (BET) method from N_2 adsorption isotherms with Quantachrome NOVA 1200e series analysis at 77 K using Nova win & Nova win P. version 10.1. The BET approach was used to evaluate specific surface area from nitrogen adsorption data in the relative pressure (P/P°) range from 0.05 to 0.2. The pore size was calculated using the BJH method from the desorption branch of the isotherm.

The surface morphology of TiO_2 was determined using a FEI Quanta 250 Scanning Electron Microscope (SEM). Solar energy conversion efficiency (the photocurrent-voltage (I-V) curve) was measured by using two computerized digital Keithley multimeters under simulated sunlight (AM 1.5, 1000 mW/cm^2).

2.3. DSSCs Assembling
DSSCs were assembled following the procedure described in the literature [6], the catalyst-coated counter electrode was placed on the top so that the conductive side of the counter electrode faces the TiO_2 film. The iodide electrolyte solution was placed at the edges of the plates. The liquid was drawn into the space between the electrodes by capillary action. Two binder clips were used to hold the electrodes together.

3. Results and Discussion

3.1. Characterization of Crystalline TiO_2 XRD
Study of synthesis of TiO_2 by sol-gel process through the addition of Pluronic F127 as a template is done on the condition with several calcination temperature. The purpose of this stage is to investigate the effect of temperature on the degree of crystallinity and the crystal size of anatase TiO_2 (Figure. 1).

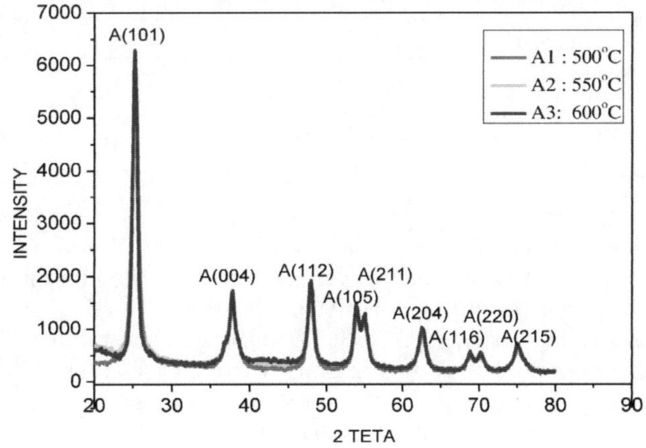

Figure 1. XRD Characterization on TiO_2 Powder Diffraction Patterns (A1, A2 and A3)

All samples synthesis of TiO_2 powder (A1, A2 and A3) have good crystallinity with high intensity and pure anatase structure shown in Figure 1. Good crystallinity helps the diffusion of electrons in the TiO_2 to be faster; thereby will improve the efficiency of solar cells performs. Synthesis of TiO_2 powder as samples A1, A2 and A3 produce pure anatase phase.

All crystalline TiO_2 obtained below 15 nm in size (Table 1). Annealing temperature affects the growth of TiO_2 crystals. The increasing calcination temperature up to 600 °C only able to grow crystals up to a size of 11-12 nm. Triblock copolymer acts as a template as well as crystal growth inhibitors. The additions Pluronic F127 can withstand rapid crystal growth so affect the crystallinity of TiO_2 obtained. Hindered the growth of crystal nucleus can prevent the formation of rutile. It also plays roles in modulating the mesostructure, surface area, pore size and wall thickness, as well as thermal stability of the resultant mesoporous TiO_2 [15].

3.2. Surface area analysis
Upon the synthesis of mesoporous TiO_2, crystallinity and surface area are directly related to the calcination temperature. Annealing temperature also affects the formation of mesoporous structure, increasing annealing temperature can increase the pore size is likely due to a more optimum pore opening and more complete combustion of the template. But increasing annealing temperature decreases the pore volumes due to collapse of mesoporous structure. In general, high crystallinity

requires high calcination temperature. However, the employment of high temperature usually leads to undesirable crystal grain growth, a total collapse of mesoporous framework, and a remarkable decrease of surface area (Table 1). Thus, only semicrystalline TiO_2 with small nanocrystals embedded in an amorphous matrix can be obtained, which greatly impedes its applications. The amorphous or semicrystalline TiO_2 easily suffers from fast electron−hole recombination owing to a large number of defects. Therefore, high crystalline TiO_2 with fewer surface defects is urgently desired to improve the conversion efficiency of solar energy and optimize the fast mass and charge transport [15].

TiO_2 from calcination of 500 °C has the highest surface area, whereas TiO_2 synthesized by calcination treatment at 600 °C has the lowest surface area, but it has the highest pore radius (Table 1). It is based on the crystal structure of TiO_2; higher calcination temperature resulted in the collapse of the mesoporous structure and lead to the formation of rutile phase with a larger crystal size than anatase phase and to increase the volume of the crystal grain size.

Table 1. Crystals Size and Surface Area Analysis Results of anatase TiO_2 Sample

	FWHM	D (nm)	Surface Area (m^2/g) BET	Pore radius (nm) BJH
A1	1.15448	7.52508	112.723	6.3
A2	1.0031	8.65969	98.723	4.9
A3	0.77327	11.2228	71.354	9.0

3.3. Surface Morphology

Figure 2 shows the scanning electron micrograph of a typical TiO_2 (anatase) film deposited by slip-casting on a conducting glass FTO that serves as current collector. The film thickness is typically 10–15 μm and the TiO_2 mass about 1-3 mg/cm^2. Analysis of the layer morphology shows the porosity to be approximately 45%. The slip-casting TiO_2 on FTO substrate appeared to be smooth under visual inspection although the film showed some inhomogeneity at the edges. The SEM micrograph of TiO_2 film (Figure 2) shows a rough surface layer containing large TiO_2 chunks in which the individual TiO_2 particles are hardly visible. The chunk structure is likely formed through the aggregation of bar-shape TiO_2 arranged in a side-by-side configuration. Another possible reason for the appearance of irregular chunks on the TiO_2 layer is the stress-induced surface rumpling caused by the fast cooling after 550 °C calcinations.

3.4. The absorbance characteristics of Anthocyanin Dye Purple sweet potato (Ipomoea L.,)

The pH values of anthocyanin dye were obtained through different preparation method. Acetic acid concentrations were 0.1% (v/v), 0.15% (v/v), and 0.2% (v/v). The result anthocyanin dye solutions have pH value of 5, 3, and 1 respectively. The lower pH condition, the more anthocyanin can be extracted. Therefore, the lower pH condition, the higher anthocyanin dye concentration reached indicating by the higher absorbance value.

The result of absorption character of anthocyanin dye at various pH (Figure 3) have the similar wavelength region of 450 nm to 600 nm. The maximum wavelength around at 530 nm was shifted by pH value. The sample pH-1 has the highest absorbance, indicate the highest anthocyanin content. At acidic pH, the dye exists mostly in the flavylium form in solution, but upon adsorption to TiO_2, the equilibrium between the adsorbed forms is thought to be shifted toward the quinonoidal form since the complexed cyanin appears purple [14]. There is an equilibrium between flavylium cation and quinonoidal form. A wide repertoire of colors in the red-blue range is available to anthocyanins as a result of their complexation with other polyphenols, pectins, and metal ions [14]. In acidic solution, the purple sweet potato anthocyanin dye appears red and has a strong absorption band at 530 nm. This visible absorption band is pH sensitive, causing the dye to appear red (flavylium form) in acidic solution and purple (quinonoidal form) as pH increases, and it is deprotonated. The visible absorption

IOP Conf. Series: Materials Science and Engineering **176** (2017) 012013 doi:10.1088/1757-899X/176/1/012013

band also shifts to the red upon complexation with titanium dioxide. It is thought that the titanyl groups of titanium dioxide surface compete with the protons, displacing them and shifting the flavylium-quinonoidal equilibrium toward the quinonoidal form in a mononuclear bidentate coordination complex with metal ions.

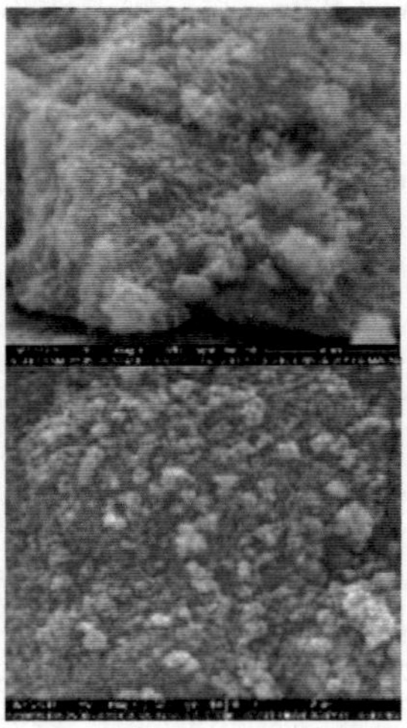

Figure 2. Morphology of TiO_2 film used in the DSSC with magnification 25,000 (top) and 50,000 (bottom).

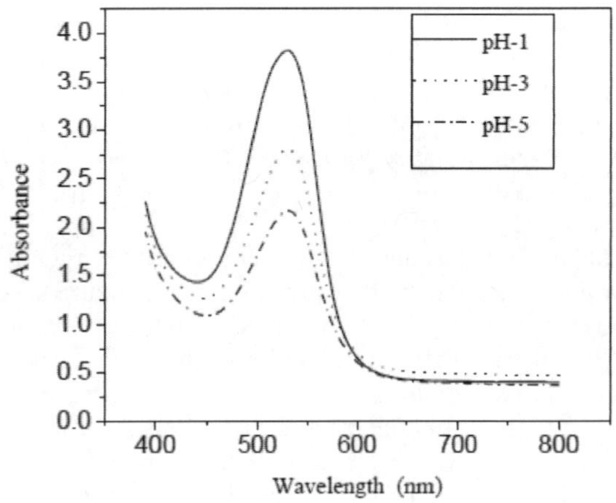

Figure 3. The visible spectrum of anthocyanin dye purple sweet potato (*Ipomoea batatas* L.,) at various pH conditions.

3.5. The Current-Voltage Characteristics of DSSC

The calcination temperature variations on the synthesis of anatase TiO_2 has an effect on the characteristics of TiO_2 powder and the performance of solar cells (Figure 4). Increasing the annealing temperature increases the pore size of TiO_2 material but lose surface area, thus affecting the adsorption of the dye. Immersion dye on TiO_2 surface requires a certain optimal time, and increased with time added. Irradiation of the DSSCs increased by 2006 W/m^2 from 1009 W/m^2 also increased the efficiency of the solar cell (Figure 4). The efficient of the solar cells decreased after 7 days, this is due to the evaporation of the electrolyte KI + I_2.

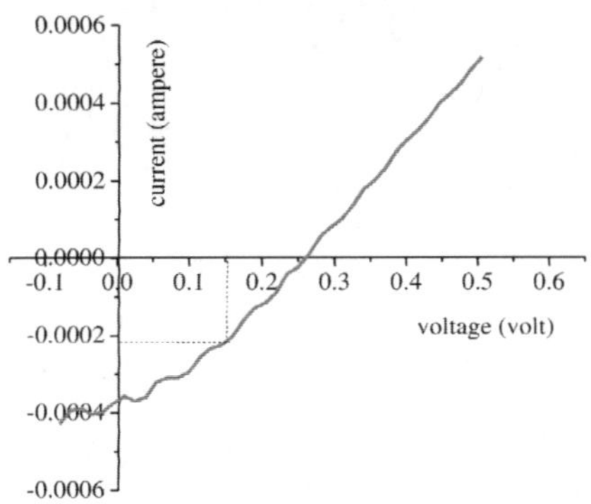

(a)

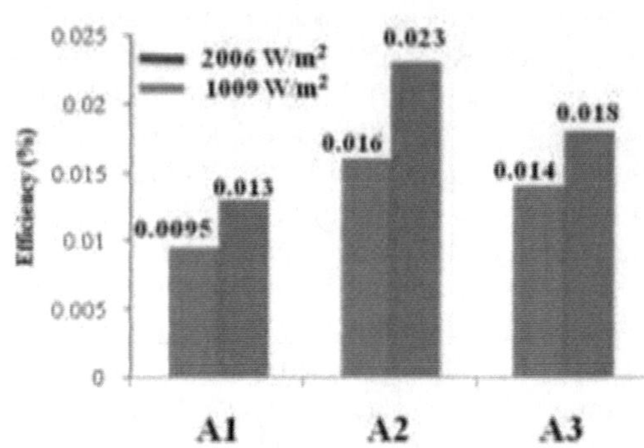

(b)

Figure 4. (a) I-V curves of A2 and (b) Effect of intensity irradiation on all DSSC samples.

The pH 1 anthocyanin dye has a characteristic absorbance and the best conductivity value, thus also improving the performance of DSSCs. Anatase TiO_2 with high intensity character showed high crystallinity. The high crystallinity increases the electron diffusion capability and is able to reduce the recombination of electrons and holes after photo excitation process.

Finally, its performance will increase the efficiency of DSSCs. The sample A2 (550 °C calcination treatment) had gave the best performance DSSC's with the highest efficiency due to the crystal size of (8.3 nm), large surface area (87.2 m^2/g), and high absorption intensity. Performance measurement with Keithley I-V meter with 1000 W/m^2 light intensity resulting DSSC efficiency of 0.0095 % (A1), 0.016 % (A2), and 0.014 % (A3). By the controlling of the pH condition in the sol gel titania preparation at pH 3 and the anthocyanin dyed on to TiO_2, we obtained anthocyanin-sensitized anatase TiO_2 have a good enough performance for DSSC's (Figure 4).

4. Conclusion

Synthesis of TiO_2 through the sol-gel process with addition of Pluronic F127 at pH 3 produced pure anatase TiO_2. Treatment of calcination temperature affects the crystallinity, surface area and pore performance of TiO_2. The higher the calcination temperature gives the better crystallinity of TiO_2, and the larger the pore distribution but lower the surface area. The anthocyanin dye at pH 1 has the highest conductivity. The conductivity properties were influenced by the structure of the dyes. The sample A2 (550 °C calcination treatment) had gave the best performance DSSC's due to the crystal size (8.3 nm), large surface area (87.2 m^2/g), and high absorption intensity. By the controlling of the pH condition in the sol gel titania preparation at pH 3, and the anthocyanin dyed on to TiO_2 was obtained a good enough performance for DSSC.

Acknowledgment

The authors would like acknowledge the Directorate General of Higher Education (DIKTI) the Ministry of Education and Culture (KEMENDIKBUD) The Republic of Indonesia for financial support this project.

References

[1] Assmann S E, Widoniak J, and Maret G 2004 *Chem. Mat.* **16** 6-11
[2] Sivakumar S, Sibu C P, Mukundan P, Krishnapillai P, and Warrier K G K 2004 *Mat. Lett.* **58** 2664-2669
[3] Lee L H and Chen W C 2001 *Chem. Mat.* **13** 1137-1142
[4] Allen N S, Edge M, Ortega A, Liauw C M, Stratton J, and McIntyre R B 2002 *Polym. Degrad. Stab.* **78** 467-478
[5] Chang H, Kao M J, Chan T L, and Kuo H G 2011 *Am. J. Engg. & Applied Sci.* **4** (2) 214 – 222
[6] Ramelan A H, Harjana, and Sakti L S 2012 *Europe Photonics* **435** 81-86
[7] Wongcharee K, Meeyo V, Chavadej S 2006 *Sol. Mat.* **11** 005
[8] O'Regan B and Grätzel M 1991 *Nature* **353** (6346) 737–740
[9] Meng S, Ren J and Kaxiras E 2008 *Nano Letters* **8** 3266-3272
[10] Wang G, Wang L, Xing W and Zhuo S 2010 *Mat. Chem. Phys.* **123** 690-694
[11] Adachi M, Jiu J, and Isoda S, 2007 *Current Nanosci.* **3** 285-295
[12] Sirimanne P M and Perera V P S 2008 *Phys. Stat. Sol.* **245** 1828 – 1833
[13] Zhan J, Peng Sun H J, Xiaohang S 2006 *An Investigation of The Performance of Dye Sensitized Nanocrystalline Solar Cell with Anthocyanin Dye and Ruthenium Dye as The Sensitizers* [Thesis] Roskilde University
[14] Cherepy N J, Smestad G P, Gratzel M, and Zhang Z 1997 *J. Phys. Chem. B* **101** 9342-9351
[15] Li W, Wu Z, Wang J, Elzatahry A A, and Zhao D 2014 *Chem. Mater.* **26** 287–298

Design of a Fiber Optic Biosensor for Cholesterol Detection in Human Blood

M Yunianto[1], A N Permata[1], D Eka[1], D Ariningrum[2], S Wahyuningsih[3] and A Marzuki[1]

[1] Physics Department, Sebelas Maret University
[2] Medicines Department, Sebelas Maret University
[3] Chemistry Department, Sebelas Maret University

E-mail : mohtaryunianto@staff.uns.ac.id

Abstract. A fiber optic sensor is to detect the cholesterol content in blood serum-based biosensor using plastic optical fiber that has been designed. The fiber optic sensor designed with sensing area was a fiber optic grated by 5 scratches then given bending treatment by 5 cm. The first test was by UV-Vis spectrometer with linearity of 0.96. The second test was by light spectrometer with linearity in white LED of 0.94. Optical fiber sensors were made to work well on a range of blood serum concentration of 140 mg/dL to 250 mg/dL.

1. Introduction

The importance of knowing the content of the concentration of cholesterol in the body is because high cholesterol in the body does not show any symptoms, it is important to check the content of cholesterol in the body on a regular basis. Diseases due to high cholesterol that exceeds the normal threshold may increase the risk of atherosclerosis (disease on the arteries or blood vessels that spread blood to the body) and cardiovascular disease (a disease caused by malfunctioning of heart and blood vessels, such as heart disease, hypertension and stroke), hypercholesterolemia (high cholesterol levels) and heart disease [1]. Therefore, a biosensor is developed to determine the content of cholesterol using blood serum that can be done in real-time. In the field of optics, to determine the concentration of cholesterol in human blood, it can be done by using optical fiber of Long Period Gratings (LPG) [2].

Research on enzyme-based biosensor, what is used to detect the cholesterol level in the blood is an electrochemical biosensor designed and fabricated by screen printing technique or thick film technology [3]. In addition, it is by the amperometric biosensors with the base of titanium oxide nanoparticles by accelerating the Reduced Instruction Set Computing (RISC) [4]. And to detect the use of cholesterol, amperometric enzyme which uses polymer-based amperometric enzyme electrodes (Pt/popd/Chox) [5] is used.

2. Experimental

In this study, to determine the concentration of cholesterol in human blood serum, it could be performed using the optical fiber sensor of Plastic Optical Fiber (POF) type scratched by 5 scratches. It was performed by examining the content of cholesterol in human blood serum using optical fiber scratched at the center of the optical fiber as the sensing area. Then the optical fiber was given bending treatment. On the sensing area, the interactions with blood serum occurred. Optical fiber connected on

the light source at one of the end of optical fiber and the other end of the optical fiber were connected to the detector, then from the detector to the interface. The measurement results would then be read and displayed on the PC. In this study, utilizing the change in the intensity of light which passed through the optical fiber sensor was the result from changes of the optical property, that was a refractive index or absorption spectrum of the medium surrounding the core optical fibers [6].
The design of the spectrometer was divided in two main parts: the manufacture of hardware and software. The hardware consisted of a convex lens, grating, light source, propulsion systems, and control system. This tool was designed to measure light spectrum using fiber optics, and then for the place of optical fiber was micropositioner. In this study, the sensing area was the optical fibers scratched by 5 scratches and then given bending treatment by 5 cm.

3. Result and Discussion

Light spectrometer calibration is using monochromatic light in the form of He-Ne laser which wavelength has been known, that is 632.80 nm. From the results of calibration, it is obtained the peak result by (631.18 ± 0.71) nm with the accuracy of 99.89% as showed at Figure 1.

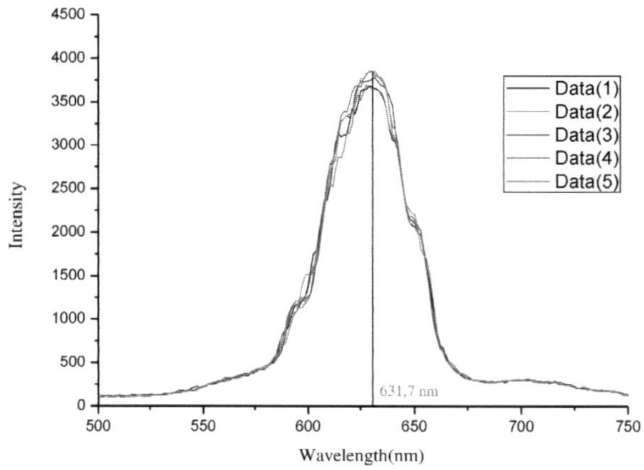

Figure 1. The resulting graph calibration using laser HeNe

The sample used in this research is blood serum obtained from Budi Sehat Clinic, Surakarta, Central Java, Indonesia. The blood serum has had information such as the different content of cholesterol concentrations, this becomes a variation in a study conducted, in which the data is presented in Table 1.

Table 1. Information of cholesterol content in blood serum

Number of Blood's data	Colesterol concentration (mg/dL)
1	140.8
2	191.3
3	232.1
4	234.2
5	243.5

The blood is serum measured using light spectrometer, each sample is measured with different LEDs. LED variation used is a white LED, a green LED, blue LED and yellow LED. The purpose of the LED color variation is due to the wavelength range of each LED is different, meaning that it will produce a different absorption as well. Figure 2 shows the results of the linearity from testing blood serum with white LED, green LED and blue LED. In the yellow LED, it is not obtained the linear graph because in the yellow LED there is no specific absorption at certain wavelengths as the light is passed. The light is passed since the blood serum samples have clear yellow color.

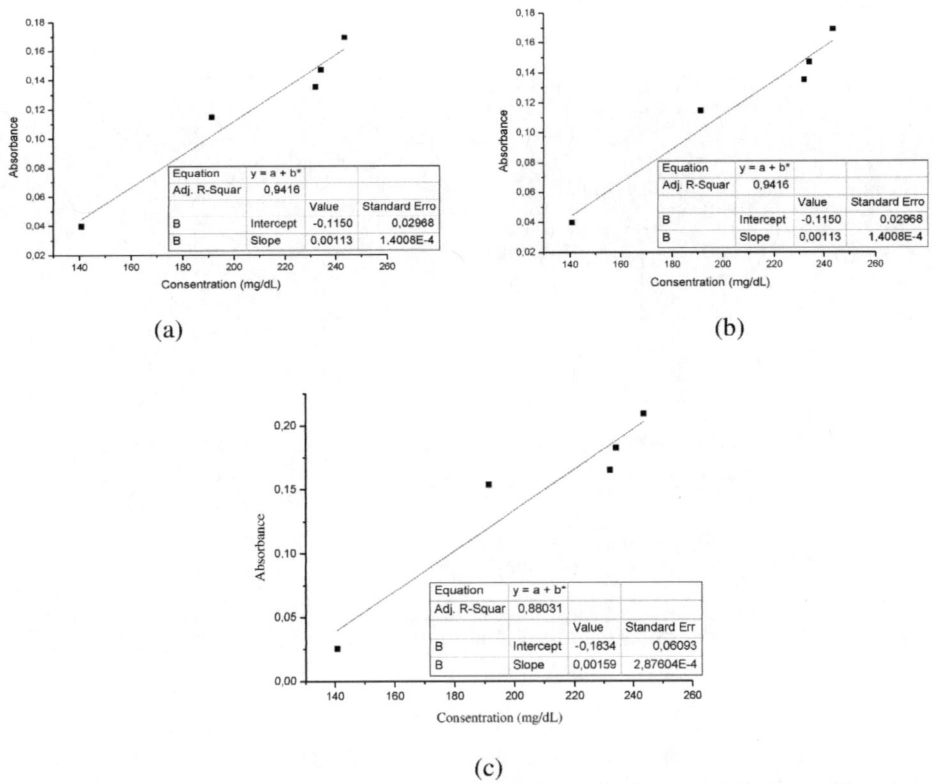

Figure 2. Results of the linearity from testing blood serum with (a) white, (b) green and (c) blue LED

Typical uptake results on white LED is at a wavelength of 514.20 nm with linearity of 0.94, on green LED at a wavelength of 527.90 nm with linearity of 0.87 and on blue LED at a wavelength of 475.70 nm with linearity of 0.88. Meanwhile, in the test using a UV-Vis spectrometer, it has a wavelength of 577.00 nm with linearity of 0.96. Figure 3 shows the results of blood serum test using a UV-Vis spectrometer and the light spectrometer graph of absorbance as the concentration function of (a) white LED (b) green LED (c) yellow LED.

IOP Conf. Series: Materials Science and Engineering **176** (2017) 012014 doi:10.1088/1757-899X/176/1/012014

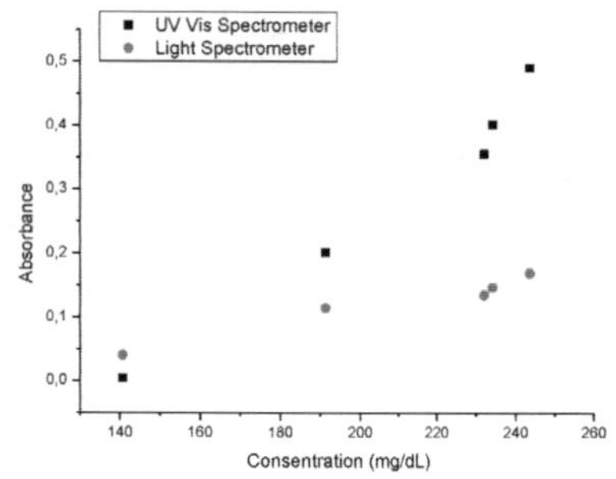

(a)

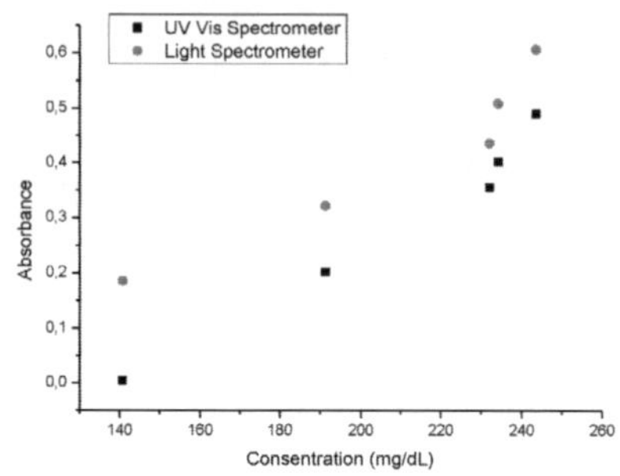

(b)

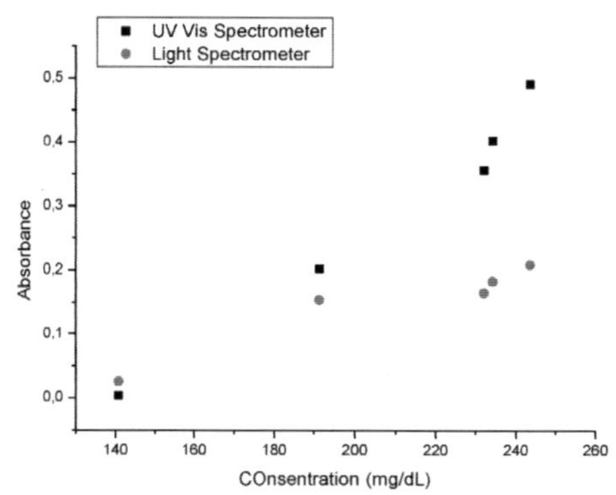

(c)

Figure 3. Graph of absorbance as the concentration function of (a) white (b) green and (c) yellow LED.

In this study, the blood serum is not mixed with other substances such as enzymes and reagents on the cholesterol test in the health field. Diagnostic study used for cholesterol with absorption wavelength of the best reagent kit is at a wavelength of 505 nm, then what is used in the estimation of cholesterol diagnostic study is at a wavelength of 505 nm [7]. Measurement of cholesterol levels purely measured using a scratched optical fiber as the sensing area which is given bending treatment using a spectrometer light tool. Before the testing using a spectrometer light, the blood serum is measured using a UV-Vis spectrometer. The results of measurements of UV-Vis spectrometer and the light spectrometer are then compared.

4. Conclusion

The study of cholesterol content uses optical fiber sensors-based biosensor using scratched optical fibers. Light source is passed in the optical fiber that has been scratched, in which the light is absorbed by the blood serum in accordance with the principles of light propagation. The first test is by UV-Vis spectrometer with linearity of 0.96. The second test is by light spectrometer with linearity in white LEDs of 0.94. Optical fiber sensors are made to work well on a range of blood serum concentration of 140 mg/dL to 250 mg/dL.

References

[1] Ahmet K, Esma A and Yeliz K İ 2012 Biosensors and Their Principles, A Roadmap of Biomedical Engineers and Milestones, Prof. Sadik Kara (Ed.), InTech, DOI: 10.5772/48824.
[2] Mathews, Bobby C, Libish T M, Linesh J, Biswas P, Bandyopadhyay S, Dasgupta K and Radhakrishnan P 2013 *Sensors & Transducers* **149(2)** 83
[3] Debataraja A, Soelaiman N F and Hiskia 2011 *Jurnal Ilmiah Elite Elektro* **2(1)** 55
[4] Jarallah N A, Baradie R A, Aljamal M, Bose A, Anazi H A, Alatram A, Aboody M A and Algohary A 2014. *Proc. Int. Conf. on Sensing Technology (Liverpool)* Liverpool, UK. p 381
[5] Zeybek D K, Zeybek B, Pekmez N Ö, Pekyardımcı Ş and Kılıç E 2012 *J. Braz. Chem. Soc.* **23(12)** 2222
[6] Crisp J and Elliott B 2006 *Serat Optik: Sebuah Pengantar* (Jakarta: Erlangga)
[7] Soni S, Gamatra S and Soni A 2014 *International Journal of Research in Engineering and Applied Sciences* **2(1)** 46

Study of Evanescence Wave Absorption in Lindane

A Marzuki*[1], E Prasetyo[1], M P Gitrin[1], V Suryanti[2]

[1]Department of Physics, Faculty Mathematics and Natural Sciences, Sebelas Maret University, Jl. Ir. Sutami No 36A Surakarta 57126 Indonesia
[2]Department of Chemistry, Faculty Mathematics and Natural Sciences, Sebelas Maret University, Jl. Ir. Sutami No 36A Surakarta 57126 Indonesia

*E-mail: amarzuki@mipa.uns.ac.id

Abstract. Evanescent wave field has been studied for the purpose of tailoring fiber sensor capable of detecting lindane concentration in a solution. The mounted fiber was optically polished such that part of the fiber clad is stripped off. To study the evanescent wave field absorption in lindane solution, the unclad fiber was immersed in the solution. Light coming out of the fiber was studied at different wavelength each for different lindane concentration. It was shown that evanescent wave field absorption is stronger at wavelength corresponding to lindane absorption band as has been shown from absorption studies lindane in UV-VIS-NIR spectrophotometer.

1. Introduction

Farming efficiency have introduced not only modern farm machinery but also the use of fertilizer and pesticides. Applying this modern method, farmers have become capable of increasing the land productivity using less labor and land. The biggest issue is improper and excessive use of chemical fertilizers and pesticides can endanger our environment. Among many pesticides used by famers, lindane is one that have been used excessively in rural country. To ensure that the disposed lindane in our environment is still under the allowed level, frequent monitoring of lindane concentration level in our environment is required.

So far, lindane and other persistent organics pollutants (POPs) concentration in a groundwater is obtained from expensive lab work. Many sensors capable of measuring POPs concentration have been developed. Among them are optical fiber sensors. In last few decades, these sensors have received much attention because of many reasons. Among them are small size, low cost, no electrical risk, and immune from interference. Taking these advantages in mind, many optical fiber sensors have been developed. Few of them are mechanical fiber sensors [1, 2, 3], thermal fiber sensors [4, 5], and biochemical fiber sensors [6 - 9]. The main advantage of using optical fiber for biochemical sensors is the capability of light to reveal the finger print (unique absorption or emission spectrum) of a compound. Aimed at designing fiber sensor to measure lindane concentration, this paper presents the analysis of light-lindane interaction in term of the amount of the evanescent wave field light absorbed as unclad fiber optic is immersed in different concentration of lindane solutions.

2. Experimental

Lindane used in this experiment was that purchaced from Sigma-Aldrich with purity of 99 %. Lindane was then dissolved in methanol 70% so that different concentrations of lindane solutions were obtained ($2.06x10^{-3}$ M, $1.06x10^{-3}$ M, $5.15x10^{-4}$ M, $3.43x10^{-4}$ M, and $2.58x10^{-4}$M). Standard absorption spectrum of each lindane solution was recorded at room temperature using UV-VIS-NIR spectrophotometer. In order to study the absorption spectrum of lindane resulted from its interaction with the light propagating along the fiber sensor, polymer optical fiber was mounted on a resin block surface. Polishing this surface gently will strip part of the fiber clad away. Figure 1 is the experimental setup. The optical fiber sensor was immersed in lindane solution. As the polychromatic light was launched from one end of the fiber, light will propagate in the fiber core with multiple total internal reflections (TIR). Evanescent wave field resulted on unclad fiber might be absorbed by lindane solutions. The absorbtion strength will then be detected by spectrometer and displayed by personal computer (PC).

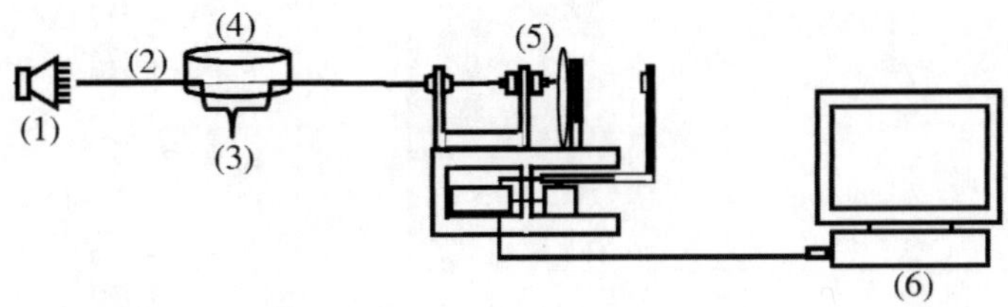

Figure 1. The experimental setup used to study evanesecent wave field of fiber sensor immersed in lindane solution. (1) light source, (2) POF, (3) unclad *fiber sensor*, (4) liquid cell, (5) spectrometer (6) PC

3. Results and Discussion

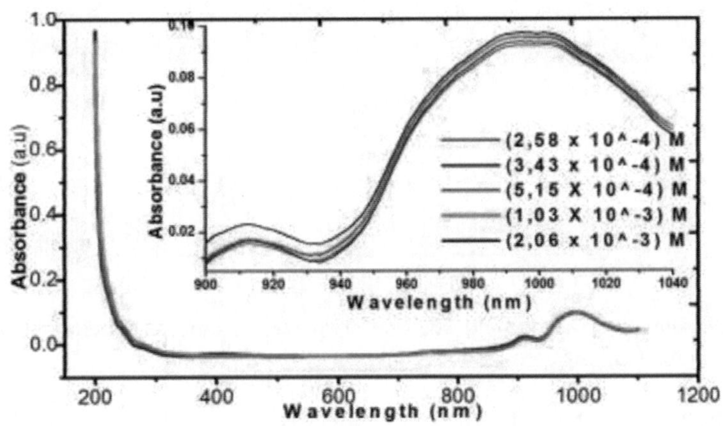

Figure 2. Absorption spectra of lindane solutions at UV-VIS-NIR recorded at room temperature

Figure 2 is absorption spectra of many different concentrations of lindane solutions recorded at room temperature. Two unique absorption bands corresponding to absorption peaks at 913 nm 1003 nm are observed. Absorbance peaks at 913 nm corresponding to lindane solutions of $2.06x10^{-3}$ M, $1.06x10^{-3}$ M, $5.15x10^{-4}$ M, $3.43x10^{-4}$ M, and $2.58x10^{-4}$M are 1.55×10^{-2}, 1.67×10^{-2}, 1.71×10^{-2}, 1.74×10^{-2} and $2,30\times10^{-2}$ arbitrary units, respectively. While for band peaks centred at 1003 nm, their corresponding absorbances are 9.25×10^{-2}, 9.37×10^{-2}, 9.53×10^{-2}, 9.58×10^{-2}, and 9.71×10^{-2}, respectively. It is clearly seen from figure 3 that the absorption peaks increase linearly with the increase of lindane concentrations. Both curves give good linearity (R^2 = 0.91 for absorption band at 913 nm and R^2 = 0.93 for that at 1003 nm). No such relation can be observed for absorbance beyond these two absorption bands. Good correlation between absorption peak and its corresponding concentration of a solution has also been observed by other researchers [10, 11] implying that a unique absorption band strength can be used as a measure a compound concentration presents in a solution.

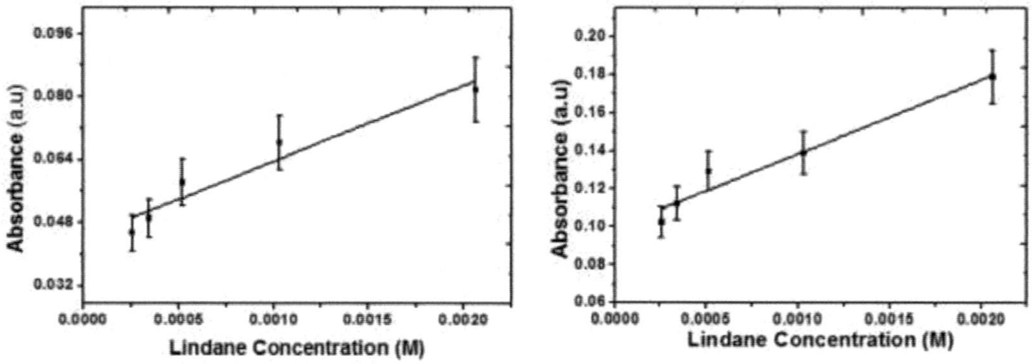

A b

Figure 3. Absorbance peaks of lindane solution at wavelength (a) 913 nm and (b) 1003 nm

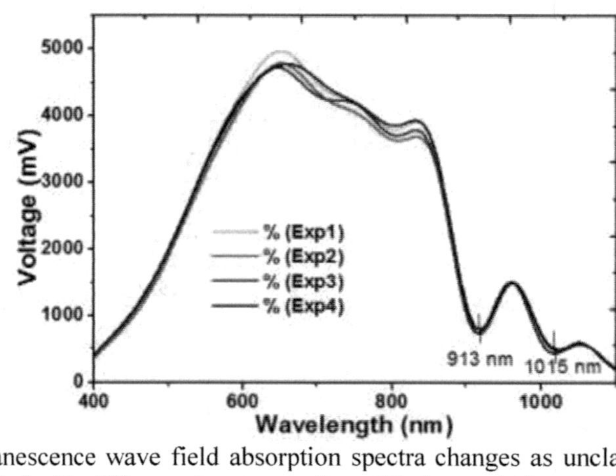

Figure 4. evanescence wave field absorption spectra changes as unclad fiber sensors were surrounded with differenent concentration of lindane solutions

Figure 4 is light intensity spectra expressed in voltage detected from the end of the fiber sensor as the unclad part of the fiber were immersed in different concentrations of lindane solutions. Higher voltage as shown in this figure means that more light is transmitted through fiber optic while less voltage means more light is absorbed. Absorption at 913 nm and 1015 nm as shown in figure 4 corresponds to absorption at 913 nm and 1003 nm, respectively. There was a slightly shift of the peaks positions of the light spectra recorded by our evanescence wave field spectrometer compared to that recorded by direct measurement of absorption by UV-VIS-NIR spectrophotometer. This shift however is still within the tolerated error.

Strong absorptions observed at 918 nm and 1015 nm as shown in figure 4 imply that part of the evanescent wave field penetrate out of the fiber core within the unclad fiber is strongly absorbed by lindane solution using mechanism outlined in figure 5. Light is launched from one end of the optical fiber and propagating toward the other end with total internal reflection (TIR). As the light strikes the core-cladding interface, light will be reflected back to the core. At the striking point, part of the light penetrates in the cladding area, but because the cladding material is not designed to absorbed light, the light will keep moving forward. In term of field distribution, strong field is confined within core region with weak tail penetrates within very depth skin in clad region. As the clad is removed and replaced with lindane the evanescent wave field might or might not be absorbed. If light with wavelength beyond the absorption band is launched into the fiber optics, light will not be absorbed. While if it is within the absorption band, light will be absorbed.

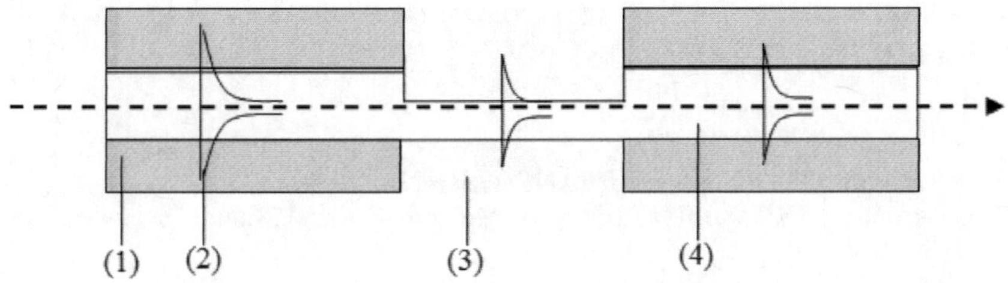

Figure 5. Absorption mechanism of evanescent wave field within unclad fiber by lindane solution. (1) fiber clad, (2) evanescent wave field penetrating in fiber clad, (3) evanescent wave field penetrating in lindane solution, (4) light propagation in fiber core.

4. Conclusions

Absorption spectra of different concentration of lindane at room temperature have been measured using UV-VIS-NIR spectrophotometer. Two absorption peaks corresponding to wavelength of 91 nm and 1003 nm have been shown to linearly increase with the increase of lindane concentration in solution. Absorption of evanescent wave field observed from the immersed unclad fiber into lindane solution also showing similar results. This suggest that optical fiber sensors to measure lindane concentration in a solution that work based on absorption evanescent wave field might be designed

References

[1] Liu X and Zhu W 2017 *Robot Comput Integr* **43** 12-17.
[2] Marzuki A, Heriyanto M, Setiyadi I D, Koesuma S 2015 *J Phys Conf Ser* **622** 012059.
[3] Kang J, Dong X, Zhu Y, Jin S, Zhuang S 2014 *Opt Com* **322** 105-108.
[4] Villalba A and Martín J C 2017 *Opt Fiber Technol* **33** 36-38.
[5] Du C, Wang Q, Zhao Y, Li J 2017 *Opt Fiber Technol* **34** 12-15.
[6] Rithesh Raj D, Sudarsanakumar C 2017 *Sensor Actuat A Phys* **253** 41-48.
[7] Yu C, Wu Y, Liu X, Fu F, Gong Y, Rao Y J, Chen Y 2016 *Sensor Actuat B Chem* (paper in Press).
[8] He Y J 2015 *Sensor Actuat B Chem* **206** 212-219.
[9] Chen Y, Yu Y, Li X, Zhou H, Hong X, Geng Y *Sensor Actuat B Chem* **226** 412-418.
[10] Chen J, Li C, Zhou M, Liu J, Kan R, Xu Z 2017 *Infrared Phys Techn* **80** 131-137
[11] Wang L, Zhang Y, Zhou X, Qin F, Zhang Z 2017 *Sensor Actuat B Chem* **241** 146-150

UTILIZATION CHITOSAN-*p*-*t*-BUTYLCALIX[4]ARENE FOR RED MX 8B ADSORBENT

D S HANDAYANI, T KUSUMANINGSIH, L A HAK
Department of Chemistry, Faculty of Mathematics and Natural Sciences, Sebelas Maret University

E-mail: dsuci72@gmail.com

Abstract. Adsorption of *Procion Red MX 8B* using chitosan dan chitosan-linked *p-t*-butylcalix[4]arene has been done. The research aimed to understand the adsorption of *Procion Red MX 8B* using chitosan *p-t*-butylcalix[4]arene compared to ordinary chitosan. The research was conducted in a batch process varying in pH, contact time and initial concentration of the *Procion Red MX 8B*. The amount of dye adsorbed was determined using UV-Vis spectrophotometer and the adsorben was characterized using FTIR and SEM-EDX spectrophotometer. The result showed that the optimum condition was achieved when the pH was set at 4, contact time 135 minutes and initial concentration at 200 ppm. The kinetic analysis showed that the adsorption followed Ho kinetic model and pseudo second order with the adsorption rate constant was 3.69×10^{-3} g/mg.minute and 2.03×10^{-3} g/mg.minute. The isotherm analysis showed that the adsorption process tend to occur following the Langmuir model with maximum capacity for chitosan and chitosan-linked *p-t*-butylcalix[4]arene 136.09 mg/g and 147.35 mg/g respectively. The adsorption energy of chitosan and chitosan-linked *p-t*-butylcalix[4]arene at 30.53 kJ/mole and 33.65 kJ/mole.

1. Introduction

The development of industrialization, besides bringing benefits also cause environmental problems, namely the emergence of environmental pollution and destruction of one of them is the emission of pollutants into the environment. Industrial liquid waste is a pollutant that causes negative environmental impact. The liquid waste containing hazardous materials and toxic contained in the water, can block the sun shone through the aquatic environment, thereby disrupting the biological processes [1].

One of the industry, namely the liquid waste of batik industry in the form of waste dye. Waste water of Batik industrial was originating from the dyeing, washing, printing and finishing proses. Waste dyed of batik industry contains several components of which the rest of the dye, salt and additive materials such as urea, sodium alginate, sodium bicarbonate and water residual staining and washing. Reactive dyes are often used in the batik industry include Procion, Cibracon, Drimaren, Lavafix, Remazol, forecasts allowing, and Primazin [2].

Many methods have been developed for waste treatment dye in the waters that coagulation, adsorption [3] and ion exchange [4]. Coagulation method will produce sludge in relatively large amounts. The resulting sludge was eventually going to create new problems for waste processing unit [5], ion exchange method and reverse osmosis are generally effective, but requires equipment and operating costs are relatively high. Effective method used to treat waste is the adsorption of the dye, because the excess of adsorption methods, among others, the process is relatively simple, relatively high effectiveness and efficiency, and do not give side effects such toxic substances [6]. Additionally adsorption method is one method that can work at low concentrations, can be recycled, and costs are relatively cheap [7].

Several studies of wastewater treatment dye adsorption proved to effectively reduce the concentration of the dye in the water. The adsorbent used comes from synthetic and natural. Synthetic adsorbent has been used, among others PSF-EGDE-CTS (polisulfon-ethylene glycol diglycidyl ether-chitosan) [8] to adsorb Procion Red MX 8B dyein textile waste batik and clay pillared CTMA-Al (cetiltrimetil ammonium-aluminium) [9] to adsorb dye indigo karmina. The adsorbent of nature, among others Aryunani [3] have investigated the the adsorption of the dye textile Remazol Yellow on the batik waste using water hyacinth activated NaOH. Setyoningsih [1] was using a pineapple leaf fibers to adsorb dye Procion Red MX 8B. The use of natural adsorbent that is cheap and plentiful in nature certainly is expected by the researchers, but to find the active compound can be used as an effective adsorbent is not easy, it can be difficult to explain the adsorption phenomena occurring in the adsorbent used. It is therefore necessary to find another adsorbent the structure, chemical and physical properties are known with certainty, and this can be obtained from the results of previous studies. Terms absorbent good, among others, have absorption large, is a solid that has a large surface area, should not be soluble in the substance to be adsorbed, must not engage in a chemical reaction with the mixture to be purified, can be regenerated back easily and non-toxic [10].

Chitosan is adsorbent with structure, chemical and physical properties are already known with certainty. It is abundant in nature and a good adsorbent criteria. Chitosan is the result of deacetylation of chitin which has undergone the removal of acetyl groups. Acetyl group is missing from the process, leaving deacetylation free amino group that causes the chitosan to be polycationic [11]. Research on the application of chitosan as an adsorbent has been done by some researchers that once used to adsorption of Procion Red MX 8B dye include Kusumaningsih et al [8] and Astuti [12], but has adsorption capability and law adsorption capacity (19.02 mg/g and 12.52 mg/g). Chitosan had specific properties such as biocompatibility, non-toxic and renewable makes a biopolymer chitosan serious attention to continue to be developed and researched. Improving the ability of chitosan can be fastened on a cross-linked macromolecular compounds crown ether is p-t-butylcalix[4]arene. The compound of p-t-butylcalix[4]arene has a geometry such as baskets and hollow, can be used in guest-host systems [13]. Modifications to improve the ability of chitosan as an adsorbent that is by p-t-butylcalix[4]arene attached to the chitosan has been done by Hilmiyana [14] who successfully apply to Procion Red MX 8B dye.

This research has been conducted on dye adsorption others, namely Procion Red MX 8B dye using butylcalix[4]arene tied to the chitosan results of research Handayani [15]. The Procion Red MX 8B dye is a reactive dye dichlorotriazine group that can undergo substitution reaction with the fibers and form ester bond that has the highest reactivity compared to other classes of dyes [8]. Waste of Procion Red MX 8B dye off for having reported carcinogenic azo group [16]. Therefore, the chitosan-p-t-butylcalix[4]arene expected to be effective and efficient way to reduce environmental pollution, especially in the batik industry.

2. Experimental

2.1. Ingredients:

Chitosan, Chitosan-p-t-butylcalix[4]arene (synthesized by Handayani [15]), NaOH Merck., HNO_3 Merck, *Procion Red MX 8B* dye

2.2. *Instruments:*
Stirer Adsorption (CV INTAN Yogyakarta), Fourier Transform Infra Red (FTIR) Prestige-21 Shimadzu, Scanning Electron Microscopy (SEM) JEOL JSM 6360 - EDX (JED-2200 and UV-Vis Spectroscopy Optima SP-300

2.3. Procedure:
2.3.1. Effect of solution pH Procion Red MX 8B
A total of 25 mL of Procion Red MX 8B at a concentration of 100 ppm respectively set the initial pH to 1.5; 2; 4; 6; 8; 10 with HNO_3 and NaOH. Then into each solution was added 20 mg of chitosan-p-t-butylcalix[4]arene. The mixture was stirred at a speed of 200 rpm for 2 hours. Afterwards the solution was filtered and the filtrate was analyzed by UV-Vis spectroscopy. Equal treatment performed on chitosan as a comparison.

2.3.2. Effect of Stirring Time
A total of 25 mL of Procion Red MX 8B each with a variation of time at 5, 15, 45, 135, 405 minutes was set at optimum pH. Then into a solution, 20 mg of chitosan-p-t-butylcalix[4]arenes adsorbent was added and stirred for optimum time with a speed of 200 rpm. The solution is then filtered and analyzed by UV-Vis spectroscopy. Equal treatment performed on chitosan as a comparison.

2.3.3. Influence of Solution Concentration Procion Red MX 8B
A total of 25 mL of Procion Red MX 8B each with varying concentrations of 100, 150, 200, 250, 300 ppm was set at optimum pH. Then into a solution, 20 mg of chitosan-p-t- butylcalix[4]arenes adsorbent was added and stirred for optimum time with a speed of 200 rpm. The solution is then filtered and analyzed by UV-Vis spectroscopy. The same treatment was done for chitosan as a comparison.

3. Results And Discussion
3.1. Determination of Optimum pH
The adsorption process depends on the type of adsorbent, the adsorbed species and the type of solvent. PH variation on the research done at the initial pH solution of pH 1.5; 2; 4; 6; 8; 10; 11 and 12. Figure 1 showed the curve of the adsorption of the dye Procion Red MX 8B by chitosan and chitosan-p-t-butylcalix[4]arenes at various pH. Figure 1 can be seen that the adsorption occurs at a pH of acids and bases with the highest absorption capacity exist at pH 4, so that the optimum pH is pH 4 to chitosan and chitosan-p-t-butylcalix[4]arenes with the adsorption capacity of 82.94 and 99.74 mg/g. The solution to the acid conditions of the hydrogen atoms (H^+) can occur protonated amine group ($-NH_2$) of chitosan. In aqueous solution, the dye will dissolve and sulfonate groups to the reactive dye Procion Red MX 8B occurs dissociation and transformed into ions.

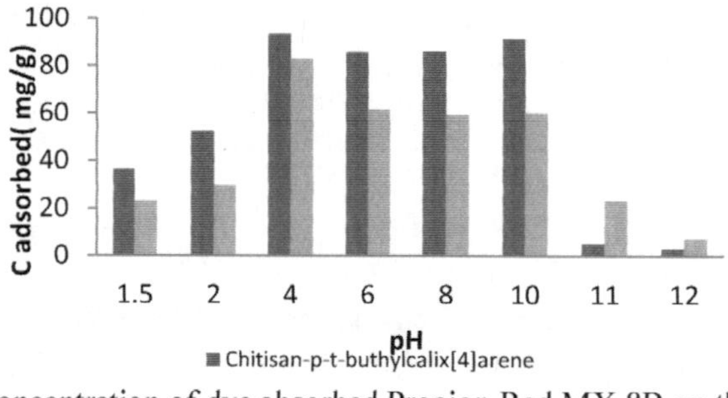

Figure 1. Concentration of dye absorbed Procion Red MX 8B on the variation of pH

When the acidic conditions, the hydrogen bonds also contribute to the the adsorption process. Hydrogen bonding occurs between nitrogen from chloride sianurate dye with hydrogen from the OH groups of chitosan [12]. When an alkaline condition, the adsorption by chitosan can still occur. This is in accordance with the opinion of Sakkayawong et al. [17].

The adsorption process at alkaline pH can be explained by the presence of ions (CH_2O-) of chitosan covalently bound to the dye replace the cluster chloride (Cl) are separated into solution. Therefore pH 8 to pH 10 increased adsorption power but not as high at pH 4 and pH down again at 11 and 12. The decrease adsorption occurs in alkaline conditions due to the hydroxyl group (OH) of the water molecules react perfectly with the dye , thus causing the interaction between the dye with an adsorbent decreases.

The adsorption capacity of chitosan-p-t-butylcalix[4]arenes at pH 4 in the amount of 93.35 mg/g. The results presented almost the same as the adsorption by chitosan, but the adsorption by chitosan-p-t-butylcalix[4]arenes showed better adsorption of the chitosan. In acidic conditions, reactive groups chitosan-p-t-butylcalix[4]arenes that has been protonated more stable and form molecules of chitosan-p-t-butylcalix[4]arenes itself shaped basket and hollow thus indicated that the dye Procion Red MX 8B trapped so that the maximum adsorption. A process of return electrostatic interaction between the amine group with sulfonate groups of dye, like chitosan, but the chitosan-p-t-butylcalix[4]arenes such as secondary amines. The process that occurs on the same basic conditions such as adsorption by chitosan, in which the hydroxyl groups of chitosan-p-t-butylcalix[4]arenes attack chloride groups of the dye and binds covalently to the dye. But in the presence of water molecules, allowing competition between the hydroxyl group (OH) of the water molecules and hydroxyl groups of the chitosan-p-t-butylcalix[4]arenes to bind with the dye as the reaction expressed by Christie [18].

3.2. Effect of Contact Time and Parameter Kinetics

Contact time was the time required adsorbent to adsorb Procion Red MX 8B dye. The time variation contacts made in this study was 5; 15; 45; 135 and 405 minutes. Effect of contact time on the amount of adsorbed dye can be seen in Figure 2.

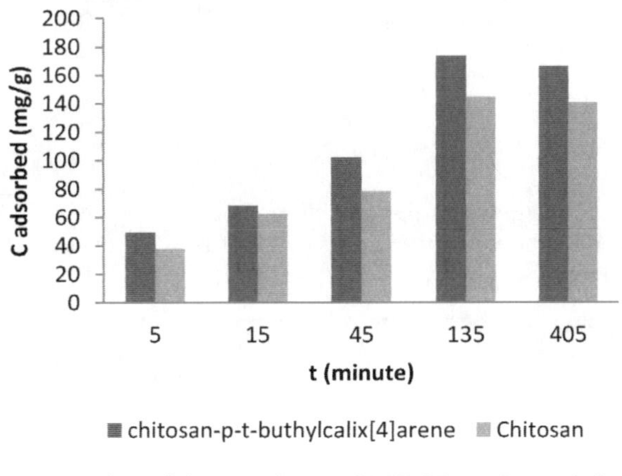

Figure 2. Concentration of dye Procion Red MX 8B on the variation of contact time

The optimum contact time for both the adsorbent was reached at minute 135. The concentration of the dye Procion Red MX 8B absorbed as much as 144.34 mg/g of chitosan and 173.79 mg/g for the chitosan-p-t-butylcalix[4]arenes. Results of the adsorption kinetics study in the two adsorbents shown in Table 1.

Table 1. Assessment of the dye adsorption kinetics Procion Red MX 8B

adsorption kinetics	Adsorbent	
	Chitosan	chitosan-p-t-butylcalix[4]arene
Lagergren:		
– K (min^{-1})	0.015	0,017
– R^2	0.804	0,896
Ho:		
– K (g mg^{-1} min^{-1})	3.69 x 10^{-3}	2.03 x 10^{-3}
– R^2	0.983	0.963

3.3. Effect of Initial Concentration Dyes and Parameter Ishoterm

Adsorption capacity Procion Red MX 8B of the adsorbent chitosan and chitosan-p-t-butylcalix[4]arene carried out under optimum pH and time that has been obtained previously. Effect of initial concentration of dye Procion Red MX 8B of the adsorbent defined in Figure 3.

Figure 5 showed the adsorption concentration tended to increase with increasing concentration of adsorbate and then remained relatively constant at a concentration of 200 ppm for both types of adsorbents. Study of adsorption isotherms determined whether the dye adsorption with adsorbent followed the Langmuir isotherm or Freundlich isotherm.

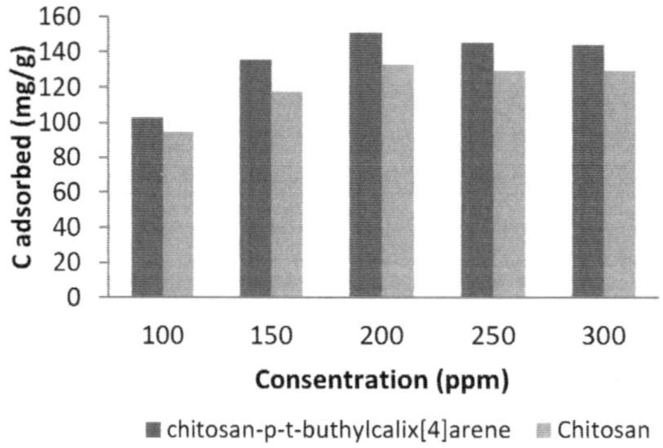

Figure 3. Variation of Initial concentration dyes Procion Red MX 8B

The study results dye adsorption isotherm Procion Red MX 8B using adsorbents chitosan and chitosan-p-t-butylcalix[4]arene are presented in Table 2.

Table 2. Parameters of isotherm adsorption of Procion Red MX 8B

Adsorbent	Parameter of isotherm adsorption						
	Langmuir				Freundlich		
	Xm (mg/g)	K x10^5 (L/mol)	ΔG (kJ/mol)	R^2	n	K(mol/g) x 10^{16}	R^2
Chitosan	136.09	2.07323	30.53	0.997	0.198	2.39	0.755
Chitosan-*p-t*-butylcalix[4]arene	147.35	7.23333	33.65	0.998	0.184	25.1	0.600

Dye adsorption isotherm Procion Red MX 8B on chitosan and chitosan-p-t-butylcalix[4]arene tend to followed the model of Langmuir adsorption isotherm, which indicates that both adsorbent occurs at monolayer adsorption. The maximum adsorption happens when all of the active sites of the adsorbent filled by adsorbat forming monolayer. In addition Langmuir models indicated that the adsorption process was chemisorpsi. Comparison of adsorption capacity optimum of adsorbents that have been used to adsorb dye Procion Red MX 8B showed that the adsorbent chitosan and chitosan-p-t-butylcalix[4]arene had a capacity of optimum adsorption was greater than the adsorbent others, as shown in Table 3.

Table 3. Adsorption of dye Procion Red MX 8B by several adsorbent

Adsorbent	qm (mg/g)
PSF-EGDE-CTS [8]	40.69
Pineapple leaf fiber [1]	6.38
Chitosan Sulfate [12]	13.76
Chitosan	136.09
chitosan-p-t-butylcalix[4]arene	147.35

3.4. Adsorbent characterization With FTIR

Characterization adsorbent using an infrared spectrophotometer was performed to identify the functional groups contained in the adsorbent. Results of FTIR spectrum of dye Procion Red MX 8B as well as the combination of chitosan before being used to adsorb dye Procion Red MX 8B and chitosan after being used to adsorb dye Procion Red MX 8B shown in Table 4.

Table 4. Comparison of the functional group of dye Procion Red MX 8B, chitosan before and after adsorbed the dye Procion Red MX 8B

Compound	wavenumber (cm^{-1})	functional groups
Dye *Procion Red MX 8B*	3440.09	Groups of hydroxy (-OH)
	1537.22	Groups of –C=C- aromatic
	1471.64	Groups of N=N azo
	1244.05; 1217.04	Groups of S=O
	752.21	Groups of C-S
	636.49	Groups of C-Cl
Chitosan before adsorbed *Procion Red MX 8B*	3445.79 ; 3421.72	Groups of hydroxy (-OH) or Groups of amine ($-NH_2$)
	2872.01	Groups of C-H sp^3
	1575.84	Groups of C=C aromatic
	1419.61; 1463.97	Groups of methylen ($-CH_2-$)
	1029.99	Groups of CN
Chitosan adsorbed *Procion Red MX 8B*	3383.14	Groups of hydroxy (-OH) atau Groups of amine ($-NH_2$)
	2956.87	Groups of $-Csp^3$-H stretching
	1481.33	Groups of N=N azo
	1201.65 ; 1240.23	Groups of S=O
	742.59	Groups of C-S
	671.23	Groups of C-Cl

IOP Conf. Series: Materials Science and Engineering **176** (2017) 012016 doi:10.1088/1757-899X/176/1/012016

The analysis of FTIR Procion Red MX 8B of chitosan-p-t- butylcalix[4]arene before it is used to adsorb and chitosan-p-t- butylcalix[4]arene once used to adsorb dye Procion Red MX 8B shown in Table 5.

Table 5. Comparison of functional groups dye Procion Red MX 8B, the chitosan-p-t-butylcalix[4]arene before and after adsorbed dye Procion Red MX 8B

Compound	wavenumber (cm^{-1})	functional groups
Dye *Procion Red MX 8B*	3440.09	Groups of Hydroxy (-OH)
	1537.22	Groups of –C=C- aromatic
	1471.64	Groups of N=N azo
	1244.05; 1217.04	Groups of S=O
	752.21	Groups of C-S
	636.49	Groups of C-Cl
Chitosan-p-t-butylcalix[4]arene before adsorbed *Procion Red MX 8B*	3419.79 ; 3346.50	Groups of Hydroxy (-OH) or Groups of Amine ($-NH_2$)
	3026.31; 3055.24	Groups of –CH- aromatic (sp^2)
	2868.15	Groups of –CH- sp^3
	1519.91; 1575.84	Groups of C=C aromatic
	1361.74	Groups of Methyl ($-CH_3$)
	675.09	Groups of C-Cl
Chitosan-p-t-butylcalix[4]arene adsorbed *Procion Red MX 8B*	3441.01	Groups of Hydroxy (-OH) or Groups of Amine ($-NH_2$)
	2949.16; 2929.87	Groups of –Csp^3-H stretching
	1469.76	Groups of N=N azo
	1325.10	Groups of Methyl ($-CH_3$)
	1201.65.1242.16	Groups of S=O
	702.09	Groups of C-S
	661.58	Groups of C-Cl

3.5. Adsorbent characterization By SEM – EDX

Morphology adsorbent chitosan before and after adsorption were characterized using Scanning Electron Microscopy (SEM) as shown in Figure 4.

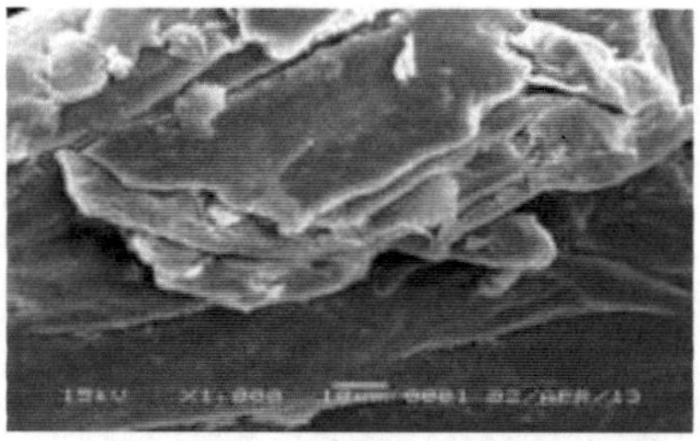

(a)

(b)

Figure 4. Results SEM of Chitosan before adsorption (a), after adsorption (b)

Figure 4 showed the difference in structure between the adsorbent chitosan before and after adsorbed. Adsorbent chitosan before adsorb dye is shown in Figure 4a, showing slotted structure which allows the adsorbate dye can meet the gap on the surface of chitosan. Chitosan adsorbent to adsorb dye that had been shown in Figure 4b, in which the seen surface structure is already covered by the dye Procion Red MX 8B. Chitosan adsorbent composition was analyzed using SEM-EDX with the results shown in Table 6.

Table 6. Composition of Chitosan adsorbent before and after adsorption

Composition	Chitosan before adsorbed (% mass)	Chitosan after adsorbed (% mass)
C	41.10	62.42
N	22.39	08.13
O	36.51	27.18
Na	-	0.14
S	-	1.11
Cl	-	1.12

While the results of SEM analysis of adsorbent chitosan-p-t-butylcalix[4]arene before and after adsorption was shown in Figure 5.

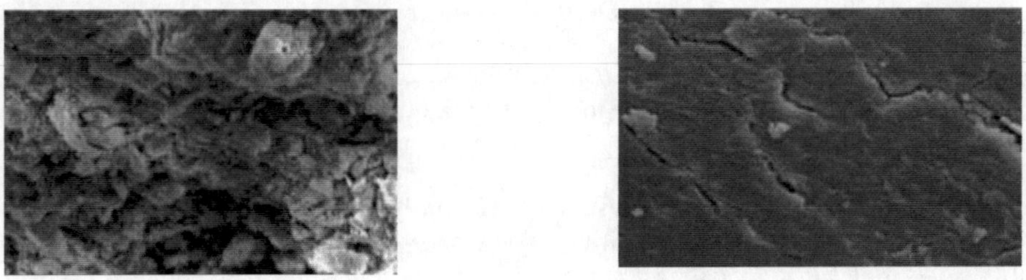

(a) (b)

Figure 5. Results SEM of chitosan-p-t-butylcalix[4]arene before adsorption (a), after adsorption (b)

Based on Figure 7 shown the difference in structure between the adsorbent chitosan- p-t-butylcalix[4]arene before and after adsorption. Adsorbent chitosan-p-t-butylcalix[4]arene before adsorption (Figure 5a) shown the structure more porous compared to chitosan, which allows adsorbate dye can meet more porous compared to chitosan. Adsorbent chitosan-p-t-butylcalix[4]arene after adsorption (Figure 5b) shown the structure of the surface has been coated by a dye Procion Red MX 8B. Chitosan adsorbent composition was analyzed using SEM-EDX with the results shown in Table 7.

Table 7. Composition of chitosan- p-t-butylcalix[4]arene adsorbent before and after adsorption

Composition	Chitosan p-t-butylcalix[4]arene before adsorbed (% mass)	Chitosan-p-t-butylcalix[4]arene after adsorbed (% mass)
C	53.95	56.64
N	18.37	10.03
O	7.55	31.62
S	-	1.35
Cl	20.13	0.20
Na	-	0.16

4. Conclusion

Adsortion of dye Procion Red MX 8B using chitosan-p-t-butylcalix[4]arene had a maximum capacity and adsorption energy of 147.35 mg/g and 33.65 KJ/mol, While adsorption kinetics study followed the pseudo second-order kinetics (Ho) with adsorption rate constants for 2.03 x 10^{-3} g $mg^{-1}min^{-1}$.

Aknowledgment

The financial support from Indonesian Directorate General of Higher Education through the research project of Penelitian Unggulan Perguruan Tinggi (PUPT) 2015 is greatly appreciated.

References

[1] Setyoningsih., 2010, *Aplication of fiber Pineapple leaves Adsorbent Dye Procion Red MX 8B, Thesis*, Surakarta: FMIPA Universitas Sebelas Maret.

[2] Huger K, 2003, Industrial Dyes: Chemistry, Properties, Applications, WILEY-VCH Verlag GmbH & Co. KGaA, Weinheim

[3] Aryunani., 2003, *adsorption of dye Remazol Yellow FG in affluent batik by water hyacinth with NaOH activator, Thesis*, Surakarta: FMIPA Universitas Sebelas Maret.

[4] Sivaiah M V, Venkatesan, K A, Sasidhar P, Krishna R M, dan Murthy G S, 2004 *J. Nuc.Radiochem. Sci.*, **5**, 1, 7-10.

[5] Manurung R, 2000, *Recast Azo Reactive Dyes In Anaerobic - Aerobic. e-USU Repository* : Universitas Sumatera Utara.

[6] Safrianti I, Nelly W, dan Titin A Z, 2012 *JKK*, **1**, 1-7.

[7] Sardjono R E, 2007, *Synthesis and Aplication of tetramer Cyclical Series calixresorsinarene, alkoxy alkylarene, and Alkenylcalixarene for Adsorption Heavy Metal Cations. Dissertation*. FMIPA, Universitas Gajah Mada, Yogyakarta.

[8] Kusumaningsih T, Handayani D S, and Lestari Y, 2004 *Alchemy jurnal penelitian kimia*, **8**, 47-56.

[9] Nur R S, 2007, *Synthesis of pillared clays Cetyltrimethylammonium-Aluminium (CTMA-Al) From Natural Clays and Utilization as Karmina Indigo Dye Adsorbent. In: Seminar Final UNDIP S1 Department of Chemistry, Department of Chemistry,* UNDIP.

[10] Rouquerol F, J Rouquerol, and K Sing. 1999. *Adsorption by Powders and Porous Solids: Principles, Methodology, and Applications.* San Diego: Academic Press. ISBN 0125989202

[11] Emma R, M T Suhartono, Dahrul S, and Sugiyono, 2004 *Jurnal Fakultas Perikanan dan Ilmu Kelautan*, Universitas Padjadjaran.

[12] Astuti P, 2007, *Adsorption of Dyes Textile Waste Procion Red MX 8B By Chitosan and Chitosan Sulfate deacetylation Results Shells Snails (Achatina fullica), Thesis*, Surakarta: FMIPA Universitas Sebelas Maret.

[13] Linane P and Shinkai S, 1994, Calixarenes : Adaptable Hosts Par Exellence, Chem. Ind., 811 - 814.

[14] Handayani D S, Purnawan C, Pranoto, Hastuti S, and Hilmiyana D, 2016IOP Conf. Series: Materials Science and Engineering **107** (2016) 012011 doi:10.1088/1757-899X/107/1/012011

[15] Handayani D S, Kusumaningsih T and Restuti A, 2014 *Prosiding ICOPIA* part of series: Advance in Physic Research Atlantis Press 21-25

[16] Izadyar S and Rahimi M, 2007 *Pakistan Journal of Biological Sciences*, **10**, 287-293.

[17] Sakkayawong N, Thiravetyan P, and Nakbanpote W, 2005 *Journal of Colloid and Interface Science*, **286**, 36-42.

[18] Christie R M, 2001, *Colour Chemistry*, RSC Paperback, The Royal Society of Chemistry, Manchester.

Comparison of shrinkage related properties of various patch repair materials

S A Kristiawan[1] & R S Fitrianto[2]
[1] SMARTCrete Research Group, Civil Engineering Department, Sebelas Maret University, Jl. Ir. Sutami No. 36A Surakarta 57126, Indonesia.
[2] Civil Engineering Department, Sebelas Maret University, JL. Ir. Sutami No. 36A Surakarta 57126, Indonesia.

E-mail: s.a.kristiawan@ft.uns.ac.id

Abstract. A patch repair material has been developed in the form of unsaturated polyester resin (UPR)-mortar. The performance and durability of this material are governed by its compatibility with the concrete being repaired. One of the compatibility issue that should be tackled is the dimensional compatibility as a result of differential shrinkage between the repair material and the concrete substrate. This research aims to evaluate such shrinkage related properties of UPR-mortar and to compare with those of other patch repair materials. The investigation includes the following aspects: free shrinkage, resistance to delamination and cracking. The results indicate that UPR-mortar poses a lower free shrinkage, lower risk of both delamination and cracking tendency in comparison to other repair materials.

1. Introduction

Numerous reinforced concrete (RC) structures have been built to provide a variety of infrastructures which serve human needs and support economic activities. Some of these RC structures prove to have an excelent performance spaning for a period of more than their intended service life. However, there is also evident that many of these RC structures tends to lose their serviceability after years in use and need a repair in order to meet their service life. Patching method is an appropriate technique to repair the degraded RC structure in the form of spalling and delamination. Patching method, when correctly applied, could recover the size and appearance of RC element, regain its strength and capacity, minimize the continuing corrosion of reinforcement and ultimately extend the service life of RC structure [1-3].

The expected additional service life of the repaired RC structures will be governed by the performance and durability of repair material being used. Previous study through case history by Mathew [4] demonstrated that the commnon forms of repair failures are cracking of repair material (32%), delamination (25%), continues corrosion of reinforcement (22%), alkali-aggregate reaction (4%) and others (17%). Cracking and delamination which count for more than 50% of repair failures could be triggered by differential shrinkage between the repair material and parent concrete. The differential shrinkage will induce tensile stress in the repair material [5]. High induced tensile stress leads to cracking when this tensile stress attains tensile capacity of repair material. In addition to tensile stress, shear and peeling stress also exist at the interface of repair material and parent concrete.

[1] To whom any correspondence should be addressed.

High induced shear and peeling stress will promote a risk of delamination when these stresses overcome the bonding capacity of repair material to parent concrete.

Curing of repair material is a good practice to improve the material's resistance againts differential shrinkage cracking. Curing will delay drying and consequently reduces the rate of shrinkage. Furthermore, curing promotes strength gain with a benefit of increasing tensile capacity of repair material. Even though curing may also reduce the stress relaxation due to an increase of repair material's stiffness, the overall effects is to reduce the risk of cracking [6]. Another approach that could improve the shrinkage cracking resistance of repair material is to incorporate fibres on the mix [7], develope a high strength-high ductility of repair material [8], make a rough surface which permit a realization of monolitic behaviour of repair system instead of smooth surface which is likely to provoke debonding [9]. The bond strength has also been identified to be influenced by the type of repair material. Greater bond strength is recognized in the following order: polymer mortar, polymer cement mortar and modified polymer cement mortar [10].

In the preceding paragraphs, the key factors influencing the risk of cracking and delamination have been highlighted and a variety of examples are shown in the development of repair materials to achieve repair materials with lower tendency of cracking and delamination. Previous research by authors [11-13] proposed a repair material made from an unsaturated polyester resin (UPR)-mortar which shows an excellent mechanincal performances. In this paper, UPR-mortar is further investigated to characterize its shrinkage related properties which include free shrinkage, delamination and cracking tendency. The properties of this material are compared to other repair materials which are available commercially in the market.

2. Experimental

2.1. Materials

UPR-mortar has been developed using an unsaturated orthophtalic type resin made from the polymerization of di-carboxylic acids with glycols as the main binder. The thermo-setting of this material is initiated by curing agent (hardener). The amount of hardener was proportioned at 3% by weight of the UPR. Aggregate used for the production of UPR-mortar was sand having a bulk specific gravity of 2.51 and fineness modulus of 3.77. Fillers were also incorporated in the development of UPR-mortar. The fillers were cement and fly ash. The final composition of UPR-mortar per m3 volume was as follows: 860.7 kg of sand, 430.4 kg of UPR, 731.6 kg of cement, and 129.1 kg of fly ash. For the purpose of comparison, three other repair materials were investigated namely normal mortar (NM), SIKA repair mortar and BASF Nanocrete. The last two materials are commercially available in the market. The proportion of NM per m3 volume was as follows: 891.6 kg of sand, 757.9 kg of cement, 133.7 kg of fly ash and 285.3 kg of water. While the commercial repair mortars were mixed with water at amount specified by the respective manufacture.

2.2. Method

Free shrinkage of repair materials were measured on cylinder specimens with diameter of 76 mm and height of 276 mm following RILEM recommendation [14]. Each repair materials were represented by two specimens. For each specimen, four pairs of demec points were attached on the surface of specimen at equidistant of 90°. The gauge length of this pair of demec points was 200 mm. The deformation of each specimen due to shrinkage was taken as an average of four changes of the distance between each pair of demec points. These changes in distance were measured using demountable mechanical strain gauge with a resolution of 1 micron. Figure 1 (a) illustrates the free shinkage specimen.

The delamination of repair materials were investigated by monitoring the tendency of the materials to detach from the SPS Plate (see Figure 1 (b)). After repair material which has been cast on top of the SPS Plate reached a hardened state (in this study 1 day after casting), one of the edge of the repair beam was clamped. The other edge was left without any sorts of clamping and so, this edge would be allowed to curl freely when the repair material exhibited shrinkage. The curling of the repair layer at this location was measured using dial gauge.

The German angle test method was adopted to measure the cracking tendency of repair materials (see Figure 1 (c)). The repair material was cast into V channel. On the following day, any visible cracks that may occur on the surface of repair materials were recorded. The observation of cracks covered the following aspects: time of initial crack, width of crack, length of crack, period of crack and propogation. Microcrack meter was used to measure the width of crack.

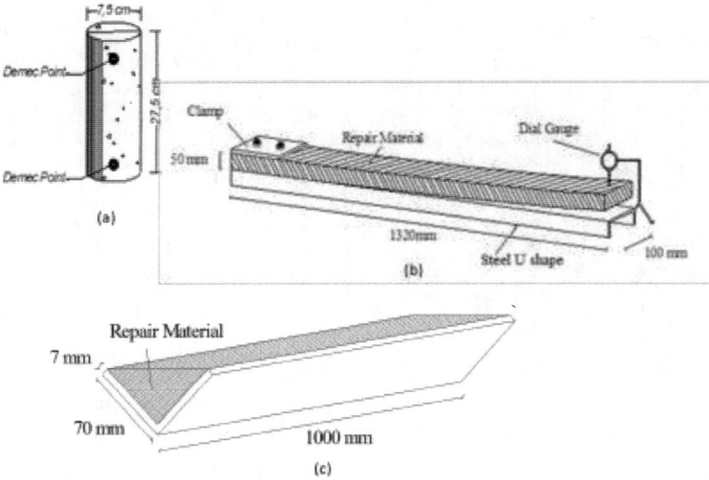

Figure 1.Specimen types for measurement of shrinkage related properties of repair material: (a) free shrinkage; (b) delamination tendency; (c) cracking tendency

3. Results and discussion

3.1. Free shrinkage

The free shrinkage of various repair materials measured for period of 30 days is presented in Figure 2. The temperature and relative humidity of the environment during test period are also presented in the figure. The fluctuation of shrinkage values is associated with the fluctuation of the environment condition especially the relative humidity. The relative humidity controls the drying process in which moisture within the pores of material evaporate to the surrounding environment. The drop of moisture in the pores induces surface tension with a consequence of shrinkage. Therefore, a fluctuation of relative humidity of the environment will create a fluctuation of drying and its subsequent shrinkage.

All repair materials show similar trend of free shrinkage behaviour i.e. shrinkage is generally increased with time at diminishing rate. The shrinkage is relatively faster at the beginning of drying and proportional to the lost of water during evaporation. The shrinkage will be restrained by the bulk material. At the beginning of hardening process, the restraint provided by the bulk material is small. In the progress of hardening, the restraint will gradually increase. Hence, both reduction of rate of water lost due to evaporation and increase of restraint provided by material in the progress of hardening could explain the observed shrinkage behaviour.

The magnitude of free shrinkage of the repair materials after 30 days of drying is in the range of 1500-3000 microstrain. This amount of shrinkage is considered to be high in comparison to the concrete. Thus, the free shrinkage of repair materials could potentially create high differential shrinkage which gives a risk of cracking and delamination. It is clearly seen in Figure 2 that UPR-mortar exhibits the lowest magnitude and rate of shrinkage. Having the lowest shrinkage property, UPR-mortar is expected to undergo the least differential shrinkage induced stresses in comparison to other repair materials.

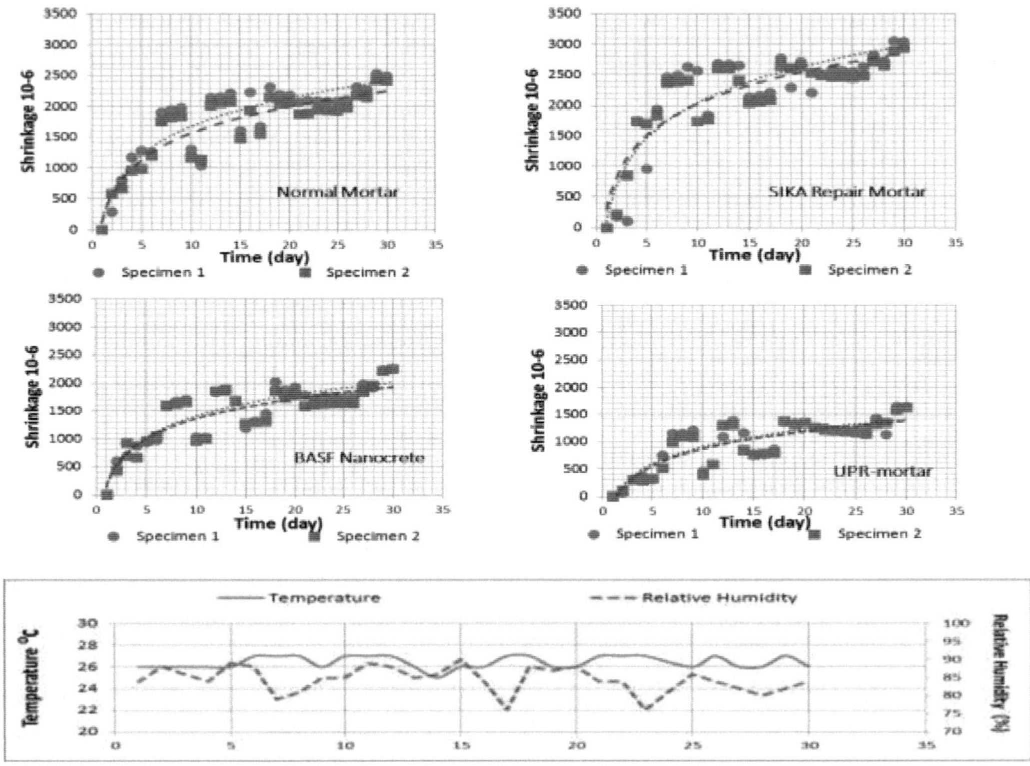

Figure 2. Free shrinkage of various repair materials and their corresponding environment condition during period of testing

3.2. Delamination tendency

Measurement of delamination tendency was carried out for period of 90 days and the results are presented in Figure 2. The delamination is represented in the form of curling of the repair material detected at the free-clamped edge. The curling is indicated by the change in elevation of surface repair material at this edge. The general trend of the delamination could be identified as follows: initially a very fast rate of curling is observed up to 10 days of drying where the curling reaches its peak value. After this period, there is a tendency of reduction in the curling. As the time is progressing, the curling continues to decrease at diminishing rate. The curling seems to attain steady condition after 40 days of drying.

The curling behaviour of repair material tested in SPS Plate is associated with restrained shrinkage action. The restraint is provided by the bond between repair material and SPS Plate. As a result of this restraint of shrinkage, there exists shear and peeling stress at the interface where the highest stresses occur at the edge. In addition to the shear and peeling stress which are the source of separation between repair material and SPS Plate, the restrained shrinkage also induces tensile stress in the repair material. This stress will trigger creep which offsets deformation of shrinkage. Recalling that shrinkage is increased at diminishing rate, the increase of creep will finally surpass the deformation of shrinkage. Consequently, the curling will drop. For the repair materials investigated in this study, this behaviour seems to occur at 10 days of drying.

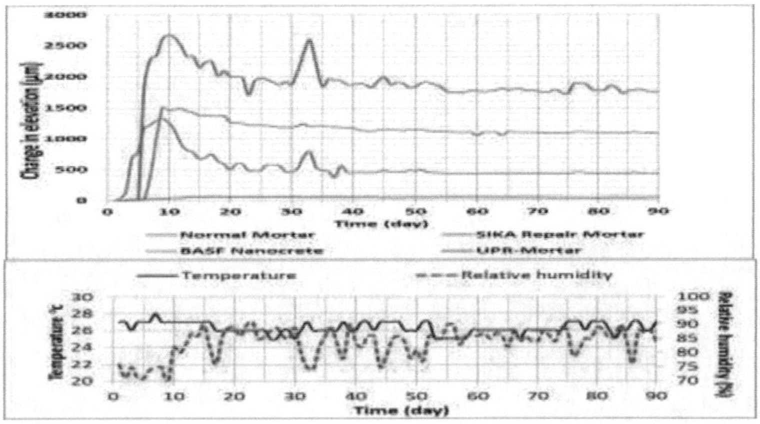

Figure 3. Delamination tendency of various repair materials due to restrained shrinkage (indicated by change in elevation) and the corresponding environment during period of testing

Repair material with higher free shrinkage will exhibit higher curling and vice versa. This is confirmed by the results of this study (see Figure 2 The low curling behaviour of UPR-mortar may also supported by the low stiffness and high bond strength of this material [11,12]. With low stiffnes, this material is able to provide high relaxation and accordingly relieve stresses that provoke separation. As for the high bond strength, this property will determine the level of accumulated stresses which are required to separate the repair material from the SPS Plate. With high bond strength, UPR-mortar could resist the stresses that may trigger separation.

3.3. Cracking tendency

Cracking tendency of various repair materials is evaluated using the German angle test method. Cracking of repair material in this test is a result of tensile stress development which surpass the tensile capacity of repair material. The build up of tensile stress is induced by restrained shrinkage in which restraint is provided by the bonding of repair material with the surface of V channel. Crack is certain when the build up of tensile stress can not be resisted by the tensile capacity of repair material. Once the repair material cracks, a portion of the induced tensile stress will be relieved. However, as long as shrinkage of repair material is still progressing, the effect is to enlarge the existing crack. The propagation of crack width follows the rate of shrinkage progress. It is why the trend of crack width propagation is similar to the rate of shrinkage progress i.e. the crack width increase at diminishing rate (see Figure 4). For the current study, the increase of crack width vanishes after 40-50 days of drying.

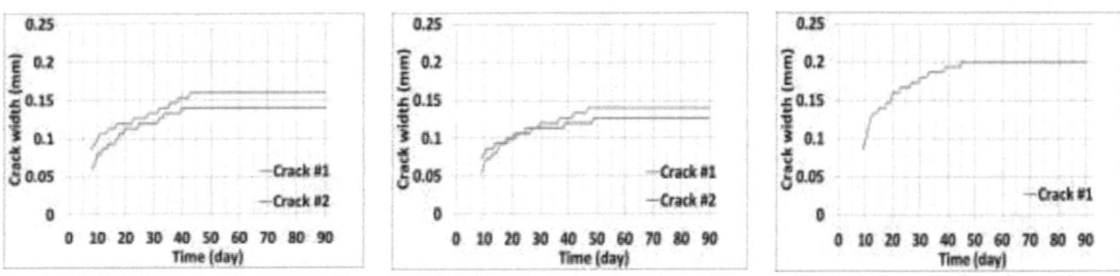

a. SIKA Repair mortar (specimen 1) b. SIKA Repair mortar (specimen 2) c. BASF Nanocrete (specimen 1)

Figure 4. Cracking evolution observed in SIKA Repair Mortar and BASF Nanocrete

Table 1. Cracking of various repair materials due to restrained shrinkage

Repair material		Initial crack width (mm)	Maximum crack width (mm)	Crack length (mm)	Crack area (mm2)	Total crack area (mm2)
Normal Mortar	Specimen 1	NA	NA	NA	NA	NA
	Specimen 2	NA	NA	NA	NA	NA
SIKA Repair Mortar	Specimen 1	0.09	0.16	108.4	17.34	32.02
		0.05	0.14	104.8	14.67	
	Specimen 2	0.06	0.14	104.8	14.67	27.53
		0.07	0.13	101.5	12.86	
BASF Nanocrete	Specimen 1	0.09	0.20	102.5	20.50	20.50
	Specimen 2	NA	NA	NA	NA	NA
UPR-Mortar	Specimen 1	NA	NA	NA	NA	NA
	Specimen 2	NA	NA	NA	NA	NA

The results of cracking observation of the current test are presented in Table 1 and Figure 4. This test shows that cracks are only observed on the SIKA Repair Mortar and BASF Nanocrete. The maximum crack width is achieved after about 40-50 days of drying. The maximum crack width of BASF Nanocrete specimen is higher than those of SIKA Repair Mortar specimens. It is common to obtain a single crack with a higher crack width while a lower crack width is usually accompanied by more cracks. Comparison of cracking performance of the repair materials shows that UPR-mortar has an excellent performance than the commercial repair materials. With no appearance of crack, the build up tensile stress in this material did not exceed the tensile capacity of UPR-mortar.

4. Conclusions

Out of four repair materials investigated in this study, UPR-mortar exhibit the lowest magnitude and rate of shrinkage. UPR-mortar shows the lowest curling behaviour indicating that this material has the lowest risk of delamination compared to other repair materials. UPR-mortar also shows a superior cracking resistance than the commercial repair materials. The relatively low shrinkage, low stiffness and high tensile strength of UPR-mortar could ensure that the induced tensile stress due to restrained shrinkage is always kept below the strength of the UPR-mortar.

References

[1] Rio O, Andrade C, Izquierdo D and Alonso C 2005 *J. Mat. Civ. Eng.* **17 12** 168
[2] Sahamitmongkol R, Suwathanangkul S, Phoothong P and Kato Y 2008 *J. Adv. Con. Tech.* **6 2** 317
[3] Supriyadi A, Kristiawan SA and Raditya S 2015 *Appl. Mech. Mat.* **754-755** 457
[4] Matthews S 2007 *J. Build Appr.* **3 2** 6
[5] Kristiawan SA 2012 *Int. J. Con. Struct. Mat.* **6 4** 221
[6] Beushausen H and Bester N 2016 *Cem. Con. Res.* **87** 87
[7] Briffaut M, Benboudjema F and D'Aloia L 2016 *Tunnel. Under. Space Tech.* **59** 215
[8] Li M and Li VC 2009 *ACI Mat. J.* **106 5** 419
[9] Perez F, Morency M, Bissonnette B and Courard L 2009 *Con. Rep. Rehab. Retrof.* **II** 347
[10] Czarnecki L 2009 *Con. Rep. Rehab. Retrof.* **II** 343
[11] Kristiawan SA and Prakoso AB 2016 *Mat. Sci. Forum.* **857** 426
[12] Kristiawan SA and Prakoso AB 2015 *Appl. Mech. Mat.* **754-755** 442
[13] Kristiawan SA, Supriyadi A and Muktamirin MK 2016 *Adv. Civ. Arch. Struct. Constr. Eng.* 225
[14] RILEM TC107-CSP 1998 *Mat. Struct.* **31** 507

Use of macrophyte plants, sand & gravel materials in constructed wetlands for greywater treatment

S Qomariyah[1,*], AH Ramelan[2], Sobriyah[1], and P Setyono[3]
[1]Department of Civil Engineering, Faculty of Engineering, Sebelas Maret University, Indonesia
[2]Department of Physics, Faculty of Mathematics and Natural Sciences, Sebelas Maret University, Indonesia
[3]Graduate School of Environmental Science, Sebelas Maret University, Indonesia

E-mail: siti.qomariyah@ft.uns.ac.id

Abstract. Greywater discharged without any treatments into drainage channels or natural water bodies will lead to environmental degradation and health risk. Local macrophyte plants combined with natural materials of sand and gravel have been used in a system of constructed wetland for the treatment of the greywater. This paper presents the results of some studies of the system carried out in Indonesia, Thailand, and Costa Rica. The studies demonstrate the success of the constructed wetland systems in removing some pollutants of BOD, COD, TSS, pathogen, and detergent. The studies resulted in the treated water in a level of treatment that fulfils the requirement of the local standards for wastewater reuse as irrigation water, fishery, or other outdoor needs.

1. Introduction

Greywater in developing countries is commonly discharged without any treatments into drainage channels, onto streets, on open grounds, or into natural water bodies. In low and middle income countries, greywater has been given the lowest priority in environmental sanitation management system compared to blackwater and solid waste management [1]. It leads to negative impacts on public health and aesthetic environment.

The wastewater in the cities in developing countries is typically discharged either into private septic tanks or directly into drainage channels. Many septic tanks may leak and are in direct contact with groundwater. Centralized wastewater treatment plants had been built in 12 cities in Indonesia. The wastewater treatment uses conventional technology systems which include aeration ponds, mechanically aerated lagoons, activated sludge systems, anaerobic sludge blanket systems, and rotating biological contact systems. The conventional wastewater treatment plant needs an extensive land, high investment and operational cost, technical trained persons, and a long time in the planning and implementation. The wastewater treatment plants are however inadequate to serve the need of urban sanitation as they only serve 10% of the Indonesian population. Consequently, untreated domestic wastewater which contributes 60% pollution will contaminate the natural waters [2].

It is understandably that sanitation is to be critical issues with polluted surface and groundwater [1]. It is getting worst with rapid urbanization and population growth, as well as the people environmentally-unfriendly activities. Meanwhile, the conventional wastewater treatment systems are costly for developing countries. Hence, an alternative low-cost technology for wastewater treatment is significantly required.

For developing countries and particularly for those in the tropics, constructed wetlands (CWs) for wastewater treatment are potentially a good solution for treating domestic and industrial wastewater due to the simple operation and low implementation cost [3]. The CWs systems do not need to install wastewater treatment systems which require investment cost, power supply, replaceable spare-parts, skilled labors, operation and maintenance cost. The CWs is simple and cost-effective technologies which can only use local aquatic macrophyte plants and local sand or gravel materials. Unfortunately, in developing countries, the CWs had not been found widespread use [4]. This paper aims to present the results of the recent studies conducted in Central Java-Indonesia and some similar studies carried out by other researchers in West Java-Indonesia, Thailand, and Costa Rica. The other aim is to inspire the use of natural materials that are locally available for the application of CWs.

1.1. Constructed wetlands

Constructed wetlands (CWs) are engineered systems which are designed and constructed to mimic the natural process of wetlands. The systems take advantages of many of the same process that occurs in natural wetlands involving wetland vegetation (macrophyte plants), substrate (soil, sand, gravel), and associated microorganisms for treating wastewater. It covers complex processes involving physical, chemical, and biological mechanisms to remove various pollutants or improve water quality [5]. The synonymous terms of "constructed" include "man-made", "engineered", or "artificial", or "treatment" wetlands [6]. The CWs with macrophyte plant roots, aquatic microbial communities, and a supporting mineral matter are effective at removing pollutants, such as suspended solids, dissolved and particulate organic matter, nitrogen, phosphorus, metals and phatogenic organisms from wastewater [7].

The first experiment of wastewater treatment using wetland plants was conducted in 1952 by the German Scientist Dr. Seidel. Recently, the use of CWs systems for wastewater treatment have been more popular in many developed countries such as Germany, UK, France, Denmark, Austria, USA, and Australia; however, in developing countries the application of CWs systems is still limited [5,6].

According to wetland hydrology, the CWs are classified into surface flow (SF) and subsurface flow (SSF). The SF-CWs are similar to natural wetlands as they occupy shallow flow with low velocity above and within the media filter (substrates). In SSF-CWs, wastewater flows horizontally or vertically through the substrates. Hence, the SSF systems are classified into vertical flow (VF) and horizontal flow (HF). The combination of VF and HF is known as Hybrid CWs [5]. The CW basin commonly contains a combination of substrates and macrophyte plants. The process of purification occur during wastewater do contact with the surfaces of the substrates and the plant roots.

1.2. Material for constructed wetlands

1.2.1. Macrophyte plants

Many studies have demonstrated that macrophyte plants contribute to water treatment. Hence, the plants are an essential part in wetlands. The most important functions of the plants are related to their physical effects in the wetlands. The plants significantly transport oxygen to the root zone to allow the roots to survive in anaerobic conditions. The roots provide a huge surface area for the growth of bacteria colonies and other microorganisms which form a biofilm attached to the surface of roots and substrate particles. The root system of the plants maintains the hydraulic conductivity of the substrate [8, 9].

Macrophyte plants are commonly found in natural wetlands or riverbanks. In tropical developing countries, the macrophyte plants grow very well. The plants function which is not related to treatment performance is to provide a local green space; hence, the plants enhance the aesthetic appearance of the wastewater systems.

The aquatic macrophyte plants can be distinguished into 4 groups based on the form or location of the growth: emergent plants, floating leaved plants, submerged plants, and free floating plants. The emergent plants grow on water-saturated or submerged soil. Most of the above ground part of the plant emerges above the water line and into the air. In subsurface flow (SSF) wetlands, their roots interact with the wastewater at the root zone only. In surface flow (SF) wetlands, the roots interact with the wastewater at the root zone, but provide an area for microbial attached growth. The floating leaved plants are rooted in submerged media. They have leaves and steams like they are floating in the water and on the water surface. The submerged plants grow with both steams and leaves within the water zone. These plants need sufficient water depth so that they can only be used within surface flow (SF) wetlands. The free floating plants are not rooted to the substrate. The plants can only be seen within the type of surface flow (SF) wetlands [10, 11].

Emerged plants such as Cattail, Bullrush, and Common Reed were recommended to be applied in SSF CWs [11]. The most frequently used plant in SSF CWs was *Pragmatis australi* (in Europe, Canada, Australia, Asia, and Africa), and the second most commonly used plant is *Typha* (in North America, Australia, East Asia, and Africa) [12]. In warm climates of South America, Africa, and Asia, [6] suggested the types of Macrophyte plants for SSF CWs that includes decorative plants (*Papyrus sedges or Cyperus papyrus, Bamboo plants, Heliconia, Canna lily, Calla lily*), very robust plants (*Umbrella sedge or Cyperus albostraitus and Cyperus alternifolius, Drwaf papyrus or Cyperus haspens*), and Broad-leaved cattail (*Typha latifola*), as well as Species of grass native to the tropical grasslands (*Napier grass or Elephant grass*).

1.2.2. Substrates

Substrates used for CWs include soil, sand, gravel, crushed rock, other natural materials, as well as organic materials. They act for growing medium of macropyhte plants, support biochemical and chemical transformation, provide sites of removed pollutants, and allow the movement of wastewater [10]. Instead of natural material substrates, there are industrial by-product substrates such as slag, fly ash, coal cinder, alum sludge, hollow brick crumbs, limestone, or oil palm shell; and artificial products substrates such as activated carbon, light weight aggregates, compost, or calcium silicate hydrate [13]. However, for the application of SSF CWs for wastewater or greywater treatment in developing countries or countries in transition (warm to moderate climate), coarse sand is considered as the most suitable substrate [6]. The coarse sand was selected in relation to the hydraulic and organic loading which is the most critical design parameter in subsurface flow CWs.

2. Experimental

Application of CWs systems for two decades of experience shows that the systems are generally efficient in removing total suspended solid (TSS) and Biological Oxygen Demand (BOD), meaning that the effluent concentration less than 20 mg/L [14]. In developed countries, the aim of wastewater treatment is elimination of all pollutants like pathogen, nutrients, organic matter, and inorganic chemicals. However, in developing countries, the main goal is protection of the public health through control of pathogens to prevent transmission of waterborne disease [4]. For this purpose, application of CWs in developing countries is suitable since they can be efficient in removal of BOD and pathogens where removal of nutrients is often limited [15]. Therefore, this paper uses the indicators of the removal of BOD, COD, and pathogens (indicated by E. coli and total coli) to evaluate the performance of the CWs. The percentage efficiencies of the removal are known by comparing the removal of the pollutant at the outlet to the initial load of the pollutant at the inlet of the CWs.

A single horizontal subsurface flow (HF) constructed wetland with dimension of 1.7m x 0.7m x 0.7m (length x width x high) was used to investigate pollutant removal efficiency of greywater from a single house in the year of 2015. The study was conducted in the City of Surakarta, Indonesia, under an average temperature of 26° C. Substrates used in this study were river sand and gravel. The HF constructed wetland was filled to the total depth of 50cm which consists of 20cm long of gravels (Ǿ 1-3cm) at the inlet and outlet zones, and 130cm long of river sand (permeability 5.48×10^{-5} cm/sec) in the plantation zone. A local macrophyte plant of Cyperus papyrus was planted in the plantation zone of constructed wetland at 25 cm interval. Before entering the constructed wetland, the greywater was mechanically pre-treated in a sedimentation tank. To acclimatize the plants to the greywater, the greywater was discharged in the constructed wetland with gradual concentration from 25 to 100% within four weeks. With a hydraulic retention time of one day, the influent and effluent were collected and analyzed twice a month during three months of experiment. The parameters to be monitored were total suspended solid (TSS), 5 days biochemical oxygen demand (BOD_5), and chemical oxygen demand (COD). Since the type of household greywater at the study site typically contains a high concentration of detergent, this study included the detergent to be tested as well. The average result of the study is presented in the Table 1 as the latest study in 2015.

3. Result and discussion

To show that the local macrophyte plants and local natural substrates can perform successfully the treatment of domestic wastewater, the results of other experiments by other researchers in West Java-Indonesia, Thailand, and Costa Rica are also presented in the following table.

Table 1. The effective removal of pollutants in CWs with natural substrates and local macrophytes.

Type & size of CW (L,W,H) & Area	Type of plant	Substrates	Country	Effective removal (%)	Source (year)
HF (1,7x0.7x0.7m) Area: 1.19 m^2	Cyperus papyrus	sand, gravel	Indonesia	BOD: 98.55; COD: 98.46; TSS: 98.06; Detergent: 99.86	Author 2015
VF (6x5x1.2m) Area: 30m^2	Phragmite Karka	sand, gravel	Indonesia	Total Coli: 99.45; BOD: 76.03; COD: 78.89; NH_4-N: 88.18; PO_4-P: 91.06; TN: 71.70	[18] 2011
HF (2x1x1m) Area: 2m^2	Canna & Helicona	Gravel	Thailand	TSS: 88-96; COD: 42-83 TN: 4-37; TP: 6-35	[16] 2009
HF (14x1.2x0.6m) & (6x3x0.6m) Area: 16.8 & 18m^2	Coix lacryma-jodi	crush rock	Costa Rica	Fecal Coli: 99.99 BOD: 94-99.4 Turbidity: 91.7-97.9	[15] 2004

The above studies use natural substrates (sand, gravel, or crushed rock) and macrophyte plants which are locally available in the study sites. From the economic point of view, these materials are considered cheaper than materials by product or artificial products substrates.

All studies demonstrated that removal of organics, in term of BOD or COD, are high in all type of the constructed wetlands. The treatment efficiency of BOD and COD varied between 76.03-99.4% and 78.89-98.46% respectively, except COD removal studied in Thailand which varied between 42-83%. According to [16], it was still efficient as the influent wastewater has low concentration of COD which might be due to high degradation rate in the wastewater collection systems and in the settling tank before entering the constructed wetlands. The latest study in 2015 resulted in an effective BOD and COD reduction which are more than 98%. The inlet BOD concentration ranged between 496-850 mg/L, while that of the outlet ranged between 2.19-17.2 mg/L. It is below the threshold of 30 mg/L as set by US Environmental Protection Agency [17] and the Indonesian standard for wastewater reuse. In addition, the greywater from a household at the study site typically contains high concentration of detergent. Fortunately, the CWs system removed the detergent successfully, having a very high percentage efficiency of 99.86%.

The above studies resulted in a high removal of suspended solid (TSS or Turbidity). Suspended solid is removed primarily through the mechanism of sedimentation and interception [6]. In the latest study, solid removal efficiency ranged from 95.47% to 99.56% with an average reduction of 98.06%. TSS at the outlet varied between 2-10 mg/L, which is below the threshold of 30 mg/L as set by US Environmental Protection Agency [17] and the Indonesian standard for wastewater reuse. The study in Thailand used two types of macrophyte plants (Canna and Heliconia) and applied gravel only as the filter media. It resulted in the well growth of both macrophyte plants [16] and very efficient removals of TSS for both plants which varied between 88-96%.

The constructed wetlands reduced both total and fecal coliforms. The average removal of total coliforms (in Indonesia) and fecal coliforms (in Costa Rica) were 99.45% and 99.99% respectively. It indicates a very high efficiency of the constructed wetland system on the removal of pathogen. The protection of the public health by means of controlling pathogens to prevent transmission of waterborne disease is the primary goal of wastewater treatment in developing countries [4]. In addition, the constructed wetlands are efficient in removal of BOD and pathogens while removal of nutrient is limited [11]. It was also stated by [5] that removal of phosphorus (TP) and nutrient (TN) in various types of constructed wetland is usually low, even recently some filtration materials by product have been tested in constructed wetland and resulted in a high removal of phosphorus; however, constructed wetlands were seldom built with phosphorus removal as a primary purpose of the treatment.

The above table shows that the study in Thailand resulted in low removal of phosphorus (6-35%) and nitrogen (4-37%). In contrast, removal of phosphorus and total nitrogen are high in the study of [18]. Substrates which contain high Ca, Al, and Fe have high phosphorus [19], so that the high removal of phosphorus in the study might due to the high amount of iron rich sand in the substrate used [18]. Nutrient removal is not one of the wastewater reuse criteria; hence, if the treated effluent is to be used for irrigation, then nutrient removal is likely to become unnecessary [15]. In addition to the study of [18], the average treatment efficiency of total coliform, NO_3-N, and PO_4-P represents the average effluent concentration of total coliform, NO_3-N, and PO_4-P of 4880 MPM 100/l, 1.36 mg/l, 0.07 mg/l respectively. These results are lower than that of Indonesia standard for irrigation water and fishery (10000 MPN 100/l, 20 mg/l for NO_3-N, 1 mg/l for PO_4-P).

The guidelines for wastewater reuse in Costa Rica place emphasis upon pathogen removal with the limit of fecal coliform of 1000 cfu/100 ml. Therefore, the study in Costa Rica used a longer hydraulic retention time (HRT) to be 7.9 days and resulted in a level of treatment that exceeds the requirement of the local standards for wastewater reuse in term of BOD average less than 10 mg/l and average fecal coliform of 122 cfu/100 ml with effective removal of fecal coli of 99.99% [15].

Houses in the cities of a developing country, like Indonesia, commonly have a limited open space. Table 1 shows the areas required for the constructed wetlands systems are relatively small (1.19 & 2 m^2) and moderate (16.8; 18; 30 m^2). Therefore, the constructed wetlands systems with the small and moderate areas are suitable to be applied as decentralized wastewater treatment systems on a single household and neighborhood respectively.

4. Conclusion

Natural materials of sand, gravel or crushed rock and local macrophyte plants used in the system of vertical or horizontal constructed wetlands have high performance for the treatment of greywater. The organic pollutants and pathogen can be removed successfully, therefore the treated water can be used safely for irrigation, fishery, or out-door uses. These materials are easily found and locally available. The CWs system offers an economic benefit for householders since there is no or less energy required for the operation of the system. In addition, the treated water can replace a part of the fresh water need supplied from the piped distribution system, and be potential to protect surface and ground water resources from pollutants. Moreover, the use of macrophyte plants creates a green space in a single house yard or green public views for neighborhood. The systems applied as decentralized wastewater systems are suitable for developing countries where the potential health benefits from pathogen removal are considerable. Hence, the CWs systems would be preferable compared to the high cost conventional wastewater treatment systems.

References

[1] Antoine M and Stefan D 2006 *Greywater Management in Low and Middle Income Countries* Sandec Report No14/6 (Switzerland: Eawag) pp 1-7

[2] Diana H, Sulistyoweni W, Setyo S M, and Robertus W T 2013 *European Sci. J.* **9** 17
[3] Patrick D 1997 *Water Sci. & Tech.* **35**. 27-34
[4] Amelia K K 2001 *Ecol. Eng.* **16** 545-560
[5] J. Vymazal 2010 *Water* **2** 530-549
[6] Heike H, Christopher P, Martina W, and Elizabeth von M 2011 *Technology review of constructed wetlands* (Germany: Gesellscaft fur Internationalle Zusammenarbeit/GIZ) pp 9-17
[7] Peter F R 2006 *J. Env. Bio.* **2.** 78-89
[8] Hans B 1994 *Water Sci. Tech.* **29**. 71-78
[9] Hans B 1997 *Water Sci. Tech.* **35**, 11-17.
[10] Robert H K and Scott D W 2009 *Treatment Wetlands* second ed. (Boca Raton: Taylor and Francis Group CRC Press) p 65
[11] Jay V 2013 *Ecol. Eng.* **61**, 582-592
[12] Jay V 2011 *Hyd. Bio.* **674.** 133-156
[13] Haiming W, Jian Z, Huu H N, Wenshan G, Zhen H, Shuang L, Jin L F, and Hai L 2015 *Biores. Tech.* **175**. 594-601
[14] Hans B, Schierup H, and Arias C A 2007 *Water Sci. & Tech.* **56.** 63-68
[15] Dennis K, Thammarat K, and Hans B 2009 *Ecol. Eng.* **35.** 248-257
[16] Stewart D, Brian S, and Goen H 2004 *Ecol. Eng.* **23**. 55-61
[17] US EPA 2000 *Constructed Wetlands Treatment of Municipal Wastewaters* (Ohio: US EPA-NRMRL) p 80
[18] Denny K 2011 *American J. of Env. Sci.* **7.** 15-19
[19] Jay V 1996 *Wat. Sci. Tech.* **34.** 159-164

IOP Conf. Series: Materials Science and Engineering **176** (2017) 012019 doi:10.1088/1757-899X/176/1/012019

The usage of activated carbon from teak sawdust *(tectona grandis* l.f.*)* and zeolite for the adsorption of Cr(VI) and its analysis using solid-phase spectrophotometry (sps)

S Saputro[1], M Masykuri[1], L Mahardiani[1], B Mulyani[1], I Qorina[1], K Yoshimura[2], K Takehara[3], S Matsuoka[4]

[1]Chemistry Education, Faculty of Teacher Training and Education, Sebelas Maret University, Jl. Ir. Sutami 36A Surakarta, Central Java-57126, Indonesia

[2]Kyushu Environmental Evaluation Association, Matsukadai, Higashi-ku, Fukuoka 813-0004, Japan

[3]Department of Chemistry, Faculty of Sciences, Kyushu University, 744, Motooka, Nishi-ku, Fukuoka 819–0395, Japan

[4]Department of Environmental Science, Faculty of Science, Niigata University, Ikarashi, Niigata 950–2181, Japan

E-mail: sulistyo68@yahoo.com

Abstract. This study aims to evaluate the usage of teak sawdust and zeolite as an adsorbent of Cr(VI) ion; optimal composition ratio of the composite adsorbent; and the sensitivity of solid-phase spectrophotometry (SPS) as a method to determine the levels of Cr(VI) ion as an adsorption results of adsorbents. The adsorbent used were teak sawdust activated carbon and zeolite as a single and composite adsorbents. The teak sawdust carbonization using *muffle furnace* and then activated with H_3PO_4 10% while the zeolite with H_2SO_4 10%. The contacting process of the adsorbents with Cr(VI) was done by varying the compositions. Analysis of Cr(VI) level was done using SPS method. Characterization of adsorbent before and after being activated is done using a FTIR. The results showed that teak sawdust and zeolite can be used as adsorbents to adsorb Cr(VI) in the simulated liquid waste with the adsorption capacity 1.19 μg/g; the optimum composition ratio of teak sawdust activated carbon and zeolite was 75%:25% with the percentage of adsorption was 62.72%. Solid-phase spectrophotometry is a sensitive method to analyze the decreased levels of Cr(VI) as an adsorption results in μg/L level with the limit of detection (LOD) was 0.03 μg/L.

1. Introduction

Industrial development in Indonesia which is fast enough can boost the economy, but the impact is waste production. One of the waste is heavy metal which caused serious pollution to the environment if it exceeds the limit and have a very dangerous toxic and will cause serious disease to humans when it accumulates in the body [1]. Indicator used to detect water pollution is the contamination of heavy metals in it. It called harmful heavy metals because it generally has a high mass density (5 gr/mL) and a small concentration can be toxic and dangerous, and one of them is Cr(VI). Water quality that contains Cr(VI) particles that exceed standard quality (0,05 mg/L), could endanger human health [2].

Determining the concentration of Cr(VI) in water is very important even in a very small concentration, because it will be absorbed by human body if the water that contains Cr(VI) is consumed. Chromium (VI) can cause liver and kidney damage, bleeding, repiratory tract damage, and lung cancer. Long-term exposure to the respiratory tract and the skin can cause inflammation of the nasal cavity, nasal and skin ulcer bleeding [3]. Thus, to determine Cr(VI) in water with a very low concentrations, *solid-phase spectrophotometry* (SPS) is used because it has a high sensitivity.

Solid-phase spectrophotometry (SPS) is a method of determining the concentration of a substance in which the light absorption by the target chemical components concentrated in a solid phase. This method is more sensitive than conventional methods that use expensive instrumentation. Additionally, it can determine up to µg/L levels [4]. If the use of existing methods are still limited to the level of mg/L, then the method solid-phase spectrophotometry (SPS), which is more sensitive to the µg/L levels, can be used in waters with lower levels of contamination.

Various methods were developed to reduce the levels of heavy metals in waters, one of the methods is the adsorption. The adsorption has proven a useful tool to control the degree of contamination of chromium in the water, low cost materials which have an adsorption capacity [5]. The adsorbent that often used is porous solids, such as charcoal, zeolite, allophane, sand, and silica gel. Most research on adsorption was performed using a single adsorbent, whereas in this study conducted adsorption with single and combination adsorbent, which are activated carbon from teak sawdust and zeolite.

The amount of waste sawdust from the sawmill industry is 15-20% of the total production [6]. This number is huge and its use is not maximized. Thus, in this study teak sawdust utilized as activated carbon using an activator H_3PO_4 10%. The largest contents of teak wood is cellulose 47.5% and 29.9% lignin. The presence of cellulose showed that teak can be used as an adsorbent.

Zeolite is a mineral group consisting of a double oxide Al_2O_3, SiO_2, Fe_2O_3, CaO and MgO. Boron can be found in the sedimentary rocks, mainly crystal of aluminum and silicate groups. Zeolites are natural materials that are widely available in Indonesia, so that knowledge and research on natural zeolite and ways of processing is very important [7]. Zeolites have a high adsorption capacity, because it can separate molecules by size and configuration of the molecule. Adsorption mechanism that may occur is the physical adsorption (involving Van der Waals force), chemical adsorption (involving electrostatic forces), hydrogen bonds, and the formation of coordination complexes [8]. The abundance of teak sawdust and zeolite needs to be maximized. Thus, in this study, both of the material used as an adsorbent for lowering the concentration of Cr(VI).

2. Experimental

2.1. Tools and materials
The tools used was a UV-Visible spectrophotometer from K-MAC, FTIR spectrophotometer Shimadzu, muffle furnace, oven, pan, analytical balance, volumetric flask, volume pipette, glass beaker, measuring cup, flask, erlenmeyer, watch glass, stirrer glass, drop pipette, aliquoting devices are assembled using a syringe, blender, mortar and pestle, a 100-mesh sieve, porcelain bowls, stirrer, and a magnetic stirrer. Materials used are teak sawdust, zeolite, $K_2Cr_2O_7$, H_3PO_4 10%, H_2SO_4 10%, resin AG 50W-X2 Muromac H^+ form 100-200 mesh, distilled water, acetone, Whatman filter paper, blue litmus, and 1,5-diphenylcarbazide (DPC).

2.2. Research procedure
2.2.1. Preparation of adsorbent
Teak sawdust is cleaned and washed with distilled water to clean, dried at 110°C then charred in a muffle furnace at a temperature of 900°C for 30 minutes. Results obtained mashed and then sieved with a 100 mesh size. Natural zeolites are cleaned with distilled water and then dried at a temperature of 105°C. Results obtained mashed and then sieved with a 100-mesh sieve.

2.2.2. Adsorbent activation
Charcoal soak in a solution of 10% H_3PO_4 for 24 hours at room temperature. Filter and rinse the residue with distilled water until the filtrate became neutral, and then dried in an oven at 110°C for 24

hours. Adsorbent was tested by FTIR before and after activation. Zeolite soak in a solution of 10% H_2SO_4 for 24 hours. Filter and rinse the residue with distilled water until the filtrate became neutral, and then dried in an oven at 105°C for 24 hours. Adsorbent was tested by FTIR before and after activation.

2.2.3. Resin preparation

Muromac resin AG 50W-X2 H^+ form 100-200 mesh is dissolved in distilled water and silence a few moments until the resin becomes more fluffy.

2.2.4. Determination of Cr(VI) calibration curve

Standard solution of Cr(VI) 0 μg/L, 2 μg/L, 4 μg/L, dan 8 μg/L respectively 20 mL plus 1 mL of H_2SO_4 0.5 M, 1 mL of 1,5-diphenylcarbazide (DPC), and 0.06 mL resin. Stir it for 20 minutes and analyzed using UV-Vis spectrophotometer at a wavelength of 540 nm and 682 nm and then absorbance difference of the two wavelengths determined, $\Delta A = A_{540nm}-A_{682nm}$, which ΔA will be made standard curve (ΔA vs concentration).

2.2.5. Determination of Cr(VI) species in simulated liquid waste

A 20 mL of simulated liquid waste solution of 50 μg/L was taken then added with 1 mL of H_2SO_4 0.5 M, 1 mL of 1,5-diphenylcarbazide (DPC) 0.25%, and 0.06 mL of resin. Then, stir it for 20 minutes and analyzed using UV-Vis spectrophotometer at a wavelength of 540 nm and 682 nm. ΔA obtained will be substituted in equation Cr(VI) calibration curve (ΔA vs concentration), so that the Cr (VI) ions in the effluent can be known.

2.2.6. Determination of optimum adsorbent composition

Inserting 20 mL of a simulated liquid waste solution of 50 μg/L in each of the 4 erlenmeyer. Insert each adsorbent composition with a ratio of teak sawdust:zeolite at 100%:0%, 0%:100%, 25%:75%, 75%:25% and stir it for 20 minutes. Each adsorbent was filtered with Whatman filter paper. The resulting filtrate was taken as much as 5 mL and then diluted with distilled water until 50 mL. Furthermore, the filtrate diluted taken as many as 20 mL added with 1 mL of H_2SO_4 0.5 M, 1 mL of solution DPC 0.25% and 0.06 mL resin then stir for 20 minutes and analyzed using UV-Vis spectrophotometer with a wavelength of 540 nm and 682 nm. ΔA obtained will be substituted in equation of Cr(VI) calibration curve (ΔA vs concentration) so species of Cr(VI) in a simulated liquid waste after the adsorption known.

2.2.7. Determination of detection limit

Taking five blank solution of 20 ml, then each added 1 mL of H_2SO_4 0.5 M, 1 mL of 0.25% DPC, and as much as 0.06 ml resin, then stir it for 20 minutes. Furthermore, it analyzed by UV-Vis spectrophotometer at a wavelength of 540 nm and 682 nm. ΔA obtained will be substituted in the equation of the calibration curve.

3. Results and discussion

3.1. Preparation of activated carbon from teak sawdust adsorbent

Preparation of activated carbon from teak sawdust adsorbent consists of three stages, including the stage of dehydration, is the removal of impurities in teak sawdust, stage of carbonization, breakdown of organic matter into carbon by heating at 900°C for 30 minutes, and the activation stage, soaking with an activator solution which is H_3PO_4 10% for 24 hours. H_3PO_4 10% is used because it is a good dehydrating agent and an oxidizing agent, so it can eliminate impurities from the adsorbent.

3.2. Preparation of zeolite adsorbent

Zeolite preparation is done by washing zeolite from impurities by using distilled water and then drying in an oven at 105°C for 24 hours, which are expected pores formed by evaporation of water through heating. At this stage of activation, zeolite soaked with a solution of 10% H_2SO_4 for 24 hours. Activation aims to clean up the surface pores, removing impurities compounds, and increase the surface area. H_2SO_4 activator used as an oxidizing agent, so as to eliminate impurities on zeolite that cause cavities formerly closed by impurities into the open, so the adsorption become more effective and can increase the adsorption capacity.

3.3. Fourier Transform Infra Red Spectra Analysis

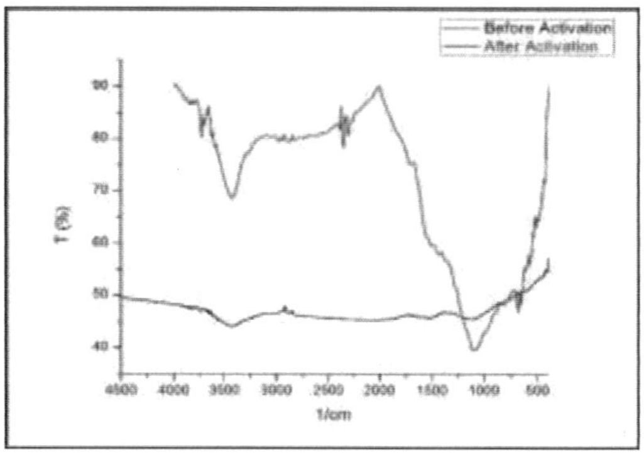

Figure 1. FTIR Spectra of Teak Sawdust Before and After Activation

Based on Figure 1, shows the presence of the -OH group at wavenumber of 3426.69 cm^{-1}, aromatic group at 1568.19 cm^{-1}, C-O at 1114.9 cm^{-1}, and O-Si-O at 679.94 cm^{-1}

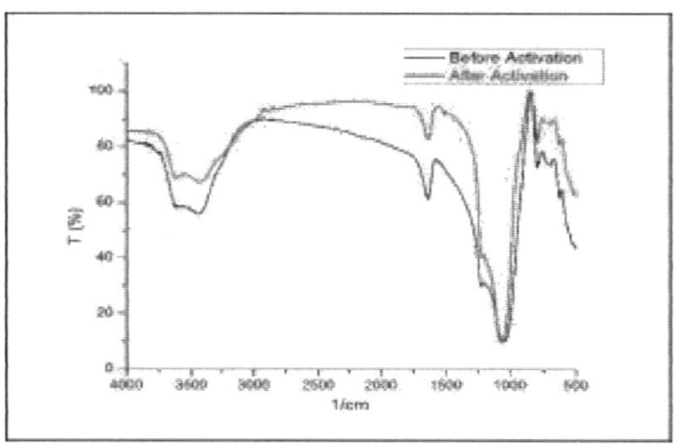

Figure 2. FTIR Spectra of Zeolite Before and After Activation

Based on Figure 2, note that the structure of the zeolite containing Al-O-Si at wavenumber of 1061.86 cm^{-1} and NH_2 groups at 3425.75 cm^{-1} and 3626.33 cm^{-1}.

3.4. Determination of Cr (VI) Calibration Curve

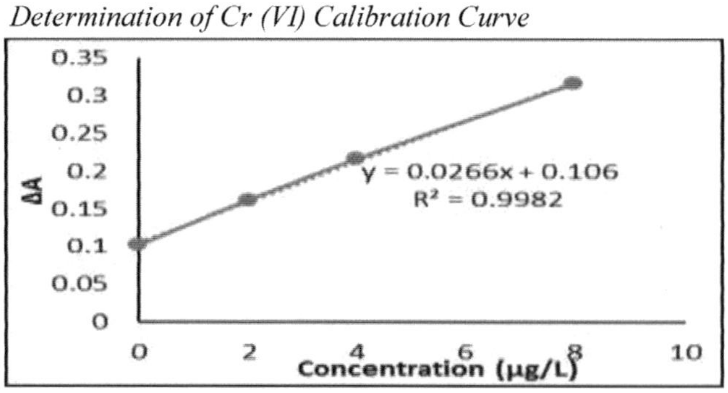

Figure 3. Cr(VI) Calibration Curve

Based on the graph in Figure 3, $y = 0.0266x + 0.106$ where y is ΔA and x is the concentration. The equation is used to determine the concentration of Cr (VI) in simulated liquid waste. By using this equation, the Cr(VI) levels in a simulated wastewater was 47.4 µg/L.

3.5. Determination of optimum adsorbent mass composition activated carbon from teak sawdust (TS) and zeolite (Z) in a decrease levels of Cr(VI)

Table 3. Determination of optimum composition

TS : Z Mass (%)	Final Concentration (μg/L)	Adsorbed Concentration (μg/L)	Adsorbed Percentage (%)
100 : 0	24.81	22.59	47.65
0 : 100	29.32	18.08	38.14
25 : 75	24.06	23.34	49.24
75 : 25	17.67	29.73	62.72
80 : 20	18.05	29.35	61.93
90 : 10	21.05	26.35	55.59

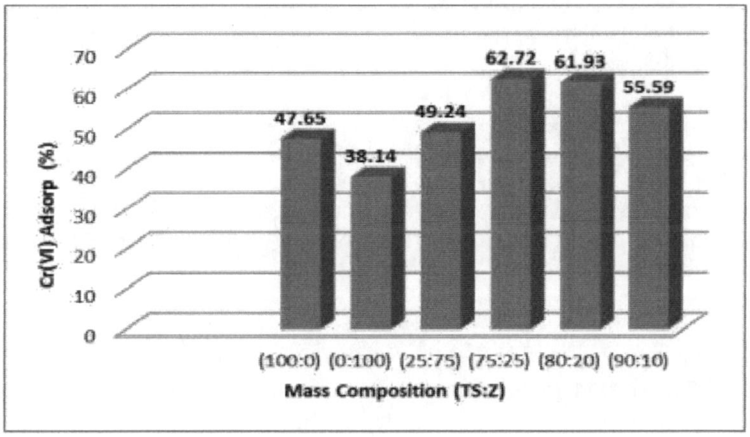

Figure 4. Adsorption percentage chart

Based on data in Table 3 and the chart in Figure 4, the mass ratio that produce the most optimum adsorption percentage is combination of adsorbent with ratio of 75%:25% are 62.72% with the adsorption capacity of 1.19 μg/g, then the ratio of 80% : 20%. Based on the analysis on a single adsorbent, the mass ratio of 100%: 0% resulted in a greater adsorption percentage than the ratio of 0%: 100%. This shows that in this study, single adsorbent activated carbon from teak sawdust is more effective at adsorption of Cr(VI) ions compared to single zeolite adsorbent. These results indicate that a combination adsorbents are more effective in the adsorption of Cr(VI) compared to a single adsorbent [9]. The process of adsorption on activated carbon or zeolites was possible because their pores and the ion exchange.

Adsorption on activated carbon from teak sawdust and zeolite may occur by way of ions of Cr(VI) enter the pores of adsorbent without forming a bond or interact with a weak binding energy (Van der Waals forces). Weak attractive force causes the metal ions of Cr(VI) can move from one surface to the other surface of adsorbent. In addition, the active groups on the adsorbent will interact with the metal ions of Cr(VI).

Greater mass ratio of activated carbon in composite adsorbents showed greater percentage of adsorption, it is possible for the adsorption process in this research is dominated by the pores, so the activated charcoal more influential in decreased levels of Cr(VI). Combination with zeolite enables the ion exchange process more optimal, because zeolites have a large ion exchange capacity [10], so that the metal ions of Cr(VI) are not only adsorbed into the pores, but rather an ion exchange process. In addition, based on the research showed that the addition of activated carbon in the composite adsorbent is directly proportional to the porosity of the adsorbent, where the greater percentage of porosity, the larger the pores, resulting in better adsorption [11].

3.6. Determination of detection limit

In this research, Limit of Detection (LOD) was 0.03 μg/L, so it can be said that the sample at a concentration of 0.03 μg/L can be read, so that the solid-phase spectrophotometry (SPS) is a sensitive and effective method for use in analysis decreased levels of Cr(VI) adsorption results of composite activated carbon from teak sawdust and zeolite in levels μg/L.

4. Conclusions

The results showed that teak sawdust and zeolite can be used as adsorbents to adsorb Cr(VI) metal ions in the simulated liquid waste with the adsorption capacity 1.19 μg/g; the optimum composition ratio of teak sawdust activated carbon and zeolite was 75%:25% with the percentage of adsorption was 62.72% in the amount of 0.5 g; and solid-phase spectrophotometry (SPS) is a sensitive method to analyzed the decreased levels of Cr(VI) as an adsorption results of a single and combination adsorbents of activated carbon from teak sawdust and zeolite in the level of μg/L with the limit of detection (LOD) was 0.03 μg/L.

Aknowledgement

Authors wishing to acknowledge assistance from Head of Chemistry Laboratory from the Chemistry Study Program, Faculty of Teacher Training and Education, Sebelas Maret University, for using the facilities of the laboratory. This work was partially supported by the PUPT Grant for S.S (2015 and 2016) and by Competency Grant for S.S (2016) from the Ministry of Research, Technology and Higher Education, Indonesia.

References

[1] Danarto Y C 2007 *Jurnal Ekuilibrium* **6** 65
[2] Silgado K J, Marrugo G D, Puello J. 2014 *Chemical Engineering* **37** 721
[3] Saputro S, Yoshimura K, Matsuoka S, Takehara K, and Narsito 2009 *Analytical Sciences* **25** 1445
[4] Matsuoka S, Nakatsu Y, Takehara K, Saputro S, and Yoshimura K 2006 *Analytical Sciences* **22** 1519
[5] Landeros C R, Barrera-Diaz C E, Bilyeu B, Guerrero V V, and Nunez F U 2013 *American Journal of Analytical Chemistry* **4** 8
[6] Pari G, Widayati D T, and Yoshida M 2009 *Journal of Forest Product Research* **27** 4
[7] Said M, Prawati A.W, and Murenda E 2008 *Jurnal Teknik Kimia* **15** 50
[8] Poerwadio A D, and Masduqi A 2004 *Jurnal Purifikasi* **5** 169
[9] Mousavi S F, Rostamian R, and Aghakhani A 2013 *E3S Web of Conferences* **1** 1
[10] Wang S, and Peng Y, 2010 *Chemical Engineering Journal* **156** 11
[11] Siagian H, 2011 *Jurnal Saintech,* **3** 6

X-RAY DIFFRACTION ANALYSIS on POST TREATMENT of Ca-Mg-Al-LAYERED DOUBLE HYDROXIDE SLURRY

E Heraldy[1,2], **K D Nugrahaningtyas**[1,2] **and Heriyanto**[1]
[1] Department of Chemistry, Mathematics and Natural Sciences Faculty, Sebelas Maret University, Jl.Ir. Sutami 36 A Kentingan, Surakarta, Indonesia 57126
[2] Solid State Chemistry & Catalysis Research Group, Chemistry Department, Sebelas Maret University, Jl.Ir. Sutami 36 A Kentingan, Surakarta, Indonesia 57126

E-mail: eheraldy@mipa.uns.ac.id

Abstract. This research objectives to study post treatment on Ca-Mg-Al-Layered Double Hydroxide (Ca-Mg-Al-LDH) slurry which was prepared from brine water by cooling treatment. The cooling rate was varied from 1 to 3 °C/min by using stirring and without stirring, and the cooling time was done at 0, 30 minutes and 24 hours. The quantitative X-ray diffraction (QXRD) was employed on Ca-Mg-Al-LDH using Le Bail refinement method. The refinement results found another Mg-Al-LDH and Ca-Al-LDH phases, such as $Mg(OH)_2$, $Al(OH)_3$ and $CaCO_3$. The highest phase composition on material Ca-Mg-Al-LDH using Le Bail refinement was showed by $Al(OH)_3$.

1. Introduction

Layered double hydroxides (LDHs), known as hydrotalcite-like (HT) compounds or anionic clays, consist of positively charged metal hydroxide sheets, interlayer anions, and water [1]. Commonly, LDH is expressed in the general formula, $[M^{2+}{}_{1-x}M^{3+}{}_{x}\,(OH)_2]^{x+}(A^{n-})_{x/n}\cdot yH_2O$, where M^{2+} is divalent cation, M^{3+} is trivalent metal cation, and A^{n-} is the exchangeable anion (usually carbonate, chloride or nitrate) and $0.2 < x < 0.33$; y is the amount of interlayer H_2O [2-7]. LDH is known have a high surface area, basicity properties, stable to thermal treatments and have a memory effect [5]. With such properties, LDHs is widely used as catalysts [6], adsorbents [7], anion exchange [8], corrosion resistance [9] pharmaceutical [10] and photocatalysis [11].

Some researchers have developed LDHs from artificial seawater as source metal ion magnesium [12] and also from seawater [13]. Zhang et al. [14], prepared LDHs using both calcium and magnesium as source divalent metal ion and aluminum as source trivalent metal ion. In addition, Heraldy et al. [15] synthesized LDHs with calcium and magnesium from brine water. Therefore, brine water which are contains magnesium, calcium [16] have great potential as the metallic source for LDHs synthesis. Moreover, Rives et al. [2] was showed that some experimental parameters can influence the LDH crystallized, such as temperature in the reactor, pH, concentration metallic salts and alkaline solution, the flow rate of reactants and aging time of the precipitate. Previous research has been demonstrated that the aging time of the precipitate [17] and cooling time treatment [18] on LDHs slurry are the main important parameters in the formation and crystallization of LDHs.

Besides that, to determine the metallic concentration in precipitates, several analytical methods were commonly used. However, these methods may not suitable for quantifying the relative phase abundances in precipitates due to the precipitates usually contain crystal and amorphous phases with similar elemental compositions [19]. Hence, quantitative phase analysis using XRD data could be further by approached with Rietveld refinement [19,20] and Le Bail method [20,21]. The aim of this paper is to discuss the crystal structure with Le Bail method on the post-treatment of Ca-Mg-Al-LDH slurry from brine water.

2. Experimental

2.1 Material

Brine Water was taken from Tanjung Jati B CFPP, Jepara, Indonesia. The reagents $AlCl_3.6H_2O$; Na_2CO_3; $AgNO_3$; HCl 37%, $LaCl_3$ and aquadest were analytical grade and used without further purification.

2.2 Synthesis of Ca-MgAl-LDHs

LDHs compounds with the molar ratio (Ca+Mg)/Al is 0.5 was prepared by coprecipitation method. Into brine water as calcium and magnesium source was added $AlCl_3.6H_2O$ solution and the alkaline solution Na_2CO_3 0,1 M until pH was adjusted $\approx$ 10. The suspension was stirred 1 h at 65 °C, and then the slurry was washed with distilled water until free ion Cl^- ($AgNO_3$ test). The slurry was separated by centrifugation at a speed of 2000 rpm for 15 min and then dried overnight at 105 °C.

Powder X-ray Diffraction (XRD) patterns of Ca-Mg-Al-LDHs material were conducted using Brucker AXS D8 Advance diffractometer at 40 kV and 35 mA with Cu Kα radiatio (λ=0.1540 nm). The diffraction intensity was measured between 10° and 80°. The crystallite size of LDHs was calculated from the main reflection (11°) using the Scherrer formula: $D = \frac{0.9.\lambda}{\beta.cos\theta}$, where D is the crystallite size (nm); 0.9 is the shape factor; λ is the wavelength of the Cu Kα radiation (0.1540 nm), β is the full width at half maximum (FWHM) and θ is the Bragg diffraction angle. In addition, the XRD could be utilized to evaluate the peak broadening with the crystallite size and the lattice strain due to the presence of dislocations in LDHs using the Williamson-Hall (W-H) analysis [22] :

$$\beta.cos\theta_{hkl} = \frac{0.9.\lambda}{D} + 4\varepsilon sin\theta_{hkl} \qquad (1)$$

where ε is strain present in LDHs crystal. The lattice parameters c and a were calculated from the diffraction plane positions of d_{003}; d_{006}; d_{009} and d_{110} [17], with lattice parameters formula $c = d_{003} + 2d_{006} + 3d_{009}$ and $a = 2d_{110}$. The product was confirmed by powder X-ray diffraction (JCPDS #890460 for MgAl-LDHs and JCPDS #870493 for CaAl-LDHs). Rietveld refinement and Le Bail refinement were perfomed using the refinement program, RIETICA. Peak profiles were modelled using a pseudo-voigt peak shape. The other experimental parameters refined were the instrument zero, scale factor, the lattice parameters and the peak shape parameters u, v, w, γ_0 and γ_1 [20].

3. Result and discussion

Fig. 1 showed that the XRD patterns of Ca-MgAl LDHs with different treatment, without (HT-1) and with stirring (HT-2) treatment. As shown in Fig 1, the diffraction patternof HT-1 and HT-2 havesharp and symmetric peaks at 2θ = 11° *(003)*; 23°*(006)*; 33°*(009)* [16]and 61°*(110)* [22], which are the characteristics of LDH compounds. When XRD patterns were compared with JCPDS Mg-LDHs (JCPDS #890460) and Ca-LDHs (JCPDS #870493), another phase has found such as $Mg(OH)_2$ (JCPDS #070239), $Al(OH)_3$ (JCPDS #070324) and $CaCO_3$ (JCPDS #862334). This was due to the radius size of Ca^{2+} ion was larger than that of Mg^{2+} ion [2,17] and because of the synthesis method could not ensure the Ca, Mg and Al cations mixed uniformly, then some Ca, Mg and Al cation agglomerated and formed the $CaCO_3$, $Mg(OH)_2$ and $Al(OH)_3$, respectively [6]. Fig. 2 showed that the 2θ (11°) has low intensity when cooling down very rapidly, so this causes a LDHs content decreased. When Thermal decreased rapidly, crystal structure organized unstable and not good. Treatment with variation cooling time 0, 30 min and 24 h in Fig. 3 shows diffraction degree 11°, 23° and 37° is a LDH phase. Peak disposed wider with increasing cooling time, so this makes a form LDHs like as amorf. Cooling time with stirring could make a collision particle one another particle perfectly, but very longer stirring could make crystal structure broken or collapse.

Table 1. Treatment at slurry Ca-MgAl-LDHs during formation LDH crystal and lattice parameter.

Material	Treatment			Lattice	
	Stirring	Cooling rate	Cooling time	*a* (nm)	*c* (nm)
HT-1	Without	1 °C/min	0 h	0.2907	2.2756
HT-2	With	1 °C/min	0 h	0.3046	2.2820
HT-3	With	2 °C/min	0 h	0.3028	2.2582
HT-4	With	3 °C/min	0 h	0.3055	2.2559
HT-5	With	1 °C/min	30 min	0.3044	2.2845
HT-6	With	1 °C/min	24 h	0.3053	2.2827
Commercial Mg-LDH	-	-	-	0.3045	2.2851

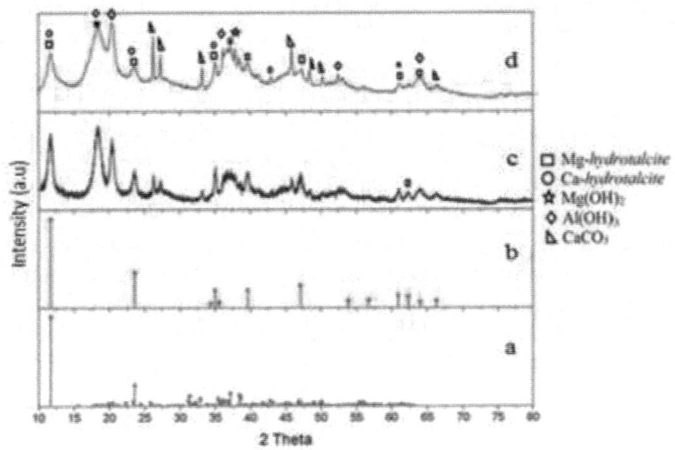

Figure 1. XRD pattern synthesized LDHs with a) Ca-Al LDHs (JCPDS #870493), b) Mg-Al LDHs (JCPDS #35096), c) HT-2 (with stirring) dan d) HT-1 (without stirring).

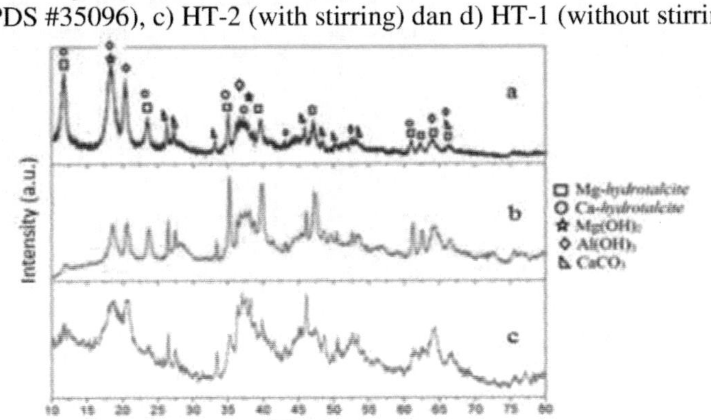

Figure 2. XRD pattern synthesis Ca-Mg-Al-LDHs with variation cooling rate a) 1 °C/min (HT-2), b) 2 °C/min (HT-3) dan c) 3 °C/min (HT-4).

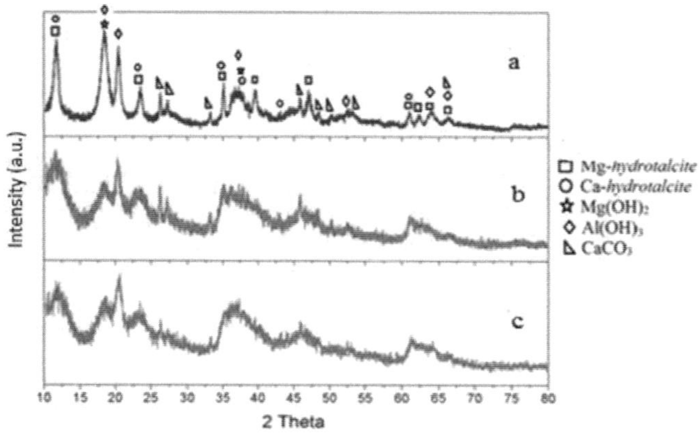

Figure 3. XRD pattern synthesis Ca-Mg-Al-LDHs with variation cooling time, a) 0 min (HT-2), b) 30 min (HT-5) dan c) 24 hour (HT-6).

Lattice parameter *a* refers distance of octahedral O-O in layer and lattice parameter *c* is the distance of interlayer [25]. This showed that HT-1 experience decreasing distance interlayer due to uneven distribution in balancing anion with the commercial Mg-LDH as the standard of comparison. Table 2 showed the lattice parameter at LDH treatment between HT-1 and HT-2 was different. HT-2 has higher lattice than HT-1, but HT-2 with other treatment has similar lattice parameter value a and c.

Table 2 showed that the LDH treatment between HT-1 and HT-2 have different crystallite size (*D*) on Scherer and Williamson-Hall Method, which HT-2 has higher crystallite size than HT-1. The crystallite size increased over the course of the experiment, suggesting that crystal growth occurred concurrently with the nucleation as the early formed LDHs slowly increased in crystallinity [24]. The decreased in crystallite size with the rising of cooling rate treatment at 1, 2 and 3°C/min might be explained that a LDHs structure lead to form small piece. In addition, the decreased in crystallite size with the rising of cooling time could be explained by partial dissolution of early formed amorphous LDHs as shown in Fig 3, which contributing to the aggregation and growth of larger crystal [17].

Table 2. The crystallite size (*D*) slurry Ca-MgAl-LDHs with Debye-Scherer and Williamson-Hall method.

Material	Scherer method (nm)	Williamson-Hall method (nm)
HT-1	4.64036	1.85614
HT-2	26.56207	0.17229
HT-3	192.58	-0.71324
HT-4	59.25385	2.30629
HT-5	256.7667	0.40821
HT-6	0.00203	0.26905
Commercial Mg-LDH	0.05538	2.46102

Le Bail refinement method was performed for both samples using as initial values the lattice parameters from standard JCPDS. The refinement had the following order: change the cell parameter value, histogram parameter, lattice parameter and instrumental peak shape. For each step were used 30 cycles. Table 3 shows the detail about result of refinement for Ca-MgAl LDHs with various treatment. The reliability index parameters i.e. the pattern factor (*Rp*), the weighted pattern factor (*Rwp*), and the goodness of fit ($\chi 2$), which is indicated the quality of refinement work in this study, are presented in Table 3. The ideal value for $\chi 2$ is 1,0 [26] or 3,0 [19], and all of the refinement was observed to be of

IOP Conf. Series: Materials Science and Engineering **176** (2017) 012020 doi:10.1088/1757-899X/176/1/012020

acceptable quality [26]. In Table 3, it was demonstrated that LDHs containing a single phase showns the $\chi 2$ value is larger than many phases. It was indicated that LDHs have contained many phases such as Mg-Al-LDH, Ca-Al LDH, $Mg(OH)_2$, $Al(OH)_3$ and $CaCO_3$ whereas $Al(OH)_3$ phase showns > 60%.

Table 3. Result Le Bail refinement with varous phase.

Material	Phase Composition					Value		
	Mg-LDH	Ca-LDH	$CaCO_3$	$AlOH_3$	$Mg(OH)_2$	*Rp*	*Rwp*	$\chi 2$
HT-1						25.38	33.92	279
	56.00	44.00				3.63	4.85	5.85
	28.86	22.97	48.07			2.47	3.3	1.99
	10.38	8.13	17.15	64.34		1.91	2.71	1.91
	10.30	8.21	16.99	63.73	0.77	1.87	2,59	1.76
HT-2						24.61	37.99	6.05
	55.95	44.05				7.16	7.45	0.23
	28.96	22.97	48.07			7.2	7.4	0.23
	10.42	8.09	17.28	64.2		7.14	7.71	0.25
	none	none	none	none	none	none	none	none
HT-3						20.35	28.5	5.67
	55.87	44.13				3.88	9.17	2.687
	29.03	22.95	48.03			3.17	5.79	1.089
	10.39	8.15	17.14	64.33		1.50	4.28	0.61
	10.30	8.12	16.98	63.83	0.77	1.95	4.81	0.78
HT-4						19.49	19.42	4.32
	55.99	44.01				3.18	5.35	4.26
	28.91	23.01	48.02			2.56	4.13	0.96
	10.38	8.19	17.13	64.31		1.81	3.88	0.86
	10.26	8.12	17.00	63.83	0.77	1.82	3.55	0.73
HT-5						12.52	12.12	0.58
	56.76	43.24				7.34	8.54	0.29
	45.57	36.17	19.07			6.44	8.1	0.26
	10.37	8.20	17.06	64.37		8.74	12.52	0.67
	none	none	none	none	none	none	none	none
HT-6						12.06	17.11	1.24
	56.17	43.83				7.51	8.42	0.30
	44.76	36.17	19.07			8.11	12.51	0.67
	10.37	8.18	17.12	64.32		5.49	7.85	0.27
	none	none	none	none	none	none	none	none

4. Conclusion

The result from this study confirmed that treatment with stirring provided higher crystallite size and lattice parameter compared the treatment without stirring. Meanwhile the cooling rate treatment lead to decrease crystallite size and the cooling time treatment has increase crystallite size. It is indicated by crystal growth occurred in nucleation and slow formation. On X-ray diffraction analysis has found another phase Mg-Al LDH and Ca-Al LDH, such as $Mg(OH)_2$, $Al(OH)_3$ and $CaCO_3$, which is supported by result analysis of refinement Le Bail, whereas $Al(OH)_3$ phases shown the highest phase.

Reference

[1] Selvam T, Inayat A, Schwieger W, 2014 *Dalt. Trans.* **43** 10365–10387.
[2] Rives V (Ed.), 2001 *Layered Double Hydroxides : Present and Future*, Nova Science Publishers, Inc., New York.
[3] Paikaray S and Hendry M J, 2012 *Appl. Surf. Sci.* **263** 633–639.
[4] Cavani F, Trifirò F, and Vaccari A,1991 *Catal. Today.***11** 173–301.
[5] Florea M and Za R, 2008 *Appl. Catal. A Gen.***341** 50–57.
[6] Gao L, Teng G, Xiao G and Wei R, 2010 *Biomass Bioenergy.* **34** 1283–1288.
[7] Heraldy E, S.J. Santosa, Triyono, and Wijaya K, 2015 *Indones. J. Chem.***15 (3)** 234–241.
[8] Roto R, Nindiyasari F and Tahir I, 2000 *J. Phys. Sci.* **20** 73–84.
[9] Wang L, Zong Q, Sun W, Yang Z and Liu G, 2015 *Corros. Sci.* **93** 256-266.
[10] Heraldy E, Nurcahyo I F and Ainurofiq A, 'Pembuatan Senyawa Hydrotalcite-like dari Brine Water untuk Eksipien Industri Farmasi' 2012, *Proc. InSINas* 2012, 37–44.
[11] Sarkarat M, Komarneni S, Rezvani Z, Wu X, Yin S and Yan Z, 2013 *Appl. Clay Sci.* **80-81** 390–397.
[12] Kameda T, Yoshioka T, Uchida M, and Okuwaki A, 2000 *Mol. Cryst. Liq. Cryst.* **341** 407–412.
[13] Palmer S J, Grand L M, and Frost R L, 2010 *J. Therm. Anal. Calorim.* **103** 473-478.
[14] Zhang, Gao Y and Li L F,2011 *Adv. Mater. Res.* **287-290** 569–572.
[15] Heraldy E, Nugrahaningtyas K D, Sanjaya F B, Darojat A A, Handayani D S, and Hidayat Y, 2016 *IOP Conf. Ser. Mater. Sci. Eng.* **107** 012025.
[16] Heraldy E, Nugrahaningtyas K D, Sanjaya F B, Handayani D S, and Hidayat Y, Influence of Variation (Ca-Mg)/Al Molar Ratios on Phase Composition of Ca-Mg-Al Hydrotalcite, 2013, *Int. Conf. Chem. Agric. Med. Sci.* 74–77.
[17] Paikaray S and Hendry M J, 2014 *Appl. Clay Sci.* **88-89** 111–122.
[18] Valente J S, Cantu M S, and Figueras F,2008 *Chem. Mater.* **20** 1230–1232.
[19] Lu X, Shih K, Li X, Liu G, Zeng E Y and Wang F, 2016 *Water Res.* **90** 9–14.
[20] Peterson V K, 2005 *Powder Diffr.* **20** 14–17
[21] Sanches E A, Carolino A D S, Dos Santos A L, Fernandes E G R, Triches D M, and Mascarenhas Y P, 2015 *Adv. Mater. Sci. Eng.* **1**1-8
[22] Gao L, Teng G, Lv J, and Xiao G, 2010 *Energy & Fuels.* **24** 646–651.
[23] Williamson G K and Hall W H, 1953 *Acta Metallurgica*, **1** 22-31
[24] Paikaray S, Gomez M A, Hendry M J and Essil J, 2014 *Appl. Clay Sci.* **101** 579–590.
[25] Sánchez-cantú M, Pérez-díaz L M, Rubio-rosas E, Abril-sandoval V H, Merino-aguirre J G, Reyes-cruz F M, and Orea L, 2014 *Chem. Pap.* **68** 638–649.
[26] Yan H and Shih K, 2016 *Water Res.* **95** 310–318.

IOP Conf. Series: Materials Science and Engineering **176** (2017) 012021 doi:10.1088/1757-899X/176/1/012021

Early Detection of Breast Cancer by Using Handycam Camera Manipulation as Thermal Camera Imaging with Images Processing Method

R Riantana[1], B Arie[2], M Adam[2], R Aditya, Nuryani[1], I Yahya[1]
[1]Physics Department of Sebelas Maret University
[2]Informatics Department of Sebelas Maret University

E-mail: rioriantana@yahoo.com

Abstract. One important thing to pay attention for detecting breast cancer is breast temperature changes. Indications symptoms of breast tissue abnormalities marked by a rise in temperature of the breast. Handycam in night vision mode interferences by external infrared can penetrate into the skin better and can make an infrared image becomes clearer. The program is capable to changing images from a camcorder into a night vision thermal image by breaking RGB into Grayscale matrix structure. The matrix rearranged in the new matrix with double data type so that it can be processed into contour color chart to differentiate the distribution of body temperature. In this program are also features of contrast scale setting of the image is processed so that the color can be set as desired. There was Also a contrast adjustment feature inverse scale that is useful to reverse the color scale so that colors can be changed opposite. There is improfile function used to retrieves the intensity values of pixels along a line what we want to show the distribution of intensity in a graph of relationship between the intensity and the pixel coordinates.

1. Introduction

Breast cancer is a disease where there is excessive growth or uncontrolled growth of cells of the breast tissue [1]. Breast cancer is a type of cancer that is often found by most women. Even in Indonesia, which suffered the highest cancer Indonesian women is breast cancer [2].

For women under 30 years old, the doctors advised doing the MRI test. MRI (Magnetic Resonance Imaging) does not use X-ray radiation and is suitable for detecting soft tissue. However, imaging with MRI is more expensive that very few women who do the MRI test.

In addition to performing the MRI test, mammogram tests (Mammography) is mostly done by the majority of the women to detect breast cancer. Mammography is the examination of the human breast using low-dose X-rays (usually around 0.7 mSv). However, the various parties still doubt the use of mammography because the error rate is still high and because of radiation used can cause hazards.

In detecting early symptoms of breast cancer, one of the important things to watch out for is the change in temperature of the breast. Rising temperatures breast was an indication of the symptoms of breast tissue abnormalities [8,10].

Therefore, we had the idea to manipulate the camera camcorder as a thermal image camera for the early detection of breast cancer with the method of processing images using Matlab. This system is simpler and more secure than mammography or MRI system.

Measurement or processing and acquisition of temperature information without contact with the object to be in the process [3,4]. The measurement is based the infrared radiation of electromagnetic waves with wavelengths longer than visible light. The object that has a temperature of more than the absolute temperature emit infrared waves [5]. Infrared measurements require the infrared radiation from the object and being transformed into electrical signals. The converters are the basis pyrometer which generates an output signal using a single sensor. More sophisticated tools can generate rows of the measurement results in detail of the image produced. Differences in visible light and infrared readings are from the light source that accompanies it. Figure utilizing infrared cameras change into color images based on the level of infrared energy. Images of the infrared reading are calorimetry [6]. Measurements using radiation to produce temperature readings from radiation generated. The temperature reading has the advantage of direct, no damage, and safe for repeated every time [7].

2. Experimental

The mechanism of manipulation of camera camcorder as a thermal image camera for the early detection of breast cancer with the method of processing images using Matlab little tricky. This is because the process to manipulate the thermal camera Handycam be not easy. Changes function does not do physically, but uses a manipulation on the software, so the infrared image on the.

.

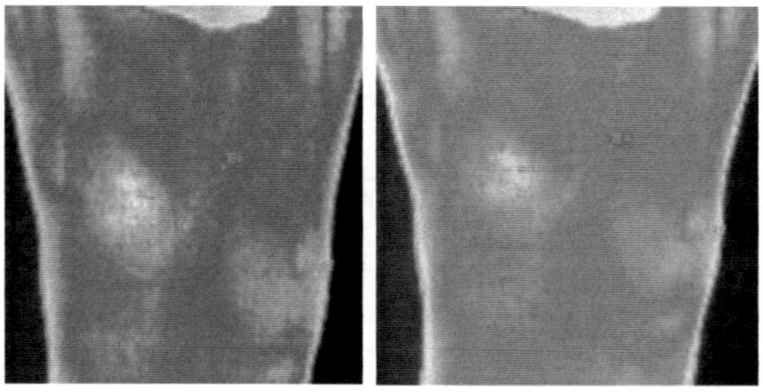

(a) **(b)**

Figure 1 (a) grayscale (b) polychromatic

The principle of this project camcorder's image can be converted into thermal (temperature) on a new program is to change the image of the infrared form of greyscale (image gray) to image thermal (temperature) in the form polychromatic (colorful), by breaking the grayscale image into a matrix of code per pixel and combines back and displayed in scale color. The process of shooting is done in a dark room in order to avoid interference of light other than infrared light. Patients are not allowed to wear clothing because it will hinder the main object. Once the images are acquired then do image processing.

In principle image of infrared images that form has a level greyscale grayish, so that each point has a value of how the gray pixel. Value grayish was converted into numeric form. So that image data is not garbled, then the identity is given pixel position in the format {x, y, z} where x is the position of the pixel on the x-axis, y is the position of pixels on the y-axis, and z is the value of grayish the pixel point. After that, the identity data rearranged in position by providing a color scale based on the value of z[9]. Figure 1 shows the image transformation from gray to polychromatic. The scale can be set based on the desired accuracy. Figure 2 shows a flowchart of this application.

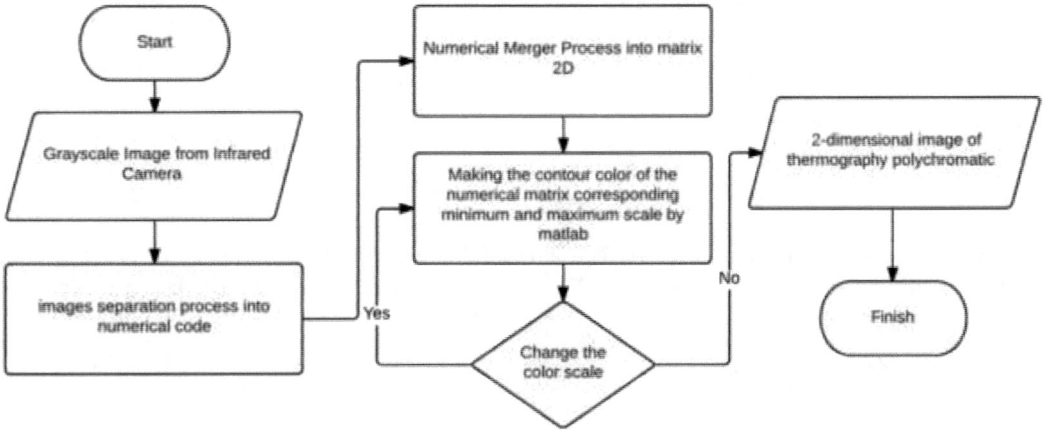

Figure 2. Flowchart of system application

3. Results and Discussion

3.1 Grayscale picture taking with night vision camera

At this stage taking pictures using a Handycam camera in night vision mode, accompanied by an infrared LED. The image is an image formed of gray. In a captured image can also be seen that the infrared can penetrate the outer skin so it looks bundles of the vein after the shooting with a camcorder and a night vision mode with the addition of LED infrared interference. Figure 3 shows the differences in the pictures taken from the camera and the usual mode with an additional mode of night vision infrared LEDs interference.

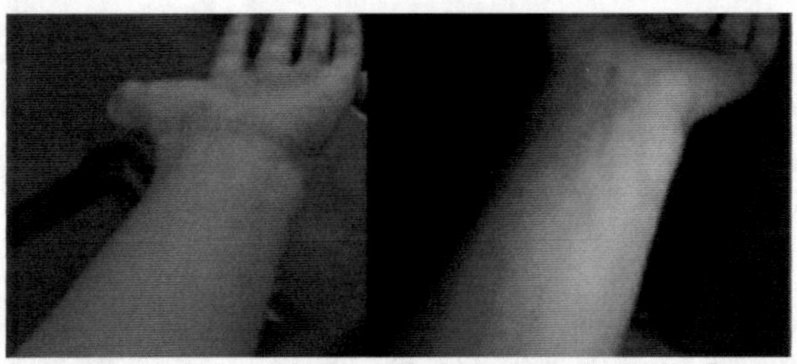

Figure 3. Image results without night vision and infrared mode (left). Image results using night vision and infrared mode (right).

3.2 Making the image processing program using Matlab

Basically, the night vision image taken from the Handycam camera is still a polychromatic image which is a 3-dimensional matrix because they have the format RGB (Red, Green, Blue). Furthermore, the image of polychromatic is converted to a form of grayscale by changing the matrix of three dimensions into two dimensions wherein image RGB totaling three layers will be converted into BW (Black and White) layer which has one layer alone, So the original image is a blend of colors Red, Green, Blue will turn into a two-dimensional image containing greyish level at any point.
Here is a scheme change RGB matrix to Grayscale.

$$RGB = \begin{bmatrix} R_{11} & \cdots & R_{1j} \\ \vdots & \ddots & \vdots \\ R_{i1} & \cdots & R_{ij} \end{bmatrix}, \begin{bmatrix} G_{11} & \cdots & G_{1j} \\ \vdots & \ddots & \vdots \\ Gi1 & \cdots & G_{ij} \end{bmatrix}, \begin{bmatrix} B_{11} & \cdots & B_{1j} \\ \vdots & \ddots & \vdots \\ B_{i1} & \cdots & B_{ij} \end{bmatrix} \quad (1)$$

$$\mathrm{BW} = \left(\sum_1^3 \begin{bmatrix} RGB_{11} & \cdots & RGB_{1j} \\ \vdots & \ddots & \vdots \\ RGB_{i1} & \cdots & RGB_{ij} \end{bmatrix} \right) : 3 \qquad (2)$$

$$Grayscale = \begin{bmatrix} BW_{i1} & \dots & BW_{ij} \\ \dots & \ddots & \vdots \\ BW_{11} & \dots & BW_{1j} \end{bmatrix} \qquad (3)$$

In the process of changing the image RGB to Grayscale matrix changes vertically inverted position. To position the matrix can be returned as the starting position, the matrix is reversed vertically using flipup() function in Matlab.

Grayscale = flipup(Grayscale); (4)

$$Grayscale = \begin{bmatrix} BW_{11} & \dots & BW_{1j} \\ \dots & \ddots & \vdots \\ BW_{i1} & \dots & BW_{ij} \end{bmatrix} \qquad (5)$$

Furthermore, the matrix will be converted into an image depicting polychromatic contour color based on the level of gray in the image that illustrates the distribution of body heat. Because it is still in the form of a matrix, then the thing to do is change the format of the matrix into a double. This is done so that each coordinate value can be determined by changing the data type to double. Once the data type matrix of gray image has become an array of type double then the next is plotting the array into a color contour graph depicting the distribution of body heat. Figure 4 shows the program display images processing using Matlab.

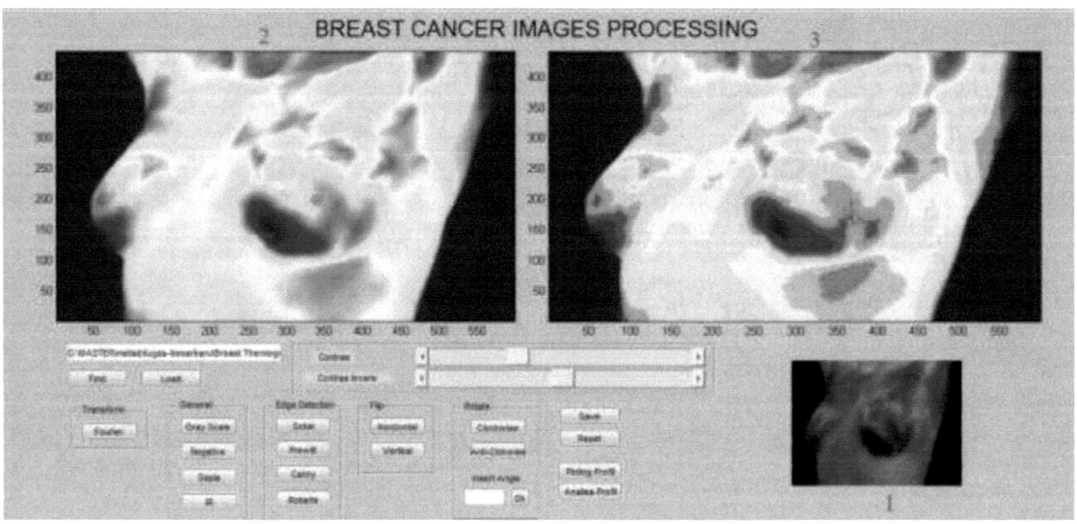

Figure 4. Display Breast Cancer Images Processing.

In Figure 4, a section marked with the number 1 is the original image of the grayscale image obtained from a camcorder with night vision mode. In the section marked with the number 2 is the result of processing images from Matlab which converts the image into a gray image polychromatic to show clearly the distribution of a person's body heat. In the section marked with the number 3 is the result of the image of the image of part 2, which has gone through the process of changing the contrast so that the color scale can be changed. Change the contrast serves to reinforce the color so as to facilitate the analysis of doctors about the potential breast cancer.

3.3 The process of contrasting scale changes

In contrast scale, change process is done by changing the intensity of the image matrix which has been converted into contour color chart. The color changes using a slider that will automatically change the image results when the slider is moved.

$$Contrast_{value} = 200^{(\frac{CV}{100})} \quad (6)$$

CV = changes value from slider

$$New\ Contrast\ Image = Grayscale\ x\ Contrast_{value} \quad (7)$$

In addition contrast changes can also be done with inverse contrast, or alter the contrast by reversing the color scale so that the image of the blue color will be red and vice versa. This is done to resolve when the gradation of an image is not very clear so it is necessary to clarify the scale of another color gradation of the image. In Figure 5 intended result of image processing with the inverse contrast

$$New\ Contrast\ Invers\ Image = 128 - (Grayscale\ x\ Contrast_{value}) \quad (8)$$

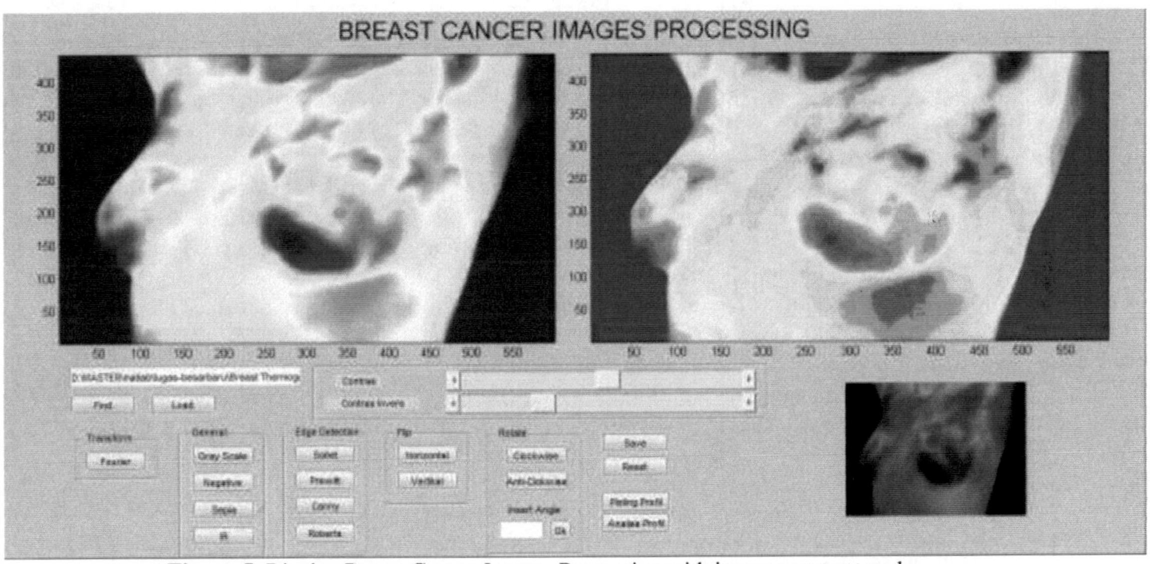

Figure 5. Display Breast Cancer Images Processing with inverse contrast scale

3.4 Improfile process

In this process, we try to display data from a row of pixels in the image that has been processed through processing images. The principle of this process is to take the value of a line between the two points. The distribution of line indicated by the following equation.

$$Hx = \{p | p \in \int_{(x,y)}^{(x',y')} a_{ij}\} \quad (9)$$

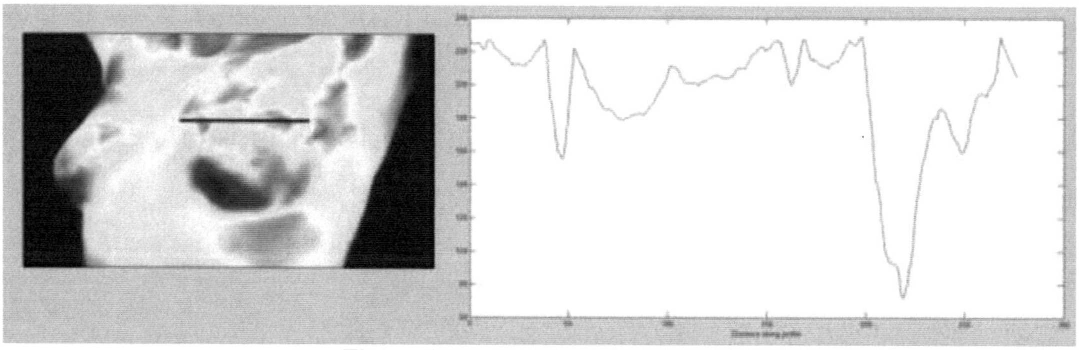

Figure 6. Result of Improfile process

Here used improfile functions on Matlab where the pixel value of the shortest distance the two lines will be displayed in the graph. Figure 6 displays a graph of the intensity distribution of pixels coordinate. Improfile retrieves the intensity values of pixels along a line or a multiline path in the grayscale, binary, or RGB image in the current axes and displays a plot of the intensity values. improfile(n) retrieves the intensity values, where n specifies the number of points to include. improfile(I,xi,yi) retrieves pixel intensity values, where I specifies an image, and xi and yi are equal-length vectors specifying the spatial coordinates of the endpoints of the line segments [11].

4. Conclusion

Handycam in night vision mode by the interference of external infrared can penetrate into the skin better and can make an infrared image becomes clearer. Images from a camcorder can be processed by the CASE Program (Detection of Breast Cancer) with the results of color contrast is pretty clear.

This program is able to change the black and white images from a camcorder into a thermal image by breaking RGB into Grayscale matrix structure. Once the matrix is rearranged in the new matrix with double data type so that it can be processed into contour color chart to differentiate the distribution of body temperature in the processed image. The graphics processing by differentiating levels of gray of the image to obtain different colors. In Case Program are also features of a scale setting the contrast of the image is processed so that the color of detail can be set as desired. There was also a contrast inverse scale feature settings are useful to reverse the color scale so that colors can be changed into its opposite. There is Improfile function used to retrieves the intensity values of pixels along a line that we want to show the distribution of intensity in a graph of relationship between the intensity and the pixel coordinates.

References

[1] R. Rastghalam, H. Pourghassem. *Pattern Recognition* **51** (2016) 176–186.
[2] E. Mahmoudzadeh, M.A. Montazeri, M. Zekri, S. Sadri. *Infrared Physics & Technology* **72** 19–28.
[3] S. V. Francis, M. Sasikala, G. B. Bharathi, S. D. Jaipurkar. *Infrared Physics & Technology* **67** 490–496.
[4] Maldague, X. *Theory and Practice of Infrared Technology for Nondestructive Testing*; Wiley: New York, NY, USA, 2001.
[5] T. B. Borchartt , A. Conci, R. C.F. Lima, R. Resmini, A. Sanchez. *Signal Processing* **93** 2785–2803.
[6] Modest, M.F. Radiative Heat Transfer; Academic Press: Waltham, MA, USA, 2013.
[7] Gaussorgues, G. *Infrared Thermography; Springer*: Berlin/Heidelberg, Germany, 1994.
[8] M. EtehadTavakol, E.Y.K. NG. J. Mech. Med. Biol. **13** 1330001.
[9] A. *Dhungana. Segmentation of Infrared Images*. Texa Tech University, 2002 (M.Sc.).
[10] N.R. Pal, S.K. Pal. *Pattern Recognit.***26** 1277–1294
[11] https://www.mathworks.com/help/images/ref/improfile.html 10/09/2016

IOP Conf. Series: Materials Science and Engineering **176** (2017) 012022 doi:10.1088/1757-899X/176/1/012022

The Characteristic and Activation of Mixed Andisol Soil/Bayat Clays/Rice Husk Ash as Adsorbent of Heavy Metal Chromium (Cr)

Pranoto[1], Sajidan[2] and A Suprapto[3]

[1] Chemistry Department, Mathematic and Natural Science Faculty, Sebelas Maret University, Surakarta, Indonesia
[2] Biology Department, Teacher Training and Education Faculty, Sebelas Maret University, Surakarta, Indonesia
[3] Post Graduate Program of Master in Environmental Science, Sebelas Maret University, Surakarta, Indonesia

E-mail : pak_pran@yahoo.com

Abstract. Chromium (Cr) concentration in water can be reduced by adsorption. This study aimed to determine the effect of Andisol soil composition/Bayat clay/ husk ash, activation temperature and contact time of the adsorption capacity of Cr in the model solution; the optimum adsorption conditions and the effectiveness of ceramic filters and purifiers to reduce contaminant of Cr in the water. The mixture of Andisol soil, Bayat clay, and husk ash is used as adsorbent of metal ion of Cr(III) using batch method. The identification and characterisation of adsorbent was done with NaF test, infrared spectroscopy (FTIR), X-ray diffraction (XRD). Cr metal concentrations were analyzed by atomic absorption spectroscopy. Sorption isotherms determined by Freundlich equation and Langmuir. The optimum conditions of sorption were achieved at 150°C activation temperature, contact time of 30 minutes and a composition Andisol soil / Bayat clay / husk ash by comparison 80/10/10. The results show a ceramic filter effectively reduces total dissolved solids (TDS) and Chromium in the water with the percentage decrease respectively by 75.91% and 9.44%.

1. Introduction

The increasing of industry gives negative impact in the form of an increase in terms of number and type of waste generated. Pollutants that enter the water (groundwater or surface water) can affect the parameters of aquatic environment. Heavy metals is one of the pollutants because of toxic and it is not degraded in the nature [1]. The adsorption method is very effective for reducing pollutants [2].

Andisol soil is soil that occurs from the weathering of volcanic rocks, which both of rock that has been frozen, or from volcanic ash. The soil is composed of andesite is Andisol soil, while allophane an aluminosilicate contained in Andisol soil. Allophane adsorbent has good characteristics, such as porosity, high absorption and cation exchange. Using the natural allophane of mountains in Java Island (Papandayan, Arjuna and Wilis) for adsorbting heavy metals (Cr, Fe, Cd, Cu, Pb and Mn) with batch method [3].

Bayat Clay has been choosen as adsorbent because its existence many found in District Bayat, Klaten Regency, Central Java, Indonesia and is currently only used by mostly Bayat community for pottery production. Clay is an aggregate of minerals that form of land consisting mainly of hydrous aluminum silicates, has active sites on the surface, it is hard and stiff when it is dry, stable at high temperatures and is plastically when it is crushed and moistened [4].

Rice husk many found as agricultural waste in Indonesia. During this time the rice husks are usually only used as a fuel or even just burned away. By making it as adsorbent, it is expected to add value to the waste. Some researchers have been using rice husk ash in processing cooking oil and it gives great results.

Based on numerous studies by using Andisol soil, natural clay and rice husk ash, so this study conducted the development adsorbent of a mixture Andisol soil, Bayat clay and rice husk ash to increase the effectiveness of the entrapment of the chromium (Cr) heavy metal and its development as a ceramic filter and pipe purifiers to reduce the content of heavy metals chromium (Cr) in groundwater.

The purpose of this study was to determine the effect of Andisol soil composition: Bayat clay: husk ash, activation temperature and contact time on the sorption of chromium (Cr) heavy metal ions in a model solution, determine the optimum conditions adsorbent of mixture Andisol soil: Bayat clay: husk ash as adsorbent chromium (Cr) heavy metal in a model solution, examine the effectiveness of ceramic filters and purifiers pipe in reducing chromium (Cr) heavy metal ion content in water.

2. Experimental

2.1. Adsorbent Preparation

2.1.1. Rice Husk Ash

Rice husks that had been washed with water for 3-4 times dried, then has been ashed at temperatures of 300°C for 2 hours. Rice husk ash obtained is then stored in sealed bottles for later use.

2.1.2. Clay

The clay used in this study came from Bayat, Klaten, Central of Java Province, Indonesia. The clay obtained is cleaned of impurities and dried with aerated in the open air to make it dry, then crushed clay until it is smooth. Clay then sieved with a 150 mesh sieve. The powder that passed 150 mesh was soaked in distilled water and filtered, then dried at 105° C for 4 hours [5].

2.1.3. Andisol Soil

Andisol soil used in this study came from the Cemoro Kandang, Lawu, Central Java, Indonesia. Andisol soil obtained cleaned of impurities, washed with water and dried with aerated in the open air to dry, then Andisol soil is crushed until smooth. Furthermore, Andisol soil has sieved with a 150 mesh. The powder that passed 150 mesh was soaked in distilled water and filtered, then dried at 105°C for 4 hours [5].

2.2. Adsorbent Identification and Characterization

The identification of the adsorbent was done by testing the pH of Sodium Fluoride (NaF), XRD, FTIR.

2.3. Adsorbent Activation

Adsorbent activation is done chemically and physically. Chemical activation is only done for Andisol soil, as many as 50 grams of ground Andisol NaOH is added to 250 mL with a concentration of 3 M. Then the mixture was stirred at a temperature of 70°C with a stirring for 5 hours, then cooled. Once the mixture is cool then filtered and washed with distilled water until the pH is neutral. After that, the Andisol soil were dried in an oven for 4 hours or until dried at a temperature of 105°C [5].

Later on, a composition variation of adsorbent was made among Andisol soil, Bayat clay, and rice husk ash, such as 100/0/0, 80/10/10, 60/20/20, 40/40/20, 20/40/40, 20/20/60, 10/80/10, 10/10/80, 0/0/100 dan 0/100/0.Each composition of the mixture of rice husk ash and Andisol soil then performed physical activation at temperature variations of 100, 150 and 200°C for 3 hours. The adsorbent used for the performance test for optimum conditions to the adsorption of metal ions of chromium (Cr^{3+}) in solution models.

2.4. Adsorbent Performance Test

2.4.1. Sorption of Ion Cr (III) in Solution Model
Metal ion adsorption process of chromium was conducted by immersion (batch), namely by as much as 0.05 gram of adsorbent mixture put in a 100 mL glass beaker containing 15 mL solution of Cr (III) 8 ppm. Then it stirred at a constant speed (150 rpm) at room temperature for 30, 60 and 90 minutes. Then filtered with Whatman No. 40 and the filtrate was measured by Atomic Absorption Spectroscopy (AAS) to determine the metal ion concentration of Cr (III) which is not adsorbed by adsorbent. Furthermore, the results of an adsorbent against metal ion Cr (III) (mg / g) to determine the contact time, which has a maximum power adsorb. Ion Cr (III) adsorbed is calculated from the concentration of Cr (III) initially reduced by concentration after sorption process. Then graphed concentration of chromium (Cr^{3+}) adsorbed with time variation. Ion Cr (III) analysis using atomic absorption spectroscopy.

2.4.2. Determination type of sorption isotherm
The best adsorbent that has been obtained then carried sorption variation adsorbed ion concentration to determine the type of isotherm. A total of 0.05 grams of adsorbent put in a 100 mL glass beaker and added to each 15 ml of Cr (III) with various concentrations of 2, 4, 6, and 8 ppm and then stirred at the optimum time. The sorption results filtered with Whatman No. 40, then the filtrate obtained, was measured by AAS. The results are then analyzed to determine the type of Langmuir or Freundlich isotherm.

2.5. Ceramic Filter Manufacture and Purifier Pipe

2.5.1. Ceramic Filter.
The process of making ceramic filter made by craftsmen of pottery in Bayat, Klaten, Central Java, Indonesia to adjust the composition of raw materials and the desire of researchers with dry mixing between Andisol, clay and rice husk ash adsorbent by comparison based on optimum conditions. The material mix gypsum cylindrical mold with a diameter of 4 cm inside, the outside diameter of 5 cm, thickness of 0.5 cm and a length of 20 cm.

2.5.2. Purifier Pipe
Water purifiers pipe use Poly Vinyl Chloride (PVC) pipe with diameter ½ inch (1.25 cm), 20 cm long with a mixture contained Andisol, clay and rice husk ash is spherical diameter of 1.2 cm and a variation of the addition of deodorizing shell charcoal.

3. Result and Discussion

3.1. Adsorbent Identification

3.1.1. NaF Test
The allophane existence of minerals in the soil can be identified by test NaF. The NaF test is done, by measuring the pH of 1 gram of soil in 50 ml of NaF 1 solution for 2 minutes. Soil pH value greater than 9.4

3.1.2. FTIR Analysis

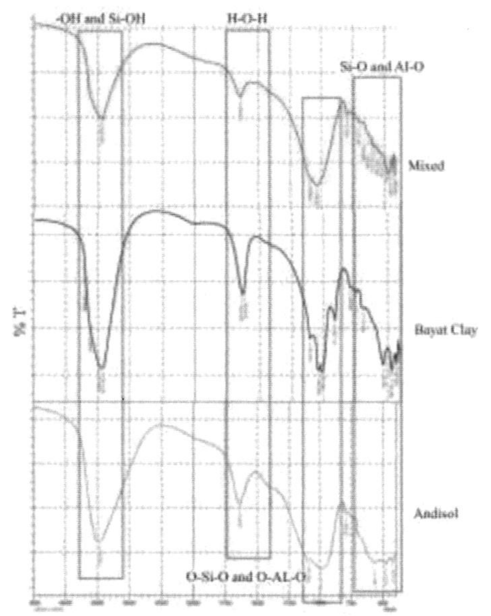

Figure 1. Sample Spectrum of andisol soil, bayat clay and mixed

As shown in Figure 1, the spectra for all natural samples are similar one to another, and they are typical for allophane.

Table 1. Comparison between the spectra of samples

Functional Groups	Allophane Uptake (Cm^{-1})		
	Allophane/ Andisol	Bayat Clay	Mixed
Vibration –OH (Al-OH/Si-OH)	3455	3444	3437
Vibration AlOh/SiOH	1108;973	912.33	960.55
The Existence of Si-O bond	579;485	428	432

3.1.3. XRD Analysis

XRD for the samples were recorded and shown in Figure 2. They are similar each other.

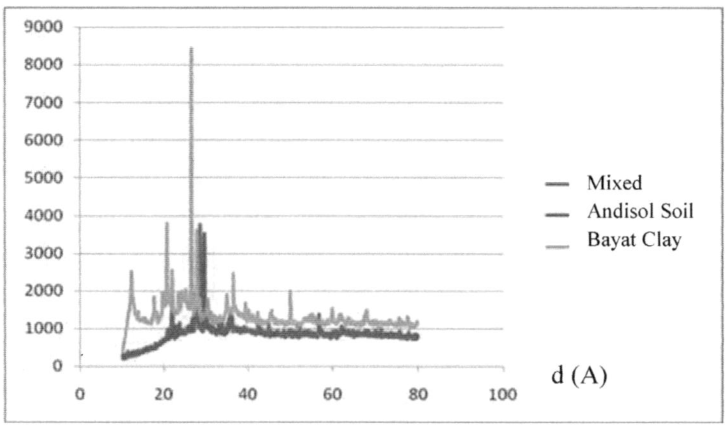

Figure 2. Difraktogram Sample of andisol soil, bayat clay and mixed

The details of particular d-spacings are confirmed with JCPDS for allophane as shown in table 2.

Table 2. typical d-spacings data of three samples compared to JCPDS

JCPDS (38-0444)	d-Spacing (Å) of allophane				
	3.300	2.250	1.860	1.400	1.230
Andisol Soil	3.305	2.249	1.861	1.400	1.230
Bayat Clay	3.305	2.249	1.861	1.400	1.230
Mixed	3.305	2.249	1.861	1.400	1.230

3.2. Adsorbent Activation

Activation of Andisol soil, Bayat clay and rice husk has done to improve the chemical and physical character as adsorbent. The chemical activation has done by soaking Andisol soil in 3 M NaOH for 5 hours. Activation using a strong base 3 M NaOH for 5 hours refers to the research that has been done [6], which resulted in an increase in specific surface area on the adsorbent of Andisol activated by 1.62% and amounted to 90.87% total acidity. As for clay and rice husk ash is not performed chemical activation. Physical activation performed on a mixture adsorbent of Andisol soil, Bayat clay and rice husk ash with their weight ratio 100/0/0, 80/10/10, 60/20/20, 40/40/20, 20/40/40, 20/20/60, 10/80/10, 10/10/80, 0/0/100 and 0/100/0.

Table 3. Data of Surface Area

No	Sample	Surface area (m^2/g)
1	Andisol	66,42
2	Bayat Clay	34,53
3	Rice husk ash	15,18
4	Mixture	35.85

Source: Primary Data

Table 4. Data of Acidity

No	Sample	Acidity (mmol/g)
1	Andisol	6,75
2	Bayat Clay	3,27
3	Rice husk ash	4,02

Source: Primary Data

3.3. Adsorbent Performance Test

Andisol soil absorbent, Bayat clay and rice husk ash, have a high affinity for metal ions. All three have active groups Si-OH and -OH, whereas Andisol soil and clay Also has an active Al-OH groups, so that all three has the electronegative surface charge. The exchange of cations in the process of metal sorption of chromium (Cr) in solution [6]. Sorption ability of silicate minerals derived from the number of negative charge on the structure of silicate minerals. The negative charge is neutralized by adsorption adsorbed positively charged ions, such as heavy metal cations. Based on the data from sorption that the three absorbent namely Andisol soil, Bayat clay and rice husk ash. All three have the ability to adsorb metal ions with different adsorption capacity[7].

Adsorbent with composition of 100% ground Andisol (100/0/0) has a sorption capacity greater than 100% clay dsorbent (0/100/0) and rice husk ash adsorbent 100% (0/0/100). Adsorbed amount of metal ions decreases with increasing percentage clay adsorbent and rice husk ash in the composition of the adsorbent mixture. This shows that the presence of clay and rice husk ash in the adsorbent mix less support adsorbent ability to adsorb the metal ion Cr (III). Reduced ability Andisol sorption of a mixture of soil, clay and rice husk ash can be explained based on the data of surface area and the number of acidity of each adsorbent.

Heating the mixture of clay and Andisol adsorbent also intended to free water molecules, both the water bound physically or chemically bound in the form of hydrated.

In this study conducted temperature variations of 100, 150 and 200°C. Through the third activation on these temperature variations, it would be known to influence the temperature difference in the ability to absorb metal chromium.

The sorption capacity of the soil mixture adsorbent of Andisol, clay and rice husk ash increased with the length of contact time. When its adsorption has reached the maximum, and the adsorption capacity to be reduced with increasing contact time. At a temperature of 100 and 150 ° C of all adsorbent compositions increased adsorption capacity with increasing contact time of 30 to 60 minutes and sorption capacity decreased when the contact time of 90 minutes. Based on data from sorption capacity results prove that the contact time influence in the process of sorption.

The research showed the best condition sorption of metal ions Cr (III) is an adsorbent with composition of Andisol soil, clay and rice husk ash (80/10/10), the activation temperature of 150° C and a contact time of 30 minutes. It is also expected to provide an overview adsorbent effectiveness Andisol soil mixture, clay and rice husk ash treatment if carried metal ion adsorption. Furthermore, the final stage of this research is the manufacture of water purification using raw material mixture of adsorbent.

3.4. Determination type sorption isotherm

Freundlich isotherm explains that the adsorption occurs on more than one surface (multilayer) and adsorbent have heterogeneous surfaces with a binding energy of different [8]. Cr (III) ion adsorption by adsorbent mixture of clay and Andisol occur fisisorption. Sorption type is suitable for sorption mechanisms that require regeneration process because the substance adsorbed only weakly bound to the surface of adsorbent.

3.5. Filter Performance Pottery Test and Pipes Purifier

3.5.1. Ceramic Filter

Ceramic filter performance test were conducted by well water drained from the tank through a PVC pipe to the water purification. Water is passed three housing containing a ceramic filter, activated carbon granules, carbon block, then the head housing contains osmosis membranes and the last pass of activated carbon powder. The water treatment outcomes, was collected and measured Total Dissolve Solids (TDS), pH, conductivity, turbidity, salinity and metal ion of Cr (III).

4. Conclusion

The composition of Andisol soil, Bayat clay and rice husk ash, activation temperature and contact time affect the adsorption capacity of Cr (III) metal ions in solution models with optimum conditions of Andisol soil composition, Bayat clay and rice husk ash (80/10/10), the activation temperature of 150 $^{\circ}$C and contact time of 30 minutes. Absorbent of Andisol soil mixture, Bayat clay and rice husk ash in optimum conditions effective to reduce the content of Cr (III) metal ions in solution models with an efficiency of 97, 56%. The measurement results show a ceramic filter effectively reduced Total Dissolved Solids (TDS) and Chromium in the water with the percentage decrease respectively by 75.91% and 9.44%. While the pipe purifiers can reduce total dissolved solids (TDS) from 260 to 249 ppm and Chromium from 0.064 to 0.052 ppm.

References

[1] Zhao, G., Wu, X., Tan, X., dan Wang, X 2011 *The Open Colloid Science Journal* **4** 19-31.
[2] Manohar, D.M., Noeline, B.F., and Anirudhan, T.S. 2006 *Applied Clay Science* **31** pp.194-206
[3] Pranoto, Suranto, Sugiyarto, K. H. and Ashadi 2013 *Journal of Environment and Earth Science* **3(5)** ISSN 2224-3216 (Paper) ISSN 2225-0948 (Online)
[4] Sulastri, S. dan Kristianingrum, S. 2007 *Prosiding Seminar Nasional Kimia dan Pendidikan Kimia* ISBN: 978-979-98063-1-4.
[5] Sulistyarini, Ari 2012 Fakultas Matematika dan Ilmu Pengetahuan Alam Universitas Sebelas Maret
[6] Widjonarko, D. W., Pranoto dan Christina, Y 2003 *Journal Alchemy* **2(2)** September 2003
[7] Diantariani, N. P. 2010 Jurnal Kimia **4(1)** Januari hlm. 91-100 Universitas Udayana Bukit Jimbaran ISSN 1907-9850
[8] Alemayehu, D. D., Sing, S. J., and Tessema, D. A. 2012 *Universal Journal of Environmental Research and Technology* **2(5)** pp 411-420

RELATIONSHIPS ANALYSIS AND PUBLIC PERCEPTION OF THE HEALTHY PLASTIC AS ONE SOLUTION TO HEALTHY LIVING

Hartatik[1], **R Hartono**[1], **A Purnomo**[1], **B K Riasti**[1], **H Munawaroh**[2]
[1]Informatics Engineering Departement, Faculty of Mathematics and Natural Sciences, Sebelas Maret University, Surakarta, Indonesia
[2]Chemistry Department, Faculty of Mathematics and Natural Sciences, Sebelas Maret University, Surakarta, Indonesia

E-mail: hartatik.uns@gmail.com

Abstract. Direct Plastics are used for various human purposes, ranging from household to industry. Tableware and drink made of plastic is very practical to use, easy to clean, durable and cost far less than tableware made of the other material .However, must also be considered in terms of security in the use of plastic containers for food storage because there are adverse effects. There are seven types of plastic based material used, namely Polyethylene, Terephthalate, High Density Polyethylene, Polyvinyl Chloride or V/PVC, Low Density Polyethylene or LDPE, Polypropylene, Polystyrene, Plastics others including polycarbonate. Experts claims that the plastic code numbers 2, 4 and 5 are used for equipment safely eat/drink because it is more stable and safe if used correctly. In this study will analyze the relationship between the recent education, family income to perception and behavior in the use of plastics in food storage daily as one solution to healthy living. The population of this research is all the people in the Solo area particularly housewife and all the people in the productive age. Data were obtained through a survey with cluster random sampling method. Statistical method used is a parametric method and Chi Square test This method is used as an alternative method of parametric when some assumptions are not met. Based on the results of Chi Square test with $\alpha = 15\%$ was concluded that recent education and income related to the behavior of people using plastic products as one of the solutions to Healthy Living.

1. Introduction

Plastic polymer has been largely superseded manufacturing of home appliances stairs of metal, wood or ceramics into devices using plastic materials. One is a plastic bottle of drinking water. The nature of plastic does not rust, inexpensive, lightweight and unbreakable, economical advantages for the user. But behind these advantages, there is a danger that if the use of plastic bottles do not follow the rules set by the Ministry of Health and WHO.

In recent years, appeared their variety of new diseases caused by unhealthy lifestyles, particularly in terms of healthy food. One of the factors that affect food plastic containers we use. We must be selective in using food storage containers, because there are chemicals that trigger the disease. Safe

[1] To whom any correspondence should be addressed.

plastic with a specific code, between 2.4 and 5. The public generally, is not too concerned about the quality of the code and platik to health and a healthy lifestyle[1].

According Kertajaya (2002), now an all consumer hearts Determine consumer choice always purchase long-term considerations with more conscious and aware well as health [2]. One thing known brands society is tupperware, tupperware where is the famous name on a brand of housewares made of plastic. From Tupperware products include storage containers, container presentation and some equipment. Tupperware introduced the first time sales manager for public review in 1948 by Earl Tupper (1907-1983) in leominster, massachusetts. Tupperware pioneered the direct marketing strategy made famous by the Tupperware party [3].

Competition household plastic products lately very tight and many new emerging brands that each race shows the benefits of its products each well in terms of form, colour and durability. Based on this it will be analyzed in this study about how important the community aware of the importance of healthy plastic.

2. Experimental

The research method was conducted using cluster sampling random sampling the population of all users of plastic Solo with Total sample 55 respondents where the sampling method used was cluster random sampling method [4]. The analytical method used is descriptive analysis in order to know the characteristics respondents and also The Chi Square test to get the relationship between education and knowledge about the safety of plastic.

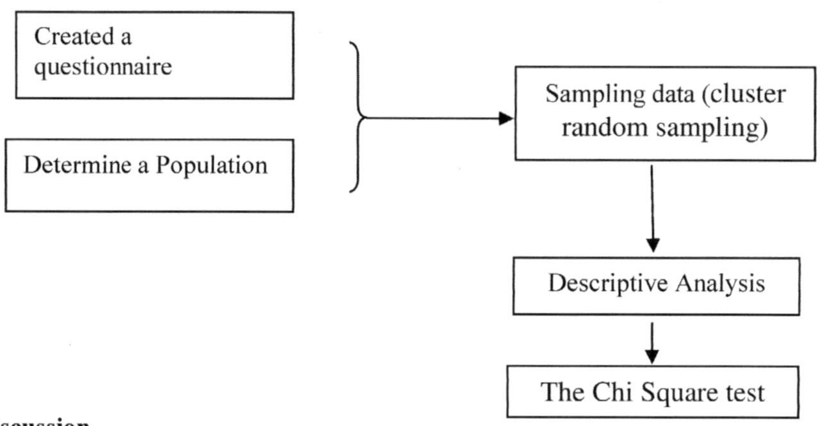

3. Result and Discussion

3.1 Public Attitudes toward Plastics products

The results showed that the communities are still not aware of the importance of the code on food storage equipment. It may be seen from the percentage of people who see the plastic code. Based on table 1, from the total of 55 respondents, some 14.5% stated that they sometimes see the code, some 34.5% said always see the code, 18.2% have never seen the code, and 32.7% Often see the code when purchasing plastic products [5].

Table 1. Percentage of code behaviour

Item	Frequency	Percent	Valid Percent	Cumulative Percent
do not care the code	10	18.2	18.2	18.2
Some consider the code	8	14.5	14.5	32.7
often consider the code	18	32.7	32.7	65.5
always consider the code	19	34.5	34.5	100
Total	55	100	100	

While based respondents (sampling), the distribution of educational level is relatively fairly even start secondary school level to college. the majority of respondents educated PT / S2 states sometimes look at the code, which is about 53.25% said always see the code while it contained 46.75% of respondents said often pay attention to the code. respondents who educated PT states sometimes look at the code, which is about 4.76% do not and sometimes look at the code, and 52.38% said often look at the code and 38.10% always look at the code. Respondents who had high school education, some 25% said never see the code, 50% said sometimes see the code, 8.33% stated often look at the code and 16.67% of respondents said always see the code in buying plastic storage equipment. While respondents junior high school education, some 75% said never noticed codes and 25% said sometimes see code.

From this discussion it can be seen that in junior and senior secondary education mostly less attention to the plastic codes they buy. In addition, the lower the level of behavior viewed codes plastics are increasingly less attention. This is evidenced by the percentage of respondents who have never seen the code is in junior high school education (see Figure 1).

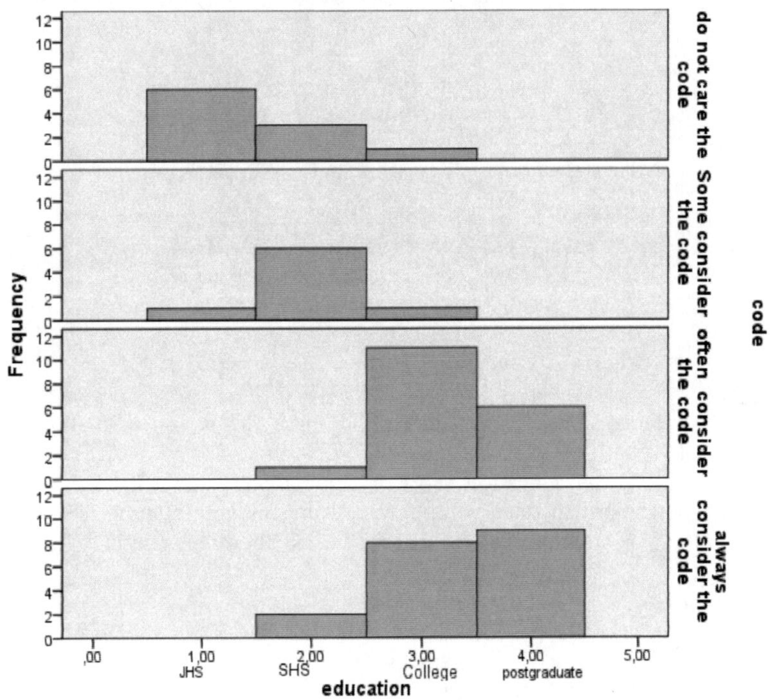

Figure 1. Education to consider the behavior of the code

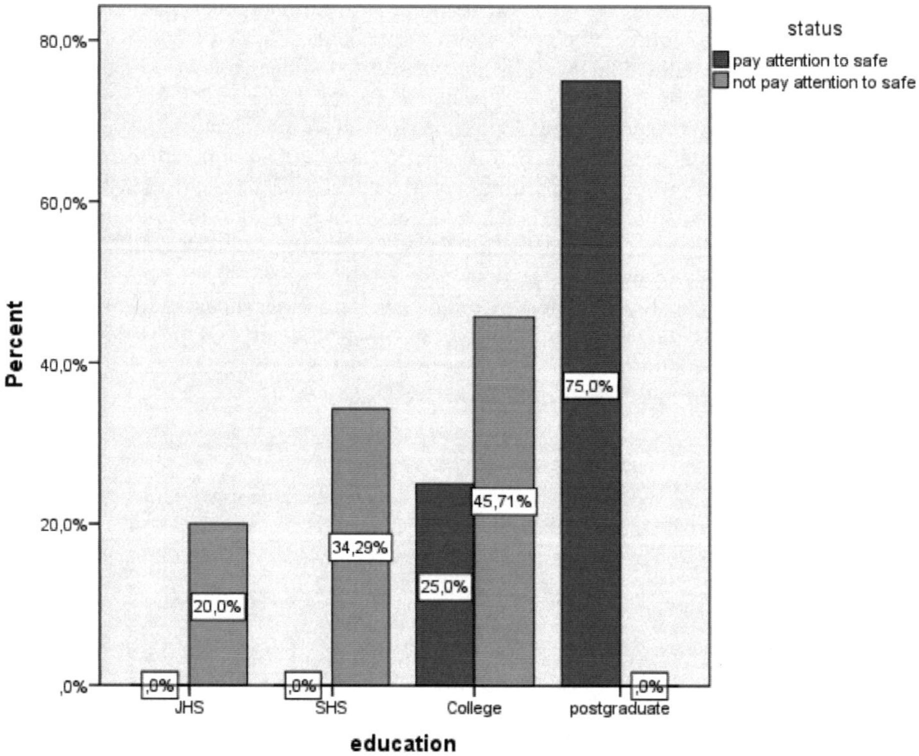

Figure 2. Education on safe usage behavior of plastic

Of the total 55 respondents, some 35.4% said that buying a plastic container According to the budget is not concerned about security of plastic and 63.6% very concerned about the safety of plastic.
To buy plastic containers, the education level of respondents, showed that the level of school are junior high school education, secondary high school and collage attention is still less attention to product safety, while the post graduates are very concerned about this that is equal to 75%

3.2 Analisis Chi Square
Hypothesis :
H_0 : education and social behavior are independent (not related)
H_1 : education and social behavior are not independent (related)
Based on analysis of the relationship with Chi square analysis with $\alpha = 5\%$ was found that education has to do with the behavior of people in view of the plastic code in any purchase or use plastic products. It is seen that the correlation values obtained spaerman sig = 0.734 and 0.00[6].

Table 2. Analyze data for education versus code and safe plastics

Variable	Correlation value (spearman)	Sig Value	Conclusion
Educational and behavioral see the code	0.734	0.00	H_0 rejected
Education and safe behavior of buying plastic	0.733	0.00	H_0 rejected

Hypothesis:
H_0 : education and safe behavior of buying plastic are independent (not related)
H_1 : education and behavior of buying plastic are not independent (related)
Based on analysis of the relationship with Chi square analysis with $\alpha = 5\%$ was found that education is related to the behavior of people in buying or using plastic products safely. It is seen that the correlation values obtained spaerman - sig = 0.733 and 0.00.

4. Conclusion

In general, there are still many people who are less attentive behavior on codes of plastic products and their use. Beside it, last education related to behavior using coded plastic products and plastic are also safe.

References

[1] Direktorat Pengawasan Produk dan Bahan Berbahaya Badan Pengawas Obat dan Makanan RI. . 2008. Materi Talkshow di RRI tentang Kemasan Pangan. Jakarta.
[2] Kertajaya, Hermawan 2002 *Marketing Plus 2000* Jakarta Gramedia.
[3] Patterson Paul G and Johnson Lester W 1997 Spreng Richard A modeling the determinants of customer satisfaction for business to business professional services.*AcadMark Sci* **25(1)** 4-17.
[4] Sugiyono 2007 *Metode Penelitian Kuantitatif Kualitatif* Alfabeta Bandung.
[5] Ghozali, Imam. 2011 Aplikasi Analisis Multivariate dengan Program IBM SPSS 19 Semarang Badan Penerbit Universitas Diponegoro.
[6] Hayter, Anthony 2007 *Probability and Statistics for Engineers and Scientists* (3rd ed.) Victoria Australia: Thomson Brooks/Cole.

IOP Conf. Series: Materials Science and Engineering **176** (2017) 012024 doi:10.1088/1757-899X/176/1/012024

Synthesis of the supported catalysts by co-impregnation and sequential impregnation methods

Khoirina Dwi Nugrahaningtyas[1,2,3], Yuniawan Hidayat[1,2], Patiha[1], Nina Prihastuti[1], Brigitta Yelvi[1], and Riza Umami Nur Kalimah[1]

[1]Chemistry Department, Faculty of Mathematics and Natural Sciences, Sebelas Maret University, Jl. Ir. Sutami 36A Kentingan, Surakarta, 57126, Indonesia

[2]Solid State Chemistry & Catalysis, Chemistry Department, Faculty of Mathematics and Natural Sciences, Sebelas Maret University, Jl. Ir Sutami No 36 A Kentingan, Surakarta, 57126, Indonesia

E-mail: khoirina@mipa.uns.ac.id and khoirinadwi@staff.uns.ac.id

Abstract. The non-sulfide catalysts Co-Mo/USY were prepared with three different methods, i.e. co-impregnation and two sequential impregnation procedures. The phases of the synthesized materials and their composition were characterized with X-ray diffraction. Furthermore, the X-ray diffraction patterns were analyzed using Le Bail methods with Rietica software. Finally, an overview of the substitution effect of cobalt and molybdenum on the metal content and acidity of the catalyst were also investigated.

1. Introduction

Transition metals are an important family of inorganic materials which have been a subject of intensive research due to their many applications, such as catalysts in a large variety of industrial processes that involve hydrogen evaluation reaction, and the activation or conversion of hydrocarbons [1–3]. For example, nickel oxide and molybdate based on alumina are important components of industrial catalysts for hydrodesulphurization [2,4]. Reference [4] investigates NiMo bimetal catalyst supported on a variety of zeolite types for HDS of thiophene. They discovered that the effect of the supported catalyst on HDS performance of thiophene over NiMo catalyst was great, and the HDS activity of NiMo could be improved by changing the supported catalyst. Their resulted research showed that USY is the best supported agent. On the other hand, the resulted of research [1] showed that a CoMo-based catalyst had good thermal stability in comparison with NiMo samples. In the other reference it was shown that CoMo-based catalysts are important components of the industrial catalysts used for hydrodeoxigenation of lignin [1].

Studies on catalysts to remove impurities from hetero_atom to allow its use in biofuel conversion are still popular. Previous researchers reported that the CoMo-based catalyst has the best catalytic performance. [5] Reported that despite its complex preparation, the best catalytic performance was shown by the sulfide catalysts of $CoMo/Al_2O_3$ as compared to Mo/Al_2O_3 and Pt/USY. In the HDO catalytic reaction, [6–8] also reported the same performance.

[3] To whom any correspondence should be addressed.

According to [5,9], sulfidation processes can be ignored to simplify the overall preparation processes. Yet, an alternative for sulfidation processes is sought. Referring to [10], calcination, oxidation and reduction processes can be alternated with the sulfidation processes to simplify the preparation processes of CoMo-based catalysts. Seeking for supports then becomes the next priority.

Supports like zeolite, provide the catalytic functions, where the cracking reaction takes place on the zeolite's strong acid sites. The most important function of support is maintenance of a high surface area for the active component and acidity for hydrocracking reactions. A support material is chosen for a given application on the basis of several important characteristics including its inertness to undesired side reactions, appropriate mechanical properties (attrition resistance, strength, etc), stability under reaction conditions, a surface area and pore size distribution appropriate for the desired reaction, and its cost [11]. Zeolites are widely used for catalyst support and play an important function in the hydrocracking process because they improve the activity, selectivity, and stability of catalysts. Zeolites are defined as crystalline alumina silicates, and have been investigated and extensively used in industry. Zeolites have been successful because of their crystalline properties, high surface area, adsorption capacity, and uniform size distribution which enable shape selectivity [12].

The hydro treatment catalyst is a typical example of a bifunctional catalyst, in which both the hydrogenation/dehydrogenation and the cracking function play crucial roles. The hydrogenation/dehydrogenation functions come from the metal side, while cracking comes from support materials such as USY.

However, the preparation of bifunctional catalyst with high performance is very important, and is related to its application for future industrial use. Molybdenum oxide and cobalt oxide precursors on zeolite are prepared by a variety of techniques, solid state reaction [13], Chemical Vapour Deposition (CVD) [14], and impregnation methods [9,10,15].The last method is the simplest one. Therefore, in this study, we prepared bimetal catalyst CoMo/USY with varied impregnated methods. The catalyst precursors were prepared by the same procedure described in our work in a previous publication [16].

Catalytic activity is determined by the chemical structure of the catalyst and its properties. In the hydrodesulphurization catalysts, the degree of acidity is one of the characteristics that determine the catalytic performance. The higher degree of acidity often results in higher catalytic activity [3]. As well as the degree of acidity, the catalytic activity is determined by many other characteristics e.g. the metal content and, the surface area. The metal content and surface area are simple and more accurate indicators of the degree of active site dispersion. That is why this research used those criteria to verify the catalysts obtained. In addition, the pattern of XRD from obtained catalyst, total pore volume and average pore radius are presented in order to complement the verification criteria of a good catalyst.

2. Experimental

2.1. Materials

The support material used in this research was NH_4-USY, which was purchased from Tosoh Inc (Japan). Other materials were ammonium heptamolybdate and cobalt nitrate hexahydrate purchased form Merck (Germany).

2.2. Methods

NH_4-USY was calcined at 550 °C for 1 hour to obtain USY. In order to address the above aforementioned research objectives, metal based-catalysts of Co and Mo supported on Ultra Stable Y-zeolite (USY) were prepared following some experimental treatments.

Three kinds of bimetal catalysts were prepared. The bimetal catalysts of (Co and Mo)/USY were prepared following the two main kind of impregnation methods - the co-impregnation method [4] and the sequential impregnation method [16]. First, in the co-impregnation method, an H-USY sample was mixed with a solution of $(NH_4)_6Mo_7O_{24}.4H_2O$, $(Co(NO_3)_2.6H_2O$, aquabidest and ammonia in a flask with a concentration of 8% w Mo metal and 2 % w Co. Then, the sample mixture was heated at a temperature of 60 ° C for 2 h. The samples were dried with a vacuum evaporator at a temperature of 85 ° C for 16 hours at a pressure of 72 mbar. The resulting catalyst was called CoMo/USY.

The sequential impregnation method was carried out through two reversed mechanisms. The first was to support the precursor of Co, followed by ammonium heptamolybdate (Co_Mo/USY) and the second was conducted through the opposite sequence (Mo_Co/USY). The next important step in the process was activation. Here, all of the produced catalysts were activated through consecutive reactions namely calcination, oxidation and reduction. Calcination was conducted by leaving the produced catalyst at a high temperature in an atmosphere of nitrogen gas. Here the temperature was set to 500 °C whereas the flowrate of nitrogen was of 5 mL/min. The oxidation and reduction processes were conducted at a temperature of 400 °C at a gas flow rate of 5 mL/min.

2.3. Characterization of the Catalyst

All catalysts obtained were then verified according to several criteria. Phase characterization was carried out by X-ray diffraction (XRD) using an XRD Philips type X'Pert with CoK_α radiation. The data were collected in a step scan from 5 to 90° in 2θ with a step of 0.04°. All the structure refinements were carried out using Le bail methods with Rietica software. The chemical composition of the catalysts was determined using a Bruker S2 Ranger wavelength X-ray dispersive fluorescence spectrometer.

Using self-construction apparatus, the gravimetric analysis method was used to measure the total acidity. The method was carried out under standard pressure and temperature. At first, the porcelain crucible heated at 110 ° C for an hour. The crucible weight was labelled as W_1. Then 0.2 g of samples were placed into the porcelain crucible and heated at the same temperature for 1 hour. The weight of the porcelain crucible containing the sample was labeled as W_2. Then, the crucible containing the sample was inserted into the desiccators. In the next step, the ammonia gas was exposed to desiccators for 24 hours (before this treatment, the desiccators were made under vacuum conditions).The weight of the porcelain crucible was labeled as W_3.

$$\text{Total acidity} = \frac{(W_3 - W_2)}{(W_2 - W_1).M} \qquad (1)$$

With M = molecular weight of NH_3 (gram/mol).

The IR spectra of samples were recorded using Shimadzu 8201 PC FTIR spectroscopy with KBr methods.

3. Results and Discussions

The characteristics of the catalyst support (USY) were reported in our previous publication. Now, a comparison of co-impregnation and sequential impregnation methods will be given.

3.1. Phase composition of catalyst sample

Characterization of crystal structure and determination of the crystallographic parameters of metal oxide catalyst were performed by XRD. Figures 1a – 1d show the diffraction patterns of samples USY, CoMo/USY, Co_Mo/USY and Mo_Co/USY respectively. Then, the parameters are listed in Table 1. X-ray diffraction pattern for several catalyst samples (a) USY, (b) Mo_Co/USY, (c) Co_Mo/USY, and (d) CoMo/USY.

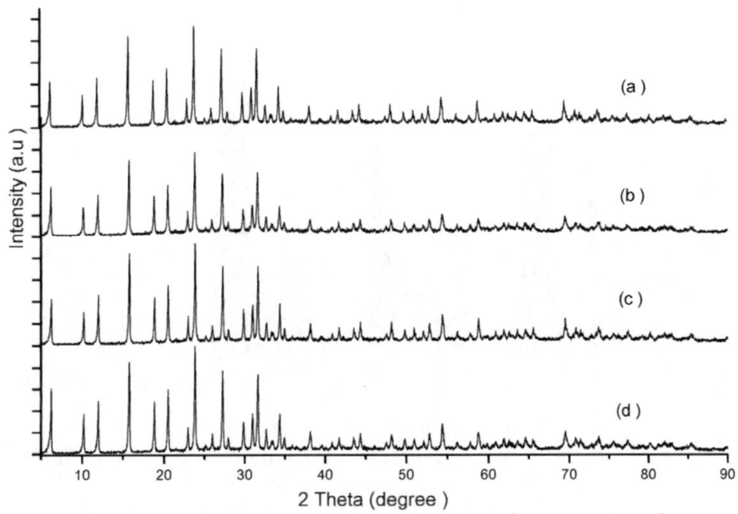

Figure 1. X-ray diffraction pattern for several catalyst samples. (a) USY, (b) Mo_Co/USY, (c) Co_Mo/USY, and (d) CoMo/USY.

Figure 1a-1d belonged to a typical diffraction pattern of the Y zeolite (Y) framework. Besides Y, we observed an extra crystalline phase. According to the ICSD #31542 data (see Figure 1a) the samples contained the type cubic Y (Fd-3mz space group, lattice parameter a = b = c = 24.539 Å) with chemical formula i.e. $Na_{0.36}H_{1.3}Al_{1.66}Si_{10.34}O_{24}$. However, we can determine the cell parameter and percentage-compounds of the samples based on the diffraction pattern data shown in Figures. 1a-1d which were refined using the Le bail method with RIETICA software (Table 1).

Figure 1 shows the three phases of metal oxide (ICSD #281235, #9865, and #3617). Although the catalysts have quite a similar character as regards the crystalline phase, they are slightly different in percentage-composition (Table 1).

Table 1. Percentage-composition of metal oxide.

Catalyst code	CoMoO4	CoO	MoO3
USY	na	na	Na
CoMo/USY	0.10	0.12	0.32
Co_Mo/USY	0.10	0.12	0.33
Mo_Co/USY	0.11	0.13	0.33

*na: not available

3.2. Metal Content

Indeed, high metal content is one of the criteria for a good catalyst. The metal content (based on the XRF data) for each impregnation method is presented in Figure 2.

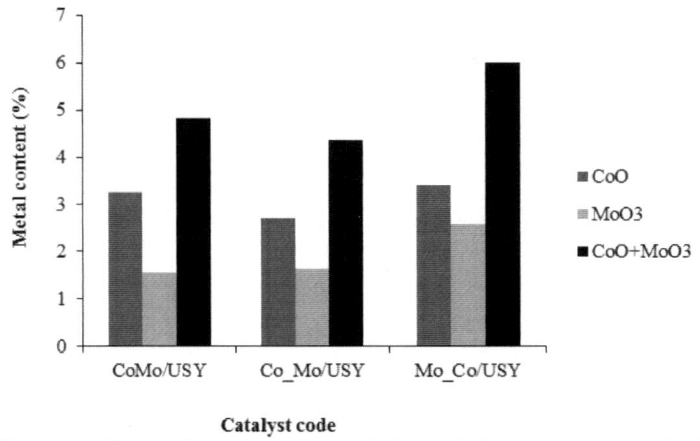

Figure 2. Totally impregnated metal for each impregnation methods.

According to Figure 2, use of the sequential impregnation method by adding Mo precursor followed by Co gives the best results as regards metal content. This method does not allow for competition between metals (Mo and Co) to adsorb into USY, but such competition between metals does exist in the co-impregnation method. A sequential impregnation method leads to greater metal content. This is because Co is smaller than Mo; hence, Co is easily impregnated upon the USY after Mo.

3.3. The acidity character

The acidity of catalysts plays an important role in the reaction behaviour of catalysts. A method often used to determine the acidity of catalysts is gravimetric and IR spectroscopy with adsorbed ammonia. The lone-pair electrons of nitrogen are involved in different types of interactions with the surface acid sites. The acidity for each applied method and each given impregnation method are presented in figure 3.

Figure 3 shows that the total acidity value is not always in line with the total metal content. There is a phenomenon of unequal distribution of the impregnated metal upon the surface of USY. The sintering phenomenon also occurs. This phenomenon causes the Co to become less capable of donating its acid sites. According to figure 4, use of the sequential impregnation method by adding Mo precursor followed by Co precursor gives the best result in terms of total acidity.

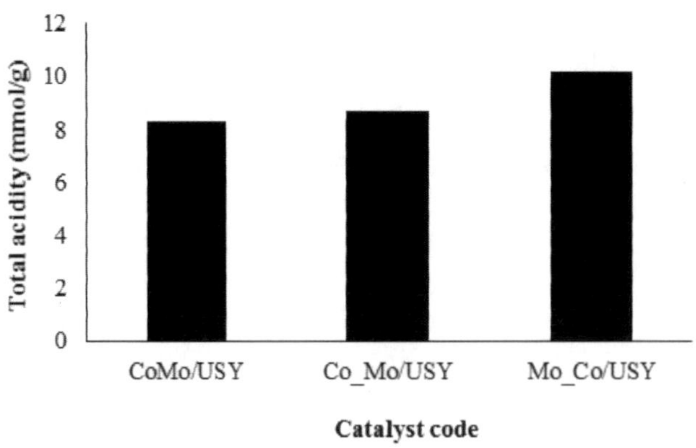

Figure 3. The total acidity for each given treatment.

Figure 4 displays the FTIR spectra of the catalyst products with various impregnation methods through careful analysis of the IR spectra. Ammonia sorption experiments, monitored with FTIR spectroscopy, were performed to determine the change in active (acid) sites with different impregnation methods. It is well known that ammonia is a useful probe molecule for Brønsted and Lewis acid sites and is characterized by the narrow and well-resolved bands in the region 1400-3750 cm^{-1}. Although the characteristic bands of the Brønsted and Lewis acid sites appear at a high wave number, all the spectra were recorded in the 400-4000 cm^{-1} range.

Three modes of adsorption have been observed: (a) proton transfer at Brønsted acidic surface sites (ammonium ions); (b) electron transfer at Lewis acidic surface sites (molecularly coordinated ammonia); and (c) hydrogen bonding to surface hydroxyl groups. All three modes of adsorption display IR spectra bands at different wave numbers and can be identified by their IR absorption bands.

Two bands at 3600 and 3500 cm^{-1} also indicated acid sites in the super cages(the high-frequency HF band) and in the sodalite cages (the low frequency LF band) of the USY, respectively [17]. The shift of the surface vibration mode of strained Al(Si)–O–Al(Si) bridges to lower frequencies for Mo_Co/USY was attributed to an Al-O-M(metal) interaction (i.e. in the 3000-3750 cm^{-1} range). It was noticed that the slight red shift in the IR spectrum of the first type of sequential impregnation method's product (namely use of Mo_Co/USY catalyst) in comparison with the products of co impregnation and second type sequential impregnation methods (namely use of CoMo/USY catalyst and Co_Mo/USY catalyst), could also be caused by the stretching vibration frequency of aqua hydroxyl radicals. Therefore, hydrogen bonds were formed between the catalyst structures, resulting in the self-assembly of a 3D cage-like structure that improved the stability of the catalyst structure. The phenomenon is an indication that the Mo_Co / USY catalyst is more stable than other types.

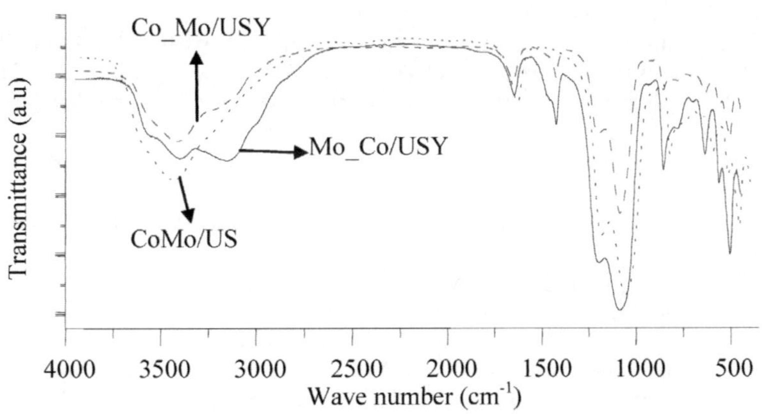

Figure 4. FTIR spectra of synthesized catalyst.

Ammonia, coordinated with the Lewis center, yields a band at *ca.*1500 cm^{-1}, and the protonated ammonia on the Brønsted center, yields a band at *ca.*1600 cm^{-1}. Both complexes also yield a peak at *ca.*1500cm^{-1}. The 1600 cm^{-1} band is generally assigned to hydrogen-bonded ammonia. In the FTIR spectra of synthesized catalyst, a small absorption band at *ca.* 1600 cm^{-1}, typical of the Brønsted acid site, was found for all samples. The presence of Lewis acid sites is indicated by a band at *ca.* 1500 cm^{-1}. The CoMo/USY catalyst did not exhibit any Lewis acidity as evidenced by the missing ammonia absorption band at around 1500 cm^{-1}, and we found that only a few Lewis acid sites could be observed for Co_Mo/USY. In CoMo/USY catalyst, the band in the 400 – 1100 cm^{-1} range was overlapped by the Mo_Co/USY catalyst.

The results of FTIR spectra analysis lead to the same conclusions as gravimetric methods. Both methods conclude that the catalyst which has the highest acidity is Mo_Co/USY.

4. Conclusion

Based on all results, the best catalyst preparation is the sequential impregnation method, first adding first ammonium heptamolybdate followed by cobalt nitrate hexahidrat. CoMo and USY are expected to be used as bifunctional catalysts and for hydro-treatment.

Acknowledgment

This research is funded by The Ministry of Education and Culture through The Directorate of general of research and community service (DP2M DIKTI) under research competent Grant. Therefore, I would like to thank the Ministry of Education and Culture, especially DP2M DIKTI, for funding this research.

References

[1] Jongerius AL, Jastrzebski R, Bruijnincx PCA, and Weckhuysen BM 2012 *J. Catal.* **285** 315-23
[2] Kiurski J, Obadovi DŽ, Kapor A, Kiš E, and Marinkovi R 1998 *Chem. Technol.***1** 135-42
[3] Rinaldi N, Yoshioka M, Kubota T, and Okamoto Y 2010 *J. Japan Pet. Inst* **53** 292–302
[4] Li D, Xu H, and Guthrie Jr. GD 2000 *J. Catal.* **189** 281–96
[5] Steiner P 2002 *Kinetic and Deactivation Studies of Hydrodesulfurization Catalysts* (Norwegia, Departament of Chemical Engineering, The Norwegian University of Scinece and Technology) p. 166
[6] Mochizuki T, Toba M, and Yoshimura Y 2011 *J. Japan Pet. Inst.* **54** 222–3
[7] He Z, and Wang X 2012 *Catal. Sustain. Energy* **1** 28–52
[8] Mortensen PM, Grunwaldt JD, Jensen PA, Knudsen KG, and Jensen AD 2011 **407** 1–19
[9] Mortensen PM, Grunwaldt J-D, Jensen AD, and Jensen PA 2013 *ACS Catal.* **3** 1774–1785
[10] Augustine RL 1996 *Heterogeneous Catalysis For The Synthetic Chemist.* (New York: Marcel Dekker In) p 672
[11] Satterfield C N 1991 *Heterogeneous Catalysis In Industrial Practice* (New York: McGraw Hill Book Co.)
[12] Weitkamp J, and Puppe L 1999 *Catalysis and Zeolites: Fundamentals and Applications* (New York: Springer)
[13] Okamoto Y 1997 *Catal Today* **39** 45–59
[14] Okamoto Y 2014 *Bull. Chem. Soc. Jpn.* **87** 20-58
[15] Muralidhar G, Massoth FE, and Shabtai J 1984 *J. Catal.* **85** 44–52
[16] Nugrahaningtyas KD, Sunaryanti WT, Triyono, Nuryono, Widjonarko DM, and Yusnani A 2010 *Proc of the 2009 Int Conf on Chemical, Biological and Environmental Engineering* (Singapore: World Scientific Pub. Co.)
[17] Taufiqurrahmi N, Mohamed AR, and Bhatia S 2011 *IOP. Conf. Ser. Mater. Sci. Eng.* (Kualalumpur: IOP Publishing)

Open Graded Asphalt Mixture Design for Environmentally Friendly Road

A Setyawan[1], Suryoto, A Sumarsono, S Widyastuti
Roadmate Research Group, Civil Engineering Department, Sebelas Maret University, Jl. Ir. Sutami 36 A Solo 57126, Indonesia

E-mail: arysetyawan@staff.uns.ac.id

Abstract. The road is an important infrastructure to support various economic development. By increasing the population, the constructions of road infrastructure are obviously required and cover more land use area, consequently narrowed the areas for water absorption into the land. Therefore, it is necessary to improve more environmentally friendly of road constructions, in this case is by using open graded asphalt. Open graded asphalt is the mixture of asphalt with low fine aggregate content to produce a high percentage of air voids, which expected to be able to absorb water into the land underneath. This observation uses laboratory experiment method. Materials that are used are natural gravel and the volcanic gravel, 60/70 asphalt penetration and stone dust filler. The tests performed for the samples are Marshall test, falling head water permeability, indirect tensile strength and unconfined compressive strength. The optimum bitumen content for each type of aggregate have been found to be 2.95% for natural gravel and 4.65% for volcanic gravel. Open graded asphalt porosity is 26.52% for natural gravel, whereas the volcanic gravel is 22.31%. The vertical and horizontal permeability for natural gravel are 1.25 cm/second and 0.92 cm/second, respectively. Whereas, for the volcanic gravel is 1.43 cm/second and 1.10 cm/second, respectively. Indirect tensile strength (ITS) value for natural gravel is 59.49 KPa whereas the volcanic gravel is 74.84 KPa. The unconfined compressive strength (UCS) value for natural gravel is 1070.63 KPa, whereas the volcanic gravel is 1120.69 KPa. It is concluded that the open graded porous asphalt is only compatible for low volume traffic road and need improvement to be able to be used as standard road.

1. Introduction

Porous asphalt is asphalt with a low fine aggregates content to get a high air voids to be able to absorb water [1]. Types of coating for porous asphalt are very innovative way, because it is easy to pass the water into the pavement vertically and horizontally through the pore to pore capillary or by using a side channel as drainage systems. Evidences show that porous asphalt is very good for improving road traffic safety in very bad weather conditions such as heavy rain and slippery, reducing hydroplaning and have a good skid resistance, so at high speeds the wheel does not slip easily. It also reduces noise and glare at night.

On the other hand to the advantages over porous asphalt it also has some drawbacks. Porous asphalt has the lowest strength or stability compared to other asphalt mixtures due to porous asphalt compositions dominated by coarse aggregate, thus becoming fragile. However, it is an

[1] To whom any correspondence should be addressed.

environmentally friendly use of materials and methods that will not adversely affect the environment, such as pollution or depletion of natural resources [2]. In this study two types of porous asphalt are investigated, the first porous asphalt used natural gravel aggregate, while the second incorporated the volcanic gravel from residuals of Merapi sand. The intention of using gravel aggregate is to avoid the process of breaking a big stone so there is no air pollution from the process at a stone crusher. The use of volcanic gravel is expected to use the waste materials from residuals of Merapi sand and it is more environmentally friendly action.

Porous asphalt mixture using a uniform graded is expected to function as drainage, anti-slip, anti-aquaplaning and noise reduction that can only be obtained through the use of a uniform graded. It also acquired a high permeability and porosity and is able to pass the surface water sufficiently [3]. The types of tests performed on a bitumen and porous asphalt mixture used are penetration, softening point, marshall properties, porosity, tensile strength indirectly, unconfined compressive strength, and the permeability of the mixture.

The objective of this study is to determine the optimum bitumen content and the mixture properties of uniform graded porous asphalt mixture contain natural and volcanic gravel such as the stability, porosity, permeability, indirect tensile strength, and unconfined compressive strength of porous asphalt mixture, which are then compared to the conventional porous asphalt.

2. **Experimental**

Aggregates are the grains of crushed stone, gravel, sand or other minerals, either natural or artificial results (Implementation Guidelines of Asphalt Concrete, Highway Agency SKBI -2.4.26.1987) [4]. Asphalt or bitumen is a viscous liquid which is hydrocarbons, contains little sulfur, oxygen, and chlorine. Asphalt as a binder in flexible pavement has viscoelastic property. Asphalt will be solid at room temperature and becomes liquid when heated. Filler is a non plastic aggregate that passes the sieve number of 200. Filler support the coarse aggregate together with fine aggregate and binder to form the skeleton of asphalt mixtures. The properties of filler depend on several factors such as grain size, grain shape and density [5]. The method used in this study is an experimental method, which is carried out by conducting trial activities to obtain the data.

(a) (b)

Figure 1. (a). Natural Gravel (b). Volcanic Gravel

The first purpose of this study was to determine the optimum bitumen content and to measure the stability, porosity, vertical and horizontal permeability, UCS and ITS of porous asphalt which use natural gravel and volcanic gravel. The two gravel types of aggregate are observed in this research as presented in Figure 1. The preparation phase, at this point all the materials and equipments needed in the research study are prepared beforehand in order to run the experiment smoothly. The materials that need to be prepared is the natural gravel and the volcanic gravel, asphalt penetration 60/70 and the stone dust filler. The identification of the aggregates, includes the Los Angeles abrasion and the specific gravity of coarse and fine aggregate. The fifteen specimen were then produced at five variations of bitumen content, three specimens for each similar bitumen content to obtain the optimum bitumen content. The total 30 specimens were produced and observed for natural gravel and volcanic gravel. Marshall Test was then performed to determine the optimum bitumen content. The specimens at their optimum bitumen contents were then objected for the vertical and horizontal permeability tests, Indirect Tensile Strength tests and Unconfined Compressive Strength tests. The data obtained from the test results are analyzed to get the conclusion according to the purpose of the research.

3. Results and Discussion

3.1 Aggregate Properties

The examination of aggregates used was conducted in the Highway Materials Laboratory of Sebelas Maret University, Surakarta according to NFPA 03-1969 - 1990 and SNI 03 - 2417-1991[6]. The results are presented in Table 1.

Table 1. Results of the aggregate examinations

The properties	Unit	Natural gravel	Volcanic Gravel	Specification
Adsorption	%	3.067	3.600	Max 3
Bulk Specific Gravity	g/cc	2.280	2.208	Min 25
SSD Specific Gravity	g/cc	2.350	2.287	Min 2.5
Apparent Specific Gravity	g/cc	2.451	2.398	-
Abrasion	%	26.18	43.96	Max 40
Aggregate Affinity	%	96	97	> 95

The filler has been examined in the laboratory and give the value of specific gravity of 2.44 g/cc.

3.2 Bitumen Properties

The bitumen used in this study is 60/70 penetration bitumen. The bitumen examination was conducted based on SNI 03-1737-1989 [7]. The results can be seen in Table 2.

Table 2. Examination of the characteristics of bitumen 60/70 penetration

Bitumen properties	Specification		Value
	Min	Max	
Penetration, 100gr, 250°C, 5 second	60	79	67.5
Softening Point, °C	48	58	48.5
Flash Point, °C	200	-	343°C
Burning Point, °C	200	-	353°C
Ductility, 25°C, 5 cm/minute	100	-	150 cm
Specific Gravity	1	-	1.0395 g/cc
Affinity, %	95	98	97.5%

3.3 The Optimum Bitumen Content

In the porous asphalt made from natural gravel (cobblestone) that comes from a Kaliboto quarry of Karanganyar district obtained the optimum bitumen content of 2.95%. The porous asphalt made from volcanic gravel obtained the optimum bitumen content of 4.65%. The difference bitumen content between the natural gravel and volcanic gravel is predicted due to the texture and the adsorption of the gravel. Natural gravel has a smooth surface texture and does not absorb much asphalt since the absorption is only 3.07% so it takes a little bitumen content. The volcanic gravel has a rougher texture and asphalt absorbs a lot of absorption of 3.60% and thus require higher bitumen content.

3.4 The Marshall Properties

Stability is the ability of asphalt mixtures to withstand deformation due to the traffic load without permanent deformation. The stability value is influenced by aggregate interlock and the bond of bitumen and aggregates. The stability value of porous asphalt mixture using gravel aggregate amounted to 481.80 kg. Porous asphalt using gravel aggregate amounted to 643.98 kg. For the value stability of porous asphalt mixture using gravel and gravel aggregate uniformly graded greater than either gradation BVR (Blackwater Valley Route) on research Suwarno with a value of 434.24 kg [8].

Flow is a state change in the form of a mixture as a result of a load to the limit of collapse. Value flow indicates the level of flexibility or stiffness of the mix. Flow demonstrated high level of flexibility is high, so that cracks arising from loading can be avoided. Conversely a low flow showed a low level of flexibility and are brittle layer, so prone to rupture due to the separation between grain particles. Comparison of flow porous asphalt mixture using gravel and pebbles can be seen in Table 3. Based on Table 3, the rated flow of porous asphalt with natural gravel obtained values of 2.24 mm. While the porous asphalt with volcanic gravel obtained values of 2.51 mm. This is due to the asphalt content of porous asphalt mixture using natural gravel more than those of volcanic gravel, so that the porous asphalt mix becomes more elastic and able to follow the deformation due to the load. From research conducted by Suwarno using the BVR (Blackwater Valley Route) gradation obtained the flow value of 3.94 mm, greater than using a uniform graded porous asphalt [9].

Table 3. Marshall and volumetric properties at optimum bitumen content

Marshall and volumetric properties	Natural Gravel	Volcanic Gravel	BVR Gradation
Stability(Kg)	481.80	643.98	434.24
Flow (mm)	2.24	2.51	3.94
MQ (Kg/mm)	215.98	265.07	n/a
Specific Gravity	2.36	2.26	n/a
Density (g/cc)	1.73	1.75	1.77
Porosity (%)	26.63	22.88	30.91

3.5 The Volumetric Properties

Density is the ratio between the weight by volume based. Comparison of density of porous asphalt mixture using natural and volcanic gravel are presented in Table 3. It could be seen that density value obtained for porous asphalt with natural gravel is 1.73g/cm^3, for porous asphalt with volcanic gravel is 1.75g/cm^3. While the study in 2006 using the gradation BVR (Blackwater Valley Route) Agustina obtained a value of 1.77g/cm^3. This is due to the ratio between the weight and volume of crushed stone is greater than the gravel aggregates, than the gradation also influences the density [9].

Porosity is the percentage of pores or air voids that are present in a mixture and is the main indicator in a mixture of porous asphalt as a pore - pore is what will be the course of the water. Comparison of porosity between the natural and volcanic gravel are presented in Table 3. It could be seen that the porosity of porous asphalt mixture using natural gravel is 26.52% as compared with the use of porous asphalt with volcanic gravel is only 22.31%. From research by Agustina using the gradation of BVR (Blackwater Valley Route) derived porosity values of 30.91%, higher than those of uniform gradation with gravel types of aggregates. This is because the porosity values obtained from the ratio of the density and SG_{mix}. The higher SG_{mix}, with equal density obtained a smaller porosity.

3.6 The Permeability Properties

The purpose of the permeability test is to show the ability of porous asphalt mixture to absorb and drain water. The two gravel types were used to obtain vertical and horizontal permeability. The permeability of porous asphalt using gravel aggregate vertically is 1.25cm/sec and horizontally is 0.92cm/sec, while those using volcanic gravel aggregate values obtained vertical permeability of 1.43 cm/second and horizontal 1.10cm/second. A higher value than those using BVR gradation on previous research by Agustina which has a vertical permeability value of 0.40cm/second and horizontal 0.43 cm/second [7]. Data presented in Table 4 show that the use of porous asphalt mix gravel aggregate has better permeability that is vertical by 1.43cm/second and horizontal amounted to 1.10cm/second.

Porous asphalt mixture using gravel aggregate is vertical at 1.25cm/second and horizontal of 0.92cm/second. Research by Agustina using BVR gradation obtained value of vertical permeability of 0.40cm/second and horizontal permeability of 0.43cm/second, are smaller than using a uniform graded aggregate [4]. This is due to pores to pore connectivity of porous asphalt mixture using natural gravel and volcanic gravel is well interconnected, so that the water can drain properly. Table 4 shows the resume of permeability test results.

Figure 5. The sample for vertical and horizontal permeability Test

Table 4. The Permeability, ITS and UCS Properties

Permeability, ITS and UCS Properties	Natural Gravel	Volcanic Gravel	BVR Gradation
Coefficient of Vertical Permeability (cm/second)	1.25	1.43	0.4
Coefficient of Horizontal Permeability (cm/second)	0.92	1.41	0.43
Indirect Tensile Strength (KPa)	59.49	78.84	394.96
Unconfined Compressive Strength (KPa)	1070.63	1120.69	2310.22

3.7 The Indirect Tensile Strength (ITS)

ITS testing to determine the tensile strength of porous asphalt mixture. Tensile force can be used to evaluate the potential occurrence of cracks (fatigue) on porous asphalt mixture. Tensile strength values of porous asphalt mixture using gravel aggregate amounted to be 59.49 kPa, while those using gravel aggregates amounted to be 78.84 kPa. Values of ITS for uniform graded porous asphalt mixture are lesser values than those using BVR gradation, the results of previous research by Widianto obtained a value of 394.96kPa ITS [8]. Table 4 resumes the results from ITS tests.

Figure 7. The setting experiment of Indirect Tensile Strengh Test

3.8 The Unconfined Compressive Strength (UCS)

Compressive strength testing aims to determine ability to withstand the traffic load of pavement layers vertically. Compressive strength can be used as a direct indication to determine how much load is handled by pavement. The UCS value of natural gravel porous asphalt amounted to 1070.63kPa, while those using volcanic gravel aggregates amounted to 1120.69kPa. UCS values of uniform graded porous asphalt mixture are lesser than those using BVR gradation, the results of previous research by Agustina obtained a value of 394.96kPa ITS as presented in Table 4.

4. Conclusion

The adsorption of aggregate and the surface texture influenced the optimum bitumen content of the mixture. The optimum bitumen content values of natural gravel porous asphalt are lower than volcanic gravel porous asphalt.

The volumetric properties: density, porosity and permeability of uniform graded gravel porous asphalt are greater than those of well graded crush stone aggregate porous asphalt, its impact on the ability of porous asphalt to drain the surface water easily. However, the strength properties: stability, tensile strength and compressive strength of uniform graded gravel porous asphalt are inferior compared to those of well graded conventional crushed stone aggregate porous asphalt.

The incorporation of uniform graded gravel porous asphalt does not meet the specifications of pavement on the road because it has low strength properties, however, it can be used as pavement for the low traffic road.

References

[1] Van Heystraeten, G. and Moraux, C., 1990. *Transportation Research Record*, (1265).
[2] Ruiz, A., Alberola, R., Perez, F. and Sanchez, B., 1990. *Transportation Research Record*, (1265).
[3] Dreelin, E. A., Fowler, L., & Carroll, C. R., 2006. *Water Research*, ***40*(4)**, 799-805.
[4] Indonesian Highway Agency, 1987, The Implementation Guidelines of Asphalt Concrete, Highway Agency SKBI Number 2.4.26-1987. Indonesia
[5] Colwill, D. M., Bowskill, G. J., Nicholls, J. C., & Daines, M. E. 1994. P. *Transportation Research Record*, (**1427**), 13-21.
[6] Indonesian National Standard, 1991, SNI 03 - 2417-1991, Indonesia
[7] Indonesian National Standard, 1989, SNI 03-1737-1989, Indonesia
[8] Widhianto, B., Setyawan, A. and Sarwono, D., 2013. *Matriks Teknik Sipil*, ***1*(2)**.
[9] Widyastuti, S., Setyawan, A., & Sumarsono, A., 2013. *Matriks Teknik Sipil*, ***1*(2)**.

Effect of Filler and Heat Treatment on the Physical and Mechanical Properties of the Brazed Joint between Carbide Tip and Steel

Y Winardi[1], Triyono[2], and A T Wijayanta[2]
[1] Graduate School of Mechanical Engineering Sebelas Maret University, Surakarta 57126, Central Java, Indonesia
[2] Mechanical Engineering Departement Sebelas Maret University, Surakarta 57126, Central Java, Indonesia

E-mail: triyonomesin@uns.ac.id

Abstract. In this study, the effect of filler and heat treatment on the physical and mechanical properties of the brazed joint carbide tip and steel was investigated. Tip carbide YG6 and low carbon steel (SS400) is joining by torch brazing with two filler metals, silver, and copper filler. Heat treatment was performed in induction furnace. Microstructure and shear strength of the brazed joint have been investigated. Many silver filler layer are formed on the surface of the base metal rather then using copper filler. The highest shear strength is achieved using a silver filler metal at temperatur 725°C. The highest shear load is 18.62 kN.

1. Introduction

Cemented carbide are widely used in manufacturing industries, mining, oil and gas drilling. It consists of tungsten carbide (WC) is known for its strength and hardness, and the Co as the binder [1]. This material is commonly used as cutting tools, abrasive wheels, metal forming equipment, drill tools and others [2,3]. To obtain the high quality of tools at low production cost, cemented carbide is usually joined with steel. Consequently, the joining cemented carbide with steel gets a real attention in recent years [4,5,6,7,8,9]. Nevertheless, the joining two these materials are not easily achieved. This difficulty caused by differences in chemical composition, properties, and physical characteristics. Moreover, as the ceramic particles, grain of WC is difficult to wetted. Research shows that by welding will produce residual stresses in the joint area. This is caused by the difference in coefficient thermal expansion [10,11,12].

Joining cemented carbide with steel required a special process, including mechanical assembly, diffusion bonding, brazing, and soldering [13]. From this process, the brazing is potential technique and able to join dissimilar metal [14,15]. To get a good brazed, some elements required, such as the selection of filler metal, base metal characteristics, heat input/ brazing procedure, and treatment after brazing. The selection of filler metal was the foundational step for brazing. Filler metals must be compatible with the base metal, and brazing procedure to be used. Especially for the characteristics of filler metal, a good filler metal has a low melting point, and the ability to form a bond with the base metal [16]. In other words, to make a joint to the base metal, filler metal should be able to wet the base metal and produces a low contact angle. As the base metal capillary action good penetration will occur if the base metal has a high capillary action, so that a good bond will be obtained.

Silver and copper filler metal has been widely applied for joining cemented carbide to other materials with brazing. This filler metal has a low melting point. Some research the use of this filler has been carried. As carried by Criss and Meyers [17], this study has successfully joining the cobalt using silver-copper filler metal. The results of the observation showed that the weld cracks can be minimized, and do not exceed the limit thickness. Delamination basic filler does not cause a failure in the area 4-point bending. Jing et al [18] investigated vacuum brazing on the cermet and steel using CuAgTi filler metal. The result of the investigation suggested that layer of CuNiTi solid solution is difficult formed between cermet and filler metal at 870°C. Sun et al [19] researching about the effect of holding time on the microstructure and mechanical properties of ZrO2/TiAl joints brazed using Ag–Cu filler metal. Kaiwa et al [20] investigating effects additions of Ni and Co to filler metals on brazed Joints of cemented carbide and martensitic stainless steel using induction brazing. the result of the investigation suggested that the joint strength increased by adding the 2 mass% Ni and 0,5 mass % C into brazing alloy.

However, from the author's knowledge, there are no experimental data of the joining between cemented carbide and steel using torch brazing. In this study, Cemented carbides have been brazed with low carbon steel by torch brazing using two filler metal, silver, and copper filler. To improve the characteristics of joint, this study proposes the heat treatment. The effect of each element on the microstructure and mechanical properties of the brazed layer was observed and analyzed.

2. Experimental

Silver (Ag) and copper (Cu) filler metals are used in this study. The chemical composition of filler metals are shown in Table 1. Tip carbide YG6 with a dimension of 25x15x8,5 mm and low carbon steel (SS400) with a dimension of 40x15x10 mm were used as test specimens. The chemical compositions of test specimens are shown in Table 2. Before brazing, the contact surfaces of specimens were polished using silicon carbide abrasive paper (number 200 to 400). Figure 1 (a) shows the schematic representation of the apparatus and types of joints used for brazing. The joint clearance was kept at 0.1 mm using a thickness gauge of 0.1 mm in thickness. To keep the specimen remains constant, use tools such as jig (fixture), as shown in Figure 1(b).

Table 1. Filler metal composition

Metal	Ag	Cu	Ni	Ti	P	Other	Melting range (°C)
Ag	Main	40	0,1	0,3	-	-	800-879
Cu	Main	-	-	-	7,5	0,15	704-815

Table 2. The chemical composition of specimen

Metal	C	Si	Mn	P	S	Ni	Cr	Fe	WC	Co
SS400	0,22	0,59	0,59	0,017	0,02	-	-	Main	-	-
YG6	-	-	-	-	-	-	-	-	94	6

After brazed, the specimens were then heated at temperature 700°C, 725°C, and 750°C by induction furnace for 30 minutes. Then cooled to room temperature with a cooling rate of 5°C/s. All of the test specimens was cleaned and ground ± 1mm using a diamond blade grinder. After grinded, the test specimens are polished with metal polish. Microstructures of the brazed layer were observed using microscope micro. Shear tests were performed on a mechanical testing machine (Model 4106 SHT) with a travel speed of 0.5 mm/min at room temperature, as illustrated in Figure 2. The results of the test specimen fractures were also analyzed.

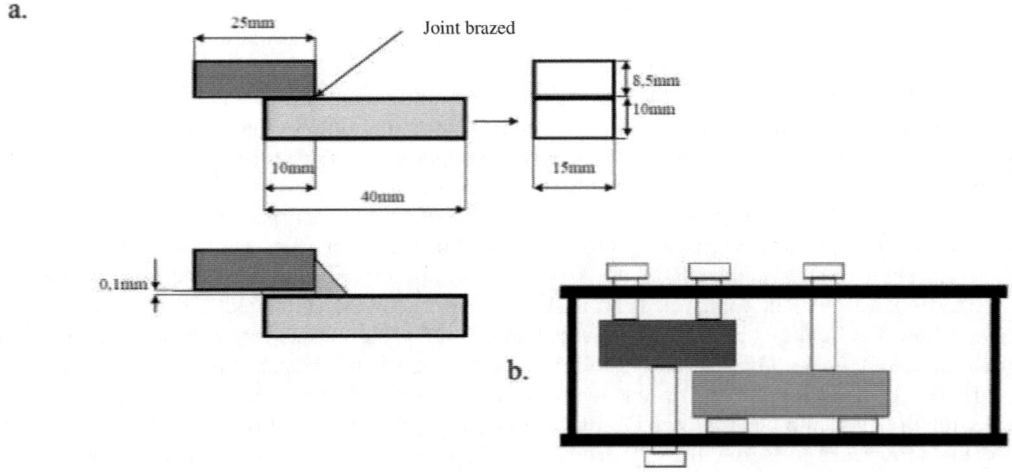

Figure 1. (a) Schematic and types of joints, and (b) Jig (fixture)

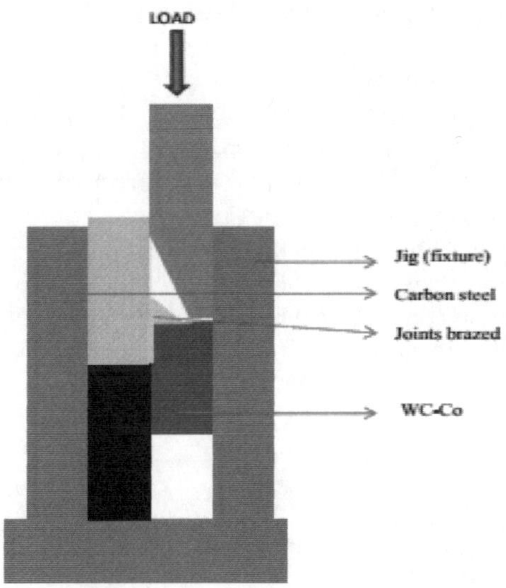

Figure 2. Schematic shear testing

3. Results and Discussion

Figure 3 shows the average shear strength of joints brazed with silver and copper filler. Which are at different temperature for 30 minutes. The test results indicate that the shear strength of joints brazed with silver filler is better than copper filler. Treatment temperature has a significant effect. The shear strength increased when the treatment temperature is increased from 700°C to 725°C, but decreases when the temperature up to 750°C. The maximum shear load of 18.62 kN at a temperature of 725°C using silver filler metal. This result is higher than the copper filler metal. While the joint with copper filler (Cu) and heated at 700 °C has the lowest shear load of 4.94 kN. Increasing the temperature can improve the fluidity of filler metals and promote the diffusion of elements to achieve a homogenous microstructure. Excessively high temperatures would induce the growth of filler metal grains and thus deteriorate the property. Moreover, at high temperature will allow the occurred of defects or void in the filler metal. Void caused by inability filler metal to remove air or gas during cooling process of joint. This void causes failure in the joint. However, the cooling process has a function against the brazed layers. The area of diffusion also gives the influence of the shear strength. Uzkut et al [21] stated to obtain high diffusion area, can be achieved with the slow cooling process after heating.

This fact relates to the results of fracture shear tests, as shown in figure 4(a) and 4(b) the result of the observation showed, that joint brazed with copper filler many defects and void. Moreover, liquid filler metal is not able to fill the gap. Consequently, a copper layer is not formed on the surface of the base metal. Observations shows, fracture mostly occurred on the surface of the base metal. It causes a low shear strength. Observations on the joint brazed with silver filler showed, silver filler metal layer is many formed on the surface of base metals. The liquid of the silver filler can flow meets the gap of base metal. These results demonstrate the fact that the silver filler has characteristics capable of wetting and good flow. The joint fractured at the Ag zone near the cemented carbide but not at the interface, show that field of Ag near on the cemented carbide is the weakest area of brazed joints [22].

Photo macro of the joint brazing at 725°C for 30 minutes after the shear test is shown in Figure 4(c) and 4(d). Wang et al expressed [23], that region of crack initiation is ductile fracture surface, which mixed with the field dimple rupture, During the cooling process, the different thermal expansion coefficients to produce the residual stress. The residual stress was dissipated by plastic deformation of Ag because it has good plasticity properties, which was good for the joint. In the shear test, Ag break along the direction of deformation, because it is not able to withstand high stress. This phenomenon is related to the formation of microstructure on the brazed layer. Figure 4 shows the microstructure of joint brazed with silver filler metals and copper in the various temperatures. When of the joint brazed with silver filler, homogeneous microstructure and there are no cracks along the line on the surface of the base metal. Silver filler metal layer is many formed on the surface of base metals, as shown in figure 4(c) and 4(d), a metallurgical bond is formed along the surface of the base metal. These results indicate that the filler metal have good wetting properties and able to fill the gap. The best condition occurred at a temperature of 725°C.

The results of the observation showed that in the solid solution filler metal is many forming islands. These islands are a solid solution phase of the composition of the filler metal. This solid solution phase will give a strengthening effect on metallurgical bond. Figure 5(c) and 5(d) show the microstructure of joint brazed with copper filler metal. Along the surface of the base metal is not formed the metallurgical bond, this will reduce the joint strength.

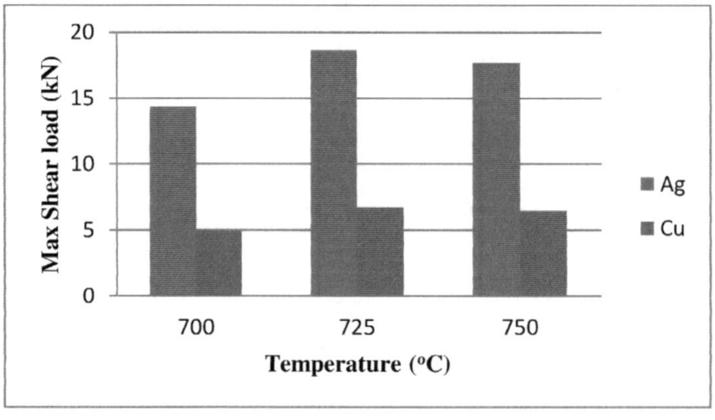

Figure 3. Shear strength of the joints brazed at different temperatures for 30 minute.

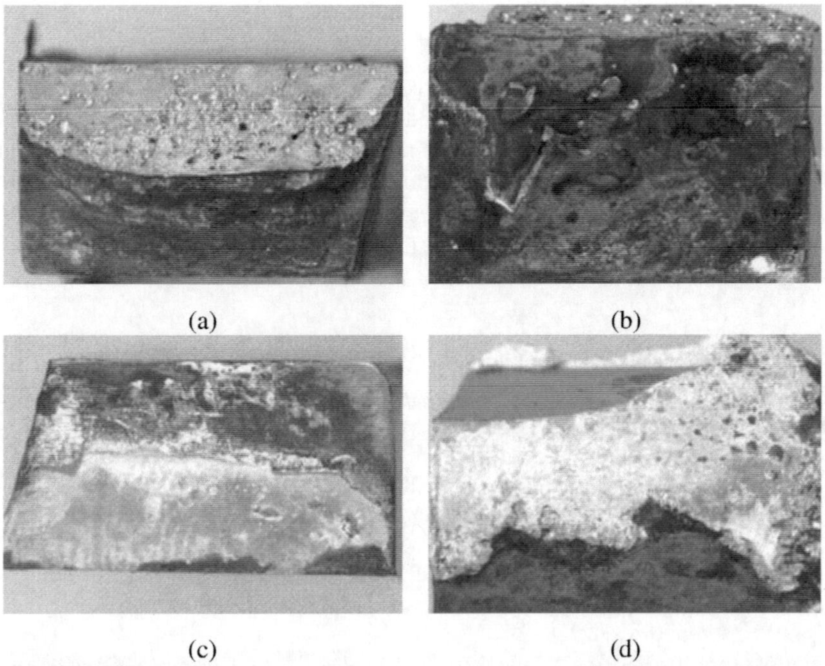

Figure 4. Fracture test specimens, (a) and (b) using copper filler metal, (c) and (d) using silver filler metal.

Figure 5. Microstructure of the joint brazed (a).700°C, (b).725°C, (c).750°C, and (d).725°C.

4. Conclusions

The aim of the present work is to investigate the effect of filler and heat treatment on the physical and mechanical properties of the brazed joint between carbide tip and steel under torch brazing. Many silver filler layers are formed on the surface of the base metal to reinforce the joint. In other hand, almost copper filler layers are not formed on the surface of the base metal which can reduce the strength of the joint. The shear strength increased when the treatment temperature is increased from 700°C to 725°C, but decreases when the temperature up to 750°C. The maximum shear load of 18.62 kN at a temperature of 725°C using silver filler metal. While the joint with copper filler (Cu) and heated at 700 °C has the lowest shear load of 4.94 kN.

References

[1] Upadhyaya G 2001 *Mater Des.* **22** 483–489
[2] Prakash L 2014 *Fundamentals and General Applications of Hardmetals.* Elsevier Ltd **1** 29-89
[3] Sheikh S, M'Saoubi R, Flasar P, Schwind M, Persson T, Yang J, and Llanes L 2015 *Int. J. Refract. Met. Hard Mater.* **49** 153–160
[4] Barbatti C, Garcia J, Liedl G, and Pyzalla A 2007 *Materwiss. Werksttech.* **38** 907–914
[5] Chen H, Feng K, Wei S, Xiong J, Guo Z, and Wang H 2012 *Int. J. Refract. Met. Hard Mater.* **33** 70–74
[6] Guo Y, Gao B, Liu G, Zhou T, and Qiao G 2015 *Int. J. Refract. Met. Hard Mater.* **51** 250–257
[7] Lee W B, Kwon B D, and Jung S B 2004 *Mater. Sci. Technol.* **20** 1474–78
[8] Xu P Q 2011 *Mater. Des.* **32** 229–237
[9] Tillmann W, Elrefaey A, and Wojarski L 2013 *14 - Brazing of cutting materials.* Woodhead Publishing Limited
[10] Al-Samhan A M 2012 *J. King Saud Univ. - Eng. Sci.* **24** 85–94
[11] Kayser W, Bezold A, and Broeckmann C 2015 *Int. J. Refract. Met. Hard Mater*
[12] Krawitz A and Drake E 2015 *Int. J. Refract. Met. Hard Mater.* **49** 27–35
[13] Miranda R M 2014 *Joining Cemented Carbides.* Elsevier Ltd **1** 527-538
[14] Martens L , Tillrnann W, Lugscheider E, and Ziegler G 2000 *J. Mater. Process. Technol.* **58** 1996
[15] Johari Miab R and Hadian A M 2014 *Ceram. Int.* **40** 8519–24
[16] Schwartz M M 2003 *Brazing - 2nd Ed* (USA:ASM International)
[17] Criss E M and Meyers M A 2015 *J. Mater. Res. Technol.* **4** 44–59
[18] Jing Y, Yang Q, Xiong W, Huang B, Li B, and Zhang M 2016 *J. Alloys Compd.* **682** 525–530
[19] Sun Y, Zhang J, Fan GH, He YM, and Shibayanagi T 2011 *J. Mater. Sci.* **46** 5830–37
[20] Kaiwa K, Yaoita S, Sasaki T, and Watanabe T 2014 *Adv. Mater. Res.* **922** 322–327
[21] Uzkut M, Köksal N S, and Unlu B S 2005 *J. Mater. Process. Technol.* **169** 409–413
[22] Jiang C, Chen H, Wang Q, and Li Y 2016 *J. Mater. Process. Technol.* **229** 562–569
[23] Wang H, Chiu L, and Chang H 2008 *Mater. Soc. Annu. Meet.* **50** 596–599

Design and Characterization of Renewable Bioasphalt Containing Damar Resin, Fly Ash, Wasted Cooking Oil and Latex

A Setyawan[1], Djumari, S J Legowo, B Widiharjo, A K S Zai, A A W Pradana, I P Rusadi, A Permana
Roadmate Research Group, Civil Engineering Department, Sebelas Maret University, Jl. Ir. Sutami 36 A Solo 57126 Indonesia

E-mail: arysetyawan@staff.uns.ac.id

Abstract. Dasphalt is one alternative of bioasphalt, made from materials that can be renewed as a substitute for conventional asphalt. Dasphalt inspired from *jabung* made of damar resin, brick powder and wasted cooking oil. Jabung have the same character with conventional asphalt. Research has been conducted by the characteristics of jabung but there are still many shortcomings, softening point and ductility values are not qualify. In this research the brick powder will be replaced by fly ash, as fly ash has a finer grain so that it can become a better absorbent. The resin will act as a natural resin for dasphalt, wasted cooking oil will be a mixed solvent. Use of additional polymers latex, is expected to improve the elasticity of dasphalt in ductility test. The purpose of this study was to determine the nature of the modification dasphalt properties in accordance with the specifications of asphalt penetration test and find the optimal composition of dasphalt. This research method is done by direct testing in the laboratory. In the present study that became the basic composition of the resin is resin (100g pure resin+ 350g resin packaging or powder), fly ash (150g) and wasted cooking oil (205g) and latex were mixed at temperatures below 150°C. While variations of latex starting from 0%, 2%, 4%, 6%, 8% and 10%. Several asphalt characterization are performed include penetration tests, test softening point, ductility test, flash point test, specific gravity test, affinity test and solubility test. Dasphalt modification achieved optimum composition of resin (100g pure resin or resin chunk + 350Gr packaging), Fly Ash powder (150g), cooking oil (205g), and latex 4%, ductility increased from 63.5 cm to 119.5 cm , the value of the flash point was originally at temperature of 240°C to 260°C, dasphalt penetration from 68.2 dmm to 43 dmm, and the value of density decreases to 1.01 g/cm^3 to 0.99 g/cm^3. Dasphalt modifications meet some of the specifications and could be categorized as bitumen penetration 40.

1. Introduction

Demand for asphalt as a pavement construction, whether for maintenance, rehabilitation or construction of road transport continues to increase along with the growth of the infrastructure construction. Asphalt required as materials for road pavement is very huge. Concerns arise as the depletion of oil reserves that are available today. As a result of oil prices tend to be more expensive. Thus requiring other alternative materials as a substitute for conventional asphalt such as asphalt derived from biomass or often called as bioasphalt.

[1] To whom any correspondence should be addressed.

One alternative that was found as a basis in the previous research was jabung. Jabung or dasphalt (damar asphalt) used by silversmith in Yogyakarta as a cushion to carve out a silver ring. Composition of jabung or dasphalt used is damar resin, brick powder and cooking oil. Visually dasphalt resembling asphalt pavement in general are dark in color, elastic, having adhesion but has the advantage that is more water resistant.

Previous research by M F Nasution et al used the best composition of dasphalt namely damar 450g, brick powder 150g and wasted cooking oil 170g [1]. However, the results showed that the softening point and ductility test does not meet the requirements of bitumen penetration 60. To follow up on it, then further research on dasphalt by changing the composition of the mixture into a damar resin [2], fly ash, oil and latex. Changes in the use of brick powder material into fly ash powder is constituted by a shortage of bricks in terms of uniformity and grain size. Fly ash has smaller grains and more uniform than brick powder, so that fly ash has a surface area greater than brick powder. The surface area of the larger and more uniform size will do better binding dirt so it will purify a dasphalt mixture. The use of fly ash as an ingredient of dasphalt will be one of the waste management solutions considering this material is a hazardous waste if not handled properly. Fly ash also have a low economic value and readily available in large quantities [3].

The addition of a polymer latex is expected to improve the shortcomings of previous dasphalt composition in terms of softening point and ductility tests. As a polymer, latex is expexted to improve stiffness and elasticity, so dasphalt will prform better resistance to deformation and rutting [4].

2. Experimental

2.1 Materials

2.1.1 Asphalt

Asphalt is a material that is a solid at room temperature until slightly dense, and are thermoplastic. Thus, the asphalt will melt when heated to a certain temperature, and re-freezes when the temperature drops. Together with aggregates, asphalt is a mixture of pavement forming material.

2.1.2 Dasphalt

Dasphalt (Damar Asphalt) is a mixture of four components include a damar resin, fly ash and wasted cooking oil with additional latex. Dasphalt included in bioasphalt due to renewable materials.

2.1.3 Damar Resin

Damar is a natural resin produced by plants Dipterocarpaceae (genus Shorea, Hopea, Balanocarpus and Vateria) and Burseraceae (clan canarium) based on several references [2].

Asphalt composition comprised of asphaltenes which is material in black or dark brown and maltenes which is a viscous liquid consisting of resins and oils. Damar is a major component in dasphalt as a natural resin, has a hardness depending on the temperature of the so-called thermoplastic properties. The function is as a resin binder of dasphalt itself. Many say the least influential resin mixture. The resin used is a resin that has been in the packaging in the form of powder [3].

2.1.4 Fly Ash

Fly ash (fly ash) is one of the residues generated in combustion and composed of fine particles. Abu who do not ride called bottom ash. In the industrial world, fly ash usually refers to the ash produced during the combustion of coal.

According to ACI Committee 226 explained that, fly ash has a fairly fine grain, which passes a sieve No. 325 (45 milli micron) 5-27%, with the specific gravity between 2.15 to 2.8 and blackish gray. Granules fine fly ash which has a large surface area so that it will be better in terms of the binding of (absorbent) impurities contained in dasphalt than brick powder used in the composition dasphalt earlier. Coal fly ash has been used as adsorbent in the preliminary parameters of pollutants in wastewater [2].

2.1.5 Wasted Cooking oil

Low quality edible oils in this research is the cooking oil that is not branded, without any *Halal* and foodgrade certification, expiration date, and other important informations. Cooking oil demand because they were cheaper and easily found in traditional markets.

Based on the information from sources that quoted from Baihaqi [3], cooking oil that serves as a controller of an elasticity of dasphalt. More or less will affect the rate of hardening of dasphalt itself. In addition it also functions as a diluent oil and solvent during the production of dasphalt.

In general, the physical properties of damar among other fragile and easily attached to the hand at room temperature, soluble in essential oils and non polar organic solvents, slightly soluble in organic solvents, insoluble in water, can not stand the heat, flammable, not volatile when not decompose and can change color if kept too long in a closed place without good air circulation [4].

2.1.6 Latex

Latex is a term used to refer issued by the sap of rubber trees. Latex contained in the bark, leaves and seed integument rubber. Latex has rubbery properties. The chewy nature related to the viscosity or plasticity of rubber. The addition of the polymer latex to the composition as an elastomer or Plastomer dasphalt to add rigidity and elasticity dasphalt. Disadvantages dasphalt composition previously located at the lack of ductility under 100 cm as a condition for penetration bitumen, so the addition of latex is needed to increase the ductility of daspahlt.

The addition of natural latex is expected to improve the resistance to asphalt cracking after experiencing long-term aging. This is because the asphalt becomes more elastic and their termination polymer molecule chains of natural rubber during the short-term aging process that reduces the brittleness of asphalt [5].

2.2 Research Method

The method used is an experimental method to the test object dasphalt with a mixture of damar resin, fly ash, and wasted cooking oil, which is used as a binder layer of asphalt concrete (laston) compared to asphalt penetration. Testing standards used are:

1.SNI 06-2456-1991 (Penetration Test).
2.SNI 06-2434-1991 (Softening Point Test).
3.SNI 06-2433-1991(Flash Point Test).
4.SNI 06-2441-1991 (Specific Gravity Test).
5.SNI 06-2432-1991 (Ductility Test).
6.PA-0312-76 (KVBB-V-19) (Affinity Test)
7.SNI 06-2438-1991 (Solubility Test)

3. Results and Discussion

3.1 Materials composition

The composition of materials to be blended for producing dasphalt is prsesented in Table 1. These composition are modified from the previous research conducted by Muhammad Fachri [1].

Table 1. Modified Dasphalt Compotitions

Code	Material			
	Damar resin (g) (powder +solid)	Fly ash (g)	Wasted cooking oil (r)	Latex (%)
S1	350+100	150	205	0
S2	350+100	150	205	2
S3	350+100	150	205	4
S4	350+100	150	205	6
S5	350+100	150	205	8
S6	350+100	150	205	10

The various composition were then subjected into bitumen tests characterization included penetration test, softening point test, ductility test, flash point test, specific gravity test, solubility and afinity test. The results are sumerized in Table 2.

Table 2. Summary of Testing Characteristics of Damar Asphalt Modified Compared with Asphalt Penetration Specifications

Test	Unit	Asphalt Specification			Results of test					
		Pen 40	Pen 60	Pen 80	S1	S2	S3	S4	S5	S6
Penetration test	10^{-1} mm	40 - 59	60 - 79	80 - 99	68	62	43	36	31	29
Softening point test	°C	51 - 63	40 - 58	46 - 54	55.5	55.5	57.5	65.5	68.5	69.5
Ductility test	Cm	100	100	100	63.5	76.5	119.5	78.5	11	6.5
Flash point test	°C	200	200	225	240	258	260	250	250	245
Specific gravity test	g/mm	1	1	1		0.99	0.98	0.97	0.97	0.96
Solubility test	%	99	99	99				95		
Affinity test	%	99	99	99				99		

Asphalt composition comprised of asphaltenes which is material in black or dark brown and maltenes which is a viscous liquid consisting of resins and oils [1]. Therefore, damar resin plays an important role in making dasphalt mixture. The function is as a resin binder of dasphalt itself.

Fly ash that serves as filler and as absorden dasphalt work by removing impurities contained in dasphalt mixture during cooking so that the results obtained mixture dasphalt purer. Fly ash is not completely lost during the cooking process but still contained in the composition dasphalt evident when solubility testing in solution trichlore ethylene were conducted, the results of the test show that dasphalt only 95% is dissolved. The 5% is fly ash which is a non-hydrocarbons do not dissolve in solution trichlore ethylene.

Wasted cooking oil as diluent of asphalt mixture also plays a role in reducing the elasticity dasphalt plastic traits possessed by the resin.

3.2 *Latex content determination*

The polymer of latex is added into the ingredient to achieve the requirement of the ductility test. Figure 1 present the role of latex into the damar asphalt penetration. The effect of latex addition into the ingredient decrease the penetration of damar asphalt.

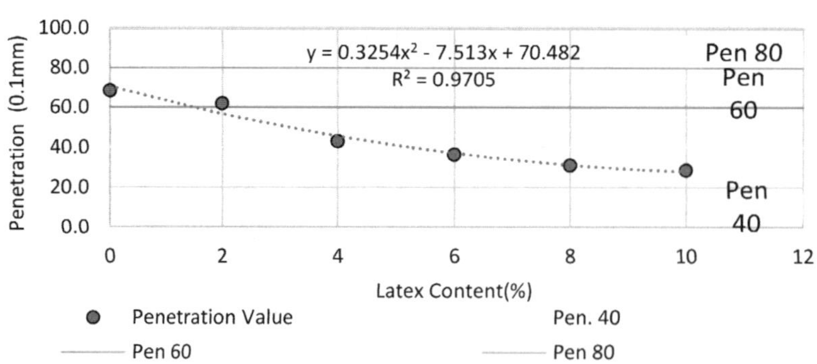

Figure 1. The effect of latex content on the penetration of dasphalt

Figure 2 shows the effect of latex addition into the softening point of damar asphalt. The addition of latex increase the softening point of the ingredient.

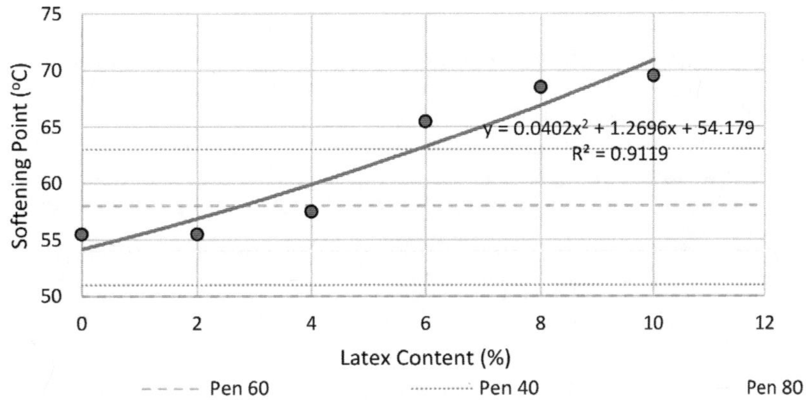

Figure 2. The effect of latex content on the softening point of dasphalt

The main objective of latex addition is to increase the ductility test, which is the shortcoming of pure damar asphalt, it could be seen in Figure 3 that by adding a latex content of 4% the ductility of dasphalt could be more than requirements of 100cm.

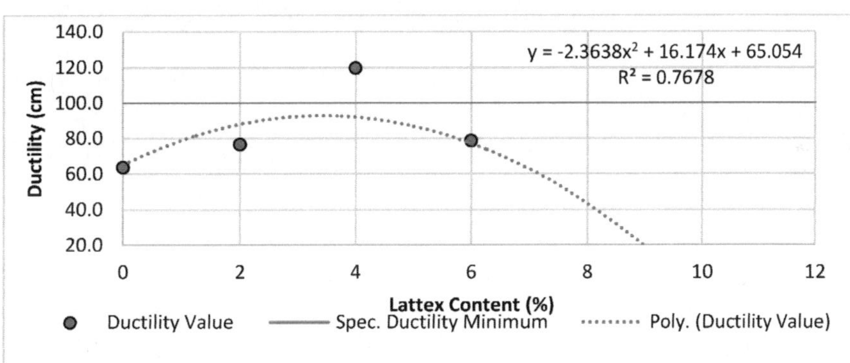

Figure 3. The effect of latex contents on the ductility of dasphalt

As expected, the addition of 4% polymer latex could improve the softening point and ductility tests results by increasing the value of softening point and ductility to full fill the requirements of 40 penetration with the sufficient value of 55°C and 119cm, respectively.

4. Conclusion

Based on the results of several investigations that have been carried out on damar asphalt composition that meets the requirements of penetration bitumen pen 60/70 is dasphalt composition with damar resin (damar resin powder 350g + solid damar 100g), fly ash 0%, wasted cooking oil 205g with latex content of 0% and 2%, but for ductility requirements of both the composition is not qualified.

The most optimal composition obtained in this test is the composition of damar resin (damar resin powder 350g + solid damar 100g), fly ash 0%, wasted cooking oil 205g with latex content of 4% with a penetration value of 43, the value of the softening point of 57.5°C, the value a flash point of 260°C, the value of the burning point 275°C, the ductility 119.5cm and an affinity of 99%. The composition does not meet the requirements for specific gravity of 0.97 and a solubility of only 95%, but due to the value of penetration of just 43pen, the most optimal composition is more similar to asphalt 40 penetration.

References

[1] Nasution, M. F., Setyawan, A., & Sumarsono, A. 2015 *Matriks Teknik Sipil* ***3*****(4)**.
[2] Gusti, R. E. P., & Zulnely, Z. 2014 *Jurnal Penelitian Hasil Hutan* ***32*****(3)** 167-174.
[3] Baihaqi, A., Setyawan, A., & Djumari, D. 2015 *Matriks Teknik Sipil* ***3*****(4)**.
[4] Afrianita, R., Fitria, D., & Sari, P. R. 2010 *Teknika Unand*.
[5] Setianingsih, N.1992 *Repository*.ipb.ac.id
[6] Emha, F. A. 2016 *eprints.uns.ac.id*

IOP Conf. Series: Materials Science and Engineering **176** (2017) 012028 doi:10.1088/1757-899X/176/1/012028

Design and Properties of Asphalt Concrete Mixtures Using Renewable Bioasphalt Binder

A Setyawan[1], Djumari, P A Irfansyah, A M Shidiq , I S Wibisono, M N Fauzy , F N Hadi
Roadmate Research Group, Civil Engineering Department, Sebelas Maret University, Jl. Ir. Sutami 36 A Solo 57126, Indonesia

E-mail: arysetyawan@staff.uns.ac.id

Abstract. The needs of petroleum asphalt as materials for pavement is very large, while the petroleum classified as natural resources that cannot be renewable. As a result of petroleum dwindling and prices tend to be more expensive. So that requiring other alternative materials as a substitute for conventional asphalt derived from biomass or often called bioasphalt. This study aims to know the volumetric and Marshall characteristics on Asphalt Cement (AC) using the *Damar* asphalt modification to substitute 60/70 penetration asphalt as a binder. The volumetric and Marshall characteristic are porosity, density, flow, stability, and Marshall quotient. The characteristic of asphalt concrete at optimum bitumen content are compared to the conditions from highway agency 1987 and the general specification of asphalt concrete Bina Marga 2010 the third revision. The research uses experimental method in the laboratory with the samples made using the dasphalt modification as binder and incorporating the aggregate gradation no. VII SNI 03-1737-1989. The research is using 15 samples divided into 5 contents of damar asphalt, they are 5%, 5,5%, 6%, 6,5%, dan 7%. Tests carried out using Marshall test equipment to get the value of flow and stability and then be searched the value of optimum damar asphalt content. The result of asphalt concrete analysis using dasphalt modification as binder gives the value of optimum dasphalt content at 5,242%. The most characteristics already met the requirements and specifications.

1. Introduction

Demand for asphalt as pavement construction, whether for maintenance, improvement, or development of accessibility of road transport continues to increase along with the growth of the construction. The needs of asphalt as materials for road pavement are very huge. Concerns rise with the increasing price of petroleum in this decade. Thus requiring other alternative materials as a substitute for conventional asphalt, there is asphalt derived from biomass or often called bioasphalt. Previous studies have been conducted on a substitute for asphalt modification of dasphalt in terms of the nature of the properties with a mixture of resin, oil, latex, and fly-ash [1,2]. Through the results of the examination conducted at the Highway Laboratory, Sebelas Maret University, the data generated are demanded to carry out for further research on the application of bioasphalt into asphalt mixture, such as asphalt concrete which is the most applied in Indonesian pavement. Asphalt concrete is a mixture of coarse aggregate, fine aggregate and filler with a binder in high temperature conditions with the composition examined and regulated by the technical specifications. Hot mix asphalt

[1] To whom any correspondence should be addressed.

pavement is a mixture of bending comprising a coarse aggregate, fine aggregate, filler, and binder with a specific composition and mixed in hot conditions [4].

2. Experimental

2.1 Materials

2.1.1 Aggregates for Asphalt Concrete

Examination of the coarse and fine aggregate is guided by the Indonesian National Standard (SNI) 1969: 2008 (Testing Methods Density and Water Absorption Coarse Aggregates) [3] and Indonesian National Standard (SNI) 1970: 2008 (Testing Methods Density and Water Absorption Fine Aggregate) [4]. The test results of the coarse aggregate and fine aggregate are presented in Table 1 and Table 2, respectively.

Table 1. The properties of coarse aggregates

Properties	Unit	Test 1	Test 2	Average
Dry weight	g	1488.7	1479.6	1484.15
SSD	g	1507.7	1515.5	1511.6
Weight in water	g	951.3	944.7	948
Bulk Density	g/cm^3	2.676	2.592	2.634
SSD Density	g/cm^3	2.710	2.655	2.682
Apparent Density	g/cm^3	2.770	2.766	2.768
Water adsorption	%	1.276	2.426	1.851

Table 2. The properties of fine aggregates

Properties	Unit	Test 1	Test 2	Average
SSD	g	500	500	500
Dry weight	g	487.7	488.4	487.95
Weight of picnometer	g	738	729.4	733.7
Weight of picnometer+ water+sample	g	1048.5	1042.8	1045.65
Bulk Density	g/cm^3	2.574	2.616	2.595
SSD Density	g/cm^3	2.639	2.680	2.659
Apparent Density	g/cm^3	2.52	2.793	2.773
Water adsorption	%	2.522	2.417	2.470

2.1.2 Bioasphalt Binder

Bioasphalt binder for asphalt mixture pavement is made from materials that are sustainable as an alternative to conventional asphalt pavement layer. In this study the bioasphalt binder made from Damar resin combined with a latex polymer, waste cooked oil, and fly-ash as a filter.

Damar is a natural resin produced by plants Dipterocarpaceae (genus Shorea, Hopea, Balanocarpus and Vateria) and Burseraceae (clan canarium) [5]. Damar also contain various compounds of the chemical element that has an important role in its function. Types of compounds found in damar are like resin acid content, complex alcohols and balm. Damar has the form of a hard, clear as plastic, easily melted and sticky.

Latex is a viscous sap, often milky, produced lots of herbs and freeze when exposed to air. Natural latex will naturally clumping or becoming solid within a few hours after collection. Natural or spontaneous clots can be caused by the onset of acids due to the disintegration of the material instead of rubber contained in the latex due to the activity of microorganisms. It was also the reason why lump natural clotting results stink. Clots terms lump has not littered with shavings tapping, wood, leaves, sand, and foreign bodies. The addition of latex is expected to improve stability.

Oil is the part of traditional damar asphalt production called jabung. Jabung often used by silversmiths to engrave silver bearing. Jabung can be created using the new cooking oil or used without significant difference. A striking difference is only on a smell of jabung produced. From previous studies it was concluded that the addition of oil may affect the content of ductility and elasticity in dasphalt [1].

Fly ash is a material which is derived from the burning of coal in power generation, so that all its properties are also determined by the composition and properties of the minerals and impurities in the coal combustion process. In this coal combustion process coal ash melting point higher than the combustion temperature. And these conditions produce ash has a very fine granular texture. Coal fly ash consists of fine grains are generally spherical solid or hollow. The particle size of the fly ash bituminous coal combustion products of less than 0,075mm. The grain size of the fly ash is very fine, ranging up to 88%.

Examination of dasphalt properties aims to determine whether it meets the criteria or similarity with bitumen based RSNI S-01-2003 (Requirements for Hard Asphalt Penetration) [6]. The properties examined are dasphalt penetration, softening point, flash point, ductility, solubility in Trichlor Ethylene, weight, and density. In this study the nature of the inspection results data modification dasphalt properties are secondary data obtained from the results of previous research conducted [1,2].

Table 3. The properties of bioasphalt binder

The properties	Condition			Properties of Dasphalt at Latex content					
	Pen 40	Pen 60	Pen 80	0%	2%	4%	6%	8%	10%
Penetration 25°C, 100g, 5sec	40 - 59	60 - 79	80 - 99	68.2	62	43	36.4	31	28.5
Softening point	51 - 63	50 - 58	46 - 54	55.5	55.5	57.5	65.5	68.5	69.5
Flash point	Min. 200	Min. 200	Min. 225	240	258	260	250	250	245
Ductility, 25°C	Min. 100	Min. 100	Min. 100	63.5	76.5	119.5	78.5	11	6.5
Solubility in Trichlor Ethylene	Min 99	Min 99	Min 99	-	-	99.95	-	-	-
Weight loss(TFOT)	Max 0.8	Max 0.8	Max 1.0	-	-	0.385	-	-	-
Specific Gravity	Min 1.0	Min 1.0	Min 1.0	1.011	1	0.97	0.981	0.971	0.956

2.2 Methods

In this study, the test is generally done consists of five stages, stage I which is the analysis of aggregates to determine the physical properties such as density and water absorption of the aggregates used in the asphalt concrete mixture, the results have been presented in Table 1 and Table 2 for coarse and fine aggregate, respectively. Second stage was producing a bioasphalt binder from damar asphalt modified with some ingredients [1,2] called damar asphalt. In the production of dasphalt, the manufacturing process is done by simple heating, while the composition is applied is a composition referring to the results of previous studies. Third stage is the job mix design using no VII aggregate gradation from SNI 03-1737-1989 (Modalities Laston Concrete Coatings for Highway) [7]. Fourth phase IV is the Marshall Test, the purpose of this test is to determine the value of Marshall stability and Marshall flow of the asphalt concrete mixture with dasphalt binder based on RSNI M-01-2003 (Test Method Asphalt Mixture heat with Marshall Tool) [8]. Phase V is the data processing of the results of volumetric properties, namely density, voids in the mixture (VIM), and Marshall properties namely stability, flow and Marshall quotien.

In this study, after the data obtained from the test results and Marshall volumetric data analysis using regression analysis in which the regression equation can describe the behavior of the test results. Regression is a line that forms a function that connects the dots with the proximity as possible. Correlation is a measure of the suitability of a regression model were used as data. The magnitude of the correlation is denoted by the letter r, where r = 0 means no correlation at all between the two variables analyzed. Conversely, if r = ± 1 then there is a strong relationship between the variables. The dasphalt optimum content was calculated from the regression equation to determine the content at the optimum Marshall stability. The asphalt concrete mixture at optimum dasphalt content were then produced to find the Volumetric and Marshall and compared to the requirements and specifications.

3. Results and Discussion

3.1 Volumetric and Marshall Properties

The series of test have been carried out to observe the volumetric and Marshall properties of 5 variables bitumen content, the average value of the properties are presented in Table 4.

Table 4. The Volumetric and Marshall properties at various dasphalt content

Dasphalt content (%)	Density (gr/cm^3)	Porosity (%)	Marshall Stability (kg)	Marshall Flow (mm)	Marshall Quotient (kg/mm)
5	2.410	2.460	1118.520	5.033	224.250
5.5	2.389	2.510	1309.307	4.013	364.650
6	2.376	2.272	990.834	5.683	177.201
6.5	2.366	1.910	908.789	7.140	128.955
7	2.348	1.867	691.691	7.980	89.178

3.2 Density

Based on a linear regression analysis on the value of density, linear equations with coefficients obtained negative x. This proves that the increasing dasphalt contents in the composition of the mixture will decrease the density values. The value of the density at optimum dasphalt content is 2.401 gr/cm^3. Based on the test requirements of Highways 1987 Marshall density value was 2 g/cm^3 - 3 gr/cm^3, so for the value of density at optimum dasphalt content already meet the requirements. Addition of dasphalt contents can cause the value of density in the test specimen more decreases. This is due to the reduced weight of the aggregate in the mix along with increasing the contents of dasphalt. The conditions set by the density value of Highways (1987) is 2-3 g/cm^3. Thus, the density values of all specimens are qualified specification of asphalt concrete.

3.3 Porosity

Based on linear regression analysis of the voids in the mixture, obtained with a coefficient of linear equations x is negative. This proves that the increasing dasphalt contents in the composition of the mixture will decrease the value of voids in the mixture (porosity). The value of porosity at optimum dasphalt contents is 2.474%. Based on Marshall Test Requirements and Specifications Highways 1987 General of Highways, 2010 (Revision 3) porosity value was 3% - 5%, so for the porosity at optimum dasphalt contents not meet the requirements and specifications. However, it could be improved by changing the aggregate gradation.

3.4 Marshall Stability

Based on regression analysis coefficient of determination (r^2) = 0.8506 This shows the influence of the independent variables (dasphalt contents) greatly affect the dependent variable (stability) with the effect of 85.06%. The value of stability at optimum dasphalt contents is 1181.564 kg. Based on the test requirements of Highways 1987 Marshall stability value is at least 550 kg and based on Highways 2010 General Specification (Revision 3) the value of stability is at least 800 kg, so for the value of stability during dasphalt optimum contents already meet the requirements and specifications. Addition dasphalt contents (%) can raise the value of stability and can also reduce the value stability. Thus, the value of Marshall stability for all specimens has been qualified to the specification of asphalt concrete mixture.

3.5 Marshall Flow

The Marshall flow at optimum dasphalt contents is 4,603 mm. Based on Marshall Test Requirements and Specifications Highways 1987 General of Highways, 2010 (Revision 3) the value of flow is equal to 2 mm - 4 mm, so for the value of the current flow at optimum dasphalt contents has not met the requirements and specifications. Addition dasphalt contents can cause the value of flow getting bigger, this is due by increasing the contents of dasphalt in the mixture, so that the test object is able to follow changes in shape until the test object is destroyed because the load. It could be improved by changing the aggregate grading.

3.6 Marshall Quotien

The value of Marshall quotient at optimum contents is 268.847 kg/mm. Based on the test requirements of Highways 1987 the Marshall quotient value is 200 kg/mm - 350 kg/mm. The Marshall quotient value at the optimum dasphalt contents already met the requirements. The influence of the independent variables (Dasphalt content) greatly affects the dependent variable (Marshall quotient) with the effect of 63.92%. The Marshall quotient value of the specimen with default dasphalt contents of 5% and 5.5% had qualified for road pavement made for the specimen while the dasphalt content of 6%, 6.5%, and 7% are not yet qualified.

3.7 Optimum Dasphalt Content

Based on the optimum stability calculation can be made a relationship to get optimum dasphalt contents. Optimum dasphalt contents is the dasphalt content that will produce the best characteristics in asphalt concrete mixture. Optimum dasphalt contents will be used as a reference on the field implementation as a full scale trial. Optimum dasphalt contents obtained from the value of the graphics stability relationship with dasphalt contents in the form of polynomial regression equation. From the relationships with high contents of stability dasphalt quadratic equation:

$$y = -165{,}53x2 + 1735{,}5x - 3367.4$$

Thus the quadratic equation, to find the extreme points on the axis x is

$$\begin{aligned} x\text{- max} &= -b / 2a \\ &= -1735.5 / (2\,(-165.53)) \\ &= 5.24 \end{aligned}$$

Thus obtained optimum dasphalt content is 5.24%

3.8 Asphalt Concrete Properties at Optimum Dasphalt Content

After getting the value of optimum dasphalt content, then be searched size of voids in the mixture (porosity), density, stability, flow, and Marshall quotient at optimum dasphalt content of 5.24%, which are then compared to the Marshall test Requirements and Specifications of Highways 1987 General of Highways, 2010 as presented in Table 5.

Based on Highways 2010 General Specification [9] for density, porosity and stability values meet the requirements. But for porosity and Marshall flow is still not eligible. While based on the Marshall test requirements of Highways 1987, the properties of mixture already meet all the characteristics except porosity and flow values.

Table 5. The Volumetric and Marshall Properties at Optimum Dasphalt Content

Properties	Equation	Value	Specification	
Porosity	y = -0.3571x + 4.3465	2.474	3.0 - 5.0	3.0 - 5.0
Density	y = -0.0294x + 2.5547	2.401	-	2 - 3
Stability	$y = -165.53x^2 + 1735.5x - 3367.4$	1181.564	Min. 800	Min. 550
Flow	y = 1.804x - 4.854	4.603	2 - 4	2 - 4
MQ	$y = -63.186x^2 + 657.06 - 1439.2$	268.847	-	200 - 350

4. Conclusion

Based on an analysis of the content of dasphalt optimum values obtained contents dasphalt optimum contents of 5.242%, with the volumetric characters and Marshall properties as follow:

The value dasphalt porosity at optimum contents is 2.474%. Based on Marshall Test Requirements and Specifications Highways 1987 General of Highways, 2010, porosity value was 3% - 5%, so for the porosity at optimum dasphalt contents has not met the requirements and specifications. The value of density at a content of optimum dasphalt is 2.401 gr/cm^3. Based on the test requirements of Highways 1987 Marshall density value (density) was 2 g/cm^3 - 3 g/cm^3, so for the value of density during dasphalt optimum contents already meet the requirements.

The value of Marshall stability at optimum dasphalt contents is at s 1181.564 kg. Based on the test requirements of Highways 1987 Marshall stability value is at least 550 kg and based on Highways 2010 General Specification, the value of stability is at least 800 kg, so for the value of stability during dasphalt optimum contents already meet the requirements and specifications. The value flow at optimum dasphalt contents is 4,603 mm. Based on Marshall Test Requirements and Specifications Highways 1987 General of Highways, 2010, the value of flow is equal to 2 mm - 4mm, so for the value of the current flow at optimum contents dasphalt not meet the requirements and specifications. The value Marshall quotient at optimum dasphalt contents is 268.847 kg/mm. Based on the test requirements of Highways 1987 Marshall quotient value is 200 kg/mm - 350 kg/mm so to Marshall quotient value at the time of dasphalt optimum contents already meets the requirements.

References

[1] Baihaqi, A., Setyawan, A., & Djumari, D. 2015 *Matriks Teknik Sipil*, **3(4)**.
[2] Nasution, M. F., Setyawan, A., & Sumarsono, A. 2015 *publikasiilmiah.ums.ac.id*
[3] Departemen Pekerjaan Umum. 2008. *SNI 1969-2008 Testing Methods Density and Water* Absorption Coarse Aggregates Badan Penelitian dan Pengembangan PU,
[4] Departemen Pekerjaan Umum. 2008. SNI 1970-2008 Testing Methods Density and Water Absorption Fine Aggregate, Badan Penelitian dan Pengembangan PU.
[5] Mulyono and Aprianto, 2005, *journal Physical, Chemical and Fungsional Damar*
[6] Departemen Pekerjaan Umum. 2003. *RSNI S-01-2003 Spesifikasi aspal keras berdasarkan penetrasi.* Badan Penelitian dan Pengembangan PU
[7] SNI 03-1737-1989, Asphalt Concrete Coatings for Highway
[8] RSNI M-01-2003, Test Method Asphalt Mixture heat with Marshall Tool
[9] Departemen Pekerjaan Umum. 2010. *Spesifikasi Umum Bina Marga 2010 Divisi 6 Revisi 3.* Badan Penelitian dan Pengembangan PU

IOP Conf. Series: Materials Science and Engineering **176** (2017) 012029 doi:10.1088/1757-899X/176/1/012029

Prediction of the optimum surface orientation angles to achieve maximum solar radiation using Particle Swarm Optimization in Sabha City Libya

F A Mansour[1], M Nizam[2], M Anwar[3]
[1] Student, Mechanical Engineering Department, Sebelas Maret University, Surakarta
[2] Professor, Electrical Engineering Department, Sebelas Maret University, Surakarta
[3] Lecturer, Electrical Engineering Department, Sebelas Maret University, Surakarta

E-mail: fathifg@yahoo.co.uk

Abstract. This research aims to predict the optimum surface orientation angles in solar panel installation to achieve maximum solar radiation. Incident solar radiation is calculated using koronakis mathematical model. Particle Swarm Optimization (PSO) is used as computational method to find optimum angle orientation for solar panel installation in order to get maximum solar radiation. A series of simulation has been carried out to calculate solar radiation based on monthly, seasonally, semi-yearly and yearly period. South-facing was calculated also as comparison of proposed method. South-facing considers azimuth of 0°. *Proposed method* attains higher incident predictions than South-facing that recorded 2511.03 kWh/m^{2}for monthly. It were about 2486.49 kWh/m^2, 2482.13 kWh/m^{2}and 2367.68 kWh/m^2 for seasonally, semi-yearly and yearly. *South-facing* predicted approximately 2496.89 kWh/m^2, 2472.40 kWh/m^2, 2468.96 kWh/m^2, 2356.09 kWh/m^{2}for monthly, seasonally, semi-yearly and yearly periods respectively. Semi-yearly is the best choice because it needs twice adjustments of solar panel in a year. Yet it considers inefficient to adjust solar panel position in every season or monthly with no significant solar radiation increase than semi-yearly and solar tracking device still considers costly in solar energy system. PSO was able to predict accurately with simple concept, easy and computationally efficient. It has been proven by finding the best fitness faster.

1. Introduction

Utilization of the solar energy depends on solar radiation that striking the surface. Solar radiation on the surface information is important to engineer for designing solar collector device installations in order to maximize solar radiation. Solar radiation on inclined surface information is needed to enrich the accurate information of solar radiations beside solar radiation in horizontal surface. It purposed to determine the optimum orientation of the solar collector device in receiving maximum solar radiation to generate maximum energy. The optimum orientation angles depend on radiation area characteristics, climate, latitude and also period utilizations during a year [1, 2, 3, 4]. However comprehensive solar radiations data is not widely available on the developing countries. Therefore, it is necessary to develop methods to predict the solar radiations [5].

IOP Conf. Series: Materials Science and Engineering **176** (2017) 012029 doi:10.1088/1757-899X/176/1/012029

Global solar radiations depend on beam and diffuse solar radiation components. Some research have been carried out to develop mathematical relations in assessing solar radiation of inclined surface gain such as Liu and Jordan, Klein, Klutcher, Hay, Riendl, Koronokis and soon. The models categorized into two, which are isotopic model and anisotropic model due to different deal about sky diffuse component. Isotropic assumes that the intensity of diffuse radiation is uniform over entire the skies, while anisotropic assumes circumsolar diffuse, horizontal brightening or both [3].
Particle Swarm Optimization (PSO) is an evolutionary computation technique that is developed to optimize continuous non-linear, constrained and unconstrained, non-differentiable multi-modal function [6]. The concept based on the behavior of organism such as bird flocking or fish schooling. This optimization has been tested as simple concept, computationally efficient, easy to implement and accurate result as well. It maintains several candidate solutions on the search space. The algorithm remembers the global best fitness and the candidate solutions. Every iteration either local or global best are updated in order to achieve better fitness value.
This study is proposed to measure the incident solar radiation on an arbitrary surface and to predict the monthly, seasonally, semi-yearly and annually optimum orientation angles in order to achieve maximum solar radiations in Sabha city, Libya. Koronakis method which is an isotropic model will be applied to predict global solar radiations on inclined surface in various angles. Further PSO will be employed to analyze and determine the best solution for optimum orientation angles.

1.1. Principle of Solar Radiation
Global horizontal solar radiation (G) is the total of solar radiation that reach the horizontal on earth surface. The power that produced by solar collectors will depends on the radiations reaching the solar panel surface which generally is not horizontal. So in each time, it must be calculated the global solar radiation incident on the surface of the panel [7]. Figure 1 illustrates beam radiation direction, which are angle of incidence (θ), zenith angle (θz), solar altitude angle (αa), tilt angle (β), surface azimuth angle (γ) and solar azimuth angle (γs).

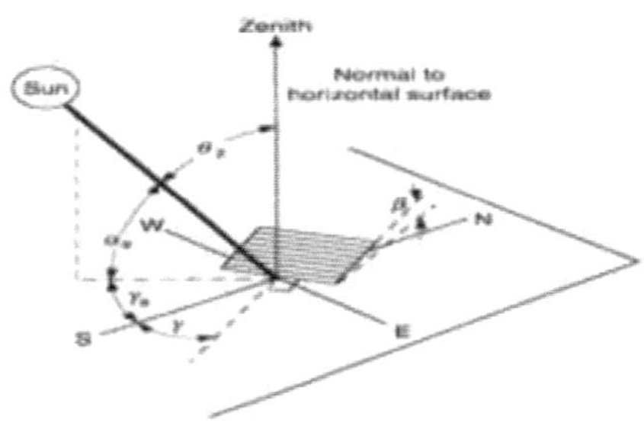

Figure 1. Angle Variety of Solar Collectors

The solar declination angle is calculated by the following equation [8]:

$$\delta = 23.45^{\circ} \sin\left[360^{\circ}\left(\frac{284+n}{365}\right)\right] \quad (1)$$

Solar hour angle can be calculated thorough following equation:

$$\omega = (t_s - 12hr).\,15^{\circ}/hr \quad (2)$$

Where t_c is the civil time in respect to the midpoint of time step (hour), λ is local longitude, Z_c is time zone in hours, east of GMT, and E is equation time.Time equation accounts due to obliquity effect, which means tilt of earth axis rotation relative to the ecliptic plane and the earth eccentricity orbit is derived as following calculation:

$$E = 3.82\begin{pmatrix}0.000075 + 0.001868.\cos B - 0.032077.\sin B \\ -0.014615.\cos 2B - 0.04089.\sin 2B\end{pmatrix} \quad (3)$$

And B here is given by:

$$B = 360^{\circ}\frac{(n-1)}{365} \quad (4)$$

n is the day of the year, starting with 1 for January 1st.Incidence Solar Radiation Angle is the angle between beam radiations and the normal on a surface for any orientation. It derives from following equation:

$$cos\theta = sin\delta\, sin\emptyset\, cos\beta - sin\delta\, cos\emptyset\, sin\beta\, cos\gamma + cos\delta\, cos\emptyset\, cos\beta$$

$$cos\omega + cos\delta\, sin\emptyset\, sin\beta\, cos\gamma\, cos\omega + cos\delta\, sin\beta\, sin\gamma\, sin\omega \quad (5)$$

Where θ is the angle of incidence, β is the slope of the surface, γ is the azimuth of the surface, $\varnothing$ is the latitude, δ is the solar declination, ω is the hour angle. Zenith angle is the angle between vertical and line to the sun, which is the angle of incidence of beam radiation on a horizontal surface. It is zero when the sun is on overhead position and 90°. Zenith angle derived be setting the $\beta = 0$ in previous equation, which is:

$$cos\theta_z = cos\emptyset\, cos\delta\, cos\omega + sin\emptyset\, sin\delta \quad (6)$$

The sunset hour angle derived from the following equation:

$$\omega_s = \pm cos^{-1}(-tan\phi\,.tan\delta) \quad (7)$$

Where the sign (+) is for the sunrise and (-) is for the sunset hour angles. $\emptyset$ is the latitude of location and δ is the declination angle.*Extraterrestrial normal radiation*G_{on} is the amount of solar radiation that arrives at the top of atmosphere over a particular point on the earth surface and can be calculated through equation as follows:

$$G_{on} = G_{sc}\left[1 + 0.033.\cos\left(\frac{360n}{365}\right)\right] \quad (8)$$

G_{sc} is the solar constant (1.367 kW/m^2), and n is the day of the year (a number between 1 and 365). *Extraterrestrial horizontal radiation* (G_o) is solar radiation that striking a horizontal surface at the top atmosphere. Its amount can be calculated as follows:

$$G_o = G_{on}\, cos\theta_z \quad (9)$$

Those calculations simulate on a time step by step basis, equation 10 integrated over the time step to determine the average extraterrestrial horizontal radiation over the time step as follow:

$$\bar{G}_o = \frac{12}{\pi}G_{on}\left[cos\emptyset\, cos\delta\,.(sin\omega_2 - sin\omega_1) + \frac{\pi(\omega_2-\omega_1)}{180^{\circ}}\, sin\emptyset\, sin\delta\right] \quad (10)$$

Where $\bar{G}_o$ is the extraterrestrial horizontal radiation averaged over the time step (kW/m^2), ω_1 is the hour angle at the beginning of the time step and $\omega_2$is the hour angle at the end of the time step. Furthermore, it can be revealed monthly average daily for extraterrestrial horizontal radiation (H_o) and global horizontal radiation (H). $H_o$and H for day derived from hourly of G_o, G respectively can be found the monthly average as follows:

$$H_{o,ave} = \frac{\sum_{n=1}^{N} H_o}{N} \quad (11)$$

$$H_{ave} = \frac{\sum_{n=1}^{N} H}{N} \quad (12)$$

N is the number of days in the month.

IOP Conf. Series: Materials Science and Engineering **176** (2017) 012029 doi:10.1088/1757-899X/176/1/012029

1.2. Solar Radiation of Surface

Enumeration of both beam and diffuse radiation defines as global solar radiation which is expressed as following equation:

$$\bar{G} = \overline{G_b} + \overline{G_d} \tag{13}$$

Where $\bar{G}_b$is the beam radiation (kW/m^2) and $\bar{G}_d$ is the diffuse radiation (kW/m^2). Beam and diffuse classification become main role in calculation of solar radiation on inclined surface. The surface orientation has an effect on the beam radiation which comes from one part of the sky, while diffuse radiation can comes from all over sky dome. It means the calculation must solve the global horizontal radiation in every step into the beam and diffuse radiation to find radiation incident on the surface. It uses correlation which introduced by Erbs for this purpose. It gives diffuse fraction as a clearness index function as follows:

$$\frac{\bar{G}_d}{\bar{G}} = \begin{cases} 1.0 - 0.09.K_T & for\ K_T \leq 0.22 \\ 0.9511 - 0.1604.K_T + 4.388.K_T^2 - 16.638.K_T^3 + 12.336.K_T^4 & for\ 0.22 < K_T \leq 0.80 \\ 0.165 & for\ K_T > 0.80 \end{cases} \tag{14}$$

Global horizontal radiation average can be used to calculate clearness index, further the diffuse radiation. It calculates beam radiation through retrieving the diffuse radiation from the global horizontal radiation.

1.3. Total Incident Radiation on Inclined Surface based Koronokis Model

Global solar radiation on the inclined consists of three components; beam, diffuse, and reflected component. So that total radiation incident can be calculated:

$$G_T = G_{bT} + G_{dT} + G_{gr} \tag{15}$$

G_T defines as hourly total radiation on a tilted surface (KW/m^2) and G_{bT} is hourly beam radiation on a tilted surface (KW/m^2).

$$G_{bT} = (G - G_d) * R_b \tag{16}$$

R_b is the ratio of beam radiation on inclined surface with the beam radiation in horizontal surface. It can be calculated as following equation:

$$R_b = \frac{cos\theta}{cos\theta_z} \tag{17}$$

Koronokis expressed diffuse solar radiation as isotropic component which comes from all over sky dome G_{dT} is hourly diffuse radiation on a tilted surface (KW/m^2) expressed by [5]:

$$G_{dT} = \left(\frac{2+\cos\beta}{3}\right) \tag{18}$$

Hourly ground-reflected solar radiation expressed as follows:

$$G_{gr}=\bar{G} * \rho_g * \left(\frac{1-cos\beta}{2}\right) \tag{19}$$

Total incident solar radiation on inclined surface can be derived from substitution of equation (18), (19), (21) and (22) as follows:

$$G_T = (G - G_d) * R_b + \left(\frac{2+\cos\beta}{3}\right) + \bar{G} * \rho_g * \left(\frac{1-cos\beta}{2}\right) \tag{20}$$

ρ_g is the ground reflectance, which is also called the albedo.

1.4. Particle Swarm Optimization (PSO)

Kennedy and Eberhart (1995) introduced Particle Swarm Optimization (PSO) as a result of motivation by the behavior of bird flocking or fish schooling. The concept consists of a number of particles omitting their position in given search space. Each particle characterized by its position and become a candidate solution towards the problem [9]. The particles change the position in a multidimensional search space exploring better fitness positions.

PSO starts with initial random population of particles where each particle is a candidate solution. The particles' velocity and position are initialized at random. During the optimization process, each particle memorizes its own best position encountered so far which is called the local best. On the other hand, the population memorizes the best position among all individual best positions obtained so far which is called the global best. To balance between the global and local exploration capabilities of the particles, inertia weight is introduced. The inertia weight is linearly decreased through the optimization process to emphasize the search globally at initial iterations and locally at final iterations. PSO has several advantages over other optimization techniques including: simple in concept, easy to implement, and computationally efficient.

2. Experimental

This study proposed to predicts the solar radiation on inclined surface for Sabha city, Libya. It is located on the latitude of 27°2.77'N and longitude of 14° 25.25'E, at elevation of 420 m. Available solar energy data is global horizontal radiation. The data of this study were hourly global horizontal radiation, and ground reflection. The data simulated using MATLAB. The range of tilt and azimuth were -90° to +90° and -180° to +180° respectively. Population of particles was 20, inertial weight was 0.9 to attempt the balance between exploitation and exploration. Iteration number was chosen 50. C1 and c2 were 1.49 where c1 is to control the step towards local best position and c2 is to control the step towards global best position.

Procedure of the present study simulation began by creating Matlab programs to calculate solar radiation components, programs of PSO and South-facing simulation based on monthly, seasonally, semi-yearly and yearly periods. Second step was calculating parameters and component for hourly solar radiation on horizontal surface (δ), $(B))$, (E), (t_s), (ω), $(\omega_s$, $(\cos\theta_z)$, $(\cos\theta)$, (G_{on}), (G_o), $(\bar{G}_0)$, (K_T), $(\bar{G}_d)$, $(\bar{G}_b)$ over the time step of simulation, each one will be calculated value at middle hour of the interval, for example that is, 9:30h for the hour 9:00 to 10:00. Third step was simulating hourly, monthly and yearly incident solar radiation on the inclined surface with the Koronokis model (equation 23) as an objective function. It calculated at the range of tilt and azimuth angles. Further, applying PSO by creating certain sub-programs with tilt and azimuth angle as the constrain variables to find optimum orientation angle and maximum incident solar radiation. The next was analyzing the simulation results to get conclusion.

The PSO algorithm can be described in the following steps:

1. *Initialization*: Initialize randomly n position vectors$\{X_k(0),\ k = 1,2,\dots,n\}$. The elements of X_k are uniformly distributed in a suitable range. Subsequently, initialize randomly n velocity vectors$\{V_k(0),\ k = 1,2,\dots,n\}$. The fitness of each particle is evaluated by using an objective function. Initialize the local best of each particle to its initial position and the global best to the best fitness among the best locals. Finally, initialize the range of the inertia weights $w(0)$.
2. *Update velocity*: Each element j of the velocity vector of the kth particle can be updated as follows:

$$\upsilon_{k,j}(t) = w(t)\upsilon_{k,j}(t-1) + C_1 r_1\left(x^L_{k,j}(t-1) - x_{k,j}(t-1)\right) + C_2 r_2\left(x^G_{k,j}(t-1) - x_{k,j}(t-1)\right) \quad (21)$$

Where t is the iteration number, C_1 is a positive constant called the cognitive parameter and controls the step size towards the particle's local best position. C_2 is a positive constant called the social parameter, and it controls the step size towards the global best position found by the entire swarm. r_1 and r_2 are uniformly distributed random numbers in [0, 1] to add randomness to the velocity updates, $x_{k,j}(t)$ represents the current position of the particle, $x_{k,j}^{L}$ is the particle's best position, and $x_{k,j}^{G}$ is the global best position, $w(t)$ is the inertia weight to control the acceleration of the particle in its original direction. Lower values of w speed up the convergence to the optima and higher values of w encourage exploration of the entire search space. The first term of the velocity update $w(t)\upsilon_{k,j}(t-1)$is the inertia component to keep the particle moving in the same direction as in the previous iteration. The second term $(C_1 r_1 \left(x_{k,j}^{L}(t-1) - x_{k,j}(t-1)\right)$ is called the cognitive component and acts as a memory of the particle causing it to return to its local best that it has encountered so far. The third term $(C_2 r_2 \left(x_{k,j}^{G}(t-1) - x_{k,j}(t-1)\right)$is called the social component as it causes the particle to move towards the global best.

3. *Update position*: After updating the velocity of each particle, the particle position is updated using the lately updated velocity as:

$$x_{k,j}(t) = v_{k,j}(t) + x_{k,j}(t-1) \tag{22}$$

4. *Update bests*: The fitness of each particle is evaluated according to the new updated position. If the updated position leads to a better objective function value, the local best and the global best are updated.
5. *Stopping criteria*: The process is repeated until the number of iterations since the last change of the best solution is greater than a pre-specified number, or the number of iterations reaches a maximum allowable number or the desired value of the objective function is reached.

3. Result and Discussion

3.1. Prediction of Monthly Incident Solar Radiation

Detailed results of monthly incident solar radiation can be seen on table 1. Incident solar radiation of all scenarios seems attaining almost same amount on May, June, July and August. However apart from those months, PSO attained much higher than South-facing. The highest solar radiation happened on July which is in the middle of summer season. Others results show monthly optimum orientation angle which consist of tilt and azimuth angle for PSO and South-facing since this method considers azimuth at 0°. Tilt angle of PSO and South-facing recorded similar trend. Optimum tilt angle turned to similar angle on August to December. Azimuth angle of PSO has been described on table 1 during a year. The tilt angle or slopes seemed to close to horizontal surface on May to August.

Table 1 shows total monthly incident solar radiation for PSO, South-facing and Horizontal surface. PSO recorded the highest total monthly incident solar radiation which was 2511.03 KWh/m^2 due to the optimal tilt and azimuth angle while South-facing recorded 2496.89 KWh/m^2. PSO gave more accurate orientation in collecting incident solar radiation compared with other scenarios because it didn't adjust tilt angle only but also the azimuth. The monthly optimization worked greatly on December and January since the incident solar radiation increased significantly than the horizontal surface compare with other months.

Table 1. Monthly result of PSO, South facing and Horizontal surface

Month	Horizontal Surface	PSO			South Facing	
	ISR (KWh/m^2)	ISR (KWh/m^2)	Optimum Angles: Tilt (°)	Optimum Angles: Azimuth (°)	ISR (KWh/m^2)	Optimum Tilt (°)
Jan	130.92	198.23	55.6503	-9.3163	196.25	55.5938
Feb	137.90	177.49	46.3746	-7.3090	176.96	46.1815
Mar	181.01	204.70	34.4587	-23.7343	202.14	32.2312
Apr	201.69	207.46	18.5012	-39.7678	205.55	14.1388

May	210.22	211.52	8.9479	-93.7058	210.24	-1.1820
Jun	222.06	224.63	10.7697	-139.4585	223.55	-8.2293
Jul	229.15	231.63	11.3927	-109.6221	229.61	-4.5247
Aug	217.64	219.94	10.9622	-43.7141	218.84	7.4662
Sep	196.45	211.91	27.6178	-20.0948	211.15	26.0278
Oct	176.45	221.41	43.2867	-6.5863	220.89	43.0775
Nov	140.52	205.97	53.6725	-1.3865	205.94	53.2158
Dec	124.11	196.14	58.0440	-4.5843	195.79	57.9611
Total	**2168.10**	**2511.03**			**2496.89**	

3.2. Prediction of Seasonally Incident Solar Radiation

The result of simulation was illustrated on the figure 2. It can be seen that similar trend happened on seasonally prediction where PSO attained the highest incident solar radiation in all season during a year. It was 569.99 on winter where falling at optimum tilt angle of 53.93° and azimuth of -7.65°. Further it attained 613.22 KWh/m^2 on spring where falling at optimum tilt angle of 20.09° and azimuth of -39.10°. Simulation predicted 673.17 KWh/m^2 and 630.11 KWh/m^2 for fall and summer respectively. Those was retrieved at the optimum tilt angle of 8.79° and azimuth of -96.30° on fall season, meanwhile optimum tilt angle of 41.52° and azimuth of -5.57° were determined during summer season.

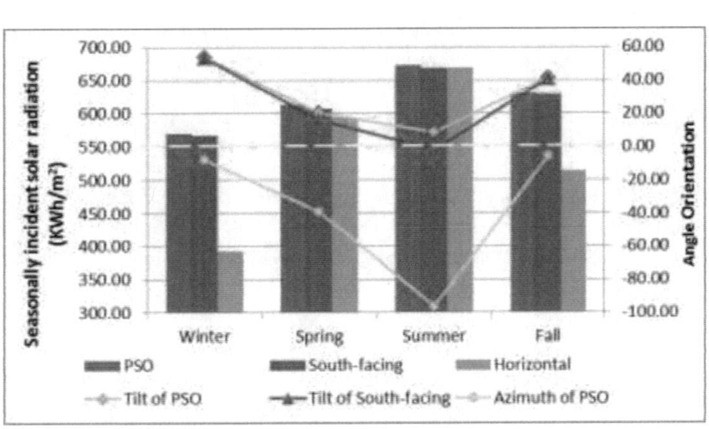

Figure 2. Comparison of Seasonally incident solar radiation and Optimum Orientation Angles

South-facing simulation calculated incident solar radiation approximately 567.40, 606.80, 669.07 and 629.13 KWh/m^2 on winter, spring, fall, and summer respectively during a year. Those fallen at optimum tilt angle of 53.66° along winter, 15.94° along spring, -1.85° along fall, and 40.96° along summer. South-facing considers 0 azimuth angle. Seasonally horizontal result shows incident solar radiations were 392.92, 592.91, 668.84 and 513.42 KWh/m^2 for winter, spring, fall, and summer season. Based on the figure 2, it can be figured out the highest incident solar radiation happened on summer season during a year, either conducted by PSO, South-facing and Horizontal surface. It also can be seen the optimum tilt angle of PSO and South-facing predicted at closed value and indicates similar trend as on the monthly prediction. Further the azimuth angle seems at same trend also with the monthly predictions. The optimization attained the best result on winter season where it increased the solar radiation better than other seasons comparing to horizontal surface.

3.3. Prediction of Semi-yearly Incident Solar Radiation

Semi-yearly simulation was conducted to determine the semi-yearly of optimum orientation angles and its maximum incident solar radiations. Semi-yearly was calculated based on first six month which starts from October to March and second six month which starts from April to September. The result illustrated on figure 4.6. PSO predicted 1192.16 KWh/m^2 for October to March which fallen at optimum tilt angle of 48.70° and azimuth of -7.50°. It was 1289.97 KWh/m^2 during April to September where fallen at optimum tilt angle of 10.87° and azimuth of -56.63°.

Table 2. Semi-yearly results of PSO and South-facing

Period	Horizontal Surface	South Facing		PSO		
	solar radiation	Solar radiation	Tilt angle (°)	Solar radiation	Tilt Angle (°)	Azimuth (°)
Oct-Mar	890.90	1187.79	48.42	1192.16	48.47	-7.50
Apr-Sep	1277.20	1280.96	5.47	1289.97	10.87	-56.63
Total	**2168.10**	**2468.75**		**2482.13**		

South-facing predicted maximum incident solar radiation of 1187.79 during October to March where fallen at optimum tilt angle of 48.42° and it was 1280.96 KWh/m^2 during April to September where it fallen at tilt angle of 5.47°. Meanwhile horizontal surface predicted about 890 KWh/m during first six month of October to March and 1277.20 KWh/m for the rest months. Solar radiation was higher during April to September where the summer season happens on this period. The semi-yearly optimization recorded the best result during October-March because solar radiation increased better than on April-September period compare to horizontal surface.

3.4. Prediction of Yearly Incident Solar Radiation

Prediction of optimum orientation angle to collect maximum solar radiation on yearly basis can be seen on table 2. It shows that proposed method was able to find better solution for yearly optimum orientation angle. It can be seen that the total incident radiation of a year recorded better than South-facing. PSO calculated 11.6 KWh/m2 higher than South-facing which was approximately 2367.68 KWh/m2. This value falls at slope of 29.50° and azimuth of -18.29°. Meanwhile South-facing recorded 2356.09 KWh/m2 which fallen at slope of 28.27° where the azimuth considered at 0°.

Table 3. Yearly Results of PSO, South-facing and horizontal surface.

Method	Total incident solar radiation	Optimum Orientation Angles	
		Tilt (°)	Azimuth (°)
PSO	2367.68	29.507	-18.289
South facing	2356.09	28.274	0
Horizontal Surface	2168.10	0	0

Based on the orientation angle results, PSO determined the accurate orientation considering the higher solar radiation attained. The best solar panel installation of PSO inclined about 29.5 ° and rotating to face the east about 18.28°.

4. Conclusion

Incident solar radiation on inclined surface has been predicted which calculated using Koronakis mathematical model were. Total incident solar radiation has been calculated based on monthly, seasonally, semi-yearly and yearly which were simulated using PSO and South-facing. Total incident solar radiation of horizontal surface was 2168.10 KWh/m2 for all periods. South-facing attained 2496.89, 2472.40, 2468.96 and 2356.09 KWh/m2 for monthly, seasonally, semi-yearly and yearly respectively. PSO recorded 2536.80, 2486.49, 2482.13, and 2367.68 KWh/m2 for monthly, seasonally, semi-yearly and yearly respectively. The best result by PSO rather than South-facing for all bases which were monthly, seasonally, semi-yearly and yearly. Monthly optimization recorded significant result in increasing solar radiation on December and January. Seasonally based optimization increased solar radiation significantly on winter season, meanwhile October-March period increased better than another period for semi-yearly based optimization.

Monthly of *PSO* orientation angles tended to rotate the east due to the negative value of azimuth for all months, moreover it rotated to the east on May and to the north-east for June to July. South-facing result indicated that the slopes were closed to horizontal line due to small degrees which was in the range of -1° to +7° on May, June, July and August. Seasonally orientation angle of proposed method tended to rotate to the east because of negative value of azimuth with the range of -5° to -96° and it was closed to the horizontal surface on summer season. South-facing was also recorded to close to horizontal surface on summer season, where the slopes were on the range of +15° to +54° on fall, winter and spring. Semi-yearly azimuth angle of proposed method tended to rotate to the east which was -7.50° on October to March period and -56.63° for another period. The optimum slopes were at 48.70° for the first period and 10.87° for another one. *South-facing* result determined the optimum slopes were at 48.42° and 5.47° for first and second period respectively. The second period angles which were April to September indicated to close to horizontal surface. Meanwhile, yearly optimum orientation angles of proposed method determined at 29.5° slope and -18.29° azimuth angle. Yearly slope of south-facing was at 28.27°.

Semi-yearly based optimization determined as the best optimization although there were higher solar radiations for seasonally and daily bases because it would only need twice adjustments of the solar panel in a year. Yet it considers inefficient to adjust solar panel position in every season or even daily with no significant solar radiation increase than semi-yearly and solar tracking device still considers costly in solar energy system. PSO was able to predict accurately with simple concept, easy and computationally efficient. It has been proven by finding the best fitness faster.

References

[1] Darhmaoui H and Lahjouji D 2013 *E. Procedia* **42** 426-435
[2] Khorasanizadeh H, Mohammadi K, Mostafaeipour A 2014 *E. Conv. and Manage.* **78** 805-814
[3] Ng K M, Adam N M, Inayatullah O, Kadir M Z A A 2014 *Inter. Journal of En. and Env. Engin.* **5** 011-13
[4] Noorian A M, Moradi I, Kamali G A 2008 *Renewable Energy* **33** 61406-1412
[5] Shukla K, Rangnekar S, Sudhakar K 2015 *Energy Reports* **1** 96-103
[6] Khare A and Rangnekar S 2013 *Applied Soft Computing* **13** 52997-3006
[7] Duffie J A and Beckman W A 2013 *Solar Engineering of Thermal Processes* (New Jersey: Wiley)
[8] Kalogirou S A 2013 *Solar Energy Engineering: Processes and Systems* (California: Academic Press)
[9] Mohandes M A 2012 *Solar Energy* **86** 113137-3145

Road Maintenance and Rehabilitation Program Using Functional and Structural Assessment

A I Setianingsih[1], S Sangaji[1] and A Setyawan[2]
[1]Magister Maintenance and Rehabilitation Infrastructure, Sebelas Maret University Solo 57126, Indonesia
[2]Roadmate Research Group, Civil Engineering Department, Sebelas Maret University, Jl. Ir. Sutami 36 A Solo 57126, Indonesia

E-mail: arysetyawan@staff.uns.ac.id

Abstract. Road sector development policy in Bangka Belitung emphasis on equitable development, which is opening up new areas for industrial development zones of potential marine and coastal tourism, so that having an impact on the budget priority to build a new road. This led to a minimal budget provided for the maintenance of the existing road. This study aimed to evaluate the condition of the pavement both functionally and structurally, the growth of traffic density and the availability of existing road maintenance costs. Then, to analyze the influence of existing road conditions, traffic density and road maintenance costs to the type of road maintenance management. The results are compared with the results of the existing maintenance conducted by the Public Works Department of Bangka Belitung province. Evaluation of pavement conditions consists of visual assessment of pavement condition using IRI, pavement condition assessment functionally with deflection method using test data tool Benkelman Beam (BB) and the actual traffic load. IRI value, deflections and traffic growth gained from years 2011-2015 subsequently created regression models to obtain the relationship and the correlation coefficient. The analysis showed that using the same relative magnitude of the budget from 2011 to 2015, giving priority to the maintenance of the road with good conditions capable of providing the road with a steady state of 100%. Recommendations can be given that maintain the road with good conditions reflecting that preservation provide maximum results with the more efficient maintenance cost.

1. Introduction

Roads are the transport infrastructures which play an important role in supporting the economic growth of a region. Therefore, it is necessary to do maintenance efforts so that the roads can function optimally.Road sector development policy in Bangka Belitung emphasis on equitable development, which is opening up new areas for development of industrial zones of potential marine and coastal tourism, so that having an impact on the budget priority to build new roads. This led to a minimal budget provided for the maintenance of the road. Assessment of road pavement conditions is the important aspects to determine the maintenance activities for road improvement. To carry out the assessment of pavement condition, the first is to determine the types of damage, the density as well as the level of severity.

[2] To whom any correspondence should be addressed.

This study aimed to evaluate the condition of the pavement both functionally and structurally, the growth of traffic density and the availability of road maintenance costs. Then analyze the influence of road conditions, traffic density and road maintenance costs to the type of road maintenance management. The results are compared with the results of the existing maintenance conducted by the Public Works Department of Bangka Belitung province. Condition assessment pavement is aimed at giving information pavement condition at a particular time. Based on the information gathered will be used to establish kinds of the study, the assessment of priorities, and maintenance programs [1]. Early research to situations the road surface is to carry out visually survey by means of seeing and analyze damage on the basis of with damage level for use as a basis for engaging in the maintenance and repair [2]. The functional pavement analysis is done on the flexible pavement based on the road conditions .The relationship obtained an exponential inclination and can be used to predict the service life of pavement of [3]. The principle road maintenance done with economic benefits effective and efficient, through budget minimum can be produced road conditions steady so that the community happy as the costs of being inferior transport [4].

2. Experimental

2.1. International Roughness Index (IRI)

Flatness level road (International Roughness Index, IRI) is one parameter used to determine the level of service a condition that affects the road driver comfort (riding quality). The main requirement is a good way strong, flat, waterproof, durable and economical over the life of the plan. To meet the requirements of monitoring and evaluation needs to be done regularly so that it can be determined that the proper method of construction improvements. To determine the level of the road surface flatness measurements can be performed using the method recommended by NAASRA.

2.2. Testing with Benkelman Beam

Benkelman beam instrument was used to measure the deflection.The turning deflection (rebounds deflection)has been the large deflection vertical synchronization the impact on the point of observation, deflection travelers commonly used to design the pavement thickness.

The procedure calculation deflection method using the BB (Benkelman Beam) is based on the standard Pd.Tt-05-2005-B as follows:

A. The burden of determining repetition of traffic plan (CESA) in ESA.

a. Determine traffic and coefficients distribution vehicles (C), Table 1 shows the determination of the number of lines based on the width of pavements. Table 2 shows the determination of C the value based on the number of line and traffic direction.

Table 1. The number of lanes and width of pavement

Wide of Pavement (L)	The number of lanes
L < 4.50 m	1
4.50 m ≤ L < 8.00 m	2
8.00 m ≤ L < 11.25 m	3
11.25 m ≤ L < 1.00 m	4
15.00 m ≤ L < 18.75 m	5
18.75 m ≤ L < 22.50 m	6

Table 2. Vehicles Distribution coefficient (C)

The number of lanes	Minor vihicle*		Heavy vehicle*	
	1 direction	2 directions	1 direction	2 directions
1	1.00	1.00	1.00	1.00
2	0.60	0.50	0.70	0.50
3	0.40	0.40	0.50	0.475
4	-	0.30	-	0.45
5	-	0.25	-	0.425
6	-	0.20	-	0.40

b. Determine the vehicle equivalent axis (E)

$$\text{Equivalent Number STRT} = \left(\frac{\text{axle load (ton)}}{5{,}40}\right)^4 \quad \dots\dots\dots(1)$$

$$\text{Equivalent Number STRG} = \left(\frac{\text{axle load (ton)}}{8{,}16}\right)^4 \quad \dots\dots\dots(2)$$

$$\text{Equivalent Numbern SDRG} = \left(\frac{\text{axle load (ton)}}{13{,}76}\right)^4 \quad \dots\dots\dots(3)$$

$$\text{Equivalent Number STrRG} = \left(\frac{\text{axle load (ton)}}{18{,}45}\right)^4 \quad \dots\dots\dots(4)$$

c. Determining factor age plans and growth of traffic (N)

$$N = \frac{1}{2}\left[1 + (1+r)^n + 2(1+r)\frac{(1+r)^{n-1} - 1}{r}\right] \dots\dots\dots(5)$$

d. Accumulation of equivalent burden the axis standard (CESA)

$$CESA = \sum_{Tractor-Triler}^{MP} m \times 365 \times E \times C \times N \quad \dots\dots\dots(6)$$

B. Count the results of the testing deflection and corrections by a factor of advance ground water (ca) and factor of temperature standard (ft) as well as bb test factors if the burden of proper test worth 8.16 tons

$$\mathrm{d_B} = 2 \times (\mathrm{d_3} - \mathrm{d_1}) \times \mathrm{Ft} \times \mathrm{Ca} \times \mathrm{FK_{B-BB}} \quad \dots\dots\dots(7)$$

C. Determine the length of section of segments having uniformity (FK) in line with the rate uniformity a license which desired in accordance guidelines.

$$\mathrm{FK} = \frac{\mathrm{S}}{\mathrm{d_r}} \times 100\% < FKijin \quad \dots\dots\dots(8)$$

$$dr = \frac{\sum_{1}^{n_s} d}{n_s} \qquad \text{......(9)}$$

$$S = \sqrt{\frac{n_s(\sum_1^{n_s} d^2)-(\sum_1^{n_s} d)^2}{n_s(n_s-1)}} \qquad \text{.......(10)}$$

D. Counting deflection vice (D_{vice}) for each road section that relies onthe class of the road

$$D_{vice} = d_R + 2s \qquad \text{.........(11)}$$
$$D_{vice} = d_R + 1{,}64s \qquad \text{.........(12)}$$
$$D_{vicel} = d_R + 1{,}28s \qquad \text{.........(13)}$$

E. Counting deflection plan permission ($D_{planning}$)

$$D_{planning} = 22{,}208 \times CESA^{(-0{,}2307)} \qquad \text{........(14)}$$

2.3. The Location of Study

The location of the study was in the segment of the National Highway from Batas Kota Pangkal Pinang to Puding Besar with the length of 22,682 Km. The segment of highway has the function of primary collector and class II type of highway in Indonesia.

2.4 The Methods

A. In this study, first collected data as follows:
   a. Functional condition
   b. Structural condition
   c. Traffic data
   d. Cost maintenance and rehabilitation pavement
B. Next up is evaluating each data from 2011 – 2015
C. The last, compared with the results of the existing maintenance conducted by the Public Works Department of Bangka Belitung province.

3. Results and discussion

3.1 The evaluation of functional pavement condition

The evaluations of functional road conditions were carried out using NASRA Equipment to measure the International Roughness Index (IRI), the results are presented in Figure 1.

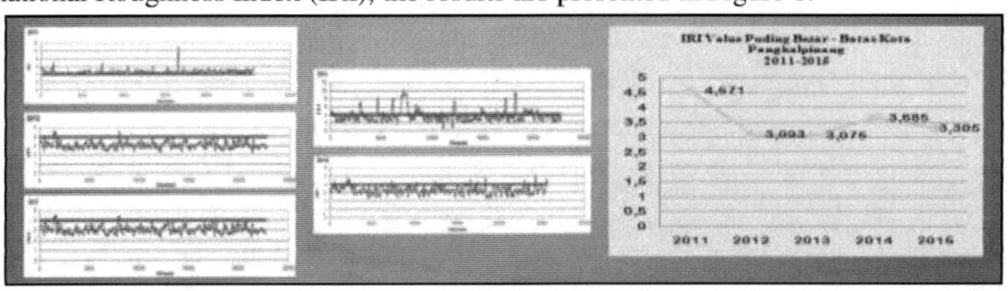

Figure 1. Result of functional road condition 2011-2015

It could be seen from the figure that the condition in 2011 is 'medium' for road condition and in the year of 2012 to 2015 performed the 'good' road condition.

3.2 The growth of traffic

The volume of traffic passing through the segment of Batas Kota Pangkalpinang to Puding Besar could be seen in Figure 4, where there was decreasing volume of traffic caused by impairment in the functions of port nearby. The traffic congestion in the road segment to Tanjung Kelian Port of Pangkalpinang city. Consequently, the traffic decreases up to the year of 2015 when the national roads have been improved. Since then, the traffic increase significantly. The data of traffic growth are presented in Figure 2.

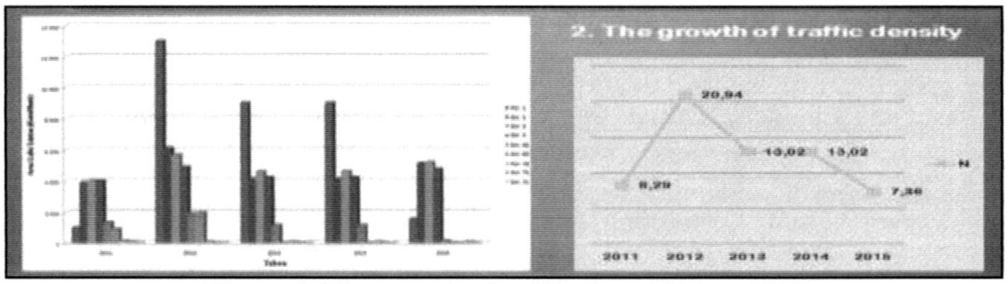

Figure 2. The growth of traffic from 2011 to 2015

3.3 The Evaluation of structural condition

The data from Bengkelmen beam were then analyzed to determine the overlay required for the segmented road. The results are presented in Figure 3. The overlay needed from 2011 to 2015 are presented in Figure 4.

The results of using benkelman beam

Year	Average Deflection (dR)	Standard Deviation (s)	Uniformty Deflection (FK)	Representative Deflection (Dwakil)	Deflection permission plan (Drencana)	Added thick layers (Ho)	Thick layers correction (Fo)	Thick layers added corrected (Ht)	Plan Add Layer Thickness
2011	0,83	0,10	12,41	0,999	0,361	17,64	1,00	17,64	3,00
2012	1,29	0,24	18,64	1,683	0,342	27,28	1,00	27,28	4,60
2013	1,23	0,29	23,42	1,699	0,387	25,37	1,00	25,37	4,30
2014	1,23	0,22	17,83	1,590	0,387	24,26	1,00	24,26	4,00
2015	1,23	0,22	17,83	1,590	0,332	26,82	1,00	26,82	4,50

Figure 3. The result of using Benkelman Beam 2011-2015

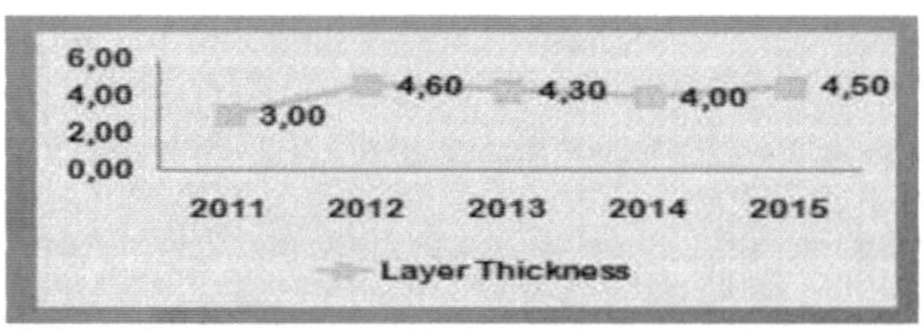

Figure 4. The thickness of overlay layers 2011-2015

3.4 Maintenance program has to be done by the Public Works Departement of Bangka Belitung Province

The type of maintenance and rehabilitation program and the respective cost for each type of treatment are presented in Figure 5.

Public Work Departemen

Program	Cost maintenance (Rp)				
	2011	2012	2013	2014	2015
Routine	508.560.000	793.870.000	771.188.000	839.234.000	861.916.000
Periodic	972.555.672	-	-	-	-
Recontruction	-	-	-	-	6.583.100.000

Figure 5. Cost maintenance 2011-2015

From 2011 to 2015 is the implementation of the program on the road. Whether it is routine program, periodical or reconstruction

3.5 Compare the results of existing maintenance conducted by the Public Works Department of Bangka Belitung Province

The results of road condition depend on the maintenance program carried out, as can be seen from Figure 6, the road maintenance program is the best choice for highway management.

IRI Value

Recomend program	2011	2012	2013	2014	2015
Routine	-	V	V	V	V
Periodic	V	-	-	-	-
Recontruction	-	-	-	-	-

Deflection Data

Recomend program	2011	2012	2013	2014	2015
Routine	-	-	-	-	-
Periodic	V	V	V	V	V
Recontruction	-	-	-	-	-

Public Work Departemen

Program	Cost maintenance (Rp)				
	2011	2012	2013	2014	2015
Routine	508.560.000	793.870.000	771.188.000	839.234.000	861.916.000
Periodic	972.555.672	-	-	-	-
Recontruction	-	-	-	-	6.583.100.000

Figure 6. Comparison result between IRI, Planning an overlay thickness and maintenance cost

Based on IRI value, deflection and growth of traffic, it can be seen parameter used as the best activities is IRI value

4. Conclusion

The results of the analysis indicate that using the same relative magnitude of the budget from 2011 to 2015 by giving the priority to the road maintenance at relatively good conditions are efficient of providing the road with a steady state of 100%. Recommendations can be given that maintain the road with good conditions reflecting that preservation provide maximum results with more efficient maintenance cost.

References

[1] Shahin, M.Y., 2005. *Pavement management for airports, roads, and parking lots* **501** New York: Springer.
[2] Suswandi, A., Sartono, W. and Christady, H., 2009, *Civil Engineering Forum Teknik Sipil*, Vol. **18(3)**, pp. pp-934).
[3] Marga, D.B. and UMUM, D.P., 1990. Petunjuk Teknis Perencanaan dan Penyusunan Program Jalan Kabupaten. *Surat Keputusan No. 77/KPTS/Db/1990.*
[4] Mahmood, M., Rahman, M., Nolle, L. and Mathavan, S., 2013. *International Journal of Pavement Research and Technology*, ***6*(5)**, pp.620-626.

The Use of International Roughness Index and Structural Number for Rehabilitation and Maintenance Policy of Local Highway

Hermawan[1], M Suprapto[1] and A Setyawan[2]
[1] Magister Maintenance and Rehabilitation Infrastructure Sebelas Maret University, Jl. Ir. Sutami 36 A Solo 57126, Indonesia
[2] Roadmate Research Group, Civil Engineering Department, Sebelas Maret University, Jl. Ir. Sutami 36 A Solo 57126, Indonesia

E-mail: arysetyawan@staff.uns.ac.id

Abstract. Rehabilitation and maintenance of road performed by technical agencies at the local government in Indonesia, are generally not based on the assessment of road conditions. The value of the structural and functional condition of the pavement is not counted on carefully. As a result, road rehabilitation and maintenance patterns tend to be similar, repetitive and improper. International Roughness Index (IRI) is a parameter for assessing the functional condition of the pavement while the Structural Number (SN) is a parameter for assessing the structural condition of the pavement. Measuring road conditions by using Roadroid applications on smartphones can provide an efficient way, scalable, and low cost to the highway authority to collect road condition data. This study was conducted to determine the conditions of the road both functionally and structurally. Results of research conducted, pavement functionally in a good condition with the acquisition of IRI value of less than 4. Structural pavement conditions indicate that the value of Structural Number Effective (SN_{eff}) is less than the value of the Structural Number Future (SN_f), thus the structural condition of the road segments has not been able to serve traffic with a design life of 20 years. Prediction of IRI value obtained to determine the type of road maintenance is functionally performed when the value of IRI exceeds the value of 4 with an overlay of material HRS WC 30 mm. Structurally road maintenance carried out various scenarios, the phased construction and direct construction. The type of material and thickness of pavement on a phased construction scenario for the design life of 15-20 years and direct construction gives better results than the phased construction scenarios for the design life of 5-20 years and 10-20 years.

1. Introduction

Handling of the flexible pavement maintenance should be done through measurable criteria for achieving the goals set. International Roughness Index is a parameter to determine the functional pavement conditions, whereas the structural condition of pavement was obtained from the value of the Structural Number [1]. SN is used to determine the structural capacity of the pavement, whether it is able to withstand traffic loads up to design life.

[2] To whom any correspondence should be addressed.

The measuring road roughness with smart phone can provide an efficient, scalable, and cost-effective way for road organizations delivers road condition data.Application web-based GIS it can use powerful filtering techniques as the IRI for road maintenance decision support [2]. Roadroid can help the road network asset management by offering a cheaper solution to monitor and report road conditions [3].

The scenario of handling and maintaining roads to serve the traffic load during the design life, using indicators IRI predicted through an empirical equation Patterson (1987) [4]. The structural condition of pavement over the design life was obtained by comparing the value SN_{eff} with SN_f. The results of the comparison value of SN_{eff} and SN_f produce overlay requirements necessary to serve the traffic load during the design life.

The results of assessment road functional and structural condition, were analyzed to determine the selection of the type of road maintenance is the thickness of an overlay and a mixture of asphalt used.

2. Experimental

2.1. Functional Condition of Pavement

Obtaining the average value of IRI on each road segment is analyzed based on the provisions of Bina Marga (Ministry of Public Work and Housing. Republic of Indonesia), as shown in Table 1.

Table 1. Classification of road condition level

Road Condition	IRI
Good	<4
Fair	4-8
Poor	8-12
Bad	>12

IRI values obtained from the application Roadroid, simulated to obtain predictions IRI or the rate of road roughness. Prediction IRI is used to determine the functional condition of the road during the design life [4]. Equations to determine the evolution of pavement roughness are presented in Paterson(1987), namely:

$$\mathrm{RI_t} = (\mathrm{RI_0} + 725\,(1 + \mathrm{SNC})^{-5} \times \mathrm{NE_t}) \times e^{0.0153t} \tag{1}$$

WhereRI_t is roughness at time, *m/km*;

RI_0 – roughnes at t=0, *m/km*;

NE_t – cumulative equivalent standard axle loadingsuntil time *t* (10^6 ESA lane^{-1});

SNC – StructureNumber Capacity, given by:

$$\mathrm{SNC} = (1/24{,}4) \times \Sigma \mathrm{a_i h_i} + \mathrm{SN_{SG}} \tag{2}$$

Where SNC is Structural Number Capacity;

a_i – the strength coefficient of the ith layer as defined;

h_i – thethickness of the ith layer provided that the sum of thicknesses $h_i$is not greaterthan 700 mm

SN_{SG} is the modified structural number contribution of the subgrade, given by:

$$\mathrm{SN_{SG}} = 3{,}51 logCBR - 0{,}85(logCBR)^2 - 1{,}43 \tag{3}$$

CBR is the California Bearing Ratio of the subgrade at in situ conditions of moisture and density [5].

2.2. Structural Condition of Pavement

Pavement structural condition assessment in this study was analyzed by using the method of Bina Marga (2002 and 2013). The value of the existing structural pavement conditions is determined by using the equation of Bina Marga adopted from the AASHTO method [6].

$$SN = a_1D_1 + a_2D_2 + a_3D_3 \quad (4)$$

Where SNis Structural Number;
$a_{1,2,3}$ – the strength coefficient;
$D_{1,2,3}$ – the thickness of layers, *inch*.

The value of the strength coefficient (a) used is the actual condition of pavement materials, so that the SN values is equal to the value of Structural Number Effective (SN_{eff}). The next phase is to determine values Structural Number Future (SN_f) obtained using the equation:

$$\log_{10} W_{18} = Z_R\, So + 9.36\log_{10}(SN+1) - 0.20 + \frac{\log_{10}\left[\frac{Po - Pt}{Po - Pf}\right]}{0.40 + \frac{1094}{(SN+1)^{5.19}}} + 2.32\log_{10} Mr - 8.07 \quad (5)$$

Where$W_{18}$is predicted number by 18-kip equivalent single axle load applications;
Z_R – standard normal deviate;
S_0 – combined standard error of the traffic prediction and performance prediction;
ΔPSI – difference between the initial design serviceability index, P_0, and the design terminal serviceability index, P_t;
MR – resilient modulus, *psi*.

2.3. *Functionally Overlay*

The use of overlay materials to repair Roughness (IRI), as shown in Table 2 [7].

Table 2. Overlay thickness to roughness repair

IRI (m/km)	Minimum Overlay Thickness *(mm)*	Materials
4	30	HRS-WC
5	45	AC-WC
6	50	AC-WC
7	55	AC-WC
8	60	AC-BC

HRS: Hot Rolled Sheet-Wearing Course
AC-WC: Asphalt Concrete-Wearing Course
AC-BC: Asphalt Concrete-Binder Course

2.4. *Structured Overlay*

Calculation of thickness layers to overlay (H_{OL}), using the equation:

$$H_{OL} = \frac{SN_{OL}}{a_{OL}} - \frac{SN_f - SN_{eff}}{a_{OL}} \quad (6)$$

Where H_{OL} is overlay thickness, *inch*;
SN_{OL} – Structural Number to overlay;
a_{OL} – the strength coefficient of the overlay layer;
SN_f – Structural Number Future;
SN_{eff} – Structural Number Effective;

The type of material and thickness overlay of asphalt mixture used to maintenance and rehabilitation of pavement, as shown Table 3 [8].

Table 3. The type of material and thickness of the asphalt mix

Minimum Thickness (mm)	Materials
30	HRS-WC
35	HRS-Base
40	AC-WC
60	AC-BC
75	AC-Base

3. Results and Discussion

3.1. Pavement Condition

3.1.1. Functional Condition

Assessment of road performance are functionally obtained from the result of survey by using the Roadroid application of the Android smartphone to get the value of IRI. The results measurement of road roughness as shown in Table 5, and the predictive value of IRI in the design life of 20 years as shown in Table 4.

Table 4. IRI prediction during the design life

Years	IRI of road segment (*m/km*)									
	I		II		III		IV		V	
	WR	R	WR	R	WR	R	WR	R	WR	R
1	2.12	2.12	1.9	1.9	1.93	1.93	1.79	1.79	2.2	2.2
2	2.31	2.31	2.03	2.03	2.02	2.02	1.89	1.89	2.34	2.34
3	2.52	2.52	2.18	2.18	2.12	2.12	2	2	2.5	2.5
4	2.76	2.76	2.35	2.35	2.22	2.22	2.13	2.13	2.67	2.67
5	3.02	3.02	2.53	2.53	2.34	2.34	2.26	2.26	2.86	2.86
6	3.3	3.3	2.72	2.72	2.46	2.46	2.41	2.41	3.07	3.07
7	3.62	3.62	2.94	2.94	2.60	2.6	2.57	2.57	3.3	3.3
8	3.98	3.98	3.19	3.19	2.76	2.76	2.75	2.75	3.56	3.56
9	4.37	**4.37**	3.46	3.46	2.92	2.92	2.94	2.94	3.84	3.84
10	4.81	2.4	3.75	3.75	3.11	3.11	3.16	3.16	4.15	**4.15**
11	5.3	2.66	4.08	**4.08**	3.31	3.31	3.4	3.4	4.5	2.13
12	5.83	2.95	4.45	2.28	3.53	3.53	3.67	3.67	4.88	2.33
13	6.43	3.27	4.85	2.5	3.77	3.77	3.96	3.96	5.31	2.56
14	7.1	3.63	5.3	2.74	4.04	**4.04**	4.29	**4.29**	5.78	2.81
15	7.84	**4.03**	5.8	3.01	4.34	2.08	4.64	2.34	6.3	3.09
16	8.66	3.01	6.36	3.31	4.67	2.25	5.04	2.55	6.88	3.39
17	9.57	3.31	6.97	3.65	5.03	2.44	5.48	2.78	7.52	3.74
18	10.58	3.64	7.65	**4.02**	5.44	2.65	5.97	3.03	8.23	**4.12**
19	11.71	**4.01**	8.41	3	5.88	2.88	6.51	3.32	9.01	2.99
20	12.96	3.13	9.24	3.28	6.37	3.13	7.11	3.63	9.89	3.27

NR : Without Rehabilitation
R : Rehabilitation

Table 5. IRI value of road segment

Road Segment	IRI	Condition
I. Kertijayan-Bligo	1.95	Good
II. Bligo-Podo	1.78	Good
III. Podo-Surabayan	1.85	Good
IV. Surabayan-Sedayu	1.7	Good
V. Sedayu-Karanganyar	2.07	Good

3.1.2. Structural Condition

The data is required to determine the value Structural Number Effective (SN_{eff}) on the existing pavement is the thickness of layer pavement (D) and value of the strength coefficient (a).The result of the calculation of Structural Number Effective (SN_{eff}), as shown in Table 6.

Table 6. Structural Number effective (SN_{eff})

Road Segment	Strength coefficient				Thickness of layer (*inch*)				SN_{eff}
	$a_{1'}$	$a_{1''}$	a_2	a_3	$D_{1'}$	$D_{1''}$	D_2	D_3	
I. Kertijayan-Bligo	0.3	0.25	0.13	0.1	1.18	2.62	2.76	7.87	**2.16**
II. Bligo-Podo	0.3	0.25	0.15	0.12	1.18	2.95	3.15	9.84	**2.75**
III. Podo-Surabayan	0.35	0.3	0.17	0.14	1.18	2.23	2.76	10.24	**2.98**
IV. Surabayan-Sedayu	0.35	0.3	0.17	0.13	1.18	2.43	4.4	7.87	**2.91**
V. Sedayu-Karanganyar	0.3	0.25	0.15	0.12	1.18	2.85	4.82	7.87	**2.74**

To determine the value of the Structural Number Future (SN_f) using the equations (5). The variables required to determine the value of the Structural Number Future (SN_f) are: Traffic forecasts of future (W_{18}/CESA), reliability, Overall standard deviation (S_0), initial design serviceability index (IP_0), design terminal serviceability index (IP_t), Design serviceability loss ($\Delta PSI = IP_0 - IP_t$) and resilient modulus (MR) of sub grade.

The Calculation of SN_f used some of the scenarios values Cumulative Equivalent Standard Axle Loadings(W_{18}) by using equation (2). The value of CESA/W_{18} and SN_f for all segments of the road with a design life (DL) of various scenarios, as shown in Table 7.

Table 7. CESA/W_{18} and Structural Number Future (SN_f)

Segment Number	W_{18}/CESA				SN_f			
	DL=5	DL=10	DL=15	DL=20	DL=5	DL=10	DL=15	DL=20
I	2,967,406	7,648,204	15,031,715	26,678,496	2.72	3.14	3.48	3.79
II	2,967,406	7,648,204	15,031,715	26,678,496	2.75	3.17	3.51	3.82
III	1,870,627	4,821,362	9,475,864	16,817,895	2.72	3.15	3.48	3.79
IV	2,338,284	6,026,703	11,844,830	21,022,369	2.71	3.14	3.47	3.78
V	2,338,284	6,026,703	11,844,830	21,022,369	2.8	3.24	3.58	3.89

Structural conditions during the design life of the pavement can be determined by calculating the difference between the value of the Structural Number Future (SN_f) with Structural Number Effective (SN_{eff}). If the Structural Number Effective (SN_{eff}) value greater than or equal to the Structural Number Future (SN_f) shows pavement in good condition, if on the contrary it is necessary to overlay structurally. The type of road rehabilitation and maintenance structurally as shown in Table 8 and Table 9.

Table 8. Calculation of Overlay

Segment Number	(SN_f - SN_{eff})				a_{OL}	Overlay thickness (cm)			
	DL=5	DL=10	DL=15	DL=20		DL=5	DL=10	DL=15	DL=20
I	0.56	0.99	1.32	1.63	0.4	**3.57**	**6.27**	**8.4**	**10.35**
II	0	0.43	0.77	1.08	0.4	**0**	**2.72**	**4.87**	**6.84**
III	-0.26	0.16	0.5	0.81	0.4	**-1.67**	**1.03**	**3.17**	**5.12**
IV	-0.2	0.22	0.56	0.87	0.4	**-1.28**	**1.42**	**3.55**	**5.5**
V	0.06	0.5	0.84	1.16	0.4	**0.4**	**3.17**	**5.35**	**7.35**

Table 9. Overlay Materials

Segment Number	Overlay thickness (cm)				Material of asphalt mixture			
	DL=5	DL=10	DL=15	DL=20	DL=5	DL=10	DL=15	DL=20
I	4	7	9	(9+3)	AC-WC	AC-BC	AC-BC	(AC-Base +HRS WC)
II	0	3	5	7	-	AC-WC	AC-WC	AC-BC
III	0	3	3	6	-	HRS-WC	HRS-Base	AC-BC
IV	0	3	4	6	-	HRS-WC	AC-WC	AC-BC
V	3	3,5	6	8	HRS-WC	HRS-Base	AC-BC	AC-BC

4. Conclusions

The simulation of IRI predictions, can determine a time plan of road rehabilitation and maintenance are functional. Layers thickness and material of asphalt mixture can be determined from the results predictions IRI in accordance with the method of Bina Marga. Road maintenance is structurally obtained from the value SN, SN_{OL} value is required to determine the thickness of the overlay and the selection of appropriate materials. So that the pavement is able to serve traffic load in the design life. Selection of road maintenance is structurally preferable, because it takes into account the capacity of the pavement, sub grade conditions and traffic load during the design life.

References

[1] American Association of State Highway and Transportation Officials 1993 *Guide for The Design f Pavement Structures* (Washington D.C.) chapter 1 pp 6
[2] Forslöf L and Jones H 2015 *J. Civil Engineering and Architecture* **9** 485
[3] Schlotjes M R, A Visser and Bennett 2014 *Proc. Int. Conf. on the 33rd Southern African Transport Conference* (Pretoria, South Africa) p 141
[4] Rachmawati D 2014 *J. Road-Bridge* **31** 95
[5] Sukirman S 1999 *Flexible Pavement Highway* (Bandung: Nova) p 114
[6] _______ 2002 *Pedoman Perencanaan Tebal Perkerasan Lentur Pd T-01-2002-B* (Ministry of Public Work and Housing Republic of Indonesia) p 16
[7] _______ 2013 *Pavement Design Manual 02/M/BM/2013* (Ministry of Public Work and Housing Republic of Indonesia) chapter 2 pp 19
[8] _______ 2014 *Standard Document Procurement and General Specifications 2010 (revision 3) for Road and Bridge Work.* (Ministry of Public Work and Housing. Republic of Indonesia) chapter 6 pp 30

IOP Conf. Series: Materials Science and Engineering **176** (2017) 012032 doi:10.1088/1757-899X/176/1/012032

Effects of Ceramic Fibre Insulation Thickness on Skin Formation and Nodule Characteristics of Thin Wall Ductile Iron Casting

D Dhaneswara[1], B. Suharno[1], N D Nugraha[1], R D S Ariobimo[2], N Sofyan[1]

[1]Department of Metallurgical and Materials Engineering, Faculty of Engineering, Universitas Indonesia, Depok, 16424 Indonesia

[2]Mechanical Engineering Department, Faculty of Industrial Technology, Trisakti University, Jakarta, Indonesia

E-mail: donanta.dhaneswara@ui.ac.id

Abstract. Skin formation has become one of the problems in the thin wall ductile iron casting because it will reduce the mechanical properties of the materials. One of the solutions to reduce this skin formation is by using heat insulator to control the cooling rate. One of the insulators used for this purpose is ceramic fibre. In this research, the thickness of the ceramic fibre heat insulator used in the mould was varied, i.e. 50 mm on one side and 37.5 mm on the other side (A), no heat insulator (B), and 37.5 mm on both sides (C). After the casting process, the results were characterized in terms of metallography by using scanning electron microscope (SEM) and tensile test for mechanical properties. The results showed that the skin thickness formed in A is 34.21 μm, 23.38 μm in B, and 27.78 μm in C. The nodule count in A is 541.98 nodule/mm^2 (84.7%) with an average diameter of 15.14 μm, 590 nodule/mm^2 (86.7%) with an average diameter of 13.18 μm in B, and 549.73 nodule/mm2 (87.2%) with an average diameter of 13.95 μm in C. The average ultimate tensile strength for A was 399 MPa, B was 314 MPa, and C was 415 MPa. Microstructural examination under SEM showed that the materials have a ductile fracture with matrix full of ferrite.

1. Introduction

In manufacturing industries, aluminium is the most preferred material for any automotive parts because of its light weight and strength. Unfortunately, aluminium has a large production cost if we compared it to iron production [1]. In term of characteristics, ductile iron can compete with aluminium but for the weight. In this instance, thin wall ductile iron (TWDI) has been developed to compete with aluminium to overcome the light weight problem.

One of the problems in thin wall ductile iron casting is the formation of the skin, which can reduce the mechanical properties of the material [2]. The main cause of this problem is due to the formation of graphite nodule in which the edge of material tends to form flakes graphite. Specifically, this problem become worse when magnesium (Mg) does not fully functioned because of rapid cooling rate at the edge of material [3].

Ceramic fibre has been known to have an excellent heat insulator that can withstand temperature even above 1000 °C [4]. By using ceramic fibre as a heat insulator inside the sand mould, it could decrease the cooling rate of the molten metal and reduce the formation of skin during the casting and hence would improve the mechanical properties of TWDI.

2. Experimental

The research was done in foundry based industry. The casting design of this research was using vertical casting design based on Stefanescu's research [5] with some modifications. The sand used for the moulding was silica sand with furan binder. In order to examine the effect of the thickness of ceramic fibre insulator in the mould on the skin formation and nodule, the ceramic fibre was placed inside the mould before pouring the silica sand inside the casting mould.

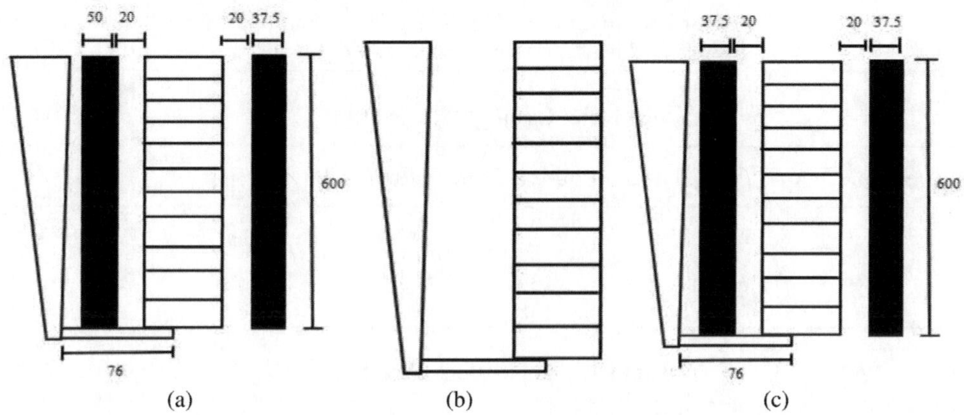

Figure 1. Casting mould with (a) ceramic fibre insulator with thickness 50 mm on left side and 37.5 mm on right side, (b) no insulator, (c) ceramic fibre insulator with a thickness 37.5 mm on both sides

In this research, certain variation of the thickness of ceramic fibre heat insulator was used, that is 50 mm on the left side and 37.5 mm on the right side (A), no heat insulator (B), and 37.5 mm on both sides (C). The distance between the heat insulator and the mould was 20 mm on both sides. After casting moulds were ready, the moulds were coated using graphite to smoothen the surface of material. The molten metal was poured with temperature between 1380-1410°C at maximum pouring time 10s to keep the quality of the end material. The raw materials for making the molten metal were steel scrap and return scrap. The melt was added with slag removal and carburizer before liquid treatment process for inoculation and nodulizing purposes. Before and after the liquid treatment, spectrometry test was carried out to examine the composition of the material. After the material was taken out, shot blasting was performed on the material to reduce surface roughness of the material.

The characterization included microstructural examination by using scanning electron microscope (SEM) and mechanical properties by using tensile test. The purpose of microstructural examination is to observe nodule diameter, nodule count, and thickness of the skin formation. For tensile testing, JIS Z 2241:2011 number 5 was used as a reference. This examination was to observe ultimate tensile strength (UTS) and elongation of the material, whereas SEM examination was performed to examine the fractography of the materials.

3. Results and Discussion

Samples codes and information are given in Table 1. Casting process data containing information on holding time, pouring temperature, pouring time during casting process, and visual information of the sample are given in Table 2, whereas composition of the sample is given in Table 3.

Table 1. Sample Codes and Information

Sample	Information
A	First pouring order; Mould was using ceramic fibre insulator with thickness 50 mm on left side and 37.5 mm on right side
B	Second pouring order; Mould did not use any insulator
C	Third pouring order; Mould was using ceramic fibre insulator with thickness 37.5 mm on left and right side

Table 2. Casting Process Data

Sample	Holding Time	Pouring Temperature	Pouring Time	Plate Thickness
A	5 min 55 s	1445°C to 1420°C and last at 1406°C	9.00 s	2.18 mm
B	6 min 4 s	1395-1390°C	11.75 s	2.08 mm
C	6 min 14 s	1385-1380°C	9.40 s	2.32 mm

Table 3. Composition of the Metal

Element	Before Liquid Treatment	After Liquid Treatment	Factory Standard
C	3.97	3.78	3.90 – 4.00
Si	1.601	2.67	1.5 – 1.7
Mn	0.496	0.51	0.30 – 0.50
P	0.007	0.0069	0.03 max
S	0.019	0.0171	0.02 max
Cu	0.0421	0.0428	0.15 max
Ni	0.0073	0.0077	0.15 max
Cr	0.0007	0.0027	0.15 max
Mo	0.0084	0.0084	-
Mg	0	0.0572	0.03 min
Sn	0.008	0.0080	-

As can be seen from Table 3, Si content after liquid treatment is much higher than that of the factory standard. This phenomenon is expected because Si helps cast iron to grow more graphite. On the contrary, Cu, Ni, Sn, and Mo are pearlite stabilizers [5], but it can be solved by giving more silicon to promote more ferrite phase. In this instance, the composition of sulphur and manganese need to be controlled because they will reduce the mechanical strength of thin wall ductile iron.

Figure 2 shows that the cooling rate of the material can affect nodule count, percentage of the nodular (nodularity) and nodule diameter. It can be seen that Sample B has the highest count of nodule compared to the others. This can be understood since sample B has the highest cooling rate as the mould lacks of the heat insulation. On the contrary, sample A has the lowest nodule count as compared to the others samples. This is because sample A has the slowest cooling rate from all of samples. Percentage of the nodule in sample C is the highest among all the samples, because sample C has the best cooling rate as compared to the other samples; it is neither too slow nor too fast. Sample A has the biggest nodule diameter among all the other samples due to the slowest cooling rate. In terms thickness of the skin formation, sample B has the thinnest skin formation because it has the highest cooling rate among all other samples. On the contrary, A has the thickest skin formation because it has the slowest cooling rate. The slower the cooling rate the thicker the skin.

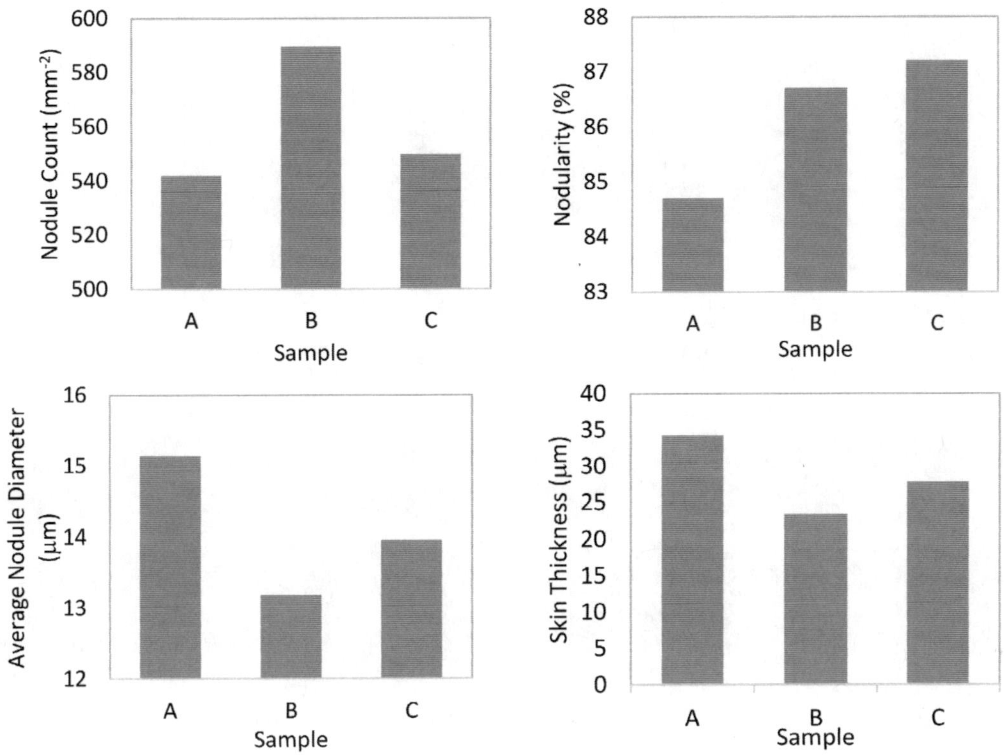

Figure 2. Characteristics of the samples for nodule count, nodularity, nodule diameter, and skin formation

Figure 3 shows nodule diameter distribution in terms of the homogeneity of the samples. As can be seen from the figure, the uniformity in nodule size is quite scattered in the range of 5 to 25 μm. Figure 4 shows the microstructure of the samples from optical microscope. From the figure, it can be seen that sample B has the most carbide on the surface compared to the other samples. This can be understood since sample B has the highest cooling rate among all other samples. The more the cooling rate the more carbide layer on the surface of the material.

Figure 5 shows the average mechanical properties of samples A, B, and C in terms of strength and elongation. As can be seen in the figure, the strength of sample C is the highest among the three despite the fact that B has the lowest skin formation. It seems that the number of nodules and percentage of nodularity cannot be related to this mechanical property. The opposite is true for nodule diameter and the skin thickness.

Figure 6 shows secondary electron images of samples fractography. From the figure, it can be seen that sample C has better fractography in terms of dimple as an indication of ductile fracture as compared to those of the other two samples. Thus, even other factor could play important role in determining the mechanical properties, in this work, nodule diameter seems to has a profound effect.

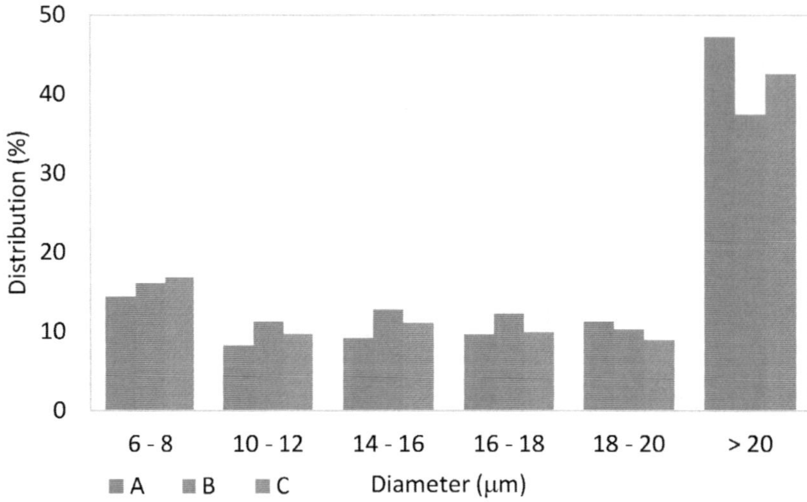

Figure 3. Nodule diameter distribution

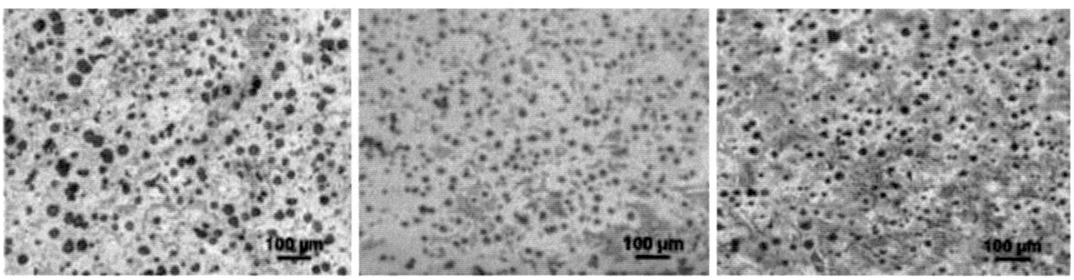

Figure 4. Optical microstructure of samples A, B, and C showing carbides on the surface

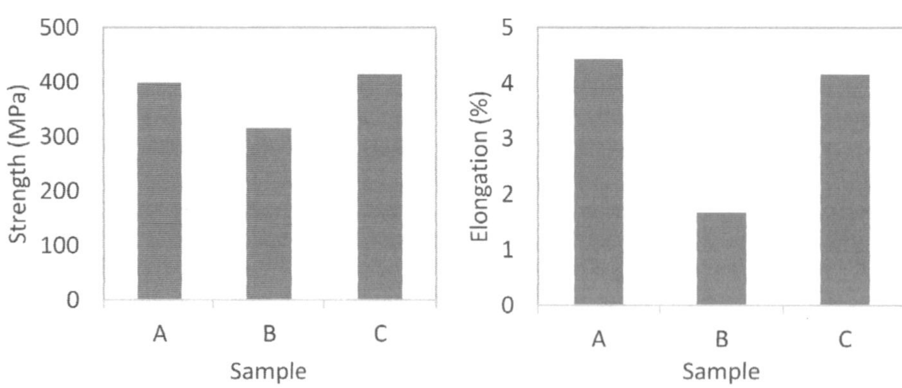

Figure 5. Mechanical properties of samples A, B, and C in terms of strength and elongation

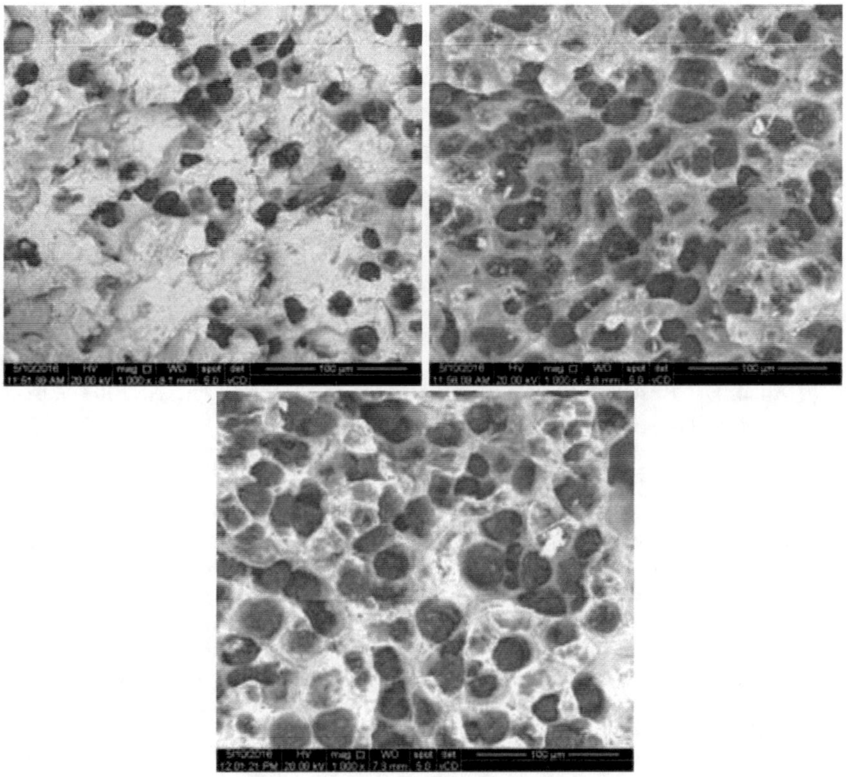

Figure 6. Secondary electron image of the fractography of samples A, B, and C, each with a scale bar of 100 μm.

4. Conclusion

The used of ceramic fibre heat insulator has the effect on cooling rate of thin wall ductile iron casting. The slower the cooling rate, the bigger nodule diameter, the lower the nodule count, and the thicker the skin layer formation. On the contrary, the higher the cooling rate, the smaller the nodule diameter, the higher the nodule count, and the thinner the skin formation.

Acknowledgment

The authors would like to thank DRPM Universitas Indonesia for the research grants through Hibah PITTA No: 2080/UN2.R12/HKP.05.00/2016.

Reference

[1] Fraś E, Górny M, Lopez H 2014 *Arch. Metall. Mater.* **59** 459
[2] Auferheide R C, Showman R. E, Hysell M A 2005 *AFS Transactions* **113** 567
[3] Boonmee S 2013 *Ductile and Compacted Graphite Iron Casting Skin-Evaluation, Effect on Fatigue Strength and Elimination* (Dissertation: The Ohio State University).
[4] Wallis R.A. 1989 *The Thermal Performance of an Ultra High Power (UHP) Electrical Furnace for The Reheating of Steel Slabs* (Dissertation: Cranfield Institute of Technology).
[5] Stefanescu D, Wills S, Massone J, Duncan F 2009 *AFS Transactions* **117** 687.

IOP Conf. Series: Materials Science and Engineering **176** (2017) 012033 doi:10.1088/1757-899X/176/1/012033

The Priority of Road Rehabilitation in Karanganyar Regency Using IRI Estimation from Roadroid

F Achmadi[1], M Suprapto[1] and A Setyawan[2]
[1]Magister Maintenance and Rehabilitation Infrastructure SebelasMaret University Surakarta, Indonesia
[2]Roadmate Research Group, Civil Engineering Department, Sebelas Maret University, Jl. Ir Sutami 36 A Solo 57126 Indonesia

E-mail : arysetyawan@staff.uns.ac.id

Abstract. The IRI (International Roughness Index) is a road roughness index commonly obtained from measured longitudinal road profiles. This is one of the functional performance a surface of road pavement. Therefore, needs to be done evaluation and monitoring periodically to getting priority of road rehabilitation right on target. The IRI standard has commonly been used worldwide for evaluating road system. The Roadroid is an application to measure road quality with a website to view road quality. It is designed for Android smartphones, so we can easily measure and monitor the road and also use the camera for GPS-tagged photo. By using the built-in vibration sensor in smartphones, it is possible to collect IRI value which can be an indicator road conditions. This study attempts to explain the priority of road rehabilitation in Karanganyar Regency. The location of the study focused on a collector street (primary, secondary and locally road). The result of IRI estimation will be combined with other aspects that influences; land use, policy, the connectivity of road and traffic average daily. Based on IRI estimation using Roadroid, the road conditions in Karanganyar Regency can be described 59,60% were good (IRI<4,5); 21,30% fair (4,5<IRI<8), 11,40% bad (8<IRI<12), and 7,70% poor (IRI>12).

1. Introduction

The road is basic infrastructure in which supported regional economy without adequate was not optimal. Not all regions having enough budget, so it might impact on the ability of allocating budget in road infrastructures. Therefore, they had to have a proper priority to available budget in order to maintenance optimal road.

Karanganyar is one of the regencies in Central Java province, about 14 kilometers from Surakarta City. It bounded with Sragen Regency in the north, Ngawi Regency and Magetan Regency (East Java) in east, Wonogiri Regency in south and Boyolali Regency, Surakarta City and Sukoharjo Regency in the west. Karanganyar has topography diverse partly in the form of lowlands and highlands in Lawumountain. Its geographical location needs to be supported facilities and infrastructure sufficient especially road management.

A preliminary step to determine the priorities of road rehabilitation is the survey road conditions. It needs as a part of rehabilitation and maintenance planning. Usually the level of road damaged determined through the visual observation. Limited equipment, minimal cost and human resource

[2] To whom any correspondence should be addressed.

ability became an absolute consideration in which visual observation a basis of the rehabilitation program. Whereas visual observation has many insufficiency.

The modern traditional techniques for measuring roughness may be categorized as either specially built trucks or wagons with laser scanners, bump-wagons or even manually operated rolling straight edges. Specially built measurement equipment is expensive, due to heavy and complex hardware, low volume of production and the need of sophisticated systems and accessories.

Roadroid is an application to measure road quality and a home page where we can see the road quality on a map. This application is designed for a smartphone with Android operating system. Roadroid is based on the result of several years of research for The Swedish National Road Administration (SNRA). It can be collecting data and give an IRI estimation for road rehabilitation references. During the IRI survey need the complex equipment, more expensive, and take a long time. Measuring roads with smartphones can provide a simple method, efficient, scalable and cost-effective way for road organization to deliver road condition data. In other side, the result of IRI estimation can be accountable and approaching the real-IRI.

The objective of this study is how making a priority system of road rehabilitation in Karanganyar Regency. The study location in road Karanganyar Regency, focused in a collector street (primary, secondary and locally road).

1.1. Evaluation of Road Conditions

The assessment of road conditions is an important step how to know the level of road damage. Depending on which characteristic is being surveyed, a pavement evaluation can be classified as functional or structural. Functional evaluation provides information about surface characteristics that directly affect the users safety and comfort or serviceability. Safety is evaluated in terms of skid resistance and surface texture, while serviceability is quantified through roughness measures. Table 1 shows the road performance condition categories.

Structural evaluation provides information on whether the pavement structure is performing satisfactorily under the traffic load and environmental conditions. This includes surveys on pavement distresses and mechanical/structural properties of pavements.

Evaluating structural conditions and functional conditions of existing, in-service pavements continuously a major part of the maintenance and rehabilitation activities undertaken by Highways Agencies. The structural and functional condition of the pavement changes with the passage of time due to the combined effects of its structural adequacy, composition and loading characteristics of traffic, environmental conditions and the maintenance inputs provided. The process of accumulation of damage is called deterioration and the failure of the pavement is said to have reached at the limiting stage of serviceability level. The physical sign of internal damage, for example, cracking, rutting, potholes, etc. are known as distress, which are indicators of the pavement condition [1].

Pavement evaluations are performed in the field through manual surveys or using specialized equipment. Evaluated characteristics of the pavement are quantified by means of indicators or condition indices. Data collection equipment should be reliable, efficient and secure. To ensure cost effective surveys and data referencing consistency, it is recommended to collect multiple pavement characteristics during a single pass of the data collection vehicle [2].

Table 1. The Indicators of road performance conditions

Evaluation	Operational Performance	Pavement Characteristic	Index & Indicators
Functional evaluation	Serviceability	Roughness	IRI
			PSI
			QI
	Safety	Texture	Macro texture
			Micro texture
		Skid Resistance	Skid resistance coefficient
			IFI
Structural evaluation	Structural Capacity	Mechanical Properties	Deflection
		Surface Distress	Cracking
			Surface defect
			Profile
			Deformation
Referencing System		(Location of Pavement Characteristic Data)	

1.2. The Influences of IRI Value on Road Rehabilitation

The ride quality is one of the most important conditions used by the travelling public to judge roadway pavements. Rough pavements have a significant impact on public satisfaction, safety, and also on the economy.The International Roughness Index (IRI) established in 1986 by the World Bank. IRI is defined as an expression or irregularities in the longitudinal profile of pavement surface that adversely affects the ride quality of a vehicle thus causing discomfort to the user. The IRI is a scale for roughness based on the simulated response of a generic motor vehicle to the roughness in a single wheel path of the road surface. IRI true value is determined by obtaining suitably accurate measurements of the profile of the road, processing it through an algorithm that simulates the way a reference vehicle would respond to the roughness inputs, and accumulating the suspension travel. It is normally reported in inches/mile or meters/kilometers[1].

The measurement of the properties of road roughness was going to be beneficial to road rehabilitation and maintenance program. In Indonesia, the local government has not done this measurement because the network of roads is so fast. In addition, equipment of roughness is very expensive and limited [4].

The road conditions survey must be done periodically, both in structural and functional survey. It's to know the level of road service. The functional road evaluations aim of checking of roughness, road texture and skid resistance. It would be useful to determine the priority of rehabilitation and maintenance program.

Roadroid was developed by the Swedish National Road Administration (SNRA). It can be an alternative for the road maintenance decision support. The asset management of a road can be optimized with the right decision and its huge money to be saved. The right decision can be taken with access to up to date data of good quality [5].

Roadroid has two options for roughness data calculation: eIRI (estimated IRI), based on a peak and RMS (root mean square) vibration analysis-which is correlated to laser measurements on paved roads, and c-IRI (calculated IRI), based on the quarter car simulation (QCR), the sensitivity of the device can be calibrated by the operator to a known reference. IRI can be expressed as a summary of a roads standard. Other input to decision support might be traffic volume and other road condition parameters, as rutting, cracking, edge brakes, culvert inspections, etc. In the correlation studies between Roadroid and IQL1/Laser Data, to the same spatial road links, we can see an illustration in Figure 1.

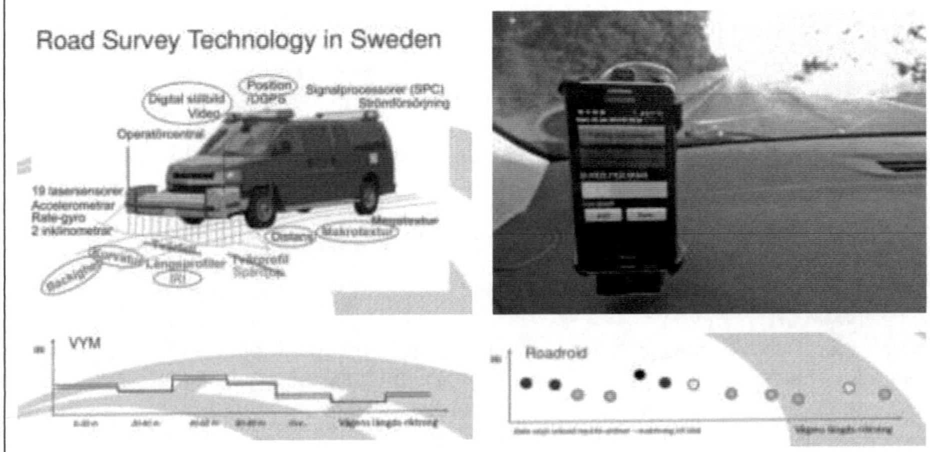

Figure 1. The comparison of IRI survey result between IQL1 and Roadroid

Based on *The Technique Management Series Guide of Regional Road Maintenance,* 2005, the road conditions minimum was being fair with the range of IRI between 4.5 m/km till 8 m/km, hanging from the function of the road. If the IRI value under 4.5, it means the road was under routine maintenance, and if the IRI value between 4.5 to 8 it means the road have to be done a periodic road maintenance or overlay. And the IRI value more than 12, the road was poor conditions and it have to be reconstructions. The explanation can seen in Figure 2.

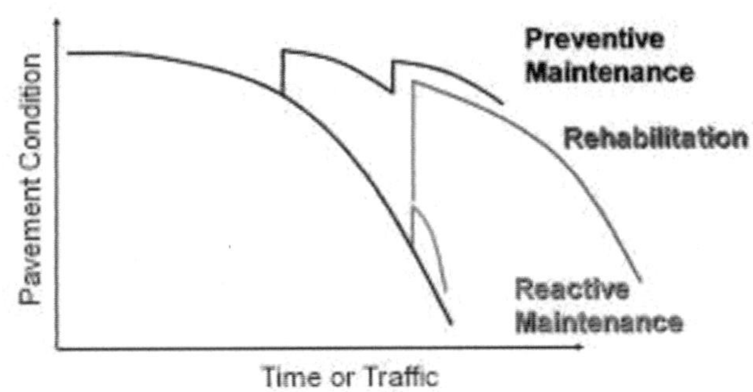

Figure 2. Road rehabilitation based on pavement condition

1.3. Other Influences Factors

For the implementation in local government, the policy of road rehabilitation priority, not only based to technical aspect like road conditions, it is also being affected by the other aspect. Based on Wahyudiana [6], the aspect that has some influences to road rehabilitation as follows:

1) Technical aspect. This factor, including road conditions and topography. The lower this aspect, the more important to rehabilitation priority.
2) Land uses aspect. This factor is used many activities, the higher of their land-use of the more important as well.
3) The policy aspect. Political factors were very needed in policy budgeting.

4) Road connectivities. This factor was needed to improve the road networking for the future.
5) Emergency aspect. Emergency factors can be a natural disaster or the other factors that cause the urge to repaired.

2. Experimental

2.1. The Location of Study
The location of the study was on the road Karanganyar Regency focussed in collector street (primary, secondary and locally road).

2.2. Instrument of Measurements
IRI estimation was obtained using Roadroid for Android smartphone. It puts on the windshield of a car or other vehicle. The next step is setting Roadroid by adjusting the type of the car, the distance of automatic documentations and a segment of IRI estimations.

3. Result and Discussion
The result of the IRI survey using Roadroid will get data as follows :

a. Date Time	f. Altitude
b. Latitude	g. Grade
c. Longitude	h. e-IRI
d. Distance	i. c-IRI
e. Speed	j. Road

The example of the IRI survey result can be seen in Table 2.

Table 2. The example of IRI survey-result using Roadroid at Colomadu-Banyuanyar road

Date Time	Latitude	Longitude	Distance (m)	Speed (km/h)	Altitude (m)	Grade (%)	eIRI	cIRI
06/11/2015 13:56	-752.991.471.832.532	11.074.937.935.612	20	21.3	140.11	0	1.88	0.35
06/11/2015 13:56	-752.990.717.215.126	110.749.728.765.123	40	21.38	140.7	2.95	2.44	0.35
06/11/2015 13:56	-752.990.079.210.511	110.749.718.910.529	60	21.21	141.16	2.33	3.03	0.35
06/11/2015 13:56	-752.988.488.668.056	110.750.041.932.126	80	21.21	141.16	0	3.03	0.35
06/11/2015 13:56	-752.988.336.063.382	110.750.076.929.464	100	21.34	143.35	10.92	2.84	0.49
06/11/2015 13:56	-752.987.317.546.356	110.750.322.838.232	120	26	144.82	7.35	1.85	0.54
06/11/2015 13:56	-752.986.783.780.621	110.750.400.358.373	140	23.31	145.28	2.32	1.92	0.54
06/11/2015 13:56	-752.983.750.853.005	110.750.711.100.958	160	20.76	145.02	-1.28	2.15	0.57
06/11/2015 13:56	-752.982.751.489.626	110.750.866.358.811	180	22.04	143.83	-5.95	3.18	0.63
06/11/2015 13:56	-752.981.566.099.216	110.751.048.589.374	200	24.1	142.69	-5.73	3.51	0.63
06/11/2015 13:56	-75.298.014.131.202	110.751.227.763.892	220	24.84	141.4	-6.45	3.03	0.63
06/11/2015 13:56	-752.978.278.157.321	110.751.419.519.228	240	23.55	140.15	-6.24	2.29	0.63
06/11/2015 13:56	-752.977.457.524.586	110.751.502.061.422	260	24.14	139.56	-2.95	2.6	0.63
06/11/2015 13:57	-752.976.879.477.996	110.751.694.902.763	280	22.84	139.31	-1.26	5.46	0.69
06/11/2015 13:57	-752.977.570.899.747	110.751.919.848.741	300	20.98	140.17	4.31	3.12	0.76
06/11/2015 13:57	-752.979.464.265.367	11.075.218.398.978	320	20.57	140.97	4	7.94	0.76
06/11/2015 13:57	-752.980.446.576.279	110.752.270.706.538	340	21.32	141.74	3.83	3.76	0.76
06/11/2015 13:57	-752.981.877.752.693	110.752.436.197.291	360	20.77	142.59	4.27	8.77	0.76
06/11/2015 13:57	-752.984.742.996.673	110.752.650.652.275	380	20.8	142.75	0.81	7.82	0.81
06/11/2015 13:57	-752.986.912.628.848	110.752.767.603.378	400	21.09	141.45	-6.53	7.72	0.96

From the table, we can see that IRI value under 4.5, it means the road was good condition, the IRI value between 4.5 to 8 it means the road is fair condition, the IRI between 8 until 12 was bad condition and the IRI more than 12 was poor condition. Have to be done at periodic road maintenance or overlay. And IRI value more than 12, the road was poor conditions.

4. Conclusion

Based on IRI estimation using Roadroid, the road conditions in Karanganyar Regency can be described as follows: the local road to the long 386.83 kilometers or 59.60% were good conditions ($IRI<4.5$); then the road about 139.98 kilometers or 21.30% fair conditions ($4.5<IRI<8$); the road about77.63 kilometers or 11.40% bad ($8<IRI<12$), and the road about 53.27 kilometers or 7.70% poor ($IRI>12$).

References

[1] Minu, P.K., Sreedevi and Babu R., 2014 *International Journal of Engineering Research* **3** 908

[2] Bennet C., 2007., *The Road and Rural Transport Thematic Group Washington DC* (Washington DC Transport Note) p 2

[3] Department of Public Works 2005*The Technique Management Series Guide of Regional Road Maintenance* Research and Development Agency Research Center Development of Transportation Infrastructure chapter A p 5-10

[4] Suwardo, 2004., *Simposium VII FSTPT Parahiyangan Catholic University* 2

[5] Hans, J. and Lars, F., 2015 *Journal of Civil Engineering and Architecture* **9** 485

[6] Wahyudiana., 2009., *The Determination of Priority Regional Road Maintenance Based on The Availability of The Allocation of Funds* (Surakarta : Surakarta Sebelas Maret University) p 67

IOP Conf. Series: Materials Science and Engineering **176** (2017) 012034 doi:10.1088/1757-899X/176/1/012034

Assessing the durability of North Buton Asphalt seal with Polymer Modified and Rejuvenation in warm mixture design

Zalman[1], B Yulianto[1] and A Setyawan[2]
[1]Magister Maintenance and Rehabilitation of Infrastructure, Sebelas Maret University, Jl. Ir. Sutami 36 A Solo 57126, Indonesia
[2]Roadmate Research Group, Civil Engineering Department, Sebelas Maret University, Jl. Ir. Sutami 36 A Solo 57126, Indonesia

E-mail: arysetyawan@staff.uns.ac.id

Abstract. Utilization of Buton Asphalt has been expanded with various optimizations through modification of the base material so that it can be used either as an additive in granular form or as modified asphalt. While in North Buton using Asphalt Buton that uses a special material specification Buton Asphalt cold mixture of Butur Seal, a thin layer of Buton Asphalt B 50/30 above the base course or existing asphalt pavement layer, which has been prepared in accordance with the General Specifications. Technically utilization of Butur Seal is still very sensitive to human resource capacity in understanding the physical condition of Asbuton, and severely affects construction to failure. Buton asphalt cold mix in the field also showed some kind of damage caused by difficulties or not fitting rejuvenation materials used. Making the challenge to do research on the characteristic properties of Asphalt Buton B50 / 30 with modifications, to get Asbuton that has higher durability in use in the field. Quality performance of the Asbuton cold mixtures is observed through a series of tests in the laboratory. This test includes testing the stability and the compressive test. Tests conducted in the laboratory are expected to be directly applicable as it is done in the field. This research aimed to investigate the characteristics of Butur Seal in warm mixture that can be used in the construction and maintenance of roads in North Burton and to investigate the characteristics of Butur Seal with the addition of elastomeric polymers and rejuvenation materials in warm mixing temperature of 30, 40, 60 and 80° C.

1. Introduction

Efforts to save the use of petroleum asphalt, began to develop research on road material, especially research for optimum use of local materials including Buton asphalt. In North Buton using Asphalt Buton that uses special material specification "Butur Seal". The definition of butur seal is asbuton B 50/30 spread on the base course or old existing asphalt pavement.

Technically utilization of Buton Asphalt (Butur Seal) is still very sensitive to human resource capacity in understanding the physical condition of Asbuton, and severely affects construction failure. Buton asphalt cold mix (Butur Seal) in the field also showed some kind of damage caused by hard or not fitting fluxing materials used.

Making the challenge to do research on the characteristic properties of Asphalt Buton (Butur Seal) B50 / 30 with modifications, to get Asbuton that has higher durability in use in the field. Quality

[2] To whom any correspondence should be addressed.

performance Asbuton mixture (Butur Seal) can be known through observation and a series of tests in the laboratory. This test includes testing of Marshall. Tests conducted in the laboratory are expected to be directly applicable as it is applied in the field.

Asphalt demand in Indonesia continues to increase, but these needs are not met due to the limited capacity of producing asphalt. In addition to materials of petroleum-based asphalt, there is a genuine asphalt or natural bitumen, as can be mined on Buton Island, Indonesia. Type natural asphalt derived from Buton Island is divided into two kinds, namely the hard asphalt was found in the area Kabungka and soft asphalt that can be mined in the area Lawele. Natural asphalt comes from the island of Buton has bitumen content varies from 10% to 40%, with an average of about 21.8%. Total natural bitumen deposits are estimated at 350 million tons [1].

Asbuton stands for "Asphalt Buton" or "Rock Asphalt Buton". Asbuton as natural asphalt deposit is located in Buton Island Southeast Sulawesi Province. Asbuton first discovered by a Dutchman named Hetzel in 1920. Subsequently, in 1936 Hetzel has been able to map the deposit of Asbuton on Buton Island [2]. Asbuton, or natural asphalt from Buton Island with the highest in the world when compared with other natural bitumen deposits, is one of Indonesia's natural wealthiest potential as material substitution, especially on the asphalt pavement asphalt mixture [3].

2. Experimental

2.1. Butur Seal (North Buton Seal)

Utilization of Buton Asphalt has been expanded with various optimizations through modification of the base material so that it can be used either as an additive in granular form or as modified bitumen [2]. Asphalt Buton technical nature based on the location of the mine is presented in Table 1. These have Buton Asphalt bitumen content variation between (20-30)% by weight, with values varying penetration. Asphalt is derived from the Lawele mine has a value of penetration is much higher than the asphalt coming from the mine site Kabungka.

Table 1. The propertis of rock asphalt from Kabungka and Lawele quarry

Properties	Unit	Kabungka Quarry	Lawele Quarry
Asphalt content	% weight	20	30,8
Penetration (25°C,100g, 5 second)	0,1mm	4	36
Softening point	°C	101	59
Flash point	°C	-	198
Ductility(25°C, 5 cm/min)	cm	<140	>140
Specific Gravity		1.046	1.037
Penetration (LOH) (25°C,100g, 5 second)	% weight		94
Softening point (LOH)	°C		62
Loss of weight (163°C, 5 hour)	% weight		0.31
Solubility in trichloroethylene	% weight		99.6

Sources : Road and Bridge Research & Development [4]

The Asphalt Buton from the mine site of Lawele or known Asbuton B 50/30 is selected for this research. In the area of North Buton, Buton B 50/30, better known by the name of North Buton Seal (Butur Seal). The property requirement for a North Buton seal are presented in Table 2.

2.2. Research Method

This research was conducted at the Highway Laboratory, Civil Engineering Department, Engineering Faculty, University of Sebelas Maret. The investigation utilizes the Marshall method. The materials used in this study are Asphalt Buton (Butur Seal) B50 / 30 and Polymer Elastomer (Latex).

The research process begins by carrying out the job mix design of Asbuton Butur Seal in warm mixing temperature at 30, 40, 60 and 80°C, then producing the Marshal sample with a mixture of Nort Buton seal. In addition the polymer latex was added to vary between 1% to 4%.

Table 2. The property requirements for North Buton Seal 50/30

No	The properties	Method	Requirements
A	Original shape		
1.	Granular size, mm	SNI 1969	9,5 mm
2.	Water content, %	SNI 2490	Max. 2
B	Asbuton Granular shape B 50/30		
1.	Bitumen content, %	SNI 03 3640	25 – 40
2.	Penetration 25°C, 100g, 5 second, 0.1 mm	SNI 06 2456	40 - 60
3.	Softening point, °C	SNI 06 2434	Min. 55
4.	Ductility, 25°C, Cm	SNI 06 2432	≥100
5.	Specific Gravity	SNI 06 2441	Min. 10
6.	Flash point, °C	SNI 06 2433	Min. 232
C	Residual after TFOT		
1.	Loss of Specific Gravity (LoH),%	SNI 06 2441	≤ 3
2.	Penetration 25°C, 100g, 5 second, 0.1 mm	SNI 06 2458	≥ 54

Sources : Centre for Road and Bridge, DPU, 2012 [5]

3. Results and Discussion

The Test results on Asbuton Butur Seal in warm mixing with a temperature range of 30, 40, 60 and 80°C coupled with latex content at varing between 1 to 4%, the best results were obtained at temperature of 80°C with high levels of latex content. Table 3 to Table 6 show the Marshall and volumetric properties at the temperature of 30°C, 40°C. 60°C and 80°C, respectively.

Table 3. The Marshall and volumetric properties at 30°C

Latex Content (%)	Stability (Kg)	Flow (mm)	MQ (kg/mm)	Porosity (%)	Density (gr/cc)
1	63.649	12.50	5.114	53.913	1.2
2	104.516	9.67	10.818	53.014	1.3
3	107.782	9.28	11.622	51.383	1.3
4	87.973	6.80	12.928	48.511	1.2

Table 4. The Marshall and volumetric properties at 40°C

Latex Content (%)	Stability (Kg)	Flow (mm)	MQ (kg/mm)	Porosity (%)	Density (gr/cc)
1	112.686	9.00	19.113	48.249	1.38
2	134.367	0.00	19.372	45.932	1.37
3	233.785	6.35	19.487	45.599	1.52
4	158.620	5.75	37.878	44.169	1.40

Table 5. The Marshall and volumetric properties at 60°C

Latex Content (%)	Stability (Kg)	Flow (mm)	MQ (kg/mm)	Porosity (%)	Density (gr/cc)
1	259.330	7.15	40.801	44.409	1.48
2	407.066	6.35	53.873	43.082	1.49
3	429.088	6.05	70.106	42.454	1.55
4	385.230	4.40	103.809	40.679	1.51

Table 6. The Marshall and volumetrik properties at 80°C

Latex Conten t (%)	Stability (Kg)	Flow (mm)	MQ (kg/mm)	Porosit y (%)	Densit y (gr/cc)
1	727.827	4.70	154.964	40.689	1.61
2	590.501	5.05	118.219	41.065	1.56
3	484.150	5.25	103.924	41.541	1.52
4	437.285	5.80	76.981	39.575	1.53

The better workability of the North Buton mixture is expected by increasing the mixing and compaction temperature and the addition of polymer latex. Porosity and density values could be used as an in indicators of workability. The correlations between density and latex content at various temperature are presented in Figure 1. It could be seen that increasing temperature resulted in higher density value. The data of porosity in Figure 2 gives similar trend of better porosity at a higher temperature of compaction. The higher the temperature the more workability of the mixture [6], so that the better compaction could be achieved. The variation of latex contents shows that the optimum percentage of latex depend on the compaction temperature, the higher the temperature the higher the latex content required for better compaction efficiency.

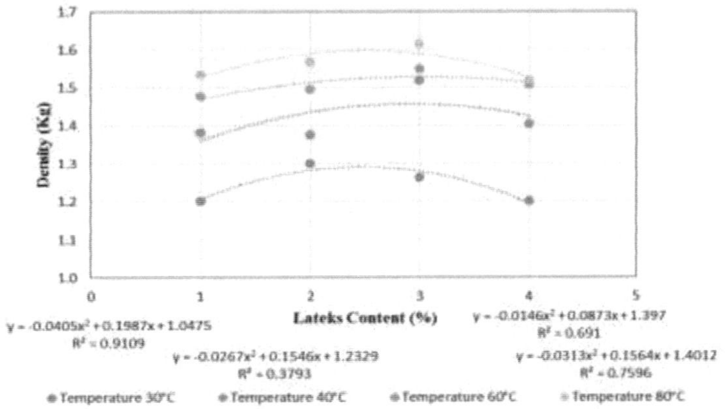

Figure 1. The correlation between Density and Latex content at various temperatures

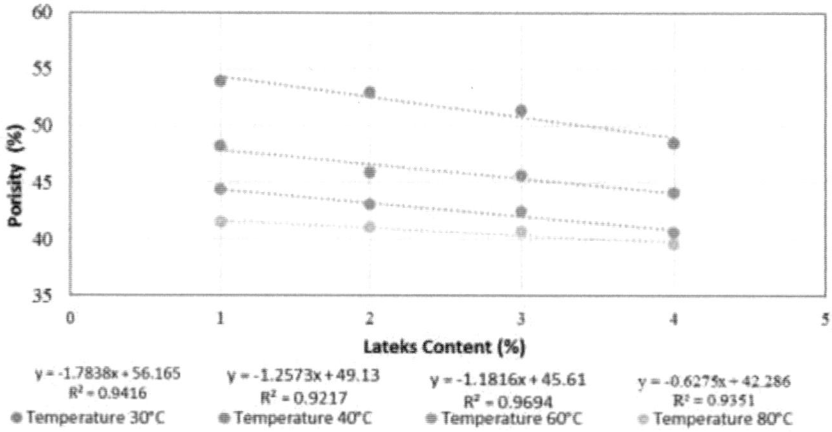

Figure 2. The correlation between porosity and latex content at various temperatures

The Marshall properties as the indicators of the strength of the materials could be seen from the Marshall stability values. As presented in Figure 3, the higher the stability value could be achieved at higher compaction temperatures. The better workability has an influence on the strength of the compacted materials. The addition of polymer as rejuvenation also achieves the optimum at 3% of latex content. It could be concluded that the problem of the North Buton seal application is at the compaction temperature and the need of rejuvenation to improve the workability. Further research is required to investigate the optimum compaction temperature and the optimum rejuvenation content to achieve ideal construction of North Buton seal [7].

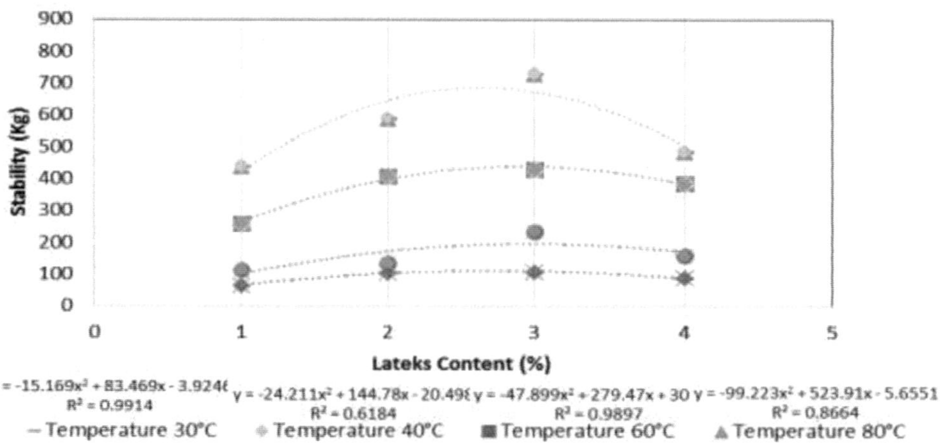

Figure 3. The correlation between latex content and Marshall Stability

4. Conclusion

Increasing the compaction temperature resulted in higher density and consequently lower porosity of the mixture, indicates that better workability has been achieved at higher combustion temperatures. Incorporating rejuvenation such as latex could improve the workability but the latex content need to be considered as the specific compaction temperature.

The Marshall stability values as indicator of the material's strength could achieve higher value at the better workability of the mixture during the compaction. Further research is required to investigate the optimum temperature compaction and rejuvenation content for better properties of the North Baton seal application.

References

[1] Hadiwardoyo, S. P., Sinaga, E. S., & Fikri, H. (2013). *Construction and Building Materials*, *42*, 5-10.

[2] Hermadi, M. and Sjahdanulirwan, M., 2008. *Jurnal terpublikasi. Bandung: Puslitbang Jalan dan Jembatan.*

[3] Davidson, J. W., 1991., *The geology and prospectivity of Buton island, SE Sulawesi, Indonesia.* 209-233.

[4] Departemen Pekerjaan Umum, 2010, Pemerintah Tingkatkan penggunaan AsButon untuk Penaganan Jalan, Jakarta

[5] Dirjen Bina Marga, 2012; Dokumen Spesifikasi Khusus Lapis Pondasi Batu kapur (Butur Seal Asbutonseks 6.6 kh, Jakarta

[6] Chowdhury, A., & Button, J. W. , 2008. *A review of warm mix asphalt* (No. SWUTC/08/473700-00080-1). Texas Transportation Institute, the Texas A & M University System.

[7] Guo-yuan, H. W. T. X., 2012. *Journal of South China University of Technology (Natural Science Edition)*, *2*, 018.

Assessing The Durability of Polymer Modified Asphalt Emulsions Slurry Seal

C Singgih[1], D Handayani[1], A Setyawan[2]
[1]Magister Maintenance and Rehabilitation Infrastructure, Sebelas Maret University Surakarta,
[2]Roadmate Research Group, Civil Engineering Department, Sebelas Maret University, Jl. Ir. Sutami 36 A Solo 57126 Indonesia

E-mail: arysetyawan@staff.uns.ac.id

Abstract. Slurry Seal is an application of road preservation in the form of impermeable non-structural thin layer with maximum thickness of 10 mm, which consisting of a cold laid mixture of asphalt emulsion with continuous graded fine aggregate, mineral filler, water and other added ingredients. Road preservation use slurry seal only functioning as a surface layer on the existing pavement structure. This preliminary research was conducted to determine the value of consistency, setting time, and indirect tensile strength of polymer modified slurry seal. The laboratory tests were conducted to determine the optimum residual asphalt content. The results show that the value of the optimum water content by pre-wetting 5% is getting smaller with increasing levels of residual asphalt emulsion. The addition of water 0 – 2.5% with 5% water for pre-wetting, the mixture provides a sufficient consistency in accordance with the specifications. The increasing levels of residual asphalt emulsion obtained the longer setting time at all slurry seal mixtures, but all of the mixtures still meet the specifications. The use of polymer modified asphalt emulsion on slurry seal was improved durability significantly, based on the value of indirect tensile strength.

1. Introduction

Slurry Seal is an application of road preservation in the form of impermeable non-structural thin layer with maximum thickness of 10 mm, which consisting of a cold laid mixture of asphalt emulsion with continuous graded fine aggregate, mineral filler, water and other added ingredients. Slurry seal is one kind of cold mix asphalt formulated precisely as maintenance materials, surface pavement treatment, or as an addition to the surface layer thickness. The addition of slurry seal will improve the pavement flatness by reducing roughness and rutting, coat the pavement surface, improve roughness without have to retexturing as well as its application is easy and fast because the traffic can be opened in a few hours [1].

Road preservation techniques already mentioned above is simply functioning as a surface layer of the existing structure pavement. One of the problems that occur in the surface layer is slick effects occurring because of road surface texture that is too smooth, therefore it is necessary of skid resistance which serves as a barrier so that the wheels of vehicles do not slip on the surface pavement. Skid resistance is an important parameter to functionally evaluation of the pavement performance, especially on roads that are planned by vehicles at high speed [2].

Another problem that occurs in surface layer is that the layers located on the top surface layer so it may change the characteristics mainly influenced by temperature. The aging process in asphalt mixture is one causes of the strength decline from flexible pavements, with the result that damage occurs earlier than service life of the pavements. It can be caused by some environmental factors such as air temperature and solar thermal. The cause of aging on asphalt mixture is the evaporation of the liquid portion of the asphalt for their heating so that the asphalt becomes brittle and loses adhesion. Declining in flexibility and adhesion of asphalt can damaged the road easily when receiving the traffic load [3].

Slurry seal is a mixture of asphalt emulsion without heating, with a content of finely graded aggregate, mineral filler, water and other additives that mixed evenly and overlay above the surface pavement as slurry. Asphalt emulsion is the asphalt dispersed in water. Terms of slurry seal characteristics can be seen in Table 1.

Based on research conducted by Oikonomou and Eskioglou [6] portland cement, fly ash, ladle furnace slag, cement klin dust and marble dust were tested as fillers in slurry sel and result showed that they can be used producing slurry seal according to specifications.

According to Read and David [5], setting time was defined as the time it takes since asphalt emulsion mixing then overlay until the time of asphalt starts to harden on the aggregate surface. This phenomenon is characterized by colour changes of asphalt emulsion. At the beginning of mixing indicated with brown (like mud) became blackish brown and when the settings have been completed, there are no brown spots on aggregate surface. At the time of slurry seal overlay has completed, it will be obtained black colour on road surface.

The aging of the asphalt mixture can occur due to warming caused by environmental conditions. Asphalt aging is a good parameter to determine the durability of asphalt mixture. It caused by two main factors, they are the evaporation of lighter oil fraction contained in the asphalt and the oxidation. Asphalt consists of a liquid part called maltenes and solid part called asphaltenes. The aging process that causes the asphalt become hard and brittle is caused by the evaporation of the light fractions in asphalt and changing the liquid fraction (maltenes) into a solid (asphaltenes). This aging causes hardening of the asphalt and will further increase the stiffness of asphalt mixture so that would affect the performance of the mixture.

Table 1. Terms of slurry seal characteristics

	Characteristics of Mixture	**Testing Methods**	**Type of Mixture**		
			1	**2**	**3**
1	Asphalt residue, % to the dry weight		10 - 16	7.5 - 13.5	6.5 - 12
2	Cone consistency, cm	ISSA TB NO.106		2 - 3	
3	Wet stripping, %	ISSA TB NO.114		Min. 90	
4	Cohesion	ISSA TB NO.139			
	- 30 sec, kg-cm			≥ 12*	
	- 60 sec, kg-cm			≥ 20*	
5	Setting time, sec	ISSA TB NO.139		15 - 720	
6	Curing Time, sec			< 720	
7	Wet Track Abrasion Test, 1 hour soak, g/m^2	ISSA TB NO.100		≤ 500	

* For fast traffic system
Source : Kementrian Pekerjaan Umum dan Perumahan Rakyat, 2015

Kliewer, et.al. [4] though Strategic Highway Research Program (SHRP) Project A-003A have been examined the relationship between the performance of asphalt concrete mixtures on the field and the aging characteristics asphalt concrete mixtures that is performed on the laboratory. The aging procedure of asphalt concrete mixture on a laboratory to represent short-term aging or during the implementation is done by heating a loose asphalt concrete mixture in the oven for 4 hours at a temperature of 135°C before compacted. Whereas for simulate the long-term aging in the laboratory by heating the compacted asphalt concrete specimen at a temperature of 85°C. Duration of heating for 2 days represent of aging asphalt concrete pavement for 5 years. Then heating for 5 days to simulate the of aging asphalt concrete pavement for 10 years. Furthermore, to measure the result of aging carried out by calculating the resilient modulus at the temperature of 25°C with data from indirect tensile test accordance with AASHTO T 322-07.

According to Kementrian Pekerjaan Umum dan Perumahan Rakyat [7] mentions that the asphalt emulsion can also be modified with latex, where the asphalt emulsion formed in rubber particles. Latex does not mix with asphalt emulsion, but latex and granular asphalt formed three-dimensional structures. Latex can be separated with asphalt emulsion due to the difference in specific gravity. Mixing of polymer latex was recommended during blending it with hot asphalt, so that the polymer can be mixed perfectly. Some researchers suggest for polymer material in liquid form can be mixed after the emulsifying process is complete, because it can reduce the energy used for heating and minimize the risk of damage to the polymer as a result of the heating process while mixing [8].

For that reason, research is needed to determine the durability of slurry seal against the effects of aging. Laboratory testing on aging can be done through Short Term Oven Aging (STOA) and Long Term Oven Aging (LTOA). Short-term aging to simulate aging while mixing, transporting, deploying, and compacting in the field. While the long-term aging is the aging process when the service life of roads [4].

This research was conducted by using a polymer modified asphalt emulsion with addition of polymer latex in an effort to improve the durability of slurry seal over the service life. Polymer was divided into two types, namely the elastomer (elastic) and plastomer (high modulus stiffness). Examples of elastomer are styrene butadiene styrene (SBS) block copolymers, styrene butadiene rubber (SBR) latex, and natural rubber latex. While the examples of plastomer are ethylene vinyl acetate (EVA) and polyvinyl acetate (PVA). Plastomer used to modify asphalt to make it stiffer, while elastomers are used to modify asphalt to make it more stiff and elastic [5].

This research aimed to determine the effect of polymer modified asphalt emulsion to the durability of slurry seal. Type of polymer that used in this research is polymer latex. The use of polymer latex is expected to improve the characteristics of slurry seal in value of consistency, setting time, and indirect tensile strength.

2. Experimental

The design of slurry seal begins to test the quality of the base material consisting of aggregates and asphalt emulsion as a bonding agent. Based on the damage of pavement surface and road conditions (environmental factors and traffic), the type of slurry seal that use in this research is Type 3. It was suitable to repairs raveling, oxidation and improve the road surface roughness. Testing the consistency, setting time, and indirect tensile strength is used to determine the optimum residual asphalt content. The optimum residual asphalt content is determined by considering some aspects of slurry seal characteristics that must be met in Table 1. This optimum residual asphalt content used for next research to be tested the durability against aging. This research was conducted by making 5 variety of asphalt content in mixture, there are: 5.5%, 6.5%, 7.5%, 8.5%, and 9.5%. The type of asphalt emulsion that used are a standard asphalt emulsion CSS-1h (I), CSS-1h+latex 1% (II), CSS-1h+latex 2% (III) and CSS-1h+latex 3% (IV).

3. Results and Discussion

3.1. Cone Consistency Testing Results

This test is used to determine the level of workable and as a control on making sample of slurry seal. The results for this test is the optimum water content. The results show in Table 2. It show that the value of the optimum water content by pre-wetting 5% is getting smaller with increasing levels of residual asphalt emulsion. The addition of water 0 – 2.5% with 5% water for pre-wetting, the mixture provides a sufficient consistency in accordance with the specifications (2 – 3 cm).

Greater consistency value obtained by increasing the residual asphalt content. This is because partly the composition of asphalt emulsion is in the form of water. Based on testing can also be seen that the use of polymer latex does not significantly affect to the value of consistency.

Table 2. Cone Consistency Testing Results (cm)

Asphalt Emulsion Type	Material Mixture (%)								
	Aggregate	Filler		Water	Residual Asphalt Content				
		OPC	LCFA		5.5	6.5	7.5	8.5	9.5
I	100	50	50	5 – 7.5	2.4	2.3	2.5	2.6	2.8
II	100	50	50	5 – 7.5	2.3	2.3	2.4	2.7	2.8
III	100	50	50	5 – 7.5	2.3	2.3	2.4	2.6	2.8
IV	100	50	50	5 – 7.5	2.4	2.4	2.5	2.7	2.8

3.2. Setting Time Testing Results

Setting time is the time required by asphalt emulsion from the beginning of mixing until hard condition. At the beginning of mixing indicated with brown (like mud) became blackish brown and when the settings have been completed, there are no brown spots on aggregate surface. The results show in Table 3. The result show that with increasing levels of residual asphalt emulsion obtained the longer setting time at all slurry seal mixtures. Based on testing can also be seen that the use of polymer latex does not significantly affect to the value of setting time. However more affected by filler that used. In the cement filler, factors that affect the setting time is mixture workability. With higher workability, asphalt emulsion will be easier to cover the aggregate surface, produce thinner asphalt layer and possibility to make greater contact with the aggregate, so that the asphalt emulsion will be more rapidly breaking and setting conditions [9].

Table 3. Setting Time Testing Results (minute)

Asphalt Emulsion Type	Material Mixture (%)								
	Aggregate	Filler		Water	Residual Asphalt Content				
		OPC	LCFA		5.5	6.5	7.5	8.5	9.5
I	100	50	50	5 – 7.5	115	140	150	165	200
II	100	50	50	5 – 7.5	110	125	150	160	190
III	100	50	50	5 – 7.5	110	130	145	160	185
IV	100	50	50	5 – 7.5	100	125	150	165	190

3.3. Indirect Tensile Strength Testing Results

Indirect tensile strength test of slurry seal mixture is intended to determine the value of tensile strength of the slurry seal mixture that have been made. Furthermore this test is used to determine the optimum residual asphalt content. The results show in Table 4.

Based on Table 4, then made correlation chart between residual asphalt content with value of ITS presented in Figure. Optimum residual asphalt content obtained when the test results show the optimum value of the indirect tensile strength. From Figure 1 it can be seen that optimum asphalt content for each type of asphalt emulsion that have been used was dissimilar. Optimum asphalt content for each type of asphalt emulsion (I, II, III, IV) were about 7.37%, 7.45%, 7.41% and 7.57%. The optimum asphalt content has a tendency to increase in line with the addition of a polymer latex. This can be caused by concomitant addition of the polymer, the penetration value will decrease and the softening point value will further increase [10]. Higher penetration value, accordingly easier to cover up and fill the cavity between the grains aggregate so that require less bitumen content to achieve the optimum characteristics.

Tabel 4. Indirect Tensile Strength Testing Results (kPa)

Asphalt Emulsion Type	Material Mixture (%)								
	Aggregate	Filler		Water	Residual Asphalt Content				
		OPC	LCFA		5.5	6.5	7.5	8.5	9.5
I	100	50	50	5 – 7.5	38.8	51.5	49.9	42.2	39.9
II	100	50	50	5 – 7.5	71.8	87.7	91.5	92.2	66.5
III	100	50	50	5 – 7.5	79.8	93.1	89.5	97.6	73.8
IV	100	50	50	5 – 7.5	53.4	75.9	69.4	67.9	60.2

The indirect tensile strength results tend to increase to the optimum point and then go down when the percentage of asphalt exceeds the optimum level. The use of polymer modified asphalt emulsion would increase the value of indirect tensile strength and optimum asphalt content that required on mixture. Based on these tests it was found that the use of asphalt type III (CSS-1h+latex 2%) gave the best results.

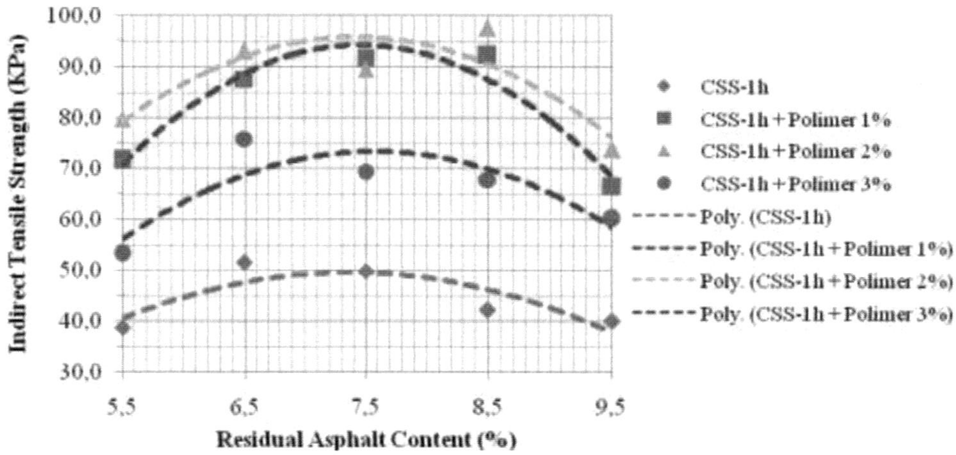

Figure 1. Correlation between residual asphalt content with value of ITS

4. Conclusion

The results show that the value of the optimum water content by pre-wetting 5% is getting smaller with increasing levels of residual asphalt emulsion. The addition of water 0 – 2.5% with 5% water for pre-wetting, the mixture provides a sufficient consistency in accordance with the specifications (2 – 3 cm). The increasing levels of residual asphalt emulsion obtained the longer setting time at all slurry seal mixtures, but all of the mixtures still meet the specifications (15 – 720 sec). Polymer modified asphalt emulsion show to improve the durability of slurry seal significantly. Based on the indirect tensile strength test showed that the use of asphalt type III (CSS-1h + latex 2%) gave the best results.

References

[1] Nono 2013 *Jurnal Jalan dan Jembatan* Volume **30** No. 2 Agustus 2013 hal. 71-79 Puslitbang Jalan dan Jembatan Badan Litbang Kementerian Pekerjaan Umum

[2] O'Flaherty C A 2002 *Highways, The Location, Design, Construction and Maintenance of Road Pavement 4th Edition* Oxford-UK: Butterworth-Heinemann

[3] Widodo S, Senja R H and Erni W 2012 *Seminar Nasional Teknik Sipil Universitas Muhammadiyah Surakarta.*

[4] Kliewer J E, Bell C A and Sosnovske D A 1995 ASTM STP 1265 Gerald A. Huber and Dale S. Decker Editors pp. 3-20 American Society for Testing and Materials Philadelphia

[5] Read J and David W 2003 *The Shell Bitumen Handbook Fifth Edition* Thomas Telford Publishing

[6] Oikonomou N and Eskioglou P 2007 *Global NEST* Journal Vol **9** No 2 pp 182-186

[7] Kementrian Pekerjaan Umum dan Perumahan Rakyat 2015 *Pedoman Perancangan dan Pelaksanaan Lapis Penutup dengan Bubur Aspal Emulsi (Emulsified Asphalt Slurry Seal)* Lampiran Surat Edaran Menteri Pekerjaan Umum dan Perumahan Rakyat No. 18/SE/M/2015

[8] Shafii M A, M Y Abdul Rand, J Ahmad 2011 *International Journal of Civil & Environmental Engineering IJCEE-IJENS* Vol: **11** No: 06

[9] Mulyono A T, Suryo H T U and Y Nunung D S 1999 *Forum Teknik 1999 XXIII (1)* Universitas Gadjah Mada Yogyakarta

[10] Suparma L B, Yosevina, Dania S L 2015 *The 18th FSTPT International Symposium Unila Bandar Lampung* August 28, 2015

Evaluation of urban drainage network based geographycal information system (GIS) in Sumenep City

F Agrianto[1]**, R Hadiani**[2]**, Y M Purwana**[3]

[1] Department of Civil Engineering, Sebelas Maret University, Indonesia
[2] Faculty of Civil Engineering, Sebelas Maret University, Indonesia
[3] Faculty of Civil Engineering, Sebelas Maret University, Indonesia

E-mail: ferryagrianto@yahoo.com

Abstract. Sumenep City frequently hit by floods. Drainage network conditions greatly affect the performance of her maid, especially those aspects that affect the capacity of the drainage channel. Aspects that affect the capacity of the drainage channel in the form of sedimentation rate and complementary buildings on drainage channels, for example, the presence of street inlet and trash rack. The method used is a drainage channel capacity level approach that level assessment of each segment drainage network conditions by calculating the ratio of the channel cross-sectional area that is filled with sediment to the total cross-sectional area wet and the existence of complementary buildings. Having obtained the condition index value of each segment, the subsequent analysis is spatial analysis using ArcGIS applications to obtain a map of the drainage network information. The analysis showed that the level condition of drainage network in the city of Sumenep in 2016 that of the total 428 drainage network there are 43 sections belonging to the state level "Good", 198 drainage network belong to the state level "Enough", 115 drainage network belong to the state "Mild Damaged", 50 sections belonging to the state "Heavy Damage" and 22 drainage network belong to the state of "Dysfunction".

1. Introduction

Drainage network conditions greatly affect the performance, especially those aspects that affect the capacity of the drainage channel. Aspects that affect the capacity of the drainage channel in the form of sedimentation rate and complementary buildings on drainage channels, for example, the presence of street inlet and filter garbage (trash rack). The greater the rate of sedimentation in the channel it will reduce the capacity of the channel which leads to reduced performance of serviceability, as well as if the channel is not equipped with complementary buildings strap water and filter junk then potentially decreased the capacity of the channel.

This study aims to obtain an evaluation value of each segment level condition in Sumenep drainage network, drainage network obtained distribution maps along with the type of building in Sumenep, and get a map of the drainage network in Sumenep based on the level and condition-based maintenance actions on geographical information systems.

If the condition has been arranged to evaluate the extent of drainage network that stretches in Sumenep, the Local Government will be easier to monitor the channels are bad for immediate maintenance action. Especially when the database is presented in the spatial data in a Geographic Information System (GIS) will facilitate the activities of asset development and maintenance [3], [5].

2. Experimental

2.1. *Urban Drainage Network Assessment*

The government has not issued a regulation on guidelines for condition assessment drainage network. Therefore, in this study the drainage network condition assessment process as a form of evaluation is done with the approach of assessment guidelines existing irrigation networks. Aspects of the assessment will be reviewed from sedimentation rate and completeness of infrastructure and facilities drainage network.

2.2. *Sedimentation Aspects Rate*

Rate aspect of the rate of sedimentation in the drainage channel made through a Circular Letter Ditjen SDA according to Minister of Public Works No. 02 / SE / M / 2011 on Guidelines for Performance Assessment Network Swamp Reclamation. Sourcebook for assessing the condition of the wet cross-section shown by the index value ranging between 1-5 [1]. The smaller the value of the index shows that the cross section of the wet conditions, the better, which means also that the function of the better wet cross section [2].

Based on the above assessment approach, in this study the function value (%) sedimentation rate assessed by comparing the existing high sedimentation of the channel capacity. Channel capacity is assumed proportional to the maximum channel cross-sectional area, the channel height (H) is the height of water (h) plus high-surveillance (w). Thus, for calculating the maximum cross-sectional area trapezoid-shaped channel (type A) is:

$$A = bH + mH \qquad (1)$$

$$H = h + w \qquad (2)$$

$$A_{sed} = bH_{sed} + mH_{sed} \qquad (3)$$

A = The maximum cross-sectional area (m^2),
A_{sed} = sectional area due to sedimentation (m^2),
b = width of the base line (m),
$1/m$ = slope of the embankments,
H = total channel height (m),
H_{sed} = high channel due to sedimentation (m).

As for calculating the maximum sectional area of a rectangular duct (type B) is:

$$A = bH \qquad (4)$$

$$A_{sed} = bH_{sed} \qquad (5)$$

Thus, the sedimentation rate can be measured by:

$$Sed = \left(100 - \frac{A_{sed}}{A} x\ 100\right)\% \qquad (6)$$

Then the calculation results of sedimentation into the condition input values in Table 1:

Table 1. Condition Assessment Based on Sedimentation Rate Drainage Network

Index	Condition	Function (%)	Condition
1	Sedimentation rate is between 76-100%.	76 - 100	Good
2	Sedimentation rate is between 51-75%.	51 - 75	Enough
3	Sedimentation rate is between 26-50%.	26 - 50	Mild Damaged
4	Sedimentation rate is between 1-25%.	1 - 25	Heavy Damage
5	Sedimentation rate to cover the entire cross section of the channel (0%).	0	Dysfunction

2.3. Assessment of Completeness Drainage Network Infrastructure

As aspects of sedimentation, the completeness of infrastructure and facilities drainage network graded into five indices as follows:

Table 2. Assessment of Completeness Drainage Network Infrastructure

Index	Condition	Function (%)	Condition
1	There is a street inlet and trash rack. The building is in good condition, not encountered significant damage.	76 - 100	Good
2	There is a street inlet and trash rack. Buildings under moderate conditions and functioning, found no damage, but can be overcome and functioned.	51 - 75	Enough
3	There is a street inlet or trash rack. Buildings in a damaged condition, encountered damage, do not function properly.	26 - 50	Mild Damaged
4	There is a street inlet or trash rack. Buildings in severely damaged condition, the damage can not be repaired, lost, collapses, and others.	1 - 25	Heavy Damage
5	Buildings or building components do not exist.	0	Dysfunction

2.4. Total Assessment of Drainage Network

Ratings total drainage network is done with the approach Circular of the Minister of Public Works No. 02/SE/M/2011 on Guidelines for Performance Assessment Network Swamp Reclamation. The condition of drainage network is a combination of the index is reviewed aspects of sedimentation condition and completeness of infrastructure and facilities. The index value waterworks condition being simulated is calculated by weighting (SE Ditjen SDA Men PU No. 02/SE/ M/2011), with the following formula:

$$Ik_{\text{drainage}} = \frac{\{(Ik_{sed} \times W_{sed}) + (Ik_{PS} x W_{PS})\}}{(W_{sed} + W_{PS})} \qquad (7)$$

Empirically, sedimentation have an influence on a greater capacity than the completeness of infrastructure and facilities. Thus, the weight of sedimentation (W_{sed}) will be greater than the weighting of infrastructure and facilities (W_{PS}). This study used a weight W_{sed}=3 and W_{PS}=1.

If the index value has been obtained drainage network in Sumenep, then the action recommendations can use the approach in the Circular Minister of Public Works No. 02/SE/M/2011 on Guidelines for Performance Assessment Network Swamp Reclamation as follows:

Table 3. Relationship between Line Condition Index and Building, Channel Function and Building, and Recommended Actions

Index	Condition	Recommended Actions
1	Functioning of between 76% to 100%	Routine Maintenance
2	Functioning of between 51% to 75%	Periodic Maintenance
3	Functioning between 26% to 50%	Rehabilitation
4	Functioning between 1% to 25%	Rehabilitation
5	There is no channel and / or buildings which should be available or 0%	Study Design

Source: SE Director General of Water Resources Ministry of Public Works No.02/SE/M/2011

3. Result and Discussion

3.1. Data Analysis Technique

Data analysis was carried out by providing an assessment of the physical condition of drainage network on each segment based on primary and secondary data. Assessment methods physical condition of the drainage network using condition assessment index sedimentation and drainage completeness of infrastructure and facilities, as described in Table 1 and Table 2. The drainage network assessment performed on each channel that is at the crossroads between the roads, but if encountered height differences of different sediment in drainage network segment every crossroads, the division of sections taken at locations or points that indicate the height difference of the sediment.

In addition, the process geographical analysis using ArcGIS version 9.3 software that starts with mapping the drainage network in all regions of Sumenep based spatial data to produce thematic maps drainage network [4].

Then this thematic maps integrated with the results of the physical assessment of drainage networks where this data as attribute data or additional information in thematic maps drainage network. Physical assessment network will appear as the value of the conditions in each network segment depicted in symbols or legends that differ in the thematic maps. These thematic maps can be useful in monitoring and maintenance, precisely a key consideration in decision making maintenance actions.

3.2. Percentage Level Sediments Drainage Network in Sumenep

Based on the measurement results of high sediment and sediment percentage calculation on each side drainage networks exist in Sumenep the obtained results of the investigation, as in Table 4.

Table 4. Percentage Rate of Sediment in Drainage Network

No	Street Name	Location	Code	B [a] (m)	H [b] (m)	M [c] (m)	Building Type	High of Sediment (H_{sed}) [d] (m)	m_{sed} [e] (m)	Percentage Sediment (%)
(1)	(2)	(3)	(4)	(5)	(6)	(7)	(8)	(9)	(10)	(11)
1	Dr. Wahidin	Pajagalan	Wahidin-1	0.50	0.70	0.20	A [f]	0.06	0.02	93.67
2	Dr. Wahidin	Pajagalan	Wahidin-2	0.50	0.70	0.20	A [f]	0.10	0.03	89.21
3	Dr. Wahidin	Pajagalan	Wahidin-3	1.05	0.80	0.00	B [g]	0.20	0.00	75.00
4	Dr. Wahidin	Pajagalan	Wahidin-4	0.90	1.15	0.00	B [g]	0.15	0.00	86.96
5	Dr. Wahidin	Pajagalan	Wahidin-5	0.70	1.15	0.20	A [f]	0.14	0.02	90.20
6 to 428	*(etc)*	*(etc)*	*(etc)*	*(etc)*	*(etc)*	*(etc)*	*(etc)*	*(etc)*	*(etc)*	*(etc)*

[a] Width of the base line (m),
[b] Total channel height (m),
[c] Slope of the embankments,
[d] High channel due to sedimentation (m),
[e] Slope sediments on embankments (m),
[f] Channel with trapezoidal cross-sectional shape,
[g] Channel with a rectangular cross-sectional shape.

IOP Conf. Series: Materials Science and Engineering **176** (2017) 012036 doi:10.1088/1757-899X/176/1/012036

Table 5. Levels and Recommended Actions Condition Drainage Network

No	Code	Sediment Factor	Infrastructure & Facilities Factor	Sediment Condition Index Values	Values Infrastructure & Facilities Condition Index	Total Value Index	Condition (%)	Condition Level	Recommended Actions
(1)	(2)	(3)	(4)	(5)	(6)	(7)	(8)	(9)	(10)
1	Wahidin-1	3	1	1	2	1.25	68.75	Enough	Periodic Maintenance
2	Wahidin-2	3	1	1	2	1.25	68.75	Enough	Periodic Maintenance
3	Wahidin-3	3	1	2	3	2.25	43.75	Mild Damaged	Rehabilitation
4	Wahidin-4	3	1	1	4	1.75	56.25	Enough	Periodic Maintenance
5	Wahidin-5	3	1	1	3	1.50	62.50	Enough	Periodic Maintenance
6 to 428	*(etc)*	*(etc)*	*(etc)*	*(etc)*	*(etc)*	*(etc)*	*(etc)*	*(etc)*	*(etc)*

Figure 1. Map of The Condition Level of Drainage Network in Sumenep City Year 2016

4. Conclusion

The value of each segment condition of drainage network in Sumenep vary according to the aspect of the sedimentation and completeness of infrastructure and facilities that demonstrate the level of condition, namely: good, enough, mild damage, heavy damage and dysfunction. Drainage network segment considered good if the network is functioning between 76%-100%, segments categorized sufficient drainage network when the network is functioning between 51%-75%, segments categorized mildly damaged drainage network when the network is functioning between 26%-50%, segment drainage network categorized as severely damaged if the network is functioning between 1%-25%, and joint dysfunction are categorized drainage network if the network can not be enabled or 0%. In each segment of the drainage network can also be determined the type of maintenance action recommendations are: routine maintenance, periodic maintenance, rehabilitation and design study which is strongly associated with the level of drainage network link conditions.

ArcGIS applications can be used as a computer-based tools in the process of evaluating the condition of the drainage network. This is done by integrating or combining the results table level assessment of drainage network link conditions with a layer of drainage network maps using the join function and relates to the application. Results of the integration in the form of drainage network map Sumenep 2016, a map of the type of building drainage network Sumenep 2016, a map of the level of drainage network conditions Sumenep 2016 and map recommendations drainage network maintenance actions Sumenep Year 2016. In addition, ArcGIS applications can do the selection of appropriate specific criteria so by using the function select by attributes can know the number of locations along the drainage network segments according to the level of condition. From the analysis of selected segments of drainage network according to the level of the conditions obtained are of a total 428 network there are 43 sections belonging to the state level "Good", 198 network belong to the state level "Enough", 115 network belong to the state of "Mild Damaged", 50 sections belonging to the state" Heavy Damage "and 22 network belong to the state of "Dysfunction".

References

[1] Director General of Water Resources Ministry of Public Works. 2011. *The Circular of the Minister of Public Works No. 02/SE/M/2011 on Guidelines for Performance Assessment Swamp Reclamation*. Jakarta.

[2] Edisono, S. 1997.*Urban Drainage*. Gunadarma, Jakarta.

[3] Prahasta, E. 2001. *Basic Concepts of Geographic Information Systems*. Informatics: Bandung.

[4] Hariyanto, T. 2013. *Modelling Decision Support System Based Irrigation Asset Management SIG*. Jakarta.

[5] United Nations Development Programs, National Technical Team. 2007. *Basic Training Module ArcGIS*. Jakarta.

The Formation of Sodium Stannate from Mineral Cassiterite by the Alkaline Decomposition Process with Sodium Carbonate (Na_2CO_3)

L Andriyah[1], **L H Lalasari**[1] **and A Manaf**[2]
[1]Research Center of Metallurgy and Materials, LIPI, Puspiptek Gedung 470, Indonesia
[2]Physics Department, Faculty of Maths and Sciences, Universitas Indonesia

E-mail: andriyah_yong@yahoo.com

Abstract. Extraction of cassiterite using alkaline decomposition of sodium carbonate (Na_2CO_3) has been studied. Cassiterite (SnO_2) is a mineral ore that contains tin (Sn) about 57.82 wt% and impurities like quartz, ilmenite, monazite, rutile and zircon. The initial step for the process was to remove the impurities in cassiterite through washing and separation by a high magnetic separator (HTS). The aim of this research is to increase the added value of cassiterite from local area Indonesia that using alkaline decomposition to form sodium stannate (Na_2SnO_3). The result shows that cassiterite from Indonesia can form sodium stannate (Na_2SnO_3) which soluble with water in the leaching process. The longer the time for decomposition, the more phases of sodium stannate that will be formed. Optimum result reached when the decomposition process was done in 850 °C for 4 hours with a mole ratio Na_2CO_3 to cassiterite 3:2. High Score Plus (HSP) was used in this research to analyze the mass of sodium stannate (Na_2SnO_3). HSP analysis showed that mass of sodium stannate (Na_2SnO_3) is 70.3 wt%.

1. Introduction

One of minerals that abundandly available in Indonesia is cassiterite, the raw material for tin (Sn). It is widely spread on the islands of Bangka, Belitung, Kundur, Singkep, Karimun and Kampar [1]. Mineral cassiterite has a chemical formula SnO_2 which mainly consisting of tin (Sn) and oxygen (O) [2]. Cassiterite ore has been processed commercially through pyrometallurgy to produce metal or high purity tin ingot. The energy extraction process using high temperature (melting) and reducing agents such as anthracite (coal) made using reverberatory furnace [1]. Lead optimization efforts from upstream to downstream is prospective to do that is by turn lead into new compounds which have a higher economic value, such as tin dioxide as a catalyst.

Various processes have been developed to obtain valuable materials from the mineral cassiterite, the process of which is a) the decomposition process cassiterite with sodium carbonate at 900 °C for 120 minutes, which then dissolved using 25% HCl to eliminate the main impurity elements, including Fe, As, S, Pb, and Sb. From the results of these studies found that cassiterite (SnO_2) slightly react with impurities during the decomposition process and does not dissolve in hydrochloric acid significantly [3]. b) Concentrate cassiterite is decomposed with sodium carbonate (Na_2CO_3) were then fed with gas mixed CO/CO_2 and dissolved into a solution of NaOH, then filtered to obtain the filtrate which is rich

[1] To whom any correspondence should be addressed.
[2] To whom any correspondence should be addressed.

in sodium stannat and will dissolve in the water when done heating and crystallization, from the research results obtained leaching efficiency of 85.6% and the product had a purity of 95.8% SnO_2, with optimal conditions the decomposition temperature of 875 °C, 15% CO gas content, the decomposition time 15 minutes, mole ratio Na_2CO_3/cassiterite 3: 2, leaching temperature of 40 °C, leaching time of 60 minutes, stirring speed of 300 rpm and the ratio L/S 4 cm^3/g [4]. c) The process of extracting tin from tin concentrate using low levels of HCl 150 g/L, 2 hours, 35 °C which is able to separate the iron up to 99.4%, but the process of decomposition continued with a mixture of 55% Na_2CO_3, 12% NaOH, 30% residue leaching and 3% of coal at a temperature of 870 °C. Tin recovery results obtained at 99.66% [5].

The alternative process that applied in this research is extracted through decomposition with sodium carbonate to produce sodium stannat, where sodium stannat can be dissolved in water [4-6] to obtain filtrate enriched with tin (Sn). The mole ratio of Na_2CO_3 to cassiterite, temperature and reaction time are critical parameters for the sodium stannat phase formation of the yield value of the final product as a tin dioxide (SnO_2). In this paper, we report results on the effect of mole ratio Na_2CO_3 to cassiterite, temperature and reaction time to the process of formation of sodium stannat compounds.

2. Experimental

2.1. Materials and reagents

The raw material cassiterite for experiment was received from Bangka Island-Indonesia. Another raw materials for experiment were an analytical grade sodium carbonate (Na_2CO_3). The detailed chemical composition of the cassiterite was examined by XRF (X-Ray Fluorescence) and the analytic results are listed in Table 1.

Table 1. Chemical composition of cassiterite

No.	Element	Percentage (% wt)	No	Element	Percentage (% wt)
1.	Sn	57.82	9.	Ca	0.63
2.	O	23.39	10.	Mg	0.62
3.	La	3.43	11.	Al	0.52
4.	Ce	3.37	12.	Si	0.45
5.	Nd	2.18	13.	Cl	0.37
6.	Fe	1.79	14.	Zr	0.29
7.	P	1.68	15.	S	0.11
8.	Ti	1.45	16.	Pb	0.04

2.2. Eperimental Methods

The prior step was grinding cassiterite to particle size -100 mesh and characterizing with X-Ray Diffraction (XRD), and XRF (X-Ray Fluorescence) apparatus. Then second step was decomposed cassiterite using various sodium carbonate (Na_2CO_3) with the mole ratio is 1:1, 3:2, and 2:1 in atmospheric conditions at various temperature of 800, 850, 870 and 900 °C for 2, 3 and 4 hours. Finally, product of decomposed was characterized using XRD analysis. Experiments were performed in a carbolyte furnace using graphite crucibles. The concentrated Na_2CO_3 and cassiterite were then homogeneously mixed in graphite crucibles and put into the carbolyte furnace when the temperature reached the present value, with free access of air. The temperature of the carbolyte furnace was controlled by a programmable temperature controller.

The reaction of cassiterite with sodium carbonate (Na_2CO_3) forms sodium stanate (Na_2SnO_3) and could be described as follows [4] :

$$SnO_{2\,(s)} + Na_2CO_{3(s)} \rightarrow Na_2SnO_{3(s)} + CO_{2\,(g)} \quad (1)$$

3. Results and discussion

3.1. The Characteristic of Bangka Cassiterite

A chemical composition of Bangka Cassiterite was analyzed by using X-Ray Fluorescence (XRF) and X-Ray Diffraction (XRD). The result is presented at Table 1 and Fig. 1, which shows that the dominant compound exist Tin (Sn) and Oxigen (O). Another compounds which are also exist in cassiterite from Bangka Island are La, Ce, Fe, Ti, Si, Mg, Ca, but their quantities are lower than Sn.

This result of chemical analysis are in a good agreement with the result of phase identification by using XRD at Fig. 1, which showed that the dominant phase exist in cassiterite from Bangka Island is tin dioxide (SnO_2), the diffraction pattern in accordance with PDF 2 No. 01-0657. The X-ray diffractograms of cassiterite Bangka Indonesia showed sharp diffraction peak intensity at 2θ of 26.64°, 33.94°, 38.02°, 51.84°, 54.84°, 61.94°, 64.82° and 66.04°. La_2Sb and CeSe phase was also found in Bangka cassiterite, but their X-ray diffraction intensities were relativelly lower than the X-ray diffraction intensities of tin dioxide (SnO_2). Another phases were not detected due to their quantities were relativelly lower as it is shown by the result of chemical analysis by using X-Ray Fluorescence (XRF) at Table 1.

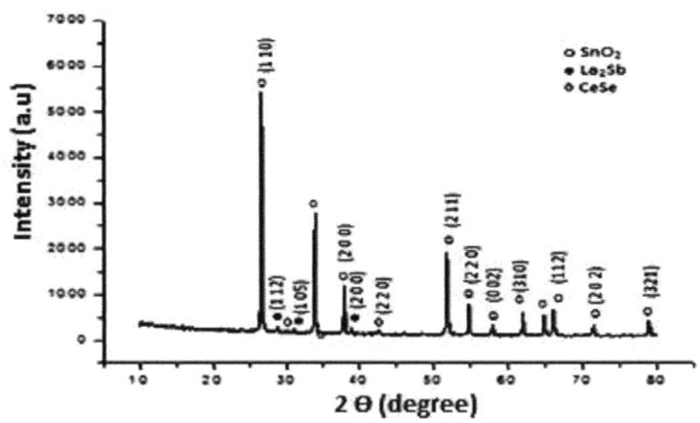

Figure 1. XRD pattern of cassiterite

3.2. Effect of mole ratio Na_2CO_3 to cassiterite

The result of X-ray diffraction of the decomposed cassiterite at Fig. 2 shows that sodium stannat (Na_2SnO_3) phase was found at cassiterite decomposition products. This result of phase identification same with the results of Zhang, Y [4] and Bunnakkha, C [6] experiments. Fig. 2 also shows that when the mole ratio of Na_2CO_3 was increased up to 2 : 1, the X-Ray diffraction peaks intensities for cassiterite decreases and intensities for sodium stannat increases are unsignificantly changed. This result of experiment shows that the decomposition reaction of cassiterite with Na_2CO_3 will be more effective when the mole ratio of Na_2CO_3 increases from 1 : 1 to 2 : 1.

The similar tendensies were found by Zhang, Y et.all [4] during decomposition of cassiterite by concentrated sodium carbonate (Na_2CO_3). He found that when the mole ratio of Na_2CO_3 were increased rapidly until it tends to be balanced at the Na_2CO_3 to cassiterite mole ratio higher than 1.50. Thus, Na_2CO_3 to cassiterite mole ratio of 1.50 is the most favorable value.

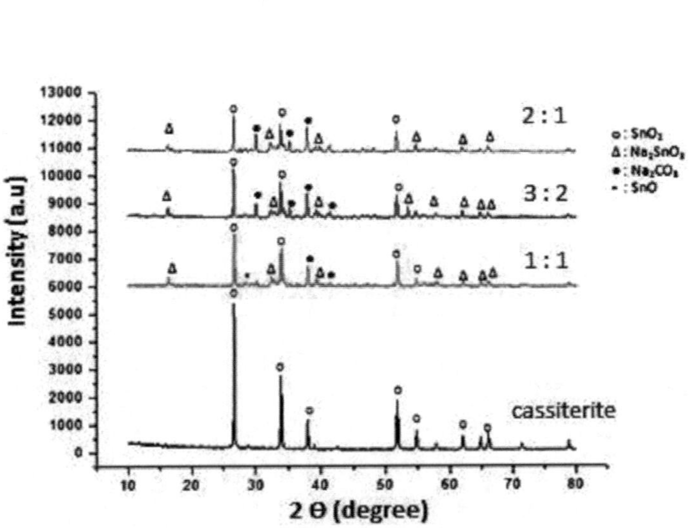

Figure. 2. X-Ray diffraction of decomposed product in various concentrations at temperature 850 °C for 2 hours

Table 2 shows the results of quantitative or weight percent of each phase were calculated by the method High Score Plus, where there is an increase in weight % Na_2SnO_3 phase of 8.1 wt % to 9.3 wt % . The greater the amount of Na_2CO_3 is added into the cassiterite will lead to increased formation of sodium stannat phase.

Table 2. Quantitative Analysis of phase in various concentrations

Mole ratio	Sample	Cassiterite (wt %)	Na_2SnO_3 (wt %)	Natrite (wt %)	GOF
1:1	A-850-2	34.6	8.1	57.3	1.52
3:2	B-850-2	15.7	8.4	75.9	1.38
2:1	C-850-2	10.9	9.3	79.8	1.53

3.3. Effect of Reaction Time

The effect of reaction time on the formation of sodium stannat phase during decomposition process with Na_2CO_3 were observed to the -100 mesh cassiterite that was decomposed with mole ratio Na_2CO_3 to cassiterite : 3:2, temperature 850 °C for decomposition time 2 hours, 3 hours, and 4 hours respectively. The result is presented at Fig. 3 which shows that the intensities of sodium stannat phase started to forms at decomposition time 2 hours, and their intensities increases significantly when decomposition time was extended up to 4 hours. Table 3 shows the results of quantitative or weight percent of each phase were calculated by the method High Score Plus, where there is an increase in weight % Na_2SnO_3 phase of 8.4 wt % to 70.3 wt % . The longer the time given in the decomposition process, reaction of sodium stannat phase formation that occurs will be increase [6]. So for this experiment, the optimum time of decomposition seen at 4 hours.

Table 3. Quantitative Analysis of phase in various reaction time

Time (hour)	Sample	Cassiterite (wt %)	Na_2SnO_3 (wt %)	Natrite (wt %)	GOF
2	B-850-2	15.7	8.4	75.9	1.38
3	B-850-3	4.8	29.1	66.0	1.41
4	B-850-4	0.8	70.3	28.9	1.45

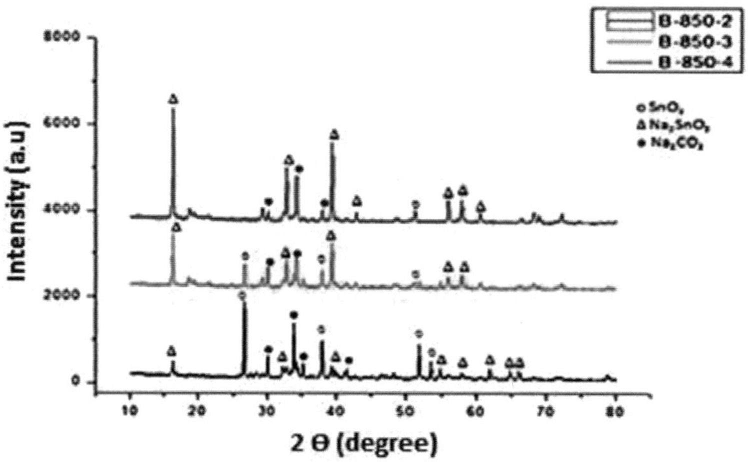

Figure. 3. X-Ray diffraction of decomposed product in various reaction time at temperature 850 °C for mole ratio Na_2CO_3/cassiterite : 3:2

3.4. Effect of Temperature

The effect of temperature on the formation of sodium stannat phase during decomposition process of cassiterite with Na_2CO_3 were observed to the -100 mesh cassiterite by using mole ratio Na_2CO_3 to cassiterite : 3:2, at decomposition temperatures 800 °C up to 900 °C, for 4 hours. The result of decomposition reactions were then analyzed by using XRD. Fig. 4 shows the result of X-Ray diffraction analysis to the decomposed cassiterite. From the result of analysis by using XRD at Fig. 4 and Table 4, it is shown that Na_2SnO_3, cassiterite and Na_2CO_3 were found at cassiterite decomposition products. The X-ray intensities of cassiterite increased and intensities of sodium stannate decreased when the decomposition temperature were increased from 850 °C to 900 °C and so sodium stannate phase optimum was found at decomposition temperatures 850 °C.

At a lower temperature of 800 °C phase sodium stannat unobserved. This is evident from the results of TG/DTA in cassiterite + Na_2CO_3 that the temperature required for the endothermic reaction occurs at temperatures of 830-880 °C, in other words sodium stannat formation occurs at this temperature. The same is done on previous research that Zhang, Y [4] and Bunnakkha, C [6] where the decomposition temperature used in research that is 870 °C with decomposition recovery yield of 90.66 % and 85.6 % .

Table 4. Quantitative Analysis of phase in various temperature

Temperature °C	Sample	Cassiterite (wt %)	Na_2SnO_3 (wt %)	Natrite (wt %)	GOF
800	B-800-4	18.0	-	82.0	1.35
850	B-850-4	0.8	70.3	28.9	1.45
870	B-870-4	3.7	59.3	37.0	1.37
900	B-900-4	4.2	55.6	40.1	1.46

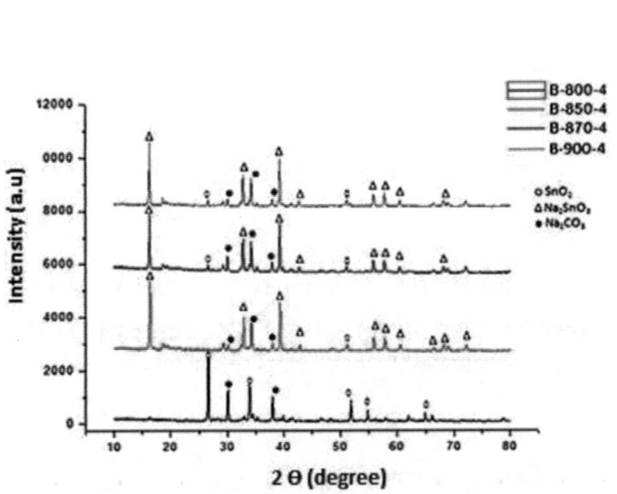

Figure. 4. X-Ray diffraction of decomposed product in various temperature at mole ratio Na_2CO_3/cassiterite : 3:2 for 4 hours

4. Conclusion

It could be concluded that the main element of cassiterite was 57.82% Sn and others were 42.18 % associated elements. The result of XRD analysis showed that the major minerals of cassiterite were indexed as tin dioxide (SnO_2) phase. It is concluded that Bangka cassiterite could be decomposed with sodium carbonat and can form sodium stannate (Na_2SnO_3) which soluble with water in leaching process and optimum result reached when the decomposition process was done in 850 °C for 4 hours with mole ratio Na_2CO_3 to cassiterite 3:2, with the percentage mass of sodium stannate (Na_2SnO_3) is 70.3 wt%.

Acknowledgment

The authors are grateful for the main financial supports in 2016 from Research Center for Metallurgy and Material-Indonesian Institute of Science (LIPI) and PT. Mitra Stania Prima for their support in preparing the raw material cassiterite.

References

[1] Djamaluddin H, Thamrin M and Achmad A 2012 *J.Teknik Geologi.* **6** 1-14
[2] Wills B A and Napiermunn T J 2006 *Mineral Processing Technology, An Introduction to the Practical Aspects of Ore treatment and Mineral Recovery* (New York: Elsevier Science & Technology Books) p 4
[3] Wright P A 1982 *Extractive Metallurgy of Tin* vol 2 (New York: Elsevier Scientific Publishing Company Press)
[4] Zhang Y, Su Z and Liu B 2014 *J.Hydrometallurgy.* **146** 82-88
[5] Yang J G, Wu Y T and Zhang X L 2014 *J. Trans. Inst. Min. Metall. C* **123** 4 228-233
[6] Bunnakkha C and Jarupisitthorn C 2012 *J.of Metals, Materials and Minerals.* **22** 1 1–6

IOP Conf. Series: Materials Science and Engineering **176** (2017) 012038 doi:10.1088/1757-899X/176/1/012038

Ion-Selective Deposition of Manganese Sulphate Solution from Trenggalek Manganese Ore by Active Carbon and Sodium Hydroxide

L Andriyah[1] and E Sulistiyono[2]
[1]Research Center of Metallurgy and Materials, LIPI, Puspiptek Gedung 470, Indonesia

E-mail: andriyah_yong@yahoo.com

Abstract. One of the step in manganese dioxide manufacturing process for battery industry is a purification process of lithium manganese sulphate solution. The elimination of impurities such as iron removal is important in hydrometallurgical processes. Therefore, this paper present the purification results of manganese sulphate solution by removing impurities using a selective deposition method, namely activated carbon adsorption and NaOH. The experimental results showed that the optimum condition of adsorption process occurs on the addition of 5 g adsorbent and the addition of 10 ml NaOH 1 N, processing time of 30 minutes and the best is the activated carbon adsorption of Japan. Because the absolute requirement of the cathode material of lithium ion manganese are free of titanium then of local wood charcoal is good enough in terms of eliminating ions Ti is equal to 70.88%.

1. Introduction

The development of battery technology has grown rapidly especially development of lithium-based battery since it has many advantages such as having light weight, large current density and quickly and easily replenished [1]. Rechargeable lithium ion battery has been commercially available in recent years. Lithium ion battery is used for component in computing, entertainment and telecommunications equipment needed to enrich the information especially mobile society. Among several choices of battery cathode materials, $LiMn_2O_4$ is an attractive candidate for a positive electrode for rechargeable lithium batteries because it is cheap, environmentally friendly and stable at low temperatures compared to other cathode materials [2]. However, these materials have problems in severe capacity fading during charge-discharge cycles. Many studies have concurred that capacity fading caused by various factors such as Jahn-Teller distortion, unstable reaction of two phases, the slow solubility of manganese into the electrolyte, instability lattice and particle size distribution [3]. Therefore, in the manufacture of lithium batteries, manganese oxide needed more pure than the manganese dioxide used for metal-based dry battery zinc-carbon and alkaline-based. One of the requirements for lithium manganese dioxide is free of Ti ion that can decrease lithium battery voltage.

The elimination of impurities such as iron is one of the main concerns in hydrometallurgical processes. One common technique used to remove impurities is precipitation. Precipitation of impurities can be done through the formation of metal hydroxides by raising the pH of the leaching solution. Some precipitations can be done by adding solution of calcium oxide (CaO), magnesium oxide (MgO) or caustic soda (NaOH) as the precipitator (Esmaeili et al., 2005) [4]. Nathsarma, 1987,

[1] To whom any correspondence should be addressed.

reported that iron impurities could be removed through the formation of iron hydroxide by raising the pH of the sulfuric acid which is used as leaching solution with calcium oxide (CaO) [5]. Calcium oxide (CaO) is preferred over other alkalies because of its abundant availability and inexpensive. The addition of magnesium oxide (MgO) and caustic soda (NaOH) are also often used for pH adjustment although more expensive than calcium oxide around two or five times [6]. When pH is adjusted, the alkaline reagent will form colloid hence the amount of solids after leaching will be slightly decrease because of the formation of metal hydroxides of impurities. The use of reagents such as MgO and NaOH with high costs should be consideration given to the amount of loss of value of the metal during the solid-liquid separation and the cost involved in the recovery process. The elimination of iron with the addition of CaO, MgO and NaOH to raise the pH of the acid solution was studied by John Jamieson, in 1995 [7]. Studies indicated that in sulfate solution, the use of calcium oxide and magnesium oxide produce multiple phase ferrihydrite. However, the formation of gypsum inhibited filterability with the addition of calcium oxide. pH adjustment with the addition of caustic soda (NaOH) produced smooth goethite crystals with slow filtration.

In the process of purification of manganese sulphate, pink manganese sulphate solution still has some impurities such as ions Ti, Fe, Cu and other metal ions. Manganese oxide that is used for battery industries required purity especially it should be free from titanium, copper and heavy metals since it might affect the performance of lithium manganese cathode in battery.

2. Experimental

2.1. Materials and reagents

In this study, the raw material was manganese sulphate solution from the leaching process of manganese ore that were reduced by activated carbon. Manganese ore was from Sukarame, Trenggalek which contains silica 48.73% by weight and an iron (III) oxide 3.77% by weight. XRF result is shown in below table :

Table 1. XRF analysis of manganese ore from Trenggalek

No.	Oxides	wt %
1	Manganese (MnO_2)	46.03
2	Silica (SiO_2)	48.73
3	Iron (III) Oxide (Fe_2O_3)	3.77
4	Calsium (CaO)	1.10
5	Barium (BaO)	0.46
6	Magnesium (MgO)	0.00
7	Phospor (P_2O_5)	0.56

Adsorption process was conducted using activated carbon. Solution of manganese sulfate that were absorbed having almost saturated concentration that resulted from the process of leaching manganese ore that has been reduced by activated carbon.

2.2. Experimental Methods

Adsorption process was carried out in batch vessel using a 250-ml erlemeyer with hot plate and magnetic stirrer. Manganese sulphate was taken from sulfuric acid leaching process result of reducted manganese ore from Trenggalek. Manganese sulphate was filtered to separate the dirt and the presence of dissolved crystals. 100 ml clear-pink manganese sulphate solution with pH 1 was taken and poured in a 250-ml erlenmeyer. Sodium hydroxide with concentration 1 N was added and mixture was stirred, pH of mixture was measured and precipitation that obtained was observed. Activated carbon was added into the mixture and stirred and immediately the time when adding activated carbon was recorded. After the adsorption process, the solution was filtrated then pH was measured. Filtrate was diluted 10 times for analyzing ion Fe, Ti and Cu and diluted 200 times for analyzing ion Mn. Concentration measurement of Fe, Ti, Cu and Mn was conducted by ICP (Inductively Coupled Plasma).

3. Results and discussion

3.1. Effect of NaOH Concentration

In this research, the adsorption was performed by activated carbon from Japan for NaOH concentration variable experiment and reaction time experiment.

3.1.1 The addition of 5 ml of 1 N NaOH solution

As shown in Fig. 1, the addition of 1 N NaOH with a concentration of about 5 ml per 100 ml solution of manganese sulphate drastically decreased the concentration of Fe and Cu ions in 10 g activated carbon. Concentration of Ti did not decrease, which means it could be concluded that the adsorption process of Ti was not running. Concentration of Ti rose later in the addition of 10 g of activated carbon starts to decline, but the decrease was not significant.

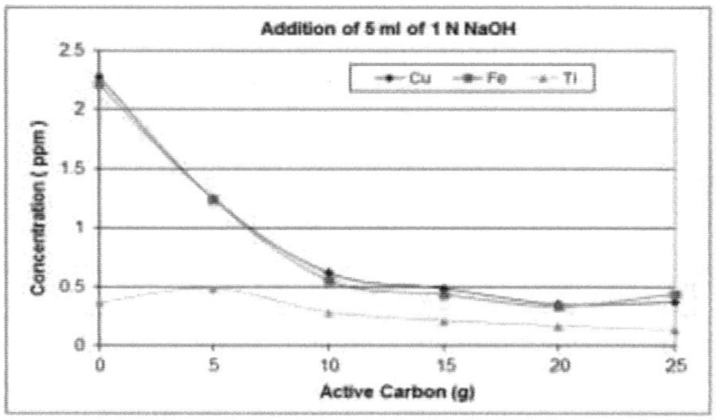

Figure 1. Graph of the relationship between the addition of activated carbon to the concentration of ions Fe, Cu and Ti on the addition of 5 ml of 1 N NaOH

From these results it can be seen that the optimum adsorption of Fe and Cu were in the addition of 10 g of activated carbon where the concentration of Fe and Cu ions drastically reduced from about 2-2.5 ppm to 0.5 ppm. Addition of 20 g activated carbon decreased concentrations of Cu / Fe from 0.5 ppm to 0.4 ppm.

3.1.2The addition of 10 ml of 1 N NaOH solution

Fig. 2 shows addition of NaOH increase the adsorption of Fe while adsorption of Cu was almost constant. Similarly, increment of addition NaOH from 5 ml to 10 ml did not give significant effect on Ti adsorption. From the experimental results shown that the addition of NaOH from 5 ml to 10 ml, the addition of 5 g activated carbon is able to adsorb Fe until the concentration is about 0.3 ppm. Meanwhile with the addition of NaOH only 5 ml were able to adsorb about 0.3 ppm of activated carbon required four times more that to 20 g.

The addition of NaOH which is just 5 ml per 100 ml solution of manganese sulfate activated carbon be able to save a quarter are from 20 g to about 5 g. By looking at these results the optimum activated carbon adsorption process on the addition of 1 N NaOH 10 mL and 5 g of activated carbon per 100 ml solution of manganese sulfate processed.

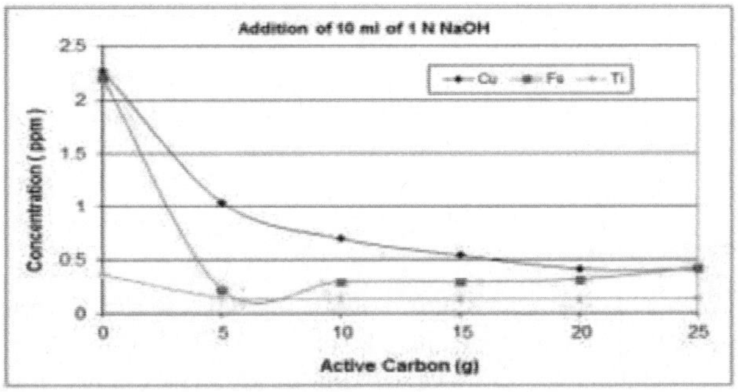

Figure 2. Graph of the relationship between the addition of activated carbon to the concentration of ions Fe, Ti and Cu in the addition of 10 ml of 1 N NaOH

3.2. Effect of Reaction Time
3.2.1 Adsorption Optimization

Optimal condition was reached on the addition of 5 ml of NaOH 1 N for 100 ml solution of manganese sulfate. The active carbon adsorption process was first carried out for 30 minutes. Therefore experiment to determine the optimum processing time was needed by varying the adsorption process from 15 to 90 minutes using 15 minutes interval. Fig. 3 shows that optimum processing time for Fe and Cu ion adsorption was 30 minutes and for Ti ion was 45 minutes. Extending the processing time much further will resolve in the detachment of Fe, Cu and Ti ion in the solution fluctuatively.

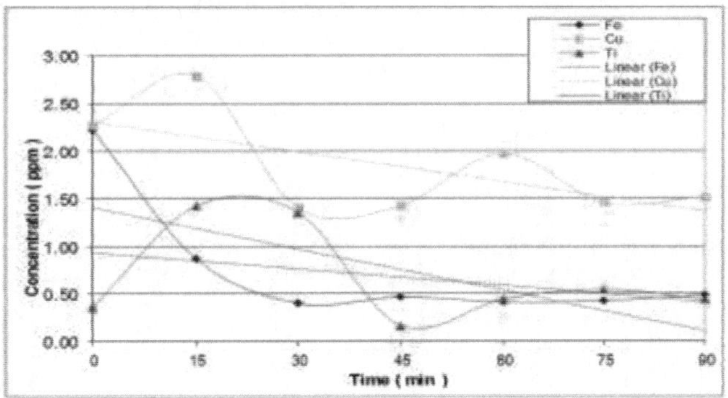

Figure 3. Active carbon adsorption processing time vs. ion concentration of Fe, Cu, and Ti in the addition of 5 g of active carbon and 10 ml of NaOH 1 N

3.2.3 Manganese Residual

Measurement of manganese residual concentration was carried out to see the effect of processing time to the reduction of manganese. There were some concentration of manganese that was carried inside the active carbon. Fig. 4 shows that the initial concentration of manganese was 42 g/l and then drop to 38 g/l in the first 15 minutes. The increase of processing time did not result in the significant manganese reduction, only 4.5% to the initial concentration. From this result, we conclude that the optimum adsorption processing time was 30 minutes.

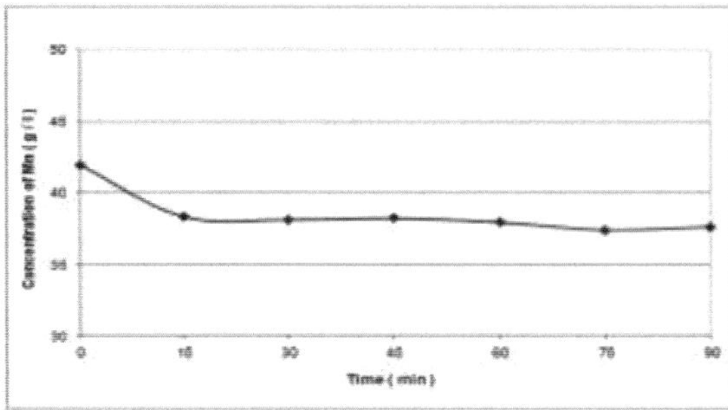

Figure 4. Active carbon processing time vs. manganese ion concentration in the 5 g active carbon and 10 ml NaOH 1 N addition

3.3. Effect of Type Absorbent

After optimal processing time was obtained then the selection of active carbon was carried out. There were 3 species of active carbon that was compared, active carbon from Philippine (which is usualy used in Gold mine adsorption process), active carbon from Japan, commercially available active carbon (Merck, Germany and Brataco, Indonesia), Indonesian locally produced active carbon, and wood charcoal for manganese reduction process.

Fig. 5 shows that Fe ion was adsorbed 89.74% using Japan made active carbon and 81.13% using local active carbon. The lowest adsorption was using Brataco and Merck active carbon resulting 61.70% and 75.14% adsorption respectively. For Cu ion adsorption the best result was obtained using Philipine made active carbon and wood charcoal with adsorption percentage 44.49% and 40.33% respectively. Wood charcoal resulted in 70.88% and the Japan made active carbon resulted in 44.54% adsorption on Ti ion, while others resulted in 0% adsorption.

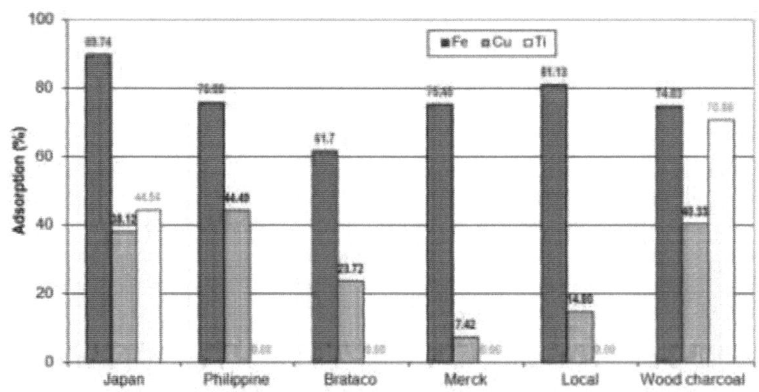

Figure 5. The adsorption of each active carbon species on Fe, Cu, and Ti ion in the 5g adsorbent and 10 ml NaOH 1 N addition

Overall the adsorption using wood charcoal was resulted in a high percentage of adsorption for all Fe, Cu, and Ti ion 74.83%, 40.33%, and 70.88% respectively.Therefore the application of wood charcoal in the reduction process would directly adsorbed the Fe, Cu, and Ti ion and the use of wood charcoal in the pilot plant need not to be replaced.

3.4. Surface Imaging of Active Carbon

From the experiment we can conclude the Japan made carbon active have the highest adsorption than other active carbon. The result from wood charcoal also showed a high percentage of adsorption. Fig.

6 shows the SEM imaging at the surface of Japan made active carbon and wood charcoal. The SEM imaging shows that both Japan active carbon and wood charcoal are porous, but the Japanese active carbon is more solid and smooth than the wood charcoal so that the adsorption of wood charcoal is higher than the others. Despite resulting in lower percentage of adsorption, the wood charcoal application in the reduction process is giving more advantage in terms impurities elimination, so that the more expensive active carbon would not be needed and lowered the cost for the leaching process.

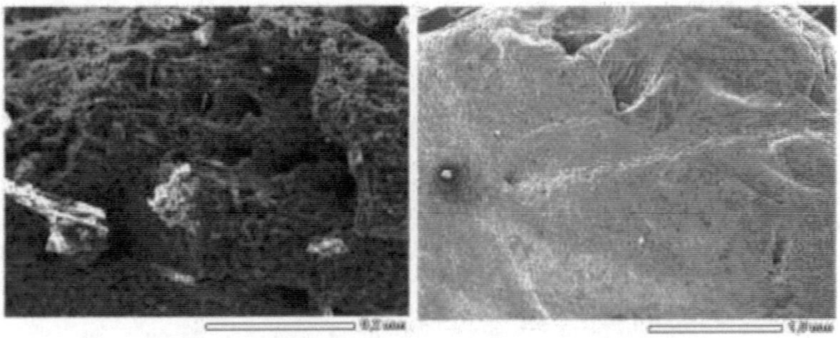

(a) (b)

Figure 6. SEM imaging of (a) wood charcoal and (b) Japanese made active carbon

4. Conclusion

Optimum adsorption process was reached for 5g carbon active and 10 ml of NaOH addition per 100 ml of solution with processing time of 30 minutes. Best result for Fe ion adsorption was using Japanese active carbon. Adsorption of Cu ion was best obtained using Philippines active carbon, and the best result for Ti ion adsorption was using wood charcoal. As required that the content of Titanium should be minimum for the material of manganese lithium catode, then the best active carbon to be used is from wood charcoal.

Acknowledgment

The authors would like to acknowledge Dr. Bambang Prihandoko dan Indonesian Concortium of Lithium Battery for their support, so this research can be conducted and presented.

References

[1] Avijit B and Barsha D 2013 *J.Hydrometallurgy*. **140** 151-162
[2] Wensheng Z and Chu Yong C 2007 *J.Hydrometallurgy*. **89** 137-159
[3] Silvio Roberto T and Jorge Rubio R 2009 *J.Minerals Engineering*. **22** 336-343
[4] Esmaeili, Abass, Mesdaghi Nia, Alireza, Vazirinejad, and Reza 2005 *J. Appl. Sci.* **2** **10** 1471–73
[5] Nathsarma K C and Bhaskar Sharma 1987 *J.Hydrometallurgy*. **17** **2** 239–249
[6] Wensheng Z and Chu Yong C 2007 *J.Hydrometallurgy*. **89** 160-177
[7] John Jamieson 1995 *Precipitation and Characterisation of Fe(III) Oxyhydroxides from Acid Liquors* (Perth Australia: Murdoch University Press)

Impurities Removal in Seawater to Optimize the Magnesium Extraction

N C Natasha[1] , F Firdiyono[1] and E Sulistiyono[1]
[1]Research Center of Metallurgy and Materials, LIPI, Puspiptek Gedung 470, Indonesia

E-mail: nadia_natasha90@yahoo.com

Abstract. Magnesium extraction from seawater is promising way because magnesium is the second abundant element in seawater and Indonesia has the second longest coastline in the world. To optimize the magnesium extraction, the impurities in seawater need to be eliminated. Evaporation and dissolving process were used in this research to remove the impurities especially calcium in seawater. Seawater which has been evaporated from 100 ml to 50 ml was dissolved with variations solution such as oxalic acid and ammonium bicarbonate. The solution concentration is 100 g/l and it variations are 2 ml, 4 ml, 6 ml, 8 ml, 10 ml, 20 ml, 30 ml, 40 ml and 50 ml. This step will produce precipitate and filtrate then it will be analysed to find out the result of this process. The precipitate was analysed by X-ray Diffraction (XRD) and Scanning Electron Microscope (SEM) but the filtrate was analysed by Inductively Coupled Plasma (ICP). XRD analysis shows that calcium oxalate and calcium carbonate were formed and ICP analysis shows that the remaining calcium in seawater using oxalic acid is about 0.01% and sodium 0.14% but when using ammonium bicarbonate the remaining calcium is 2.5% and sodium still more than 90%. The results show that both oxalic acid and ammonium bicarbonate can remove the impurities but when using oxalic acid, not only the impurities but also magnesium was precipitated. The conclusion of this research is the best solution to remove the impurities in seawater without precipitate the magnesium is using ammonium bicarbonate.

1. Introduction

Magnesium is an alkaline earth metal and highly abundant in the earth's crust that has many functions in automotive, aerospace, medical, electronic and etcetera [1]. To improve its properties, magnesium needs to be designed and alloyed with other metals [2]. Magnesium can be found not only in dolomite, magnesite but also in seawater and brine water [3]. Seawater is a solution that contain of many elements and organisms. The most abundant of chemical dissolved in seawater can be seen on Table 1.

Table 1. Chemical dissolved in seawater [4]

Chemical ion	Concentrations (parts per thousand)	Proportion of total salinity
Sodium	10.752	55.03
Magnesium	1.295	3.68
Calcium	0.416	1.18
Potassium	0.390	1.11
Chloride	19.345	55.03
Sulfate	2.701	7.68

[1] To whom any correspondence should be addressed.

Based on table 1, magnesium is the second abundant element in seawater after sodium so that magnesium extraction from seawater is promising way. Unfortunately, there are many impurities in seawater such as sodium, calcium, and potassium so that the removal impurities needed to be done. Magnesium extraction was done by many researchers such as Robinson in 1943 gave patent about recovering magnesium as $Mg(OH)_2$ from seawater [5], Yildirim in 2010 extracted the MgO from dolomite ore by removing the calcium first as carbonates [6], Khuyen Thi Tran in 2013 extracted magnesium oxalate from Uyuni Salar brine by removing the impurities such as calcium, potassium and lithium [3] and Khuyen Thi Tran in 2015 extracted the magnesium from Uni Salar brine as magnesium carbonate by removing the calcium first using oxalic acid then use NaOH and Na_2CO_3 solution [1].

From previous works, authors conclude that to optimize magnesium extraction, the impurities especially calcium in seawater needed to be eliminated. Calcium needed to be eliminated because it can reduce the magnesium purity in the process of magnesium extraction [3]. The effect of oxalic acid and ammonium bicarbonate addition to calcium removal in seawater from local area Indonesia was studied. This study was done to know the best solution to remove the impurities without remove the magnesium.

2. Experimental

2.1. Materials and reagents

Seawater from beach in Tangerang was taken and brought to laboratory at Research Center of Metallurgy and Materials. Chemical composition of seawater was analysed first using Inductively Coupled Plasma (ICP). Oxalic acid and ammonium bicarbonate were used in this study as reagents to remove the impurities.

2.2. Experimental Methods

Seawater was filtered to remove the dirt and analysed by ICP to know its initial composition. Seawater was heated with the ratio 1 : 2 of seawater evaporated then analysed with ICP. The concentration of oxalic acid and ammonium bicarbonate is 100 g/l. Volume variations of oxalic acid and ammonium bicarbonate were done from 2 ml, 4 ml, 6 ml, 8 ml, 10 ml, 20 ml, 30 ml, 40 ml and 50 ml. Dissolution of oxalic acid and ammonium bicarbonate in seawater was done at ambient temperature for 24 hours without stirring. The precipitate from this step filtered to separate the precipitation and filtrate. Precipitation was analysed by SEM to know the morphology and XRD to find its phase. Filtrate was analysed by ICP to get the final concentration of seawater.

3. Results and discussion

3.1. Filtrate

Before doing the experiment, ICP analysis was done to know the initial concentration of seawater. ICP analysis shows that seawater consisted of sodium (10129.3 ppm), magnesium (1672.09 ppm) and calcium (553.21 ppm). It tells that sodium dominates and magnesium is the second abundant element in sample. Heating the seawater until 50% of sample evaporated can affect the concentration of seawater either it will be increased or decreased. ICP analysis was done to find the effect of heating process. Analysis results in filtrate are sodium (20175.5 ppm), magnesium (1576.2 ppm) and calcium (826.204 ppm). It indicates that heating process effect the seawater more saline because the concentration of sodium and calcium increased.

This filtrate then added with two different solutions that are 100 g/l oxalic acid and 100 g/l ammonium bicarbonate. Volume variations were done to find the optimum volume to remove the impurities. This process produced liquid residue and precipitate. Liquid residue was analysed with ICP to find the final concentration of seawater. The analysis can be seen at Table 2.

Table 2. ICP analysis of seawater with 100 g/l oxalic acid and 100 g/l ammonium bicarbonate

Volume (ml)	Sodium (%)		Magnesium (%)		Calcium (%)	
	Oxalic Acid	Ammonium Bicarbonate	Oxalic Acid	Ammonium Bicarbonate	Oxalic Acid	Ammonium Bicarbonate
2	10.98	103.35	14.04	102.76	21.11	31.13
4	11.93	102.24	15.36	104.91	2.62	5.61
6	3.63	98.14	15.51	104.13	2.21	4.46
8	10.52	92.98	16.02	105.74	2.40	4.08
10	3.69	96.93	18.02	108.13	0.65	3.38
20	21.1	101.80	15.24	113.78	2.34	2.98
30	16.11	98.35	22.45	120.44	3.66	2.50
40	0.92	102.65	7.84	122.95	0.20	3.29
50	0.14	100.08	0.00	131.39	-0.01	3.56

Table 2 shows that by using oxalic acid the content of sodium and calcium are reduced until 0.14% and 0.01% remained while using ammonium bicarbonate only calcium that reduced rapidly until 2.5% remained but sodium only 2% removed from seawater. Unfortunately, more than 90% magnesium in seawater also reduced when using oxalic acid. Calcium with oxalic acid will react and form white precipitation as calcium oxalates. Calcium oxalates will be produced immediately in concentrated solution but in aqueous solution it will be produced slowly [7] so that calcium can be eliminated from seawater using oxalic acid. According to Khuyen Thi Tran, magnesium in oxalic acid will be precipitated at pH > 1. Oxalic acid solution in this experiment is an aqueous solution with the concentration less than 1 M. Thus, magnesium in seawater was removed.

Magnesium in mixture of seawater and ammonium bicarbonate was not removed. It can be seen at Table 2 that more than 90% of magnesium in seawater using ammonium bicarbonate still exist. Calcium with hydrogen carbonate solution will give white precipitation but this precipitation has high solubility in water and the precipitation will dissolve in water [7]. So that if calcium in oxalic acid was compared with calcium in ammonium bicarbonate, the calcium concentration in ammonium bicarbonate higher than calcium concentration in oxalic acid. It caused by the precipitate re-dissolved in seawater. Magnesium with ammonium bicarbonate will not produce a white precipitate. It caused by ammonium (NH_4^+) effect the equilibrium reaction and hydrogen carbonate ion produced [7].

3.2. Precipitate

Reaction between oxalic acid and ammonium bicarbonate produced a precipitate. The precipitate was analysed by SEM to know the morphology of precipitate and XRD to confirm the phase that formed from this experiment.

Oxalic acid	Ammonium Bicarbonate
	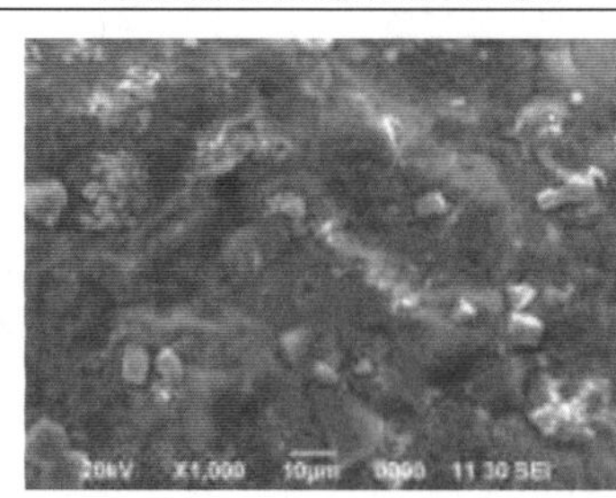
a	e

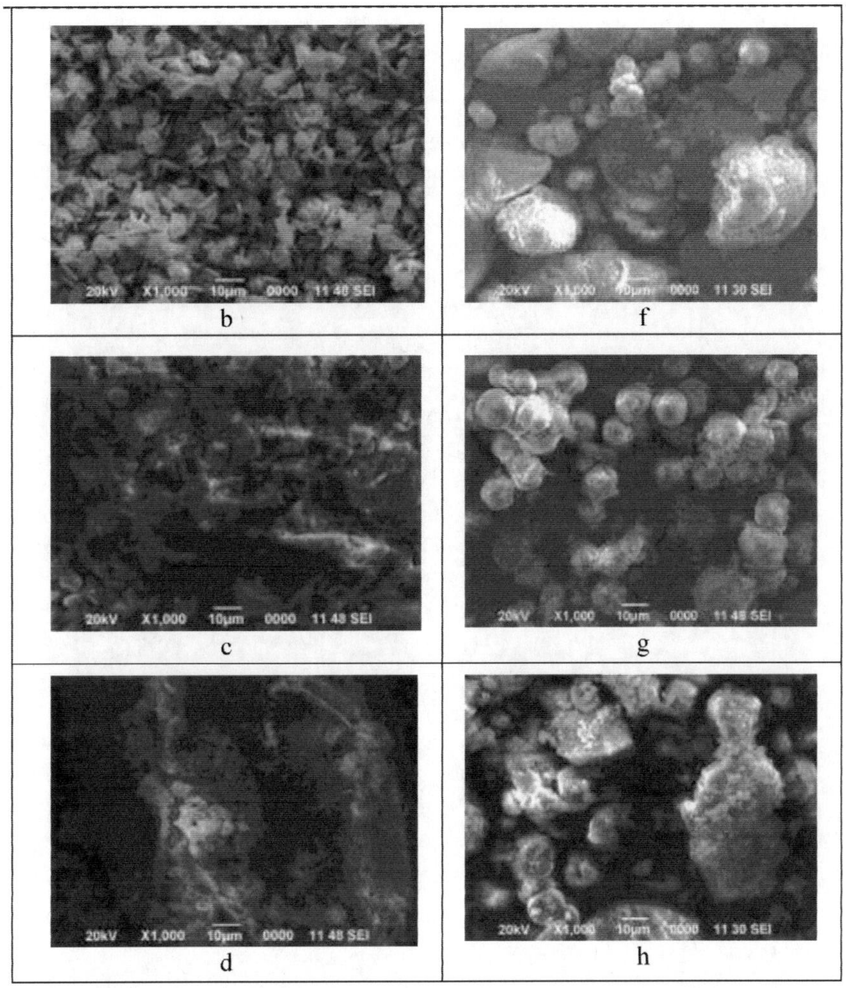

Figure 1. SEM image of precipitation in 1000x magnification using oxalic acid and ammonium bicarbonate. a) 2 ml oxalic acid in seawater b) 10 ml oxalic acid in seawater c) 30 ml oxalic acid in seawater d) 40 ml oxalic acid in seawater e) 2 ml ammonium bicarbonate in seawater f) 10 ml ammonium bicarbonate in seawater g) 30 ml ammonium bicarbonate in seawater h) 40 ml ammonium bicarbonate in seawater

Figures 1a – 1d are precipitate result from seawater with oxalic acid addition whereas Figures 2e – 2h are precipitate result from seawater with ammonium bicarbonate addition. Figure 1a – 1b shows that monoclinic and prismatic crystal system appeared clearly. It indicates that calcium oxalates monohydrated was formed in this experiment [8]. Figures 1c – 1d show that monoclinic and prismatic crystal appeared in small amounts and dominate with other forms. It indicates that not only calcium oxalate formed in this experiment but also other oxalates. While in figures 2e – 2h show that the rounded crystal dominated and become extended when the ammonium bicarbonate increased. It indicates that calcium carbonate was formed from this reaction.

IOP Conf. Series: Materials Science and Engineering **176** (2017) 012039 doi:10.1088/1757-899X/176/1/012039

To confirm the phase of the precipitate that produced from oxalic acid and ammonium bicarbonate addition in seawater, XRD analysis was done on these samples.

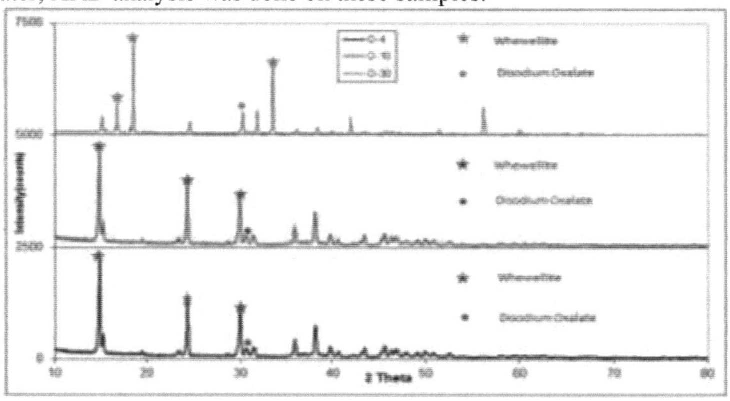

a

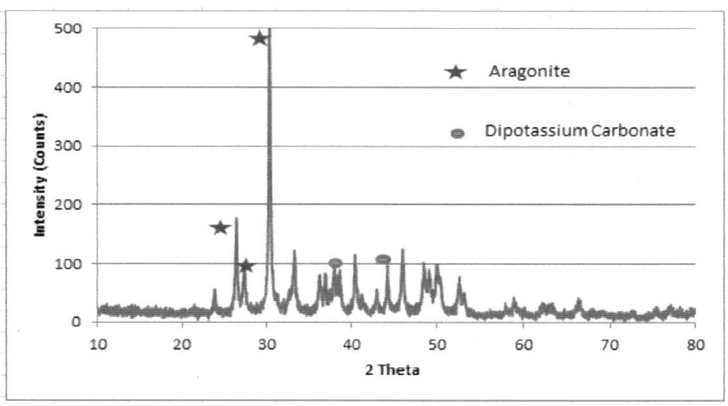

b

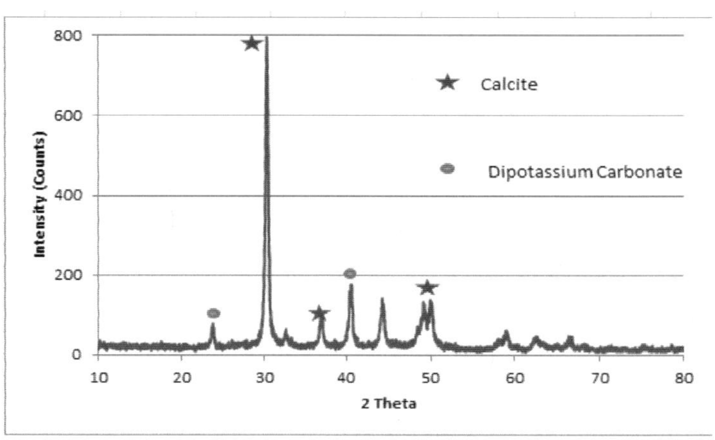

Figure 2. XRD analyse of precipitate from oxalic acid and ammonium bicarbonate addition. a) seawater with 4 ml oxalic acid, 10 ml oxalic acid and 30 ml oxalic acid. b) seawater with 10 ml ammonium bicarbonate. c) seawater with 40 ml ammonium bicarbonate

X-ray diffraction (XRD) analysis with 2 Theta 10° - 80° was done in this study. Figures 2a showed that whewellite ($CaC_2O_4.H_2O$) and disodium oxalate monoperhydrate ($Na_2C_2H_2O_6$) formed when 10 ml oxalate was added into sea water while whewellite and sodium hydrogenoxalate hydrate formed when 30 ml oxalate was added into seawater. Figures 2b – 2c showed ammonium bicarbonate in seawater addition caused aragonite ($CaCO_3$) and dipotassium carbonate (K_2CO_3) formed. Figures 2a – 2c show that the intensity of oxalic acid addition is higher than ammonium carbonate addition in seawater. It can be seen that the intensity of oxalic acid addition reached 7000 while the intensity of ammonium carbonate addition only reached 800.

This analysis also tells that both whewellite and disodium oxalate monoperhydrate have a monoclinic crystal system but sodium hydrogen oxalate hydrate has anorthic crystal system. While aragonite has orthorhombic crystal system and others have hexagonal crystal system. The analysis results confirmed that both oxalic acid and ammonium bicarbonate are able to eliminate the impurities in seawater but the best solution to eliminate the impurities without remove the magnesium is ammonium bicarbonate.

4. Conclusion

Oxalic acid and ammonium bicarbonate can remove the impurities in seawater from Tangerang, Indonesia that consisted of sodium (10129.3 ppm), magnesium (1672.09 ppm) and calcium (553.21 ppm) in order to extract the magnesium. Oxalic acid able to remove sodium and calcium until 0.14% and 0.01% remained while using ammonium bicarbonate only calcium that reduced rapidly until 2.5% remained but sodium only 2% removed from seawater. This experiment shows that oxalic acid also can remove the magnesium but ammonium bicarbonate only can remove calcium so that the best solution to remove the impurities without remove the magnesium is ammonium bicarbonate. Variations in solution concentration need to be done to know whether oxalic acid can eliminate the impurities without remove the magnesium.

References

[1] Khuyen T T, Kyu S H, Su J K, Myong J K, Tam T 2016 *hydrometallurgy* **160** pp 106 – 114
[2] Anonymous http://www.intlmag.org/magnesiumresources/process.cfm International Magnesium Association akses 22 september 2016
[3] Khuyen T T, Tri V L, Jeon-Woong A, Dong-Jun K, Myong-Jun K, Tam T 2013 *hydrometallurgy* **138** pp 93 – 99
[4] Genny A 2008 Seawater Composition http://www.marinebio.net/marinescience/02ocean/swcomposition.htm akses 22 september 2016
[5] Robinson, H.A., Friedrich, R.E., Spencer, R.S 1943 *Magnesium hydroxide from seawater. US Patent 2,405,055* (30 June 1943).
[6] Yildirim, M., Akarsu, H 2010 *Physicochem. Probl. Miner. Process* pp 257–272
[7] Ir. L. Setiono, Dr. A. Hadyana Pudjaatmaka. 1990. *Vogel Buku Teks Analisis Anorganik Kualitatif Makro dan Semimikro Bagian I*. Jakarta.
[8] Emel A, Mualla O 2014 *J. Cryst. Growth* **401** pp 260 – 265

IOP Conf. Series: Materials Science and Engineering **176** (2017) 012040 doi:10.1088/1757-899X/176/1/012040

The Synthesis of Calcium Salt from Brine Water by Partial Evaporation and Chemical Precipitation

L H Lalasari[1], M K Widowati[2], N C Natasha[1], E Sulistiyono[1] and A B Prasetyo[1]
[1]Research Centre for Metallurgy and Material, Indonesian Institute of Sciences (LIPI), Kawasan Puspiptek Serpong, Tangerang Selatan 15314, Indonesia
[2]Faculty of Mathematics and Natural Sciences, Universitas Sebelas Maret (UNS), Ir. Sutami No.36 A Surakarta 57126

E-mail: ifa_sari@yahoo.com; lati003@lipi.go.id

Abstract. In this study would be investigated the effects of partial evaporation and chemical precipitation in the formation of calcium salt from brine water resources. The chemical reagents used in the study was oxalate acid ($C_2H_2O_4$), ammonium carbonate $(NH_4)_2CO_3$) and ammonium hydroxide (NH_4OH) with reagent concentration of 2 N, respectively. The procedure was 10 liters brine water evaporated until 20% volume and continued with filtration process to separate brine water filtrate from residue (salt). Salt resulted from evaporation process was characterized by Scanning Electron Microscopy (SEM), X-Ray Fluorescence (XRF) and X-Ray Diffraction (XRD) techniques. Filtrate then was reacted with $C_2H_2O_4$, $(NH_4)_2CO_3$ and NH_4OH reagents to get salt products in atmospheric condition and variation ratio volume brine water/chemicals (v/v) [10/1; 10/5; 10/10; 10/20; 10/30; 10:50; 20/1; 20/5; 20/10; 20/20; 20/30; 20:50]. The salt product than were filtered, dried, measured weights and finally characterized by SEM/EDS and XRD techniques. The result of experiment showed the chemical composition of brine water from Tirta Sanita, Bogor was 28.87% Na, 9.17% Mg, 2.94% Ca, 22.33% O, 0.71% Sr, 30.02% Cl, 1.51% Si, 1.23% K, 0.55% S, 1.31% Al. The chemical composition of salt resulted by partial evaporation was 53.02% Ca, 28.93%O, 9.50% Na, 2.10% Mg, 1.53% Sr, 1.20% Cl, 1.10% Si, 0.63% K, 0.40% S, 0.39% Al. The salt resulted by total evaporation was indicated namely as NaCl. Whereas salt resulted by partial evaporation was $CaCO_3$ with a purity of 90 % from High Score Plus analysis. In the experiment by chemical precipitation was reported that the reagents of ammonium carbonate were more reactive for synthesizing calcium salt from brine water compared to reagents of oxalate acid and ammonium hydroxide. The salts precipitated by NH_4OH, $(NH_4)_2CO_3$, and $H_2C_2O_4$ reagents were indicated as NaCl, $CaCO_3$ and $CaC_2O_4.H_2O$, respectively. The techniques of partial evaporation until 20% volume sample of brine water and chemical precipitation using $(NH_4)_2CO_3$ reagent are recommended in the synthesis of calcium salts from brine water because are simple, flexible and economical.

1. Introduction

Calcium is a mineral that is essential for life, good for living beings as well as for the needs of the industry, agriculture, pharmaceuticals and other chemicals. Calcium is an important determinant of water harness, and it also functions as a pH stabilizer, because of its buffering qualities. Calcium also gives water a better taste. Calcium salts that are often encountered in daily life are calcium carbonate. It is an ingredient utilized as filler in paper, paint, sealants and coatings [1][2]. It also has many applications in agriculture and the food industry [3][4]. Some agricultural uses of this substance are as animal feed and agricultural lime for soil. Many foods and cooking products, such as baking powder, dough and dry mixes also contain calcium carbonate. It is also an ingredient in toothpaste.

1 To whom any correspondence should be addressed.
2 To whom any correspondence should be addressed.

Many natural resources in Indonesia, which can be used as a source to obtain calcium salts is sea water, brine water, limestone. Nowadays, calcium salts are abundantly synthesized from limestone and sea water compared to brine water resources. It becomes an interesting potential if could take advantage of brine water as raw material for the manufacture of the calcium salt. Brine water is a solution that has a high salt concentration. It comes from byproduct of reducing salt content process in seawater (desalination) so that brine water containing some of dissolved salts [5].

Few researchers have done experiments for synthesizing calcium salt from natural resources as follows: Wenhao Sun et al. [6] tried to know mechanisms of nucleation of metastable aragonite $CaCO_3$ in seawater. He observed that there are a direct relationship between the calcite surface energy and solution Mg–Ca ion concentrations. It showed the calcite nucleation barrier surpasses that of metastable aragonite in solutions with Mg:Ca ratios consistent in seawater, allowing aragonite to dominate the kinetics of nucleation. Ljerka Brecevic and Damir Kralj [7] said that calcium carbonates can be precipitated in the form of six modifications (polymorphic or hydrated). They consisted of crystal growth and transformation of amorphous calcium carbonate, calcium carbonate hexahydrate, calcium carbonate monohydrate, vaterite and calcite. Keith E. Chase and Erwin [8] Suess explained the addition of Ca^{2+} or CO_3^{2-} ions could increase supersaturation from seawater that would begin rapid precipitation of $CaCO_3$. Ritika Gupta [9] tried to synthesize of precipitated calcium carbonate nanoparticles using modified emulsion membranes. She has successful prepared calcium carbonate nanoparticles with particle size ranging from 100 nm to 1200 nm.

On this opportunity, we would investigate the effects of partial evaporation and chemical reagents in the synthesis of calcium salt from brine water resources in Indonesia. The chemical reagents used in the study was oxalate acid ($C_2H_2O_4$), ammonium carbonate ($(NH_4)_2CO_3$) and ammonium hydroxide (NH_4OH) that was the type of acidic, neutral, alkaline reagents. From this research could be obtained the following data such as cheaper technology and an effective reagent in extracting calcium from brine water to form calcium salts.

2. Experimental

The sample was originally from *brain water,* Tirta Sanita, Bogor. Whereas the chemical reagent used in the research were purchased from Merck with analytical pure grade that was oxalate acid ($C_2H_2O_4$), ammonium carbonate ($(NH_4)_2CO_3$) and ammonium hydroxide (NH_4OH).

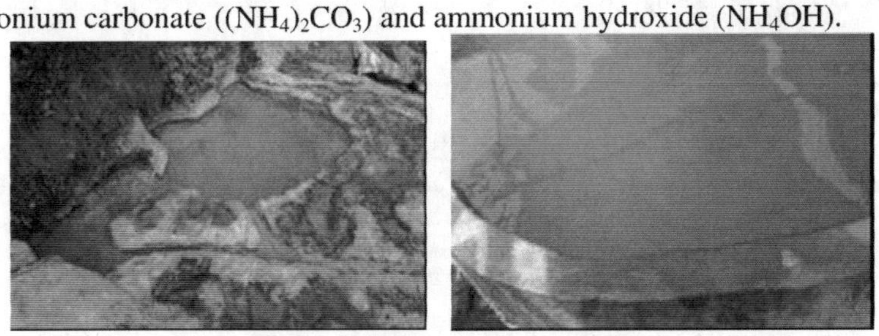

Figure 1. Brine water sample

For identifying chemical composition of brine water was carried out total evaporation processes on sample, than the resulted residue was characterized by X-Ray Florescence (XRF) and X-ray Diffraction (XRD) analysis as shown in Table 1 and Fig. 2.

Table 1. The chemical composition of brine water

Element	Ca	O	Na	Mg	Sr	Cl	Si	K	S	Al
Mass (%)	2.94	22.33	28.87	9.17	0.71	30.02	1.51	1.23	0.55	1.31

In this experiment, the formation mechanism of calcium salt was investigated. The processing methods were partial evaporation and chemical precipitation. The procedure was 10 liters of brine water evaporated until 20% volume sample and continued with filtration process to separate brine

IOP Conf. Series: Materials Science and Engineering **176** (2017) 012040 doi:10.1088/1757-899X/176/1/012040

water filtrate from residue product. Residue resulted from evaporation process was characterized by X-ray Fluorescence (XRF) to identify chemical element of residue, Scanning electron microscopy/energy (SEM) to know morphology particle and X-ray diffraction (XRD) to identify phase of residue product. Filtrate of brine water then was reacted with $C_2H_2O_4$, $(NH_4)_2CO_3$ and NH_4OH reagents to get salt products in atmospheric condition and variation ratio volume brine water/chemicals (v/v) as follows: [10/1; 10/5; 10/10; 10/20; 10/30; 10:50; 20/1; 20/5; 20/10; 20/20; 20/30; 20:50]. The salt product than were filtered, dried, measured weights and finally characterized by SEM/EDS and X-ray diffraction (XRD) to identify phase of salt products. The flowchart of research is shown in Fig. 3.

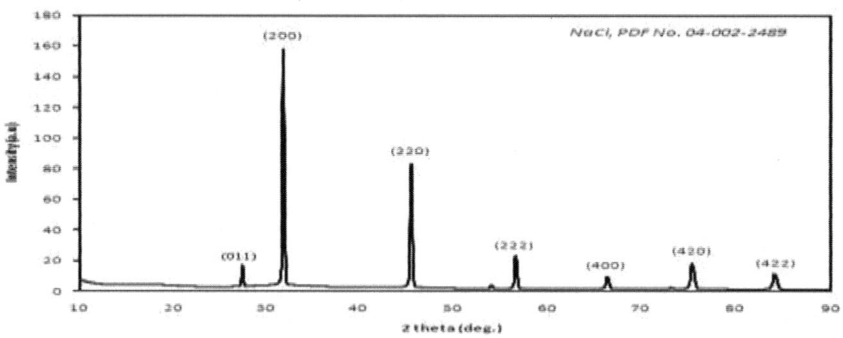

Figure 2. The XRD pattern of brine water after total evaporation process

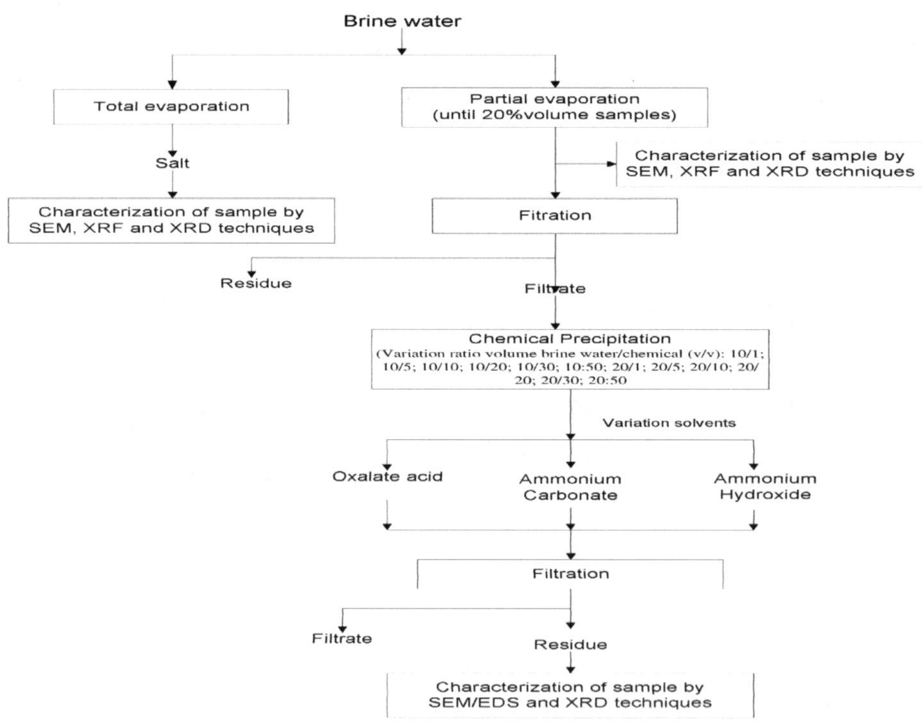

Figure 3. The flowchart of research

3. Result and Discussion

The brine water from Tirta Sanita, Bogor has murky green colorless and smelled like sulfur as shown in Fig. 1. From XRF analysis in Table 1 shows the chemical composition of brine water is 28.87% Na, 9.17% Mg, 2.94% Ca, 22.33% O, 0.71% Sr, 30.02% Cl, 1.51% Si, 1.23% K, 0.55% S, 1.31% Al. It describes the resources of brine water consisted of sodium, calcium, magnesium and chloride elements that have the percentage higher than other elements. If brine water was carried out total evaporation process, it would form precipitated salts as shown in Fig. 2. Figure 2 looks that the peaks of diffraction are clearly and sharply at 2Θ degree of 27.44°; 31.79°; 45.58°; 54.02°; 56.63°; 66.42°. Based on PDF No. 04-002-2489, It is indicated namely as sodium chloride (NaCl). This is suitable with the data of chemical composition in Table 1 that the sodium (Na) and chloride (Cl) elements has the highest percentage compared to other elements. Therefore the phase of NaCl appears more dominant than other phases.

The diffraction peaks of other salts have not looked clearly in Fig. 2 because has lower intensity than the diffraction peaks of NaCl salt. It would become economical if brine water could be utilized for synthesizing calcium salts. We know that the consumption of calcium salt is predicted a rise slightly every year in the world. The formation mechanism of calcium salts of brine water would be explained in this research.

3.1. The effect of partial evaporation on the chemical composition, morphology and phase of residue product

In this research would be investigated the effect of partial evaporation on the chemical composition, morphology and phase of residue product. For describing and comparing the chemical composition and morphology of residue product resulted from total evaporation and partial evaporation until 20% volume sample of brine water was carried out the SEM and XRF analysis as represented in Table 2 and Fig. 4.

Table 2. The chemical composition of brine water and residue by partial evaporation until 20% (v/v) sample

Element	Ca	O	Na	Mg	Sr	Cl	Si	K	S	Al
Brine water (%)	2.94	22.33	28.87	9.17	0.71	30.02	1.51	1.23	0.55	1.31
Residue from partial evaporation (%)	53.02	28.93	9.50	2.10	1.53	1.20	1.10	0.63	0.40	0.39

From Table 2 looks that the chemical composition of residue product by partial evaporation is 53.02% Ca, 28.93%O, 9.50% Na, 2.10% Mg, 1.53% Sr, 1.20% Cl, 1.10% Si, 0.63% K, 0.40% S, 0.39% Al. It appears the percentage of calcium element increased from 2.94% to 53.02% by partial evaporation until 20% volume sample of brine. Partial evaporation undertaken is able to increase supersaturation of brine water so that precipitation of calcium element could occur rapidly in hours. Based on Table 2 could be explained that calcium element is easier precipitated than sodium and magnesium elements by partial evaporation techniques.

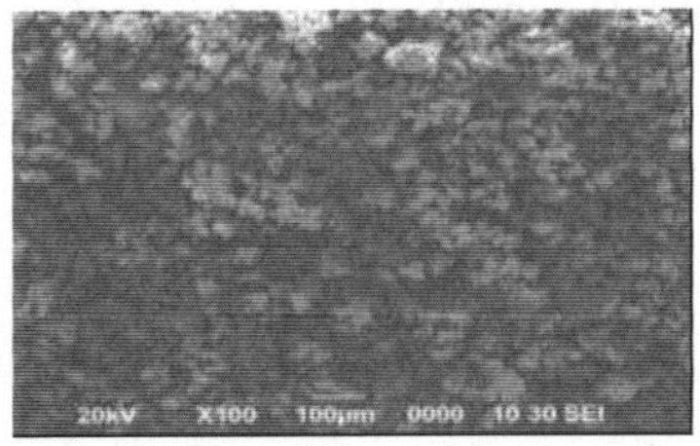

Figure 4. The morphology of residue product on evaporation: (a) total evaporation; (b) partial evaporation (evaporation until 20% volume sample)

Fig. 4 describes the morphology of residue products by total evaporation and partial evaporation. It represents particles has irregular morphology like cubic for residue product by total evaporation and almost regular morphology like spherical for residue product by partial evaporation. On observation by SEM with magnification at 100 x also looks the particle sizes of residue product by total evaporation is approximately 100 μm that were larger than particles sizes by partial evaporation. It probably caused agglomeration of particles from brine water such as salts elements (Na, Ca, Mg, K) so that appears larger and irregular by SEM observation. The growth of NaCl crystal is more dominant than other salt crystals by total evaporation. Kazunori Kadota et al. [10] reported the precipitated NaCl and KCl particles have asymmetric structure. He also explained that other surfaces in contact with the aqueous phase were a flat cubic structure.

This condition is different with residue product by partial evaporation that has smaller particle size with regular particle distribution. It is probably indicated that particles resulted by partial evaporation is dominant as calcium elements with homogeneous particles. The difference of morphology of particles depends on evaporation condition. Takiyama et al. [11] concluded that the shape of the crystals was found to differ with the operating conditions. For identifying the phase of residue product by partial evaporation was carried out the XRD analysis as represented in Fig. 5.

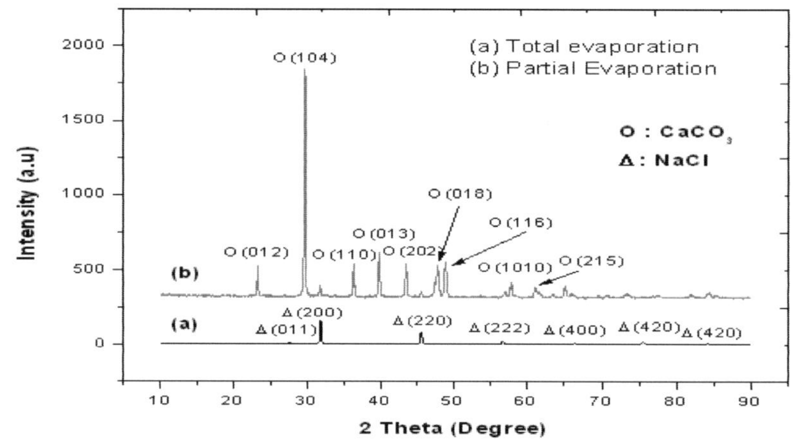

Figure 5. The XRD pattern of residue product on variation evaporation techniques: (a) total evaporation; (b) partial evaporation

Fig. 5 looks that the phase of residue product changes with different evaporation techniques. At explaining above, it is known that the phase of residue product by total evaporation is sodium chloride (Fig. 2 or Fig. 5a). Whereas the result of XRD analysis from residue product by partial evaporation (Fig. 5b) shows the peaks of diffraction is clearly and sharply at 2Θ degree of 23.3°; 29.67°; 36.3°; 39.8°; 43.54°; 47.87°; 48.91°; 57.91°; 61.3°. It is known namely as calcium carbonate ($CaCO_3$) that is approximately 90 % $CaCO_3$ from High Score Plus analysis.

The mechanism of $CaCO_3$ formation based on chemical thermodynamics that the precipitation of solid phase (CaCO3) is controlled by the law of mass action. The purpose of partial evaporation is to create a solution of brine water become supersaturation condition with ion activation product (IAP) > solubility product (Ksp). For this condition, IAP = $[Ca^{2+}][CO_3^{2-}]$ and Ksp ($CaCO_3$) = $10^{-8.48}$ (Table 3). Therefore $[Ca^{2+}][CO_3^{2-}] > 10^{-8.48}$, it causes precipitation of $CaCO_3$ in brine water. This research is such interesting because calcium element was precipitated firstly besides other elements. It could be looked from the Data of Ksp as shown in Table 3. If the value of Ksp is low, it would rapid precipitation of solid phase.

Table 3. The solubility product (Ksp) of brine salt product

Compound	pKsp	Ksp	Classification rule	Ref
$CaCO_3$	8.48	$10^{-8.48}$	insoluble	[12]
CaC_2O_4	2.27	$10^{-2.27}$	insoluble	[13]
$Na_2CO_3.10H_2O$	1.31	$10^{-1.31}$	insoluble	[12]
$Na_2C_2O_4$	-3.26	$10^{3.26}$	Soluble	[13]
$MgCO_3$	8.03	$10^{-8,03}$	insoluble	[12]
MgC_2O_4	-	-	-	[13]
$CaCl_2.6H_2O$	-4.09	$10^{4.09}$	Soluble	
NaCl	-1.58	$10^{1.58}$	Soluble	[12]
$MgCl_2$.6H2O	-4.39	$10^{4.39}$	Soluble	
KCl	-0.85	$10^{0.85}$	Soluble	

Table 3 shows that $CaCO_3$ salt has Ksp value lower than other salts such as CaC_2O_4, $Na_2CO_3.10H_2O$, $MgCO_3$ (insoluble). Therefore the $CaCO_3$ salt would begin to form a precipitation more rapid than other salts in supersaturation condition of brine water.

3.2. The effect of chemical precipitation processes on the morphology and phase of salt product

In this research, the synthesis of calcium salt was conducted by chemical precipitation techniques using several reagents, such as oxalic acid ($H_2C_2O_4$), ammonium carbonate ($(NH_4)_2CO_3$) and ammonium hydroxide (NH_4OH), which each reagent would react with elements in the filtrate of brine water to form complex precipitates. The influence of reagent and ratio volume (v/v) the concentrated brine water (filtrate)/reagents would form precipitates differently. Formation of the precipitated product such as calcium salt would be explained detail in this research to indicate the kind of reagent and ratio volume (v/v) concentrated brine water/reagents that are optimum for synthesizing calcium salt. The precipitated products resulted by chemical precipitation are presented in Table 4.

Table 4. The formation of salt from the concentrated brine water in variation ratio volume (v/v) concentrated brine water (filtrate)/reagents and variation reagents

Ratio / Reagents	10:1	10:5	10:10	10:20	10:30	10:40	10:50
Salts using $H_2C_2O_4$ (gram)	4.28	4.4	4.7	4.68	4.78	4.8	4.72
Salt using $(NH_4)_2CO_3$ (gram)	4.08	4.2	4.25	4.2	4.25	4.2	4.25
Salt using NH_4OH (gram)	-	3.62	3.70	3.87	4.05	3.78	3.84
Ratio / Reagents	**20:1**	**20:5**	**20:10**	**20:20**	**20:30**	**20:40**	**20:50**
Salts using $H_2C_2O_4$ (gram)	4.31	5.21	5.39	5.2	5.4	5.4	5.8
Salt using $(NH_4)_2CO_3$ (gram)	4.29	4.73	4.89	5.12	4.48	4.31	4.38
Salt using NH_4OH (gram)	-	4.18	4.19	4.17	4.2	3.96	3.65

According to Table 4 looks that by using both oxalic acid and ammonium carbonate reagents, it would form precipitated salts more rapid (faster) than by using ammonium hydroxide at ratio volume (v/v) concentrated brine water/reagents of 10:1 and 20:1. It could be seen in the formation of brine salt using ammonium hydroxide reagent at ratio of 10: 1 and 20: 1 that did not produce precipitates. The

IOP Conf. Series: Materials Science and Engineering **176** (2017) 012040 doi:10.1088/1757-899X/176/1/012040

brine salt by using ammonium hydroxide begins to form at a ratio of 10: 5 and 20: 5. The formation of precipitated brine salt is maximum at ratio volume (v/v) concentrated brine water/reagents of 10:10 and 20:20 using ammonium carbonate; 10:30 and 20:50 using oxalate acid; 10:30 using ammonium hydroxide. From this research could be concluded briefly that the number of reagents of ammonium carbonate used to form a precipitate is less than reagents of oxalate acid and ammonium hydroxide. It shows that the reagent of ammonium carbonate is more reactive for synthesizing calcium salt from brine salt compared to reagents of oxalate acid and ammonium hydroxide. It is also supported with the data of Ksp of salt carbonate (CaCO3) that shows lower than the data of Ksp of other salts in Table 3. Keith E et al. [8] reported addition of CO_3^{2-} ion to seawater would increase supersaturation and finally would begin rapid $CaCO_3$ precipitation.

To determine the morphology and chemical composition of salt brine was carried out by SEM-EDS analysis. The SEM-EDS analysis is not representative, because the results are based on only one point or area from a sample shot X-ray. This analysis aimed to compare the morphology and the chemical composition of salt brine with semi quantitative through chemical precipitation techniques. The morphology and chemical composition of salt brine is shown in Fig. 6 and Table 5

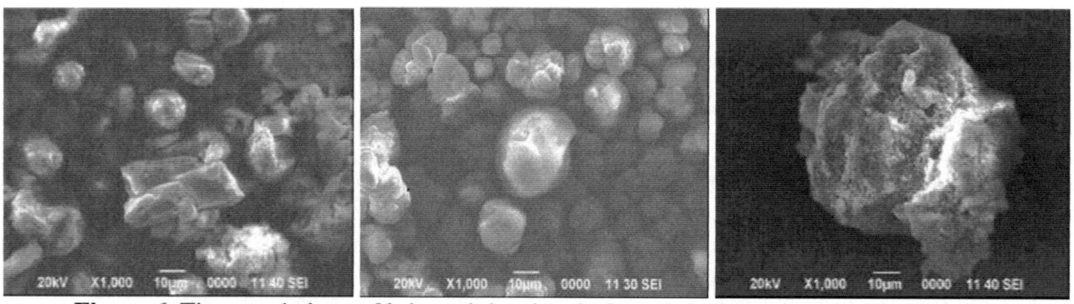

Figure 6. The morphology of brine salt by chemical precipitation using: (a) Oxalate acid; (b) Ammonium carbonate; (c) Ammonium hydroxide

The results of SEM analysis shows the morphology of salt brine precipitate is different that depend on a kind of reagents. In the salt brine formed using a reagent oxalic acid will produce irregular morphology like cubic. On the SEM results of precipitate with ammonium carbonate reagent has spherical shape morphology, while the SEM results of precipitate with ammonium hydroxide reagent has a larger size at the same magnification 1000x compared with both other precipitates.

Table 5. The semi quantitative of chemical composition from salt brine by EDS-SEM

Element	Oxalate salt	Carbonate Salt	Hydroxide Salt
Ca	17.64 %	39.91 %	3.47 %
K	0.02 %	0.41 %	0.91 %
Cl	0.26 %	0.17 %	24.28 %
Al	0.10 %	0.05 %	0.78 %
Mg	1.52 %	2.15 %	16.91 %
Na	15.22 %	0.98 %	9.95 %
O	43.85 %	46.82 %	29.25 %

Table 5 shows the chemical composition of salt precipitated by oxalate acid, ammonium carbonate and ammonium hydroxide reagents. It looks that the brine salt precipitated by oxalate acid and ammonium carbonate reagents has the calcium percentage higher than by ammonium hydroxide reagent. The salt precipitated by oxalate acid consists of calcium and sodium elements with high levels that were spread evenly in salt. It causes the morphology of salt precipitated by oxalate acid is irregular spherical (Fig. 6a). The same thing is also shown on the salt precipitated by ammonium hydroxide that salt consisted of sodium, magnesium, chloride with high levels besides other elements. On the precipitation using ammonium hydroxide, there are agglomerations of elements from brine water when begin the crystal growth together. It causes the morphology of particles appeared irregular and larger (Fig.6c)[10][11].

From EDS-SEM analysis in Table 5 could be seen that salt precipitated by ammonium carbonate has chemical composition as follows: 39.91% Ca, 0.41% K, 0.17% Cl, 0.05% Al, 2.15% Mg, 0.98% Na, 46.82% O. It shows that the calcium element is the most dominant precipitated by ammonium carbonate. Therefore the morphology of salt precipitated by $(NH_4)_2CO_3$ reagent has spherical shape and appeared more regular. For identifying the phase of brine salts precipitated by $H_2C_2O_4$, $(NH_4)_2CO_3$, NH_4OH reagents were carried out the XRD analysis as shown in Fig. 7.

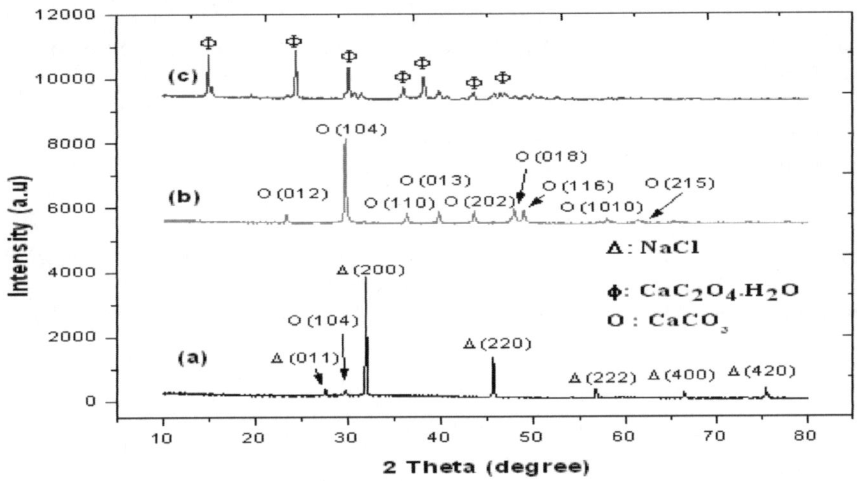

Figure 7. The XRD pattern of salts resulted by chemical precipitation using variation reagents: (a) ammonium hydroxide; (b) ammonium carbonate; (c) oxalate acid

The peaks of diffraction are clearly and sharply at 2Θ degree of 27.44°; 31.79°; 45.58°; 54.02°; 56.63°; 66.42° as represented in Fig. 7. Based on PDF No. 04-002-2489, it is indicated namely as sodium chloride (NaCl). This is supported with the data of chemical composition in Table 1 that the Na. Mg and Cl elements has higher percentage than other elements for chemical precipitation using NH_4OH. On this condition, salt of NaCl is precipitated earlier than $MgCl_2$ because value Ksp of NaCl is smaller than Ksp of $MgCl_2$ as shown Table 3. Therefore salt precipitated by NH_4OH is indicated as NaCl. From this research could be explained that the NH_4OH reagent is not suitable used for synthesizing calcium salt from brine water.

Fig. 7b shows that the peaks of diffraction from salt precipitated by $(NH_4)_2CO_3$ are clearly and sharply at 2θ degree of 23.3°; 29.67°; 36.3°; 39.8°; 43.54°; 47.87°; 48.91°; 57.91°; 61.3°. It is known namely as calcium carbonate ($CaCO_3$). The Ca^{2+} ion of brine water is almost precipitated completely using $(NH_4)_2CO_3$ reagents[8]. This technique of chemical precipitation using $(NH_4)_2CO_3$ is recommended for synthesizing calcium salt from brine water.

On observation in Fig. 7c looks that the peaks of diffraction from salt precipitated by $H_2C_2O_4$ are clearly and sharply at 2θ degree of 14.89°; 24.36°; 30.02°; 38.14°; 43.17°; 46.41°. It is indicated namely as calcium oxalate ($CaC_2O_4.H_2O$). The chemical precipitation using $H_2C_2O_4$ reagent is also recommended for synthesizing calcium salt from brine water. However the rate of $CaCO_3$ precipitation is faster than $CaC_2O_4.H_2O$ precipitation.

4. Conclusion

The result of experiment showed that the chemical composition of brine water from Tirta Sanita. Bogor was 28.87% Na, 9.17% Mg, 2.94% Ca, 22.33% O, 0.71% Sr, 30.02% Cl, 1.51% Si, 1.23% K, 0.55% S, 1.31%. The salt precipitated by total evaporation was indicated namely as sodium chloride (NaCl). The chemical composition of salt resulted by partial evaporation was 53.02% Ca, 28.93%O, 9.50% Na, 2.10% Mg, 1.53% Sr, 1.20% Cl, 1.10% Si, 0.63% K, 0.40% S, 0.39% Al. It was known namely as calcium carbonate ($CaCO_3$) with a purity of 90 % $CaCO_3$ from High Score Plus analysis. The particles have irregular morphology like cubic for salt resulted by total evaporation and regular (homogenous) morphology like spherical for salt resulted by partial evaporation. For the synthesis of calcium salt by chemical precipitation was obtained information as follows: the reagents of ammonium carbonate were more reactive for synthesizing calcium salt from brine water compared to reagents of oxalate acid and ammonium hydroxide. The salts precipitated by NH_4OH, $(NH_4)_2CO_3$ and $H_2C_2O_4$ reagents are indicated as NaCl, $CaCO_3$ and $CaC_2O_4.H_2O$, respectively. The morphology of salt resulted by chemical precipitation was irregular with asymmetric structure using NH_4OH reagent, spherical shape with homogeneous particles using $(NH_4)_2CO_3$ reagent, and irregular spherical morphology using $H_2C_2O_4$ reagent. From this research could be concluded briefly that the technique of partial evaporation until 20% volume sample of brine water and chemical precipitation using $(NH_4)_2CO_3$ reagent are recommended in the synthesis of calcium salts from brine water for domestic and industry applications. These technologies are simple, flexible and economical.

Acknowledgments

The authors are grateful for the main financial supports from the SINAS program of Lithium Battery Consortium (2014-2015), Research Center for Physic, Indonesian Institute of Science (LIPI) and thanks for contributing from Mudhita Kusuma Widowati, Department of Chemistry, Universitas Sebelas Maret (UNS) Solo in this experiment.

References

[1] S El-Sherbiny, S M. El-Sheikh and A Barhoum 2015 *Carbohydr. Polym.* **279** 290–300
[2] J Shen, Z Song, X Qian and F Yang 2010 *Carbohydr. Polym.* **81** 545–553
[3] A Declet, E Reyes and O M Suarez 2016 *Rev. Adv. Mater. Sci.* **44** 87–107
[4] R A Ataee, J Derakhshanpour MTA and A Eydi 2011 *Iranian Journal of Military Medicine.* **13** 2–7
[5] Wiwiek Karina 2015 *Skripsi*. Fakultas Matematika dan Ilmu Pengetahuan Alam. Universitas Sebelas Maret (UNS). Surakarta.
[6] W Sun, S Jayaraman, W Chen, K A Persson and G Ceder 2015 *Appl. Phys. Sci.* **112**
[7] L Bre and D Kralj 2007 *Croat. Chem. Acta.* **80** 467–484
[8] K E Chase and E Suess 1970 *Limnol. Oceanogr.* **15** 633–637
[9] Ritika Gupta 2004 *Thesis in Georgia Institute of Technology.*
[10] Kazunori Kadota, Yoshiyuki Shirakawa, Ikumi Matsumoto, Atsuko Shimosaka and Jusuke Hidaka 2007 *Adv. Powder Technol.* **18.** 775–785
[11] H T T O M Matsuoka 1998 *Chem. Eng. Res. Des.* **76** 809–814
[12] J W Ball and D K Nordstrom 1991 U.S.G.S. *Open-File Report* 90-129.
[13] J D Allison, D S Brown and K J Novo-Gradac 1991 Version 3.0. *User's Manual.* EPA/600/3-91/021

IOP Conf. Series: Materials Science and Engineering **176** (2017) 012041 doi:10.1088/1757-899X/176/1/012041

Comparison of Dolomite Crystal Structure, Calcinations Dolomite and Magnesium Hydroxide in Partial Calcinations and Slaking Process

E Sulistiyono[1], F Firdiyono[1], NC Natasha[1] and Y Amalia[2]
[1]Research Center of Metallurgy and Materials, Indonesian Institute of Sciences
[2]Department of Chemistry, Faculty of Mathematics and Natural Sciences, Sebelas Maret University, Surakarta

E-mail: eko221068@gmail.com

Abstract. Dolomite is a mineral that consists of calcium carbonate and magnesium carbonate with various mole ratio depend on the formation of mineral source. Recently, Utilization of dolomite only used as raw material for fertilizer and building materials, so that the enhancement of its added value becomes low. If the components in dolomite can be separated, magnesium carbonate and calcium carbonate will be produced then provide high added value. To separate these two components in dolomite is through partial calcinations followed by slaking process. The purpose of this paper is to prove that the partial calcinations can be used as a fundamental process for calcium and magnesium separation process in dolomite. SEM-EDX and XRD analysis proved that partial calcinations at 675°C for 6 hours is able to produce magnesium oxide (MgO) and calcium oxide (CaO). Then sea water was added to calcinations product so magnesium hydroxide and calcium carbonate that easily separated by sea water. The weakness of partial calcinations process at 675°C and processing time 2 hours is the dolomite has not perfectly reacted yet. XRD analysis showed that $MgCO_3.CaCO_3$ compounds still exist, so there is a possibility that magnesium was not fetched after the separation process.

1. Introduction

Dolomite is a mineral that consists of calcium carbonate and magnesium carbonate with various mole ratio depend on the formation of mineral source [1]. Gresik is one place that produce dolomite with a very good quality and has a potentially to be processed to become advanced materials [2-3]. One of advanced material product that developed from dolomite is magnesium carbonate with an ultrafine grain size and high purity [4]. It obtained from dolomite processing which started from dolomite calcinations, slaking process, carbonation, separation process between calcium and magnesium and the final step is magnesium carbonate formation [5].

Separation process between calcium and magnesium is an important role in making magnesium carbonate from dolomite. This process uses carbon dioxide from the result of calcinations process so it makes this process environment friendly. Separation process between magnesium and calcium from dolomite with carbonation process will be successful if in calcinations process, the calcium carbonate does not change into calcium oxide (CaO) [6]. In this research, calcinations process at low temperature

[1] E Sulistiyono, Indonesian Institute Of Sciences
[2] Y Amalia , Sebelas Maret University

will be done. This purpose of this research is to transform magnesium carbonate into magnesium oxide without transform calcium carbonate into Calcium Oxide [7].

2. Experimental

2.1. Raw Materials

Raw material that used in this study was dolomite powder from PT. Polowijo Gosari. Based on XRF analysis, other oxides that contained in the raw material are CaO, MgO, Na_2O, Fe_2O_3, K_2O, SiO_2, and Al_2O_3. XRF analysis can be seen in Table 1.

Table 1.XRF analysis of raw material.

No	**Oxides**	**% wt**
1	Calcium (CaO)	61.83
2	Magnesium (MgO)	25.75
3	Sodium (Na_2O)	7.93
4	Silica (SiO_2)	1.19
5	Aluminium (Al_2O_3)	0.84
6	Potassium (K_2O)	0.40
7	Iron (Fe_2O_3)	0.37

Table 1 show that CaO content in dolomite is higher than MgO content. Calculations by Stoichiometry on XRF analysis showed that dolomite consisted of 61.83 %(wt) CaO and MgO 25.73%(wt) and it mole ratio is 1: 0.33. It indicates that calcium oxide is released from magnesium oxide.

2.2. Experimental Procedure

This experiment was done to determine the differences between the dolomite structure, the results of dolomite calcinations and the results of the hydration process. Dolomite was crushed until it has -60 mesh and calcinations the dolomite at 625°C, 650°C and 675°C for 1 hour to 6 hours. Then sea water was added to the calcinations in slaking process in order to transform magnesium oxide (MgO) into magnesium hydroxide (MgOH). Magnesium hydroxide can be easily separated with calcium carbonate after carbonation process because in this process magnesium hydroxide will change to become soluble magnesium bicarbonate. The optimum results were analysed by XRD, XRF and SEM to know that calcium oxide was not produced. Flow diagram of this experiment can be seen in Figure 1.

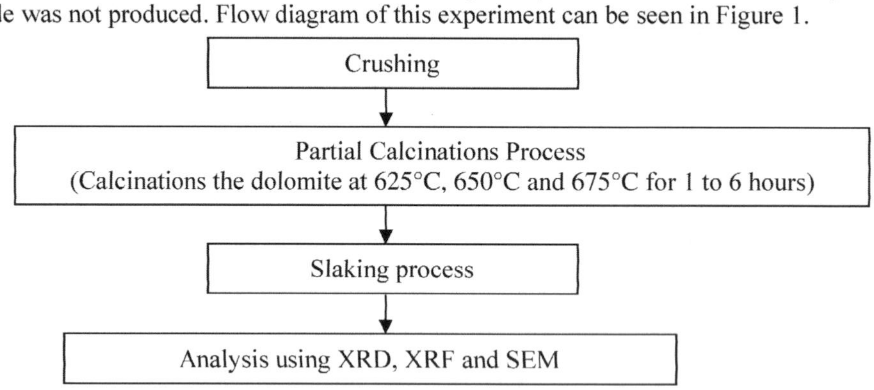

Figure 1. Experimental procedure

3. Results and Discussion

3.1. Calcinations Process

Calculation of the weight reduction in the dolomite calcinations at 625°C, 650°C and 675°C for 1 to 6 hours can be seen in Table 2.

Table 2. Sample weight lost during calcinations (% wt)

Time (Hour)	**625°C**	**650°C**	**675°C**
1	10.79	13.99	14.82
2	14.62	17.74	21.29
3	19.69	22.79	28.86
4	26.07	29.17	31.07
5	28.95	31.48	31.62
6	31.58	31.89	32.09

Table 1 show that the content of MgO is 25.75%, it can be estimated that 28.0923 %wt will be lost during partial calcinations process. Based on that weight reduction, the maximum results in theoretical occur at 625°C for 5 and 6 hours. Table 2 shows the weight lost during calcinations process is caused by temperature and when the temperature at 675°C for 3 to 6 hours the partial calcinations process perfectly occurred. The highest temperature caused 28.86 %wt, 31.07 %wt, 31.62 %wt and 32.09 %wt of sample weight lost. In this study, samples at 675°C for 4 hours were taken subsequently. Results of experiments conducted by Eni Febriana show that the mass loss at temperatures 700°C and calcinations time for 6 hours is only 27.68 wt% dolomite [7].

3.2. XRD Analysis

XRD analysis of dolomite, calcinations dolomite and slaking process using sea water can be seen in Figure 2. It shows that the calcinations process is not running perfectly. There is still a peak of $MgCO_3.CaCO_3$ from calcinations process so partial calcinations process is not perfect. Consequently of partial calcinations process was not occur so that if CO_2 was streamed into hydration products, Mg and Ca cannot be easily separated. Peak of CaO from calcinations process and peak of calcium hydroxide ($Ca(OH)_2$) from slaking process were not appear in XRD analysis. Results of experiments conducted by Eni Febriana show that peak dolomite was formed at temperatures of 700°C and was not formed at temperatures of 800°C [7].

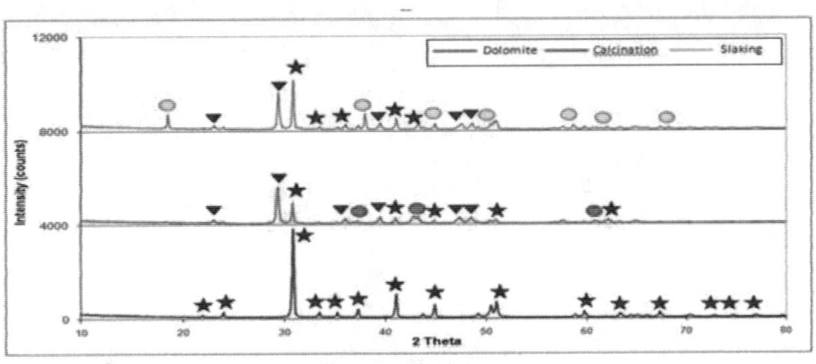

Figure 2. XRD analysis on the dolomite, calcinations dolomite and slaking process

3.3 SEM-EDX Analysis

EDX analysis of mineral dolomite shows that the amount of Ca and Mg are not always in same ratio so in dolomite not every calcium and magnesium bonded to form $MgCO_3.CaCO_3$ but there were loose in the form $CaCO_3$. Figure 3 shows that the oxygen distribution tends to spread more evenly than the carbon distribution in dolomite. This suggests that oxygen which bonded with carbon is also bound up with Ca and Mg, than it can be concluded that very difficult to make Ca and Mg in metal from dolomite.

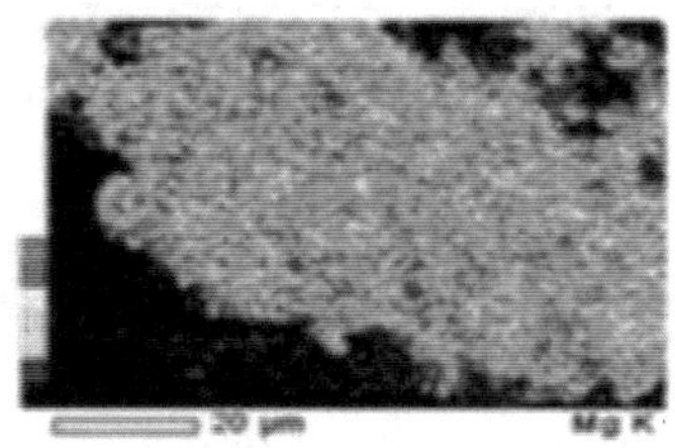

(a)

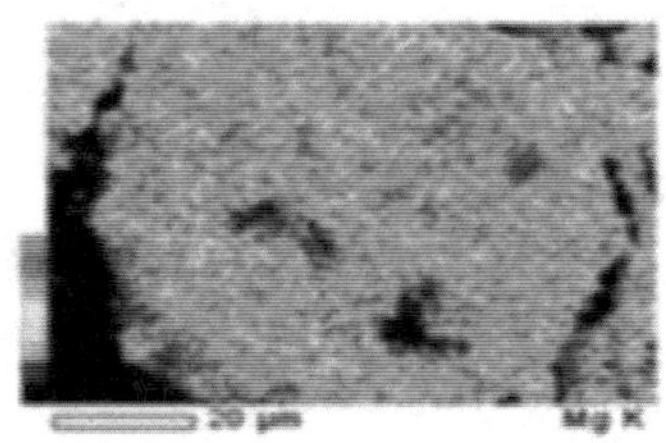

(e)

(b)

(f)

(c)

(g)

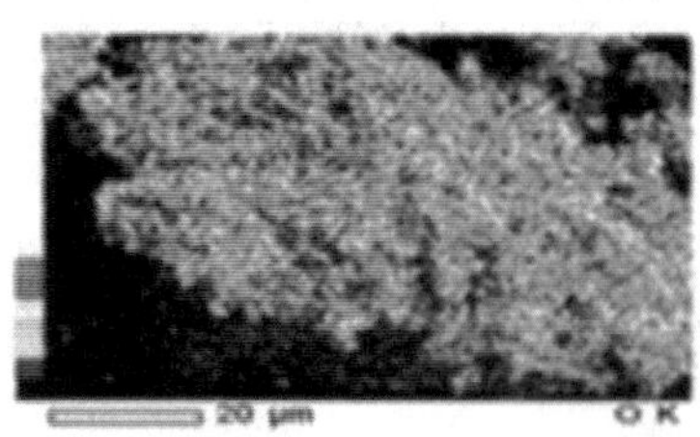

(d)

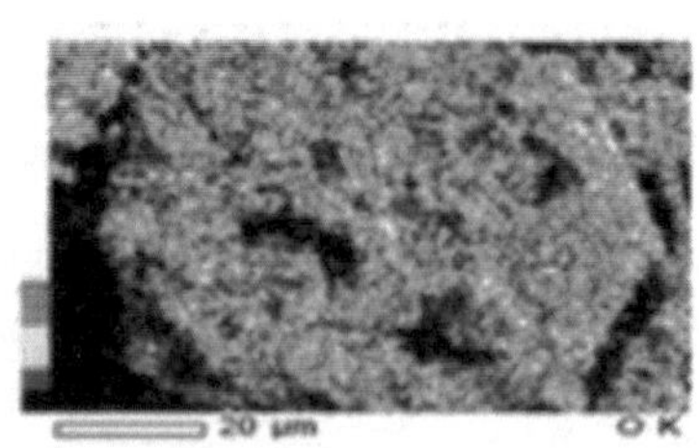

(h)

Figure 3. The results of SEM - EDX mapping for the element of magnesium, carbon, calcium and oxygen in the dolomite, (a – d) before calcinations and (f – h) after calcinations.

By comparing the right image with the left in Figure 3, there is a reduction in the concentration of carbon and oxygen are also quite a lot. This indicates that the result of the calcinations process is the release of carbon dioxide (CO_2). Based on Figure 3, the dispersion of carbon and oxygen concentration in calcinations dolomite was slightly decrease compare to dolomite ore. It tells that partial calcinations process has occurs. The carbon dioxide was released and $MgCO_3$ was transformed into MgO, but $CaCO_3$ still attached with carbon.

SEM analysis on Figure 4 shows that the dolomite after calcinations has a porous surface. It caused by CO_2 released from dolomite grain and it proves that the dolomite calcinations reaction occurs not only on the surface of the dolomite but also in all parts of dolomite. From this result can be concluded that the grain size from calcinations process will not give an effect on the reaction kinetics. In the calcinations process it is required a large grain size so circulation of CO_2 gas will easily happened.

(a) (b)

Figure 4. The appearance of granular dolomite, which (a) is dolomite before calcinations process and (b) after calcinations dolomite.

4. Conclusion

From the experimental results on calcinations process, hydration and drying on a dolomite with partial calcinations can be concluded that calcinations process at 675°C for 4 hours is still partial calcinations process. Partial calcinations process makes separation process between magnesium and calcium in dolomite easy to do. However, calcinations at 675°C for 4 hours causing an imperfect calcinations reaction and characterized by the presence of peak $MgCO_3.CaCO_3$ and it can reduce magnesium carbonate producing. Calcinations reactions of dolomite occur not only on the surface of the grains but also inside of the grain.

References

[1] Jamiludin L 2010 *Artikel Bahan Galian Industri Dolomit* (Mataram:Fakultas MIPA Universitas Mataram) pp 7-10
[2] Anonim 1996 *Memperkenalkan Bahan Galian Golongan C di Jawa Timur "Dolomit"* (Surabaya: Dinas Pertambangan Provinsi Jawa Timur)
[3] MadiadipoeraT 2006 *Bahan Galian Industri di Indonesia* (Bandung: Pusat Sumberdaya Geologi Departemen Energi dan Sumberdaya Mineral)
[4] Guo M, Li Q, Ye X, Wu Z 2010 *Powder Technology* (Beijing: Elsevier)
[5] Andliswarman 2003 *Proses Ekstraksi MgO dari Mineral Dolomit dan Analisis Techno Economic Proses Produksi* (Jakarta: Universitas Indonesia)
[6] Gelai H, Pijolat M, Nahdi K, Trabelsi AM 2007 *J. Solid State Ionics* **178** 1039-1047
[7] Febriana E 2011 *Kalsinasi Dolomit Lamongan untuk Pembuatan Bahan Baku Kalsium dan Magnesium Presipitat* (Jakarta: Universitas Indonesia) p 44-68

IOP Conf. Series: Materials Science and Engineering **176** (2017) 012042 doi:10.1088/1757-899X/176/1/012042

TiO_2 Nanorods Preparation from Titanyl Sulphate Produced by Dissolution of Ilmenite

S Wahyuningsih[1], L Rinawati[1], R M I Munifa[1], A H Ramelan[2], Eko Sulistyono[3]

[1]Inorganic Material Research Group, Faculty of Mathematics and Natural Sciences Sebelas Maret University, Jl. Ir. Sutami 36A Kentingan Jebres Surakarta 57126, Indonesia

[2]Electronic Materials Research Group, Faculty of Mathematics and Natural Sciences Sebelas Maret University, Jl. Ir. Sutami 36A Kentingan Jebres Surakarta 57126, Indonesia

[3]Pusat Penelitian Metalurgi, Lembaga Ilmu Pengetahuan Indonesia, Kawasan PUSPIPTEK Gedung 470 Serpong Tangerang 15314, Indonesia

E-mail: sayekti@mipa.uns.ac.id

Abstract. One-dimensional titanium oxides (TiO_2) nanorods have substantial applications in photocatalytic, nanoelectronic, and photoelectrochemical solar cells. These applications require large quantities of materials and a production technique suitable for future industry fabrication. We demonstrate here a new method of TiO_2 nanorods production from ilmenite sands ($FeTiO_3$). In this process, the roasted ilmenite sand was separated from the iron content and dissolved in the sulphuric acid solution. Separation process of TiO_2 from ilmenite has been carried out by roasting, leaching and precipitation processes. The roasting process was conducted by the addition of Na_2S at a temperature of 800°C that had been deomposed ilmenite into hematite (Fe_2O_3), anatase TiO_2, rutile TiO_2, Na_2SO_4, $NaFeS_2$ and $NaFeO_2$. Separation TiO_2 from titanyl sulfate ($TiOSO_4$) after leaching in H_2SO_4 solution was conducted by hydrolysis-condensation step and complexation step of Fe^{2+} content. KCNS solution was used as a complexing agent. The xerogel synthesized TiO_2 then was prepared to 1-D nanostructure of TiO_2 nanorods by hydrothermal process under alkaline condition. By the two-step method, we finally gain the 1D nanorods TiO_2 extracted from ilmenite sand. The characterization using the Scanning Electron Microscopy (SEM) and Transmission Electron Microscopy (TEM) obtained the nanorod morphology at a diameter about 9.6 nm.

1. Introduction

The high potential of alternative energy as a renewable energy like solar energy. The development of solar cell technology from silicon solar cells to Dye-sensitized Solar Cell (DSSC) was increased recently [1]. Dye-sensitized solar cells (DSSCs) made from oriented, one-dimensional semiconductor nanostructures such as nanorods, nanowires, and nanotubes are receiving attention because direct connection of the point of photogeneration with the collection electrode using such structures may improve the cell performance [2]. In the DSSCs system, absorption of light is carried by the dye molecules and the charge separation was done by inorganic semiconductor nanocrystal having a wide band gap. One of the semiconductors used recently is titanium dioxide (TiO_2).

One-dimentional (1-D) TiO_2 has become the preferred semiconductors in a variety of studies because it possesses good photo- sensibility [3] and has a large band gap energy (> 3.00 eV) so it can absorb the energy of photons on most of the spectrum of sunlight in ultra violet and near visible. TiO_2 and TiO_2-derived materials are importance for utilizing solar energy and environmental purification [4]. One-dimentional (1-D) nanostructured materials (nanotubes, nanorods, and nanowires) have received considerable attention due to their unique properties of novel applications [5-8]. Titanium oxide nanorods have been produced using wet chemistry methods including electrodeposition [9], sol-gel electrophoresis [10, 11] and hydrothermal method [12]. Ilmenite is a naturally occurring iron titanate (nominally $FeTiO_3$) and thus significantly reduces the production cost. In addition, roasted $FeTiO_3$ contains both titanium oxide and catalytic iron (Fe) required for nanorod growth. TiO_2 content in the ilmenite ore is about 34.30-60.5% [13]. Most of 95% of titanium content in minerals used to TiO_2 production so that only 5% was converted to titanium.

Large scale production of TiO_2 requires abundant resources and naturally available ilmenite is the major mineral resource of titania across the world. Indonesia has huge ilmenite resources; large deposits of ilmenite are found in the Bangka and Kalimantan Island. Since it contains 35–50% TiO_2, it has to be upgraded for its effective utilization. Removal of iron from ilmenite is the major challenge to obtain high-quality rutile phase of titania, especially to generate one-dimensional nanostructure that very perspective to applied in electronic materials. The conventional method of iron removal involves hydrochloric/sulfuric acid treatment of ilmenite.

In our previous works, nanorods TiO_2 were synthesized by hydrothermal method under basic condition using TiO_2 xerogel for hydrolisis and condensation of titanyl sulphate ($TiOSO_4$) from ilmenite source in Indonesia.

2. Experimental

Ilmenite as a precursor of $TiOSO_4$ source was roasted at a temperature of 800°C for 2 hours by Na_2S according previous procedure with several modifications [14]. The roasted-ilmenite then was characterized using XRD. About 100 grams of roasted ilmenite was washed with hot water for 2 hours for three times. The dissolution of roasted-ilmenite performed with H_2SO_4 at 3, 3.5, 4.5, 6, and 9M by refluxing process under 2 hours, according Wahyuningsih et al. (2016) [16]. The ilmenite dissolution was produced $TiOSO_4$ solutions. For preparation TiO_2 xerogel from $TiOSO_4$ solution, warm water about 60°C was added to KCNS 1M for complexing agent of Fe^{2+}/ Fe^{3+} in the $TiOSO_4$ solution. Finally, we got the white precipitated at the bottom. The precipitated was washed using HCl 0.1 M and water, then was dryed. at 100°C until gain the TiO_2 xerogel. TiO_2 nanorods was synthesized by TiO_2 xerogel throught refluxing at a high concentration of NaOH at 12 M for 24 hour at boiling point condition. The solution was put in a Teflon-lined autoclave for hydrothermal process that was carried out at 120° C for 24 hours to obtained white material such as mud at the bottom of the autoclave. The result was washed using 0.1 M HCl solution and then dried at a temperature of 60°C for 24 hours. The TiO_2 nanorods then was calcined at 400°C for 2 hours and characterized using XRD, SEM and TEM.

3. Results and Discussion

Ilmenite was characterized by XRF to determine the composition of chemical compositions in the metal oxides.

IOP Conf. Series: Materials Science and Engineering **176** (2017) 012042 doi:10.1088/1757-899X/176/1/012042

Table 1. Chemical compositions of ilmenite. Chemical compositions were analysed by XRF (% w/w)

Content	Level (%)	Content	Level (%)
TiO_2	45.35	SiO_2	3.29
Fe_2O_3	31.48	Al_2O_3	2.35
Na_2O	3.86	SO_3	2.24
ZrO_2	3.57	Cr_2O_3	1.87

As shown at Table 1, the chemical compositions of ilmenite were dominant by TiO_2 about 45.35% and Fe_2O_3 about 31.48%. While the others (23.17% remaining) are metal oxides content, such as Na_2O, ZrO_2, SiO_2, Al_2O_3, SO_3, and Cr_2O_3. Ilmenite with chemical composition TiO_2 45.35% and Fe_2O_3 31.48% were used as a precursor of nanorods TiO_2 synthesis that was prepare according previous procedure of Wahyuningsih et al (2016) [16]. Roasting process of ilmenite was conducted by added Na_2S to simplify the extraction process of TiO_2 which will facilitate decomposition process in solid phase. Na_2S addition aim to reduce the oxidation process during anneling treatment [14]. This oxidation processes may occur during roasting process shown in equation reaction 1[16].

$$2\ FeOTiO_2(s) + \tfrac{1}{2}\ O_2(g) \rightleftharpoons Fe_2O_3(s) + 2\ TiO_2(s) \quad (1)$$

The O_2 diffusion processes that may occur at high temperatures will increase the separation of Fe and Ti due to the different diffusion speeds. Fe diffuse more rapid than Ti because of the affinity difference of Fe and Ti against O_2 atmospheric. In this way, a phase separation of TiO_2-rich and Fe_2O_3-rich in roasted ilmenite ore may have been occured also according plausible equation in the following reaction:

$$2\ FeOTiO_2(s) + \tfrac{1}{2}\ O_2(g) \leftrightarrow Fe_2O_3(s) + 2\ TiO_2(s) \quad (2)$$

XRD pattern of roasted ilmenite was conducted at a temperature of 800 °C, as has been done Wahyuningsih (2016) [16]. The roasted ilmenite before and after dissolving in water are shown in Figure 1.

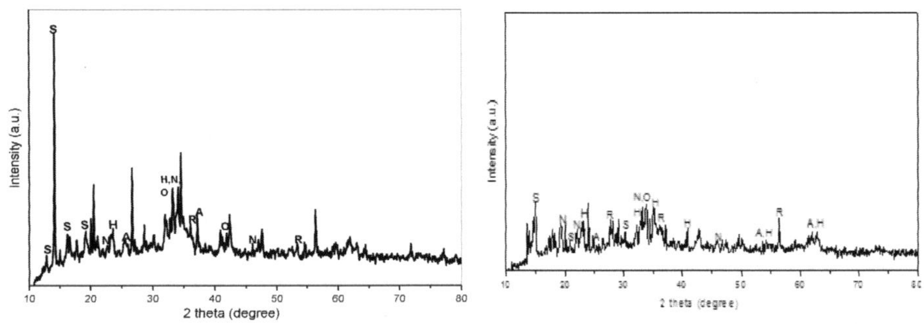

Figure 1. Diffractogram X-Ray of roasted ilmenite (a) before dissolving in water (b) after dissolving in water. N= Na_2SO_4; S= $NaFeS_2$; O= $NaFeO_2$; H= Fe_2O_3; A= anatase TiO_2; R= rutile TiO_2

The roasted ilmenite show the decomposition result of ilmenite into hematite, anatase TiO_2, rutile TiO_2 and sodium salts including Na_2SO_4 (N), $NaFeS_2$ (S) and $NaFeO_2$ (O) (Fig.1). XRD pattern at 23.85° (d_{012}) and 32.85° (d_{104}) are characteristic diffractogram of hematite (according JCPDS No. 88-

2359). The XRD pattern at 25.53° (d_{101}) and 37.79° (d_{004}) in accordance with the characteristic diffractogram of anatase TiO_2 (JCPDS No. 21-1272). The XRD pattern at 36.15° (d_{001}), 54.43° (d_{211}) are characteristic diffractogram of rutile TiO_2 (JCPDS No. 87-0710). Diffractogram peaks at 12.73° (d_{110}), 16.525° (d_{202}), 19.45° (d_{020}) and 20.35° (d_{111}) are characteristic diffractogram of $NaFeS_2$ (JCPDS No. 34-0935). Meanwhile, $NaFeO_2$ compound have characteristic diffractogram at 23.40° (d_{006}); 32.86° (d_{104}) corresponding to JCPDS No.82-1495. Diffractogram characteristic of Na_2SO_4 are at 24.50° (d_{002}), 33.25° (d_{110}) and 46.24° (d_{202}), respectively, corresponding to JCPDS No. 74-1948. The formation of Na_2SO_4 indicate that it had produced by products, as shown in equation 3:

$$Na_2S(s) + 2\ O_2(g) \rightleftharpoons Na_2SO_4(s) \quad (3)$$

The $Na_4Ti_5O_{12}$ probably was formed along with the decomposition of ilmenite followed by complexes formation of $NaFeS_2$ and $NaFeO_2$, even though less stable. The reaction in solid phase during roasting process may occurs in the following equations:

$$4\ FeTiO_3(s) + 2\ TiO_2(s) + 2\ Fe_2O_3(s) + 6\ Na_2S(s) + 7\ O_2(g) \rightarrow Na_4Ti_5O_{12}(s) + 5\ NaFeO_2(s) + NaFeS_2(s) + Na_2SO_4(s) + TiO_2(s) + Fe_2O_3(s) + 3\ SO_3(g) \quad (4)$$

If we notice adequate, there are many complexes on the roasted ilmenite after it was wash away by hot water (Fig. 1b). However, after dissolving any element process, we gained diffractogram characteristics of hematite, anatase TiO_2, and rutile TiO_2 slightly.

Dissolution of roasted ilmenite in acidic solution produce titanyl sulfate ($TiOSO_4$) and ferro sulfate ($FeSO_4$), respectively. The reactions of ilmenite decomposition are in the following reaction:

$$FeTiO_3 + Fe_2O_3 + TiO_2 + 4\ H_2SO_4 \rightarrow 2\ FeSO_4(aq) + 2\ TiOSO_4(aq) + 4\ H_2O(l) \quad (5)$$

If we compared the dissolution of roasted ilmenite with Na_2S addition and without Na_2S addition, there are not pseudobrokite (Fe_2TiO_5) formation resulted during annealing process via roasting ilmenite with Na_2S addition [14]. Dissolution of roasted ilmenite were carried out by H_2SO_4 at various concentrations. The H_2SO_4 concentrations influenced the dissolution process of roasted ilmenite. Ilmenite solubility increases with increasing concentration of H_2SO_4 and gain optimum condition at a concentration about 4.5 M to 6 M. The highest solubility has been studied that was occurred at the dissolution of roasted ilmenite at concentrations H_2SO_4 6 M [16].

The next step of TiO_2 preparation is synthesis TiO_2 from $TiOSO_4$ with the addition of KCNS as a complexing agent of Fe^{2+}. Separation TiO_2 from Titanyl sulphate solution by hydrolysis-condensation process lead to transform $TiOSO_4$ into TiO_2 and to separate the iron content left in liquid phase. The addition of 1 M KCNS aims to form $[Fe(CNS)n]^{3-n}$ complexes that soluble in water. In contrast, titanyl sulphate content will precipitate into TiO_2 via hydrolysis–condensation route. The white precipitate was obtained when pH was increased. Hydrolysis-condensation reactions for precipitation TiO_2 should occurs in the following equations:

$$TiOSO_4(aq) + 2\ H_2O(l) \rightarrow TiO(OH)_2(s) + H_2SO_4(aq) \quad (6)$$

$$TiO(OH)_2(s) \rightarrow TiO_2(s) + H_2O(l) \quad (7)$$

The addition of KCNS was capable to separate the iron content in TiO_2 to obtain high-purity TiO_2. TiO_2 as white solids of anatase type then was washed using 0.1 M HCl and water to reduce the impurities content in the solids TiO_2.

The synthesis procedure of TiO_2 nanorods was done by hydrothermal process under basic condition of 12 M NaOH. This process was obtained viscous white solution indicates that there has been a nano size formation perfectly production. By hydrothermal treatment at a temperature of 120 °C for 24 hours, finally we gain the nanorods. The white solids of TiO_2 nanorods (Figure 2) were then washed using 0.1 M HCl and aquades to reduce the impurities.

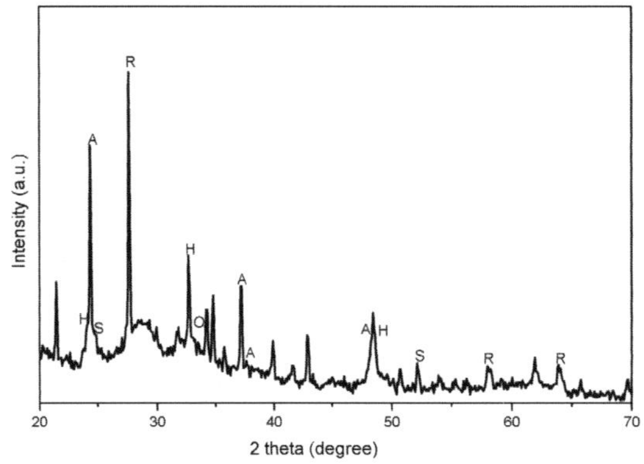

Figure 2. XRD pattern of TiO_2 nanorod. H= Fe_2O_3, A= anatase TiO_2 R= rutile TiO_2

There are two type of TiO_2 in nanorods morphology, namely anatase and rutile with a certain ratio. TiO_2 rutile was shown at a peak of 27.49° (d_{110}), while TiO_2 anatase was shown at a peak 25.28° (d_{101}). Rutile TiO_2 type on a peak of 27.49° has higher intensity than anatase TiO_2 type at 25.28°. This is a specific character of nanorods TiO_2 morphology, like the analysis result reported by Liu et al. [15].

The analysis of SEM of synthesized TiO_2 shown in Figure 3. The images showed the comparison between nanoparticle TiO_2 (it was produced with another procedure) and TiO_2 nanorods. The image of nanoparticle TiO_2 showed the spherical morphology. While the image of TiO_2 nanorods showed that the surface morphology has rods like in micrometers size. However, the nanorods material showed that the morphological form is not homogeneous; it could be caused by the pressure in systems that are not maintained during the process.

(a) (b)

Figure 3. SEM images of TiO_2 (a) nanoparticles (b) nanorods

The nanoparticles have so many advantages when it's compared to the bulk material. It has a ratio of surface area/volume greater than the similar particle in large size. Reactivity material was determined by atoms in the surface that will have interacted directly with the other materials. Furthermore, the TEM analysis aims to determine the morphology of the inside of TiO_2 nanorods that can be seen in Figure 4. The image of TiO_2 nanorods showed that the inside morphology are very similar with rod-shaped with the particle size of 9.26 nm.

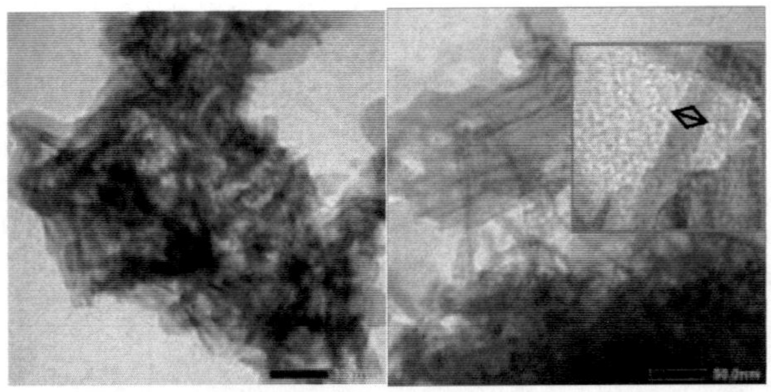

Figure 4. TEM image of TiO_2 nanorods

4. Conclusion

TiO_2 was capable synthesized from titanyl sulfate ($TiOSO_4$) resulted from dissolution ilmenite. By the addition of KCNS, the hydrolysis-condensation and complexation process will occurred to increase the separation between titania and iron content in their colloidal system. Further treatment, the synthesis of TiO_2 nanorods has been done by hydrothermal process under highly alkaline conditions. Morphology of the TiO_2 nanorods was rod-shaped samples with a diameter of 9.26 nm

Reference

[1] Green A, K Emery, Y Hishihira, W Warta, ED Dunlop 2015 *Prog. Photovolt: Res. Appl*, **23**, 1-9.
[2] Jiu J, S Isoda, F Wang, M Adachi 2006 *J. Phys. Chem. B.* ***110*** (2087-2092).
[3] Vasquez R and A Molina 2008 *Metal,* **15,** 13–15.
[4] Wahyuningsih S, AH Ramelan, E Pramono, C Purnawan, V Anjani, P Estianingsih, L Rinawati, K Fadli 2016 *IOP Conf. Series: Materials Science and Engineering,* **107**, 012019.
[5] Rao C N R and Nath M 2003 *Dalton Trans,* **1,** pp.1-24.
[6] Patzke G R, Krumeich F and Nesper R 2002 *Angew. Chem. Int. Ed.* **41,** pp 2446-2461.
[7] Huang M, Mao S, Feick H, Yan H, Wu Y, Kind H, Weber E, Russo R, and Yang P 2001 *Science.* **292,** pp 1897-1899.
[8] Pan Z W, Dai Z R, and Wang Z L 2001 *Science,* **292** pp 1947-1949.
[9] Y. Lei; L. D. Zhang; J. C. Fan, 2001 *Chem. Phys. Lett*., **338**, 231-236
[10] L. Miao; S. Tanemura; S. Toh; K. Kaneko.; M. Tanemura, 2004 *J. Cryst. Growth.,* **20,** 59-62.
[11] L. Miao; S. Tanemura; S. Toh; K. Kaneko; M. Tanemura, 2004 *Appl. Surf. Sci.* **238**, 175-179
[12] Y. Li; M. Guo; M. Zhang; X. Wang 2009 *Mater. Res. Bullet*, **44,** 1232–1237
[13] M. Sundarajan; K.H. Bhat; S. Velusamy; N. Babu; M.E.K. Janaki; S. Sasibhooshanan, and P.N. Mohan Das. 2009 *J. of Minerals & Materials Characterization & Engineering*, **8,** 427-438
[14] Wahyuningsih S., E Pramono, S Budi Rahardjo, 2013 *Asian Journal of Chemistry*, **25**(12),
[15] Liu Z, Zhang X, S Nishimoto, M Jin . DA Tryk, T Murakami , A Fujishima 2007 *Langmuir*, **23** (22), pp 10916–10919
[16] S. Wahyuningsih1, A. H. Ramelan, R. M. I. Munifa, L. N. M. Z. Saputri, and U. Chasanah 2016 *Journal of Physics: Conference Series* **776**, 012044

Porosity of Self-Compacting Concrete (SCC) incorporating high volume fly ash

S A Kristiawan[1], Sunarmasto[1] and G Y Murti[2]
[1]SMARTCrete Research Group, Civil Engineering Department, Sebelas Maret University, Jl. Ir. Sutami No. 36 A Surakarta 57126, Indonesia
[2]Undergraduate Student, Civil Engineering Department, Sebelas Maret University, Jl. Ir. Sutami No. 36 A Surakarta 57126, Indonesia

E-mail: s.a.kristiawan@ft.uns.ac.id

Abstract. Degradation of concrete could be triggered by the presence of aggressive agents from the environment into the body of concrete. The penetration of these agents is influenced by the pore characteristics of the concrete. Incorporating a pozzolanic material such as fly ash could modify the pore characteristic of the concrete. This research aims to investigate the influence of incorporating fly ash at high volume level on the porosity of Self-Compacting Concrete (SCC). Laboratory investigations were carried out following the ASTM C642 for measuring density and volume of permeable pores (voids) of the SCC with varying fly ash contents (50-70% by weight of total binder). In addition, a measurement of permeable voids by saturation method was carried out to obtain an additional volume of voids that could not be measured by the immersion and boiling method of ASTM C642. The results show that the influence of fly ash content on the porosity appears to be dependent on age of SCC. At age less than 56 d, fly ash tends to cause an increase of voids but at 90 d of age it reduces the pores. The additional pores that can be penetrated by vacuum saturation method counts about 50% of the total voids.

1. Introduction

Self-compacting concrete (SCC) is a type of concrete with special fresh properties i.e. it flows under its own weight and passes any obstacles on its way to fill every spaces within the formwork. With these properties, there is no need of vibration to obtain uniform and dense concrete in the construction of SCC. On the other hand, vibration is necessary in the production of conventional concrete to force the solid particles in freshly mixed become in a closer arrangement. Otherwise, any entrapped air will settle within concrete to form large voids resulting in a very porous concrete. Interconnected voids also facilitates an ingress of aggressive agents from the environment surrounding concrete.

Penetration of aggressive agents into the body of concrete and the chemical reactions that follow may disrupt the integrity of concrete. To increase the resistance of concrete against aggressive agents, it is important to reduce the rate of ionic transport from the environment into the concrete. This ionic transport may be driven by one or combination of the following mechanisms: pressure-induced water flow, water absorption, water vapour diffusion, wick action, ion diffusion, gas diffusion and pressure-induced gas flow [1]. Whatever the driving force is, lower porosity and finer pores with more tortuous pores structure will promote a better durability than concrete with the opposite properties. For this reason, porosity may be used as one of the durability indicators [2].

The porosity of concrete is controlled by the pores size in the interfacial transition zone (ITZ) between aggregates and paste. This is because the porosity of this zone is higher compared to the bulk paste as shown from the image analysis of microstructure [3]. It is also revealed that there are non-homogeneous distribution of pores in the ITZ and bulk paste. The pore volume is higher toward aggregate. It is also interesting to note that the pore volume beneath aggregate is higher compared to that at the top and the side of aggregate. The pore volume in the ITZ is influenced by packing density of the solid ingredients, water to cement ratio (w/c), binder composition, type of filler, etc. All these parameters are represented in the composition and proportion of materials for making concrete.

SCC has been investigated to show a finer and more tortuous pores structure than conventional concrete [4]. The different in the pores characteristic between the two concretes is originated in the proportion of the ingredients. In SCC, one of the key factors to achieve moderate viscosity is its proportion where finer components tend to be higher while the coarser components are lower compared to those of conventional concrete. With a higher finer proportion, they will be a higher surface area of the solid particles to hold free water in the mix. Moreover, the free water is not disrupted in the absence of compacting process. Thus, there is a lower tendency of bleeding and of water accumulating around aggregates and leads to lower pores in the ITZ [3, 4]. On a contrary, free water in conventional concrete is in a motion during vibration and this may provoke bleeding as well as bleeding channels. Therefore, an interconnected network of large pores is developed. In addition, part of the free water is also accumulated around coarse particles on its way to move upward and creates large pores in the ITZ. A closer arrangement of coarse aggregates in conventional concrete due to higher amount of coarse aggregates content will further introduce a closer network of large cappilary pores from one ITZ to the adjacent ITZ.

Most of SCC utilises fillers or pozzolanic materials to increase its finer components. Otherwise, the mix will consist of large amount of cement to meet a moderate viscosity. The introduction of fillers and pozzolanic materials will certainly affect the porosity and microstructure of SCC. Silva and Brito [5] point out that the predominant factor influencing the porosity of SCC with the substitution of cement by limestone filler, fly ash or combination of limestone filler and fly ash is the w/c. In a mix with higher amount of cement substitution, w/c ratio is increased resulting in a higher porosity. However, for SCC with fly ash substitution, finer pores is obtained and therefore, lower cappilary absorption is noted in comparison to SCC with limestone filler substitution. The pozzolanic reactivity of fly ash also provokes additional pores refinements as indicated by the lower cappilary absorption at later age. The beneficial effect of incorporating fillers or pozzolanic materials to reduce water absorption is achieved if the w/b (water/binder ratio) is kept similar. A higher addition of fillers or pozzolanic materials into SCC mix results in a lower water absorption [6]. This may be associated with the enhance of particle distribution, reduction of inter-particle friction and greater packing density.

Concrete with the same amount of total pores may show different ability to facilitate ionic transport. It is the size of the pores which could determine the behaviour. The pores in the concrete can appear as macro-pores, capillary pores and gel pores. The pores in the size of 0.01-1μm along with the macro-pores have a real effect on the ionic transport properties. For ionic transport by capillary suction as measured by water absorption, the water can only access the pores in the size of about 0.5μm. For finer pores, water pressure is needed to penetrate these finer pores [4]. This research aims to investigate porosity of SCC incorporating various contens of fly ash at high volume level. The porosity is determined on the basis of volume of accessible pores by water absorption and water pressure. The parameters to be investigated include density and volume of permeable pores (voids).

2. Experimental

2.1. Materials and their properties

The composition of materials and their proportion for producing SCC is presented in Table 1. The coarse aggregates were selected to limit their maximum size of 10 mm. The bulk specific gravity of coarse and fine aggregates were found to be comparable i.e. 2.51 and 2.50, respectively, when they were tested according to ASTM C128. Both coarse and fine aggregates had also been tested to determine their gradation following ASTM C136 and the results conformed to the requirement of gradation as specified by ASTM C33. The aggregates were free from any deleterious substances as specified by ASTM C33. The cement used for making the concrete was Portland Composite Cement (PCC) obtained from PT Indocement. While fly ash was supplied from Cilacap Power Plant. The chemical composition of fly ash is given in Table 2. The total composition of SiO_2, Al_2O_3 and Fe_2O_3 is more than 70% (Class F according to ASTM C618). However, the CaO content is moderate (13.32%). Considering the amount of CaO, the fly ash is categorised as Class CI according to CSA A3001.

After all the solid ingredients were mixed with the liquid ingredients and obtain a uniform mixture of fresh concrete, this mixture was tested to determine the flowability, fillingability and passingability using flow table, J-Ring, Box-Type and V-funnel test. Table 3 indicates the fresh properties of all SCC used in this research.

Table 1. Composition of SCC

ID	fly ash (kg)	cement (kg)	coarse aggregate (kg)	fine aggregate (kg)	water (kg)	superplasticizer (kg)
50%	384.30	384.30	709.8	595.35	231	7.686
55%	422.73	345.87	709.8	595.35	231	7.686
60%	461.16	307.44	709.8	595.35	231	7.686
65%	499.59	269.01	709.8	595.35	231	7.686
70%	538.02	230.58	709.8	595.35	231	7.686

Table 2. Chemical composition of fly ash and their relative amount (%)

SiO_2	Al_2O_3	Fe_2O_3	TiO_2	CaO	MgO	K_2O	Na_2O	P_2O_5	SO_3	MnO_2
45.27	20.07	10.59	0.82	13.32	2.83	1.59	0.98	0.41	1	0.07

Table 3. Fresh properties of SCC

Test Method	Parameter	50%	55%	60%	65%	70%
Flow Table	diameter (mm)	655	675	720	730	790
J-Ring	diameter (mm)	570	590	675	670	700
Box-type	h_2/h_1 (mm)	340/340	340/340	340/305	340/340	340/330
V-funnel	time (sec)	13.72	20.77	12.36	11.50	12.43

2.2. Testing of density and voids in hardened SCC

ASTM C642 was adopted in this research to determine the density and voids in hardened SCC. The specimen was cylinder of 70 mm diameter and 100 mm height. A total of 16 cylinders were prepared from each type of SCC mixture. Testing of specimens were carried out at the age of 7, 28, 56 and 90 day. The density and void of each SCC mixture were determined as an average of measurements on 4 cylinder specimens. The testing procedure for measuring density and voids in hardened SCC is summarized as follows: First, the specimens were dried in an oven at a temperature of 110 °C for 24 h. The oven-dry specimens were weighted to obtain their respective mass (A). Second, the specimens were then immersed in water for a period of 48 h. After taken from the water, the specimens were wrapped with a dry towel to remove any moistures on the surface of specimens. The saturated surface dry (SSD) mass of specimens after immersion were determined (B). Third, the SSD specimens were then put in a suitable receptacle, covered with tap water and boiled for 5 h. The boiled specimens were allowed to cool by natural loss of heat for a period of not less than 14 h to a final temperature of 20-

25°C. Subsequently, any moistures on the surface of specimens were removed and then the SSD mass of specimens after boiling were determined (C). Finally, the specimens were immersed and weighted in water to determine their apparent mass (D). The density and voids of the hardened SCC could be calculated using the following equations:

$$BD = \frac{A}{(C-D)} \tag{1}$$

$$Vo_1 = \frac{(C-A)}{(C-D)} 100\% \tag{2}$$

where *BD* is bulk density density; Vo_1 is volume of permeable pore space (voids).

2.3. *Testing of voids in hardened SCC by vacuum saturation*

This research adopted RILEM recommendation [7] for measuring the voids in hardened concrete by vacuum saturation method. Cube specimens with a size of 50x50x50 mm were used to determine the voids of hardened SCC. A total of 12 cubes were prepared to represent each type of SCC mixture. The mesurements of voids were carried out at the age of 7, 28, 56 and 90 day. To determine voids of hardened SCC, an average value of measurements on 3 specimens was used. The procedure of testing could be summarized as follows: First, specimens were stored in oven at a temperature of 110 for 24 h. The mass of the oven-dried specimens were determined (*Wo*). Second, the specimens were then put into a desiccator which was connected to a hose of vaccuum pump and to a hose of water tank. The vacuum pump was turned on to evacuate moistures within the specimens with a pressure of -1 bar for 24 h while the hose of water tank was closed (disconnected). The hose of vacuum pump was then closed (disconnected) and the dessicator was filled with tap water when hose of water tank was opened. The water would penetrate into the evacuated pores of specimens with the aids of atmospheric pressure. The specimens were left while immersed in such condition for 24 h. After which the specimens were weighted in water to determine their mass in water after immersion (*Ww*). The next step was to take the specimens and wrap with a dry towel to remove moistures on the surface. The SSD mass of the specimens were then weighted (Wa). The total voids (Vo_2) could be calculated using the following equation:

$$Vo_2 = \frac{Wa-Ww}{(Wa-Wo)} 100\% \tag{4}$$

3. Results and discussion

3.1. *Density and voids in hardened SCC measured by ASTM C642*

The bulk density of hardened SCC is presented in Fig. 1 (a). All SCC show an increase of density with time. Initially the bulk density of SCC is vary in accordance with a variation of fly ash content. The density is in the range of 2100-2200 kg/m3 when it is measured at 7 d. It is likely that at early age, SCC with a lower inclusion of fly ash content will give a greater bulk density. However, at later age the bulk density of all SCC is nearly similar at about 2300 kg/m3 after 90 d of age. The comparable value of bulk density at later age is linked with the rate of the increase in density. As shown in the figure, the increase of bulk density with time in SCC containing higher fly ash is greater than that of SCC with lower fly ash content. The increase in the density is associated with the pores refinement due to progress of cement hydration and pozzolanic reaction of fly ash.

The voids of SCC measured on the basis of water which can enter the permeable pores by capillary suction is presented in Fig. 1 (b). All SCC show a consistent behaviour where the voids drop with time. This trend is coherent with the corresponding trend of bulk density where density is increased with time. The reduction of voids which leads to a denser hardened SCC with time is associated with the continuous hydration and pozzolanic reaction which takes place at later age. These two reactions give a similar product i.e. calcium silicate hydrates which refine the pores of hardened SCC.

It is also noticed that the decline in the percentage of voids is more significant in SCC with high (above 60%) fly ash content. The highest reduction of voids is observed on SCC with 60% fly ash content. Initially the voids in this SCC count about 15.3% at 7 d, but the value drops to 2.1% at 90 d. The significant reduction of voids in hardened SCC with high fly ash content may be explained as follows. At low fly ash content, higher cement is available for hydration; more hydration products would fill the inter particles space and so a denser hardened SCC is obtained at early age [8]. However, at later age pozzolanic reaction will takes place since at this age there are sufficient calcium hydroxide, which is one of the main products of cement hydration, to be consumed by fly ash for the pozzolanic reaction. The product of pozzolanic reaction i.e. calcium silicate hydrate will fill an empty pores which are not occupied by the previous hydration products. Consequently, there is a continuation of pores refinement at later age and the pores refinement would be more effective when the original pores is coarser as in SCC with high fly ash content. Thus, even though initially the pores in SCC with high fly ash content is greater, but in a longer time pozzolanic reaction together with continuous hydration will cause a refinement of pores at a faster rate.

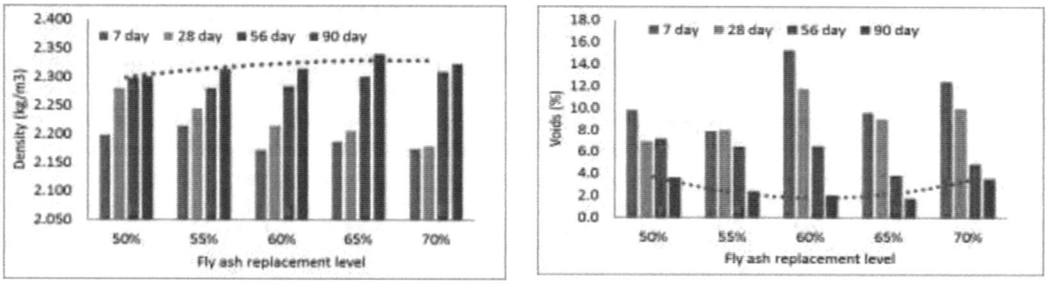

(a) Density of SCC (b) Volume of permeable pores (voids)

Figure. 1. Density and voids in hardened SCC determined according to ASTM C642

Fig. 1 (b) also indicates that the effect of fly ash on the formation of voids in hardened SCC depends on age. At early ages, fly ash tends to increase the presence of voids within hardened SCC. However, from 56 d and afterward the influence of fly ash seems to decrease voids. At 90 d of age, there is an optimum amount of fly ash content which provide a minimum voids in hardened SCC as shown in the curve of Fig. 1 (b). There are several factors that may explain the influence of fly ash content on the formation of voids in hardened SCC. First, is the particle packing density factor. It is thought that the finer size of fly ash would be beneficial to fill the inter particles of solid ingredients. A denser inter particles packing is expected in the presence of fly ash. However, there is always an optimal amount of combination that the voids are minimized. Excessive fly ash content could dilate the inter particles space leading to higher porosity at the start of the hardening process [8-10]. The particle packing density factor may be considered as the origin of the inter particles space of mixture before the concrete hardened. Therefore, the effect is foremost at early age. Comparison of the voids of all SCC at early age, it seems that fly ash content of 55% gives the best particle packing density. Second, is cement content. As previously mentioned a higher cement content provokes a higher hydration products which occupy the inter particles space. Hence, a lower voids is expected on SCC with high cement (or low fly ash) content at early age. Third, is the pozzolanic reaction between fly ash and calcium hydroxide that takes place at later age. The reaction together with the continuous hydration cause a refinement of pores at later age [10]. To sum up, the voids at early age is predominantly affected by particle packing density and rate of cement hydration factor while at later age the major factor is refinement of pores by continuous hydration and pozzolanic reaction.

3.2. Voids in hardened SCC measured by RILEM Method

The voids within the hardened SCC are formed in various sizes. Cappilary suction during water absorption test of ASTM C642 method can only draw water into capillary pores with a minimum size of 0.5μm. For finer pores, a hydraulic pressure is necessary [4]. RILEM Method provides an atmospheric pressure to aid the water to penetrate into the finer pores. As a result, the voids measured using this method is higher compared to those measured by ASTM C642. Fig. 2 clearly confirms this phenomenon. At early age, the difference of the total voids content measured by RILEM and ASTM C642 is high. However, the difference in the total voids is getting lower with time. For example, at 7 d of age the voids in hardened SCC with 70% fly ash is 22.1% and 12.3% when measured by RILEM and ASTM C642, respectively. The corresponding voids in this concrete reduce to, respectively, 7.4% and 3.6% at 90 d of age. This range of total voids difference actually represents about 50% of the voids. The difference in the voids content measured by the two methods may be considered as the amount of the finer pores that can not be penetrated by capillary suction only.

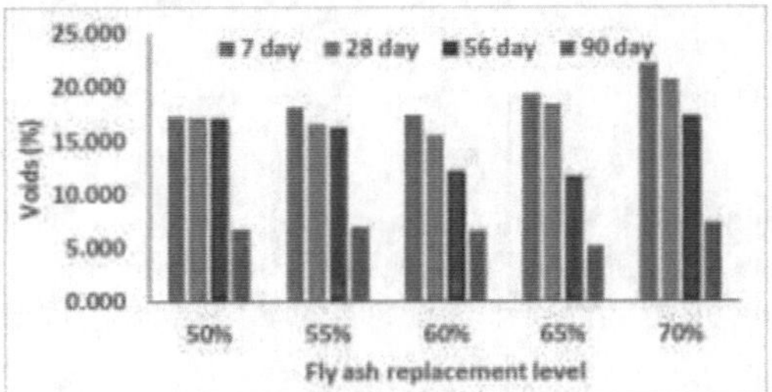

(a) Volume of permeable pores (voids) by RILEM Method

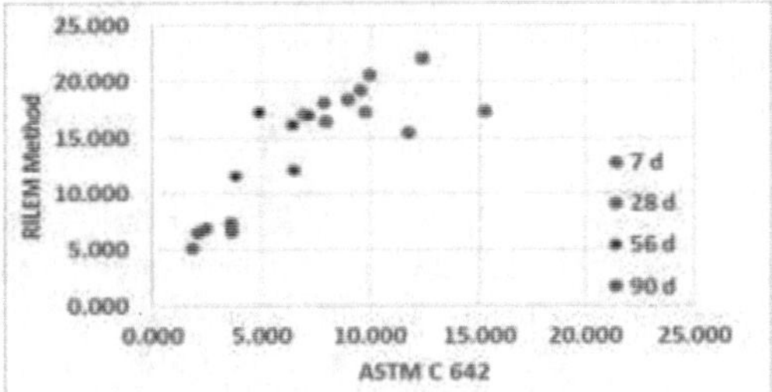

(b) Relationship between voids obtained from RILEM Method and ASTM C642

Figure 2. Voids in hardened SCC measured by RILEM Method

It is interesting to note that at 50-55% of fly ash replacement level, the voids in the hardened SCC are barely drop until 56 d of age (see Fig. 2 (a)) even though the voids measured by ASTM C642 show a reduction of voids (see Fig. 1 (b)). This finding indicates that the refinement of pores in this concrete may only occur at the inter particles space while fine cappilary pores remain unchanged.

4. Conclusions

The volume of permeable pores (voids) in hardened SCC decreases with time. The trend is coherent with the increase of density with time. The decline in the total voids is more significantly observed in SCC with at least 60% of fly ash content. The influence of fly ash content on the porosity appears to be dependent on age of SCC. At age less than 56 d, fly ash tends to cause an increase of voids but at 90 d of age it reduces the pores. The voids observed via vacuum saturation method is higher than those of immersion and boiling method. The finer pores that can be penetrated by vacuum saturation method counts about 50% of the total voids.

References

[1] Ferreira R M 2004 *Probability-based durability analysis of concrete structures in marine environment* (Guimaraes: University of Minho)
[2] Kanellopoulos A, Petrou M F and Ioannou I 2012 *Constr. Build. Mat.* **37** 320
[3] Leeman A, Loser R and Munch B 2010 *Cem. Con. Comp.* **32** 116
[4] Valcuende M, Parra C, Marco E, Garrido A, Martinez E and Canoves J 2012 *Constr. Build. Mat.* **28** 122
[5] da Silva P R and de Brito J 2015 *Constr. Build. Mat.* **86** 101
[6] Elyamany H E, Elmoaty A B M A and Mohamed B 2014 *Alexandria Eng. J.* **53** 295
[7] RILEM Recommendations 1984 *Mat. Struct.* **101** 393
[8] Liu M 2010 *Constr. Build. Mat.* **24** 1245
[9] Ghoddousi P, Javid A A S and Sobhani J 2014 *Constr. Build. Mat.* **53** 102
[10] Elrahman M A and Hillemeier 2014 *Constr. Build. Mat.* **58** 225

IOP Conf. Series: Materials Science and Engineering **176** (2017) 012044 doi:10.1088/1757-899X/176/1/012044

Variation of Strontium (Sr) in the Ferroelectric Material Barium Strontium Titanate ($Ba_{1-x}Sr_xTiO_3$) by Co precipitation Method

Y Subarwanti[1], R D Safitri[1], A Supriyanto[2], Y Iriani[2], A Jamaludin[3]
[1]Graduate Physics Department, Sebelas Maret University, Surakarta, Indonesia
[2]Physics Departement, Faculty of Mathematics and Natural Sciences, Sebelas Maret University, Surakarta, Indonesia
[3]Physics Department, FKIP, Sebelas Maret University, Surakarta, Indonesia

E-mail: yopen_2005@yahoo.com

Abstract. Barium Strontium Titanate (BST) have been made with variation strontium (Sr) 10%, 30% and 50% by co-precipitation method. This study aims to determine influence addition Sr against the crystal structure, crystallite size, lattice parameter, grain size and dielectric constant. Samples have been made by co-precipitation method and then the samples were sintered by furnace at 1100°C with holding time 4 hours. Characterization of BST use X-Ray Diffraction instrument, Scanning Electron Microscopy and Resistance Capacitance Inductance (RCL meter). Based on result obtained, the larger Sr content cause the diffraction angle shift to the right (the greater) and crystallinity increasing. But, the value of dielectric constant, crystallite size and grain size decreasing with additional Sr content. Measurement of dielectric constant (*K*) performed in the frequency range 1 kHz to 100 kHz and the highest value at Sr content 0.1 i.e. 258.35. The addition of Sr content 30% and 50% change the crystal structure from tetragonal to cubic which has paraelectric phase.

1. Introduction

The development of electric components is one of the factors that support exist as modern technology. One of the material that many applied and continues to be developed is a ferroelectric material. The ferroelectric is a material that can spontaneous polarization if applied external field. Spontaneous polarization is defined as the surface density of the bound charge on the sample surface [1]. The ferroelectric material has a perovskite structure. One of the ferroelectric material often used is Barium Titanate (BT). It has perovskite structure ABO_3.

Barium Titanate has high dielectric constant, there for its often used to electronic component as capacitor [2]. Moreover, BT has a Curie temperature 120°C and environmentally friendly. When the BT is below or above the Curie point, the crystallographic of BT will change [3]. Many research had been done in improving the electric properties as well as the morphology of BT. One of the ways often used i.e give doping. The doping can increase the dielectric constant, decrease the Curie temperature and the ferroelectric properties change. Some dopings are often used for BT i.e Pb, Zr, Ca, Sr, and Sn [4]. The strontium (Sr) is one of the doping that often used because it has an atomic radius similar to the atomic radius of Ba. The Sr caused crystal structure change and increase the dielectric constant of BT. The BT doped with Sr has a chemical formula $Ba_{1-x}Sr_xTiO_3$ (BST). The BST has high dielectric constant, chemical stability and low dielectric losses [5].

The many methods usually used to make BST i.e solid state reaction, sol-gel [4], pulsed laser deposition (PLD), metal organic chemical vapor deposition (MOCVD) [6], chemical solution deposition (CSD) [7], sputtering [8], and co-precipitation [9]. The co-precipitation method is a simple method with low reaction temperature [5]. This paper will discuss on the manufacture of $Ba_{1-x}Sr_xTiO_3$ with variation strontium 10%, 30% and 50% by co-precipitation method. The samples will be sintered by the furnace at 1100 °C with holding time 4 hours. The microstructure of the $Ba_{1-x}Sr_xTiO_3$ will be characterized by XRD to obtain lattice parameter and crystallite size. The morphology structure is characterized by SEM, while the dielectric constant is characterized by Resistance Capacitance Inductance (RCL meter).

2. Experimental

The starting materials used for the synthesis of Barium Strontium Titanate ($Ba_{1-x}Sr_xTiO_3$, x = 10%, 30% and 50%) were barium hydroxide (Sigma Aldrich, 95%), strontium nitrate (Sigma Aldrich, ≥ 99%), titanium tetrabutoxide (Sigma Aldrich, 97%), oxalate acid (Sigma Aldrich, ≥ 99%), and isopropanol (IPA). The BST powder was synthesized by the procedure described in the flowchart as given in Fig. 1. Initially, prepared tools and materials. Then the materials were weighed by digital neraca. 0.2 M solution of oxalic acid was dissolved in 35 ml IPA and taken in beaker glass and then stirred continuously. Then the solution of titanium tetrabutoxide (0.1 M) was made by dissolving titanium tetrabutoxide in 18 ml IPA. The solution of titanium tetrabutoxide had been added to a solution of oxalic acid with continuous stirring and then obtained clear solution oxalotitanic acid (HTO). Then barium hydroxide and strontium nitrate were added directly (in solid) to the clear HTO solution. After that, the distilled water was added a dropwise and kept overnight. The precipitate obtained was filtered, washed and dried at 100°C for 5 hours.

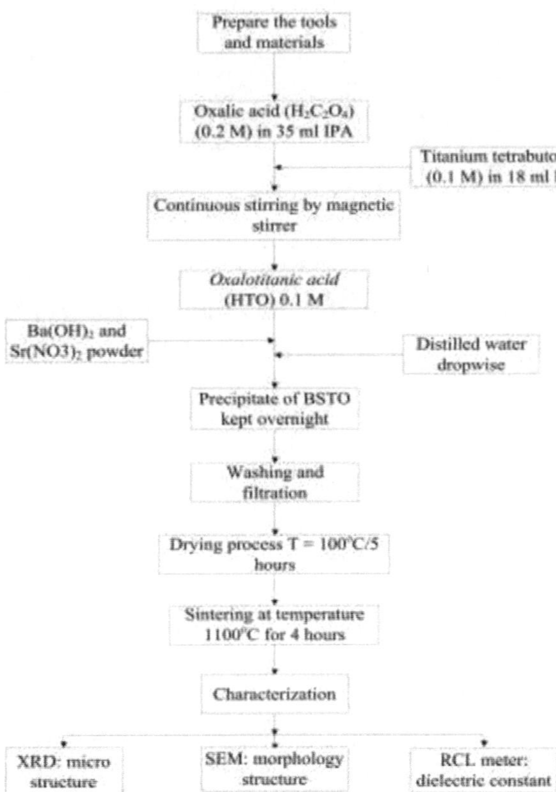

Figure 1. Flowchart procedure of experiment

After the drying process, barium strontium titanate oxalate (BSTO) powder was pressed by the hydraulic pump. The pellets of BSTO were sintered by the furnace at 1100°C with holding time 4 hours. The microstructure will be characterized by XRD with λ Cu = 1.5406Å, the data obtained will be matched with a database ICDD. The morphology structure was characterized by SEM and the dielectric constant was characterized by Resistance Capacitance Inductance (RCL meter).

3. Result and Discussion

The Result obtained from characterization using XRD is diffraction patterns with intensity value and 2θ. They were shown in Fig. 2. The diffraction patterns will be matched with database ICDD number pdf#440093 for $Ba_{0.9}Sr_{0.1}TiO_3$ sample and number pdf#340411 for $Ba_{0.7}Sr_{0.3}TiO_3$, while $Ba_{0.5}Sr_{0.5}TiO_3$ sample will be matched with number pdf#391395. The Fig. 2 shows the peaks formed were only BST peaks and the impurities (BaC_2O_4 and SrC_2O_4) were absent. So, the XRD patterns shown in Fig. 2 were only for single-phase of BST. This shows that the co-precipitation method can produce high purity yield > 99% [9]. It can be seen on the results of high crystallinity. Moreover, the result of GSAS refinement analysis proves that the peaks formed were only BST peaks.

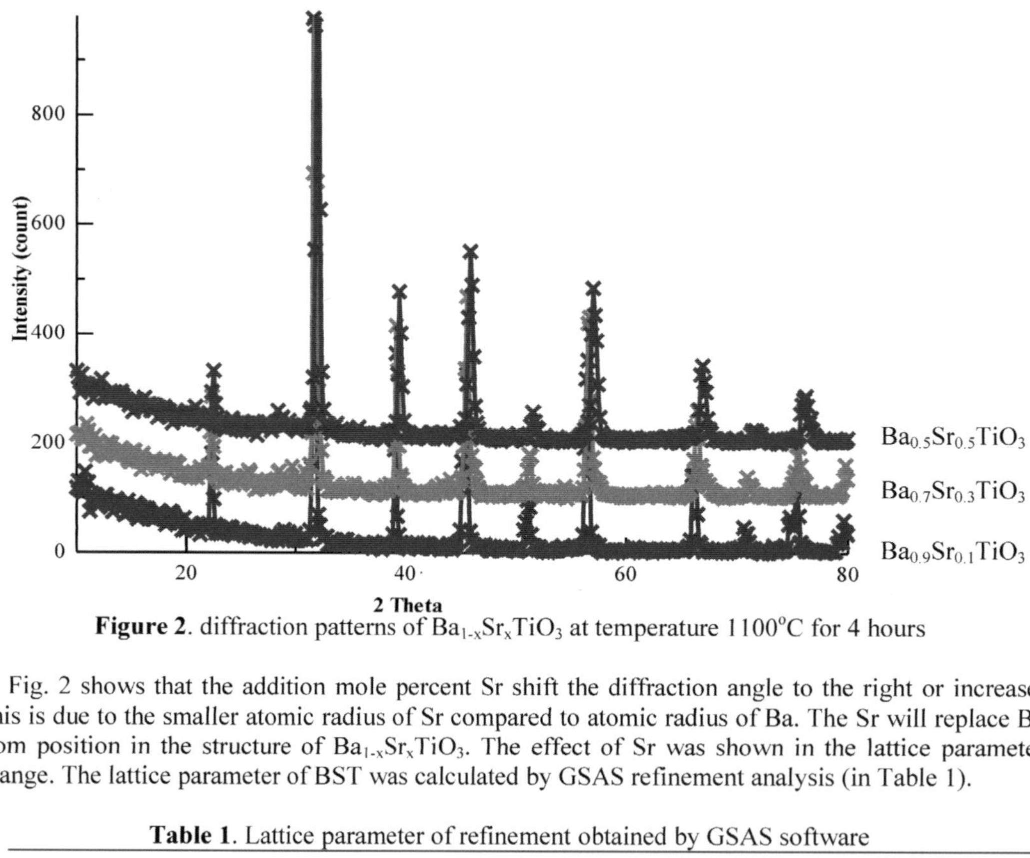

Figure 2. diffraction patterns of $Ba_{1-x}Sr_xTiO_3$ at temperature 1100°C for 4 hours

Fig. 2 shows that the addition mole percent Sr shift the diffraction angle to the right or increase. This is due to the smaller atomic radius of Sr compared to atomic radius of Ba. The Sr will replace Ba atom position in the structure of $Ba_{1-x}Sr_xTiO_3$. The effect of Sr was shown in the lattice parameter change. The lattice parameter of BST was calculated by GSAS refinement analysis (in Table 1).

Table 1. Lattice parameter of refinement obtained by GSAS software

$Ba_{0.9}Sr_{0.1}TiO_3$		$Ba_{0.7}Sr_{0.3}TiO_3$		$Ba_{0.5}Sr_{0.5}TiO_3$	
2θ	Lattice Parameter (Å)	2θ	Lattice Parameter (Å)	2θ	Lattice Parameter (Å)
31.625 45.575 56.525	a = b = 3.9762 c = 4.0026	31.775 45.575 56.675	a = b = c = 3.9693	31.925 45.875 56.975	a = b = c = 3.9510

Based on Table 1 shows that the addition mole percent of Sr affected on the crystal structure. The increase of Sr will decrease the lattice parameter. This is caused crystal structure change from tetragonal to the cubic structure. At additional mole percent of Sr 10% has tetragonal structure, space group p 4 m m and ferroelectric phase. But, when the addition mole percent of Sr 30% and 50%, the crystal structure is transformed into cubic, space group p m 3 m and paraelectric phase. The BST has a tetragonal structure and ferroelectric phase if mole percent of Ba ≥ 0.75. Mean while, the BST has a cubic structure and paraelectric phase if the mole percent of Ba < 0.75 [10].

We can obtain the crystallite size of BST through the result of XRD peaks. Table 2 shows crystallite size and crystallinity were influenced by composition mole of Sr. The increasing mole of Sr will decrease the crystallite size. The decreasing crystallite size is caused the atomic radius of Sr which smaller than Ba. Therefor the substitution of Sr in the unit cell Ba causes the volume of the unit cell and the crystallite size become smaller.

Table 2. Crystallite size and crystallinity of $Ba_{1-x}Sr_xTiO_3$

Sample	Crystallite size (nm)	Crystallinity (%)
$Ba_{0.9}Sr_{0.1}TiO_3$	36.99	95.09
$Ba_{0.7}Sr_{0.3}TiO_3$	33.95	96.96
$Ba_{0.5}Sr_{0.5}TiO_3$	21.49	97.32

Table 2 also shows the crystallinity of BST. The highest crystallinity was obtained at the sample with the composition of Sr 50%, while in the composition of Sr 10% obtained the lowest value of crystallinity. It also can be seen in the SEM photos. The value of crystallinity shows the level of regularity atomic structure on a material.

The dielectric constant value was obtained from measurement Resistance Capacitance Inductance (RCL meter) by giving input frequency 1 kHz to 100 kHz. Based on the Fig. 3, increasing frequency will decrease the dielectric constant. While the increasing the composition of Sr will decrease the dielectric constant. This is caused the greater composition of Sr will change crystal structure into cubic and become paraelectric phase. The paraelectric material has the dielectric constant which smaller than the ferroelectric material.

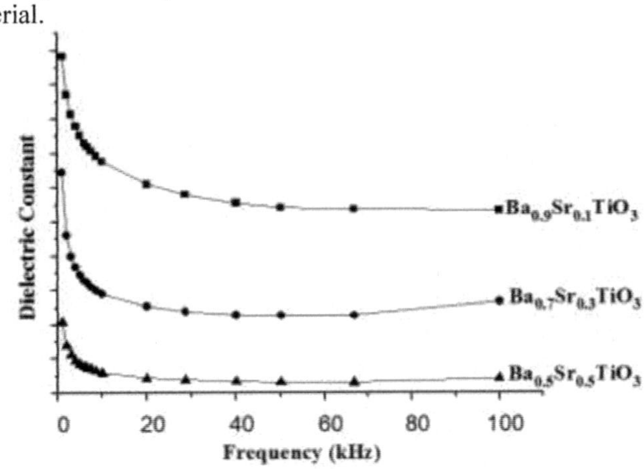

Figure 3. Graphic the dielectric constant of $Ba_{1-x}Sr_xTiO_3$

The morphology characterization results of BST is shown in Fig. 4. The addition of Sr to the BST decrease the grain size and increase homogeneous. It is also can be seen from the XRD characterization results. The grain size of $Ba_{0.9}Sr_{0.1}TiO_3$ with 80.000x magnitude, $Ba_{0.7}Sr_{0.3}TiO_3$ with 80.000x magnitude and $Ba_{0.5}Sr_{0.5}TiO_3$ with 40.000x magnitude are about 230 nm, 166 nm, and 113 nm, respectively.

Figure 4. SEM photos of $Ba_{1-x}Sr_xTiO_3$ at temperature 1100 °C (a) $Ba_{0.9}Sr_{0.1}TiO_3$, (b) $Ba_{0.7}Sr_{0.3}TiO_3$ and (c) $Ba_{0.5}Sr_{0.5}TiO_3$

All samples show uniformity and homogeneous morphologies. Moreover, the sintering at high temperature has two effects, grain growth and slight reduction of lattice parameter [11]. This one like Fig. 4 the $Ba_{1-x}Sr_xTiO_3$ samples at temperature 1100 °C shows less porosity. It shows growth the grain. The decreasing grain size was caused the addition mole percent of Sr. It also can be seen in the crystallite size decrease. Because the grain size, crystallite size and lattice parameter were comparable.

4. Conclusion

Ferroelectric BST material (x = 10%, 30% and 50%) has been made by co-precipitation method. It can be concluded that the increasing mol of Sr shift diffraction angle to the right (the greater) and the crystallinity also gets bigger. But, the dielectric constant value, the crystallite size and grain size decrease along with the addition of Sr. The addition of Sr 30% and 50% change the crystal structure become cubic which has paraelectric phase.

Acknowledgment

The project support by Penelitian Unggulan Perguruan Tinggi (PUPT) Kemenristek and Dikti 2016.

References

[1] How S C 2007 *Thesis* (Malaysia: University Sains Malaysia)
[2] Ertug B 2013 *American J. of Engineering Research* **02** (08) 1-7
[3] Vijatovic M M, Bobic J D, Stojanovic B D 2008 *Science of Sintering* **40** 155-165
[4] Iriani Y, Saputri D F, Hadiati S, Ramelan A H and Variani V I 2014 *Advance Materials Research* **91** 205-209

[5] Li M L, Liang H, Xu M X 2008 *J. Materials Chemistry and Phys.* **112** 337-341
[6] Fuentes S, Cespedes F, Padilla-Campos L and Diaz-Droguett D E 2014 *J. Ceramics International* **40** 4975-4984
[7] Iriani Y, Setyadhani R T, Jamaluddin A *Advanced Materials Research* **896** 229-232
[8] Xu Z, Yan D, Xiao D, Yu P and Zhu J 2013 *Ceramics International* **39** 1639-1643
[9] Khollam Y B, Bhoraskar S V, Deshpande S B, Potdar H S, Pavaskar N R, Sainkar S R, Date S K 2003 *Material Letters* **57** 1871-1879
[10] Dewi R, Krisman, Khaironiati, Fauziana 2014 *J. Fisika Indonesia* **XVIII** 53
[11] Suasmoro S, Pratapa S, Hartanto D, Setyoko D, Dani U M 2000 *J. of The European Ceramic Society* **20** 309-314

IOP Conf. Series: Materials Science and Engineering **176** (2017) 012045 doi:10.1088/1757-899X/176/1/012045

Preparation of Amine-Functionalized TiO_2/Carbon Photocatalyst by Arc Discharge in Liquid

Erlina Arikawati, Pranoto and Teguh Endah Saraswati[1]

Department of Chemistry, Faculty of Mathematic and Natural Sciences, Sebelas Maret University, Jl. Ir. Sutami 36A Surakarta 57126 INDONESIA

E-mail: teguh@mipa.uns.ac.id

Abstract. Amine-functionalized titanium dioxide/carbon (TiO_2/C) was prepared via the arc discharge method using graphite electrodes and a liquid medium consisting of 50% ethanol with the addition of urea. The arc discharge was conducted using a voltage of 20 to 40 V. X-ray diffraction (XRD) of prepared TiO_2/C showed a pattern of definitive peaks at 25.32°, 26.61°, and 36.14°, which are the main characteristic peaks of TiO_2, C graphite, and titanium carbide, respectively. The successful surface modification of TiO_2/C synthesized in liquid ethanol/urea resulted in better dispersion of nanoparticles in water than TiO_2/C synthesized in ethanol only. This surface characteristic was also confirmed via Fourier transform infrared (FTIR) spectra of TiO_2/C synthesized in liquid ethanol/urea, which revealed C=O, C–N, C–O, and N–H stretching vibrations at 1600–1700, 1400–1100, 1200–1300, and 3300–3400 cm^{-1}, respectively. Scanning electron microscopy (SEM) analysis showed that the nanocomposite had a spherical morphology. Transmission electron microscopy (TEM) analysis found that the structure of the nanocomposite was carbon coated with TiO_2.

1. Introduction

Photocatalysis has been considered a promising way to remedy the problem of chemical waste. Among various semiconductors, titanium dioxide (TiO_2) has been found to decompose different kinds of organic and inorganic waste in gas and liquid phases, and it has become one of the most popular photocatalysts because it is chemically and biologically inert, photocatalytically stable, commercially available, inexpensive and, environmentally friendly [1-3]. However, TiO_2 has several limitations such as a tendency to aggregate, causing TiO_2 particles to lose the effective surface area on which the catalytic process takes place. Moreover, TiO_2 has low adsorption ability for pollutants because it is non-porous [4].

One doping compound that can optimize the performance of TiO_2 is carbon [5]. The composite TiO_2/carbon is a potential photocatalyst in the purification of water because of the high adsorption ability of porous carbon and because carbon doping will narrow the bandgap, producing the active catalyst under visible light [4-5].

One top-down nanotechnology method that can be used to synthesize TiO_2/carbon nanocomposite is the arc discharge method [6,7]. Arc discharge is a simple, convenient, and economical method [8] that uses a liquid medium in place of a vacuum system [9], which is simpler and cheaper than the conventional arc discharge method [4,7,10]. While 50% ethanol can be used as a liquid medium in the

[1] To whom any correspondence should be addressed.

arc discharge process, TiO_2/C nanocomposites produced via arc discharge in 50% ethanol do not disperse well in water [5]. To enhance the surface hydrophilicity of the nanocomposite, acetic acid was added to the ethanol liquid medium to provide attachment of the oxygen-containing functional group. TiO_2/C nanocomposites produced with the addition of acetic acid to the medium have better surface characteristics than those produced in ethanol only, but the degradation efficiency is still less than 90% [7].

In the study described here, we synthesized and modified the surface of TiO_2/C photocatalyst in one step via the arc discharge method in a liquid medium. The amine-functionalized TiO_2/C prepared by adding urea ($CO(NH_2)_2$) to the ethanol medium provides attachment of the amine functional group onto the nanocomposite surface. This addition of urea is expected to improve the dispersibility of TiO_2/C nanocomposites in water compared to the dispersibility of TiO_2/C nanocomposites synthesized in ethanol alone. Moreover, amine-functionalized TiO_2/C will show enhanced photocatalytic activity when applied to the photodegradation of organic pollutants such as in wastewater treatment.

2. Experimental detail

2.1. Materials

The materials used here are carbon electrodes, titanium dioxide A-100 98% (PT. Brataco Chemica, Surakarta), urea p.a (Merck), ethanol 70%, silica glue (Autosil Black Gasket RTV silicone), and deionised water.

2.2 Synthesis and Surface Modification

Graphite electrodes were filled with a mixture of TiO_2 anatase, carbon powder, and silica glue as a binder at a weight ratio of 1:3:1. This electrode was heated at 120 °C for 6 hours. The liquid medium consisted of 300 mL of 50% ethanol and 50% urea in a volume ratio of 1:1. Two graphite electrodes, an unfilled carbon electrode, and a TiO_2-filled carbon electrode were placed very close together in the liquid medium. The synthesis process was carried out using a voltage of 20 V and current of 10 A. The resulting nanocomposite was collected from the surface and bottom of the liquid medium and then characterized.

2.3 Characterization

The collected TiO_2/C nanocomposite was characterized by X-ray diffraction (XRD) using a Bruker D8 Advance to determine the crystallinity and crystal structure. A Fourier transform infrared spectrometer (FTIR) type FT-IR-8201 PC (Shimadzu) was used to identify the surface modifications to the material. A few milligrams of the nanocomposite was sonicated in deionised water to observe the dispersion performance of the modified and unmodified TiO_2/carbon nanocomposite. Scanning electron microscopy (SEM) was used to identify the surface structure morphology of the nanocomposite and transmission electron microscopy (TEM) was used to determine the detailed structure of the nanocomposite.

3. Results and Discussion

3.1 Synthesis and Surface Modification

TiO_2/carbon (TiO_2/C) nanocomposite was successfully prepared via arc discharge in liquid. The experimental setup of the synthesis and surface modification of the TiO_2 nanocomposite via the arc discharge method is presented in Fig.1.

When a high electrical current was passed from the anode to the cathode, a flame leaped between the two electrodes, and TiO_2 and carbon in the electrodes evaporates into the liquid medium containing ethanol and urea. This interaction produced amine-functionalized TiO_2/C nanocomposite.

Figure 1. Experimental setup for the synthesis of amine-functionalized TiO_2/C nanocomposites via the arc discharge method.

3.2 Characterization

Following the synthesis process, the TiO_2/C nanocomposite was characterized using X-ray diffraction (XRD) to determine the crystallinity changes after synthesis and surface modification. The XRD patterns of TiO_2, TiO_2/C, and amine-functionalized TiO_2/C are shown in Fig. 2.

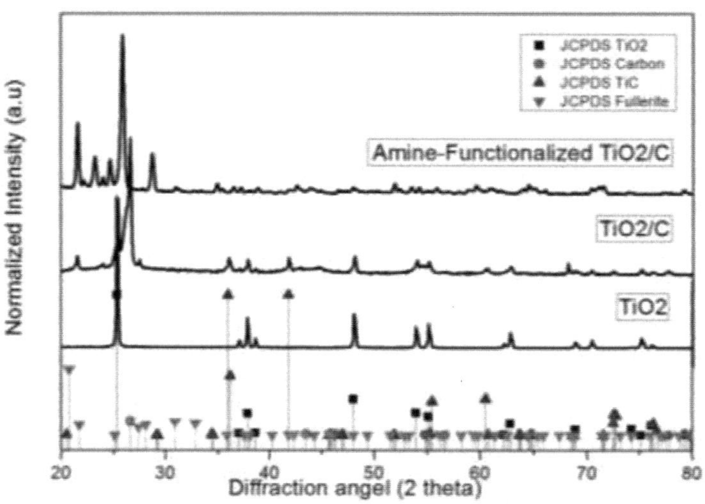

Figure 2. X-ray diffraction pattern of TiO_2, TiO_2/C, and amine-functionalized TiO_2/C.

The diffractogram of TiO_2/C shows peaks comparable to those in the TiO_2 pattern. The peak at 25.32 (101) is the main peak characteristic of TiO_2 (JCPDS No. 86-1157) and 26.62 (002) is the main peak characteristic of carbon (JCPDS No. 41-1487). The main peaks of both TiO_2 and carbon were still present in the XRD pattern of TiO_2/C, indicating that the as-prepared particles were composed of crystalline TiO_2 and elemental carbon. In addition, new peaks of lower intensity appeared at 36.02 (202) and 29.29 (104), which are assigned to titanium carbide (TiC) (JCPDS No. 72-2496). Meanwhile, the main peaks located at 20–25 represent fullerite, which mainly consists of polymerized C_{60}. These peaks mark the successful formation of the TiO_2/C structure.

The nanocomposite was also analysed using Fourier transform infrared (FTIR) spectroscopy to observe the functional groups of pristine TiO_2, unmodified TiO_2/C, and amine-functionalized TiO_2/C produced via the arc discharge method. The surface character of nanocomposites prepared in a liquid medium of 50% ethanol and in 50% ethanol with added urea show significant differences, as shown in Fig. 3.

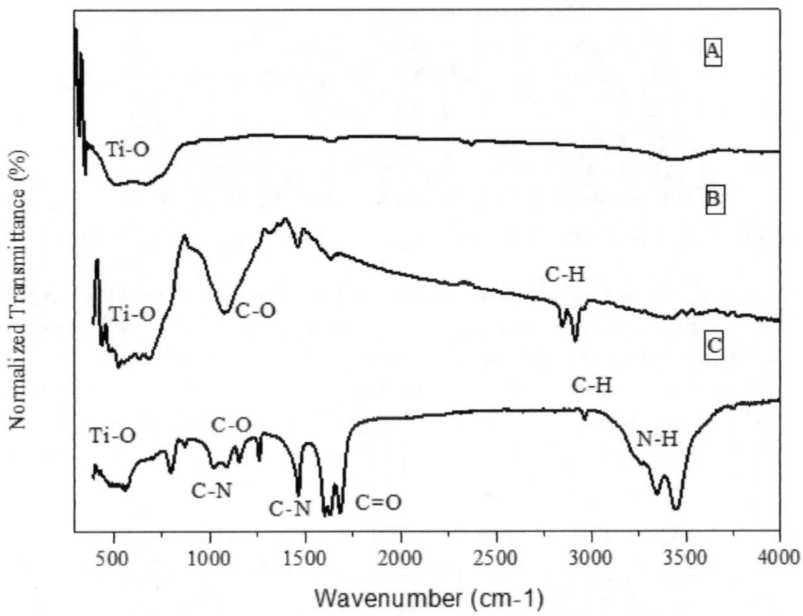

Figure 3. FTIR spectra of TiO_2 (A), unmodified TiO_2/C (B), and amine-functionalized TiO_2/C (C).

The FTIR spectra of the successfully surface modified amine-functionalized TiO_2/C show definitive peaks at 1600–1700, 1400–1100, and 3300–3400 cm^{-1}, which are assigned to C=O, C–N, and N–H absorption bands [10-14]. However, these peaks did not appear in the FTIR pattern of either pristine TiO_2 or unmodified TiO_2/C. The presence of the N–H band shows the success of the surface modification of TiO_2/C nanocomposite via the liquid medium containing ethanol and urea.

The successful amine-functionalized TiO_2/C nanocomposite showed improved dispersion in water, as shown Fig. 4. The present of amine groups on the nanoparticle surface provided a better interaction between nanoparticle surface and water via hydrogen bonding.

Figure 4. Dispersibility of unmodified TiO_2/C (A) and amine-functionalized TiO_2/C (B) in water after sonication.

Additional nanocomposite characterization by TEM and SEM confirmed the structure of the as-prepared amine-functionalized TiO_2/C. Fig. 5 shows the TEM and SEM images of the TiO_2/C nanocomposite obtained after arc discharge in 50% ethanol with 50% urea.

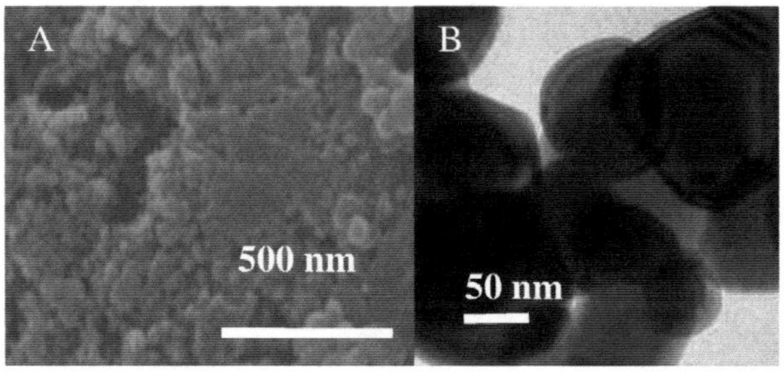

Figure 5. SEM (A) and TEM (B) images of amine-functionalized TiO_2/C.

Fig. 5 shows that amine-functionalized TiO_2/C had a spherical morphology, and two structures resulted from the synthesis process, that is, TiO_2 nanoparticles coated with carbon and carbon nanoparticles. This is in good agreement with the results of XRD, which showed that the synthesized nanocomposite consisted of incorporated TiO_2 and carbon. Analysis of the bandgap energy using ultraviolet (UV) reflectance spectroscopy showed that the bandgap energy of the TiO_2/C material and amine-functionalized TiO_2/C was narrower than the bandgap energy of TiO_2. The bandgap energy of the materials are shown in Table 1.

Table 1. Bandgap energy of TiO_2 and material synthesized.

Material	Bandgap Energy (eV)
TiO_2	3.200
TiO_2/C	3.150
amine-functionalized TiO_2/C	3.192

4. Conclusion

Amine-functionalized TiO_2/C was prepared via the arc discharge method in a liquid medium of ethanol with added urea. The XRD pattern shows the main characteristic peaks of TiO_2, C graphite, and titanium carbide. The successful surface modification of TiO_2/C synthesized in liquid medium containing both ethanol and urea demonstrated better dispersion in water than TiO_2/C synthesized in ethanol alone. FTIR spectra confirmed the presence of the amine functional group on the TiO_2/C synthesized in liquid ethanol with the addition of urea.

Acknowledgement

The author would like to thank the Minister of Research, Technology and Higher Education, Republic of Indonesia, for providing the Research Grants under project No: 632/UN27.21/LT/2016 and 458/UN27.21/LT/2016.

References

[1] Feng Y, X Yan, C Liu, Y Hong, L Zhu, M Zhou, and W Shi 2015 *Applied Surface Science* **353** 87–94
[2] Xiang N, R G Song, B Xiang, H Li, Z X Wang, and C Wang 2015 *Applied Surface Science* **347** 454–460
[3] Dai J, Yang J, Wang X, Zhang L, and Li Y 2015 *Applied Surface Science* **349** 343–352
[4] Saraswati T E, Andika, I F, Patiha, Purnawan C, Wahyuningsih S, and Anwar M 2015 *Advanced Materials Research* **1123** 285-288
[5] Subramani A K, Byrappa K, Anada S, Ray K M L, Ranganathaiah, C, and Yoshimura M 2007 *Bulletin of Materials Science* **31** 37-41.
[6] Andika, I F, Patiha, and Saraswati T E 2014 Prosiding *Seminar Nasional Kimia dan Pendidikan Kimia ISBN 979363174-0*

[7] Saraswati T E, Nandika A O, Wahyuningsih S, and Purnawan C 2015 *Proceedings of The 9th Joint Conference on Chemistry*

[8] Miranzadeh M., and Kassaee M Z 2014 *Chemical Engineering Journal* **257** 105–111.

[9] Sano N, Wang, H, Alexandrou I, Chhowalla M, and Teo K B K 2002 *Journal of Applied Physics* **92** 2783-2788.

[10] Anwar M, Argawan P R, Indro C S, Saraswati T E, Purwanto A, and Nizam M 2015 *Advanced Materials Research* **1123** 247-251

[11] Karabacak M and Cinar M 2012 *Spectrochimica Acta Part A* **86** 590– 599

[12] Liu Y D, Lee B M, Park T, Kim J E, Choi H J, and Booh S W 2013 *Journal of Colloid and Interface Sciene* **404** 56–61

[13] Stuart B 2004 *Analytical Techniques in the Sciences Infrared Spectroscopy: Fundamental and apllications* John Wiley & Sons Ltd ISBNs: 0-470-85427-8 (HB); 0-470-85428-6 (PB)

[14] Li C, J Zhang, Z Yi, H Yang, B Zhao, W Zhang, and J Zhang Li 2016 *International Journal of Adhesion &Adhesives* **66** 26–32.

[15] Fu X, H Yang, GuanghuaLu, YanmeiTu, J Wu 2015 *Materials Science in Semiconductor Processing* **39** 362–370

The Modification of Carbon with Iron Oxide Synthesized in Electrolysis Using the Arc Discharge Method

Teguh Endah Saraswati[1,4], Oktaviana Dewi Indah Prasiwi[1], Abu Masykur[1], Nestri Handayani[2], and Miftahul Anwar[3]

[1]Department of Chemistry, Faculty of Mathematic and Natural Sciences
[2]Department of Pharmacy, Faculty of Mathematic and Natural Sciences
[3]Department of Electrical Engineering, Faculty of Engineering
Sebelas Maret University, Jl. Ir. Sutami 36 A Surakarta 57126 INDONESIA

E-mail: teguh@mipa.uns.ac.id

Abstract. The modification of carbon-based nanomaterials with metals is widely studied due to its unique properties. Here, the modification of carbon nanomaterial with iron oxide has been successfully carried out. This modification was achieved using arc discharge in 50% ethanol liquid media. The anode used in the arc discharge was prepared from a mixture of carbon and iron oxide that was synthesized in electrolysis and was then calcined at 250°C with silicon binder with a mass ratio of 3:1:1, and the cathode used was graphite rod. Both electrodes were set in the nearest gap that could provide an arc during arc-discharging, leading to carbon-based nanoparticle formation. The diffractogram pattern of the X-ray diffraction of the fabricated nanoparticles confirmed the typical peak of carbon, iron oxide and iron. The magnetization value of the result analysis of the vibrating sample magnetometer was 9.9 emu/g. The bandgap energy measurement using diffuse reflectance ultra violet was estimated to be 2.18 eV. Using the transmission electron microscopy, the structure of the nanomaterial produced was observed as carbon-encapsulated iron compound nanoparticles.

1. Introduction

Carbon-based nanomaterials have a large surface area that can bind biomolecules with high capacity [1-4]. Carbon nanomaterials with graphite structures might also be used as protective shells or coatings on the outside of magnetic nanoparticles to prevent these nanoparticles from being rapidly oxidized, and the carbon layers also magnetically isolate the nanoparticles from each other [5].

At present, graphite-encapsulated magnetic nanoparticles [6-10] are promising candidate materials for drug delivery. The carbon or graphite layers can not only prevent iron nanoparticles from being rapidly oxidized but also magnetically isolate the nanoparticles from each other [5]. Graphite-encapsulated magnetic nanoparticles are conventionally prepared using the conventional arc-discharge method, which is conducted in low pressure using a sophisticated vacuum system. Two graphite electrodes are installed vertically, and the distance between the two rod tips is usually in the range of 1–2 mm. The anode and cathode are made of pure graphite (with a purity of 99.999%). After evacuating the chamber with a vacuum pump, an appropriate ambient gas is introduced at the desired pressure. When a direct current (DC) arc voltage is applied between the two graphite rods, the anode is

[4] To whom any correspondence should be addressed.

consumed and carbon-based nanoparticles are formed in the chamber soot [11]. Arc discharge has been used by Huffman and Krätschmer to prepare fullerene by evaporating graphite electrodes via restrictive heating in a helium atmosphere [12].

The conventional arc plasma growth method for carbon materials requires complex gas-handling equipment, a sealed reaction chamber, a liquid-cooled system and time-consuming purge cycles. Ishigami et al. developed a simple arc method in liquid nitrogen for high-quality carbon nanotube (CNT) synthesis [13]. Following this work, researchers generated carbon onions, nanotubes and encapsulates using arc discharge in water [14-17].

Here, we fabricated carbon-modified iron compound magnetic nanoparticles using arc discharge in a liquid medium of 50% ethanol as carbon sources using carbon electrodes and iron oxide-containing carbon electrodes. The iron oxide could be prepared using various methods. One of the easiest iron oxide preparation methods is the electrolysis method, which is a simple and economical process.

2. Experimental

2.1. Fabrication of iron oxide/carbon

The fabrication of carbon modified with iron oxide material was performed using the submerged arc discharge method. This method used two electrodes as the cathode and anode. The anode was a carbon electrode filled with a mix of graphite carbon, iron oxide (synthesized via electrolysis) and silica binder with a mass ratio composition of 3:1:1. The mix was sonicated for 480 seconds and then moulded into a rod with dimension of Ø 10 mm x 50 mm. The electrode was then heated in furnace at 180°C for 6 hours. The cathode used was a graphite electrode that had a sharp pencil tip. The cathode and anode distance were set with a very close gap to procure stepping electrons. Both the cathode and anode were set up in a 600-mL glass beaker containing 300 mL of 50% ethanol. A current of 10 A (13.75 V) was passed through electrodes. After arcing, the nanoparticles were collected, and the remaining ethanol was evaporated and dried.

2.2. Characterizations of iron oxide/carbon

The nanoparticles obtained from the arc discharge were characterized using an X-Ray Diffractometer (XRD-6000, Shimadzu) (Cu; 40.0 kV; 30.0 mA) to analyse the crystallinity and elements contained in the material. The magnetization of the material was measured using a vibrating sample magnetometer (VSM OXFORD type 1.2H). The change of band gap energy was determined using a diffuse reflectance spectrophotometer-ultra violet (UV 1700 PHARMASPEC). The structure, shape and size of the nanoparticle were studied using a transmission electron microscope (JEOL JEM-1400).

3. Result and Discussion

The nanoparticles synthesized by the submerged arc discharge were collected and named iron oxide/carbon. Fig. 1 shows a diffractogram of the iron oxide/carbon fabricated by arc discharge compared with the starting material of iron oxide synthesized via electrolysis and calcined at 250°C.

The X-ray diffraction (XRD) pattern of the initial material of iron oxide shown in Fig. 1(a) had agreement peaks to the Joint Committee on Powder Diffraction Standards (JCPDS) Fe_3O_4 or magnetite No. 89-0691 of magnetite Fe_3O_4. Some relevant peaks at 30,5296°; 35,9186°; 43,6172°; 53,9676°; 57,5602° and 74,6690° represented to the Fe_3O_4 planes of (220), (311), (400), (422), (511), and (440), respectively. After arc discharging as shown in Fig. 1(b), the existence of carbon peak on 26.5° represented to a graphitic layer of (002) and Fe_3O_4 peaks as mentioned are still confirmed. Moreover, several peaks at 2θ = 26.5303°, 35.6197°, 43.1364°, 54.5446° and 77.387°, which, according to Joint Committee on Powder Diffraction Standards (JCPDS), matched Fe_3C No. 85-0871 on hkl (002) 26.50°, hkl (020) 35.598°, hkl (022) 43.23°, hkl (220) 54.57° and graphite JCPDS No. 41-1587 on hkl (110) 77.24°.

These distinguish peaks strengthen that during arc discharging, the incorporated structures of the starting material of iron oxide and carbon was successfully condensed in the plasma zone indicationg that the product of this arc discharge is incorporated nanoparticles contained iron- and carbon-containing compounds.

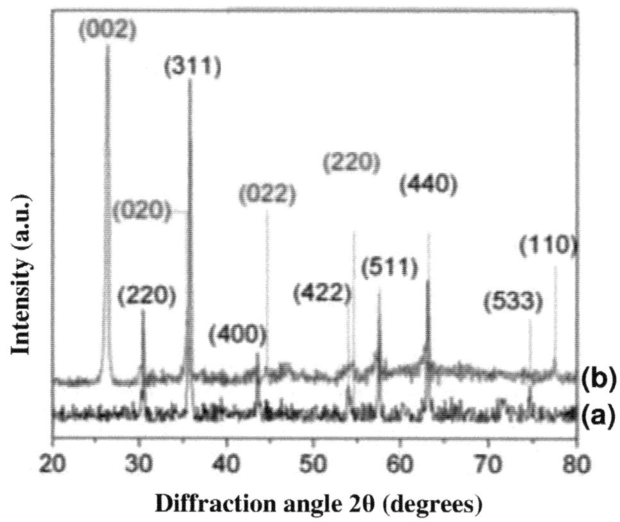

Figure 1. Diffractogram of iron oxide (a) and iron oxide/carbon (b)

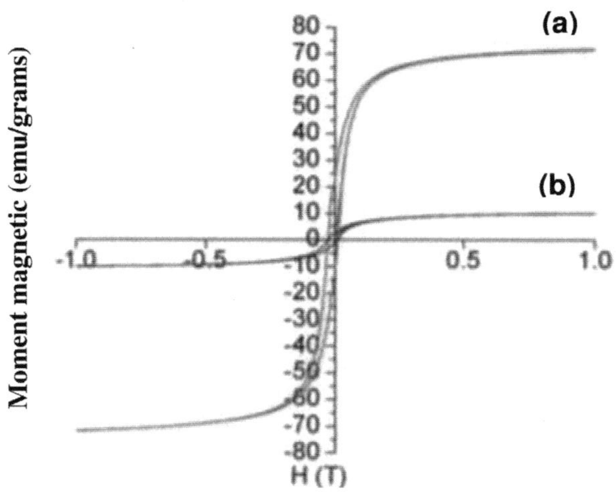

Figure 2. Hysteresis loop of iron oxide starting material (a) and iron oxide/carbon (b) fabricated by arc discharge.

Analysis was performed using the VSM to determine the magnetic properties of the material. Fig. 2 shows that the difference related to spin magnetization is irregular, which was caused by both the interaction of the superexchange between the Fe ions that were linked by O^{2-} ions (Fe^{3+}-O^{2-}-Fe^{3+}) and the coordination between atoms, which is incomplete on the surface [18]. This phenomenon caused the magnetization of the iron oxide/carbon nanoparticles to be weaker than that of the iron oxide starting material. Moreover, the addition of carbon to the iron oxide nanoparticles caused the magnetization to decrease. According to the hysteresis loop as shown in Fig. 2, both iron oxide and iron oxide/carbon have a superparamagnetic property with a magnetization value of 9.9 emu/g.

Fig. 3 shows the transmission electron microscopy (TEM) imaging of the iron oxide (Fig. 3(a)) and nanoparticle product of iron oxide/carbon (Fig. 3(b)). As shown in Fig. 3, the iron oxide particles had a diameter in the range of 30–40 nm and the nanoparticle product of iron oxide/carbon had a diameter in the range of 20–25 nm. The significant structures were clearly shown between both. The iron oxide was typically in a sphere without a coated layer, and the iron oxide/carbon had a core–shell structure. The core was highly probably occupied by iron oxide or iron carbide, and the shell was constructed from a carbon graphite layer.

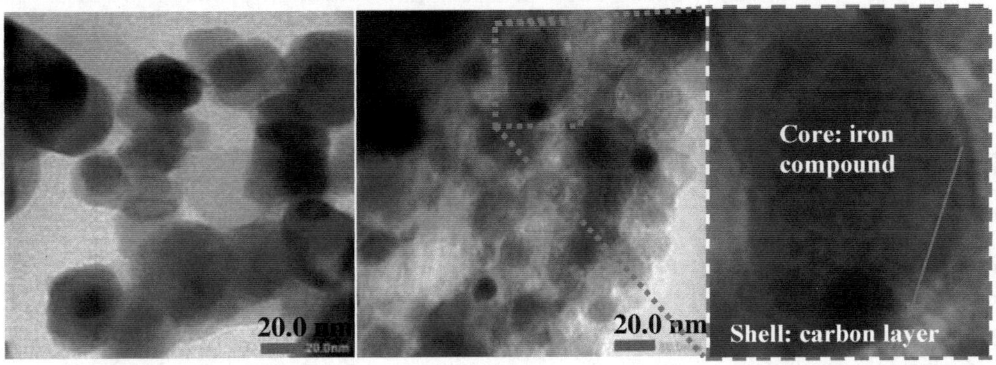

(a) (b)

Figure 3. The transmission electron microscope images (a) iron oxide and (b) iron oxide/carbon

For additional characterisation, the optical property of nanoparticle product was also studied. Table 1 shows the band gap energy of the iron oxide and iron oxide/carbon. The addition of carbon to the iron oxide material caused the band gap energy to decrease slightly from 2.25 eV to 2.18 eV. Since the nanoparticle product still had a similar band gap energy value, the carbon did not disturb the iron oxide structure; therefore, the iron oxide/carbon still had the same character as a semiconductor material as before.

Table 1. Bandgap energy of iron oxide and iron oxide/carbon fabricated by arc discharge.

Materials	Bandgap Energy (eV)
Iron oxide	2.25
Iron oxide/carbon	2.18

4. Conclusion

The character of the product of iron oxide/carbon prepared by submerged arc discharge has a crystalline structure with distinctive peaks representative of iron oxide, graphite, and a new peak of iron carbide. The nanoparticle had a nanosize and superparamagnetic property with a magnetization value of 9.9 emu/g. The bandgap energy of the iron oxide/carbon narrowed from 2.25 eV to 2.18 eV. The structure of nanoparticle product was observed as a core–shell structure with an iron compound core coated by a carbon shell.

Acknowledgement

The author would like to thank the Minister of Research, Technology and Higher Education, Republic of Indonesia, for providing the Research Grants under project No: 353/UN27.21/PN/2016 and 632/UN27.21/LT/2016.

References

[1] Vinu A, Miyahara M, Mori T and Ariga K 2006 *J. Porous Mater.* **13** 379-83
[2] Lu C H, Yang H H, Zhu C L, Chen X and Chen G N 2009 *Angew. Chem.* **121** 4879-81
[3] Kum M C, Joshi K A, Chen W, Myung N V and Mulchandani A 2007 *Talanta* **74** 370-5
[4] Vashist S K, Zheng D, Pastorin G, Al-Rubeaan K, Luong J H and Sheu F-S 2011 *Carbon* **49** 4077-97
[5] Bystrzejewski M, Huczko A and Lange H 2005 *Sensors and Actuators B: Chemical* **109** 81-5
[6] Saraswati T E, Matsuda T, Noguchi S, Ogino A and Nagatsu M 2010 *Proceeding of Symposium of Plasma Processing (SPP) 27*
[7] Saraswati T E, Matsuda T, Ogino A and Nagatsu M 2011 *Diam. Relat. Mater.* **20** 359-63
[8] Saraswati T E, Ogino A and Nagatsu M 2012 *Carbon* **50** 1253-61
[9] Saraswati T E, Tsumura S and Nagatsu M 2014 *Adv. Mat. Res.* **896** 122-5
[10] Saraswati T E, Tsumura S and Nagatsu M 2014 *Japanese Journal of Applied Physics* **53** 010205
[11] Ando Y, Zhao X, Sugai T and Kumar M 2004 *Mater. Today* **7** 22-9
[12] Kratschmer W, Lamb L D, Fostiropoulos K and Huffman D R 1990 *Nature* **347** 354
[13] Ishigami M, Cumings J, Zettl A and Chen S 2000 *Chem. Phys. Lett.* **319** 457-9
[14] Lange H, Sioda M, Huczko A, Zhu Y Q, Kroto H W and Walton D R M 2003 *Carbon* **41** 1617-23
[15] Saito Y 1995 *Carbon* **33** 979-88
[16] Saito Y, Yoshikawa T, Okuda M, Fujimoto N, Yamamuro S, Wakoh K, Sumiyama K, Suzuki K, Kasuya A and Nishina Y 1993 *Chem. Phys. Lett.* **212** 379-83
[17] Saito Y, Yoshikawa T, Okuda M, Fujimoto N, Yamamuro S, Wakoh K, Sumiyama K, Suzuki K, Kasuya A and Nishina Y 1994 *J. Appl. Phys.* **75** 134-7
[18] Barbeta V B, Jardim R F, Kiyohara P K, Effenberger F B and Rossi L M 2010 *J. Appl. Phys.* **107** 073913

High temperature Oxidation of ODS alloy with zirconia dispersions synthesized using Arc Plasma Sintering

Bandriyana[1], A Sujatno[1], R Salam[1], B Sugeng[1], A Dimyati[1]
[1]Center for Science and Technology for Advanced Materials, BATAN-Indonesia

E-mail: bandri@batan.go.id

Abstract. Microstructure formation and oxidation behaviour of the Oxide Dispersion Strengthened (ODS) steels for application as structure material in Nuclear Power Plant was investigated. A mixture composed of Fe and 12 wt. % Cr powder with addition of 0.5 and 1 wt.% ZrO_2 particles was milled and isostatic pressed to form a sample coin. The coin was then consolidated in the Arc Plasma Sintering (APS) for 4 minutes. The samples were subjected to the high temperature oxidation test in the Magnetic Suspension Balance (MSB). The oxidation test was carried out at 700°C for 6 hours to evaluate the oxide growth in the early stage of it formation by extraction the mass gain curve. The Scanning Electron Microscope (SEM) imaging and X-ray Diffraction Spectroscopy (EDX) elemental mapping were performed to study the microstructure change and compositional distribution. SEM and EDX observation revealed the time dependent development of the Fe-Cr-phases during consolidation. The oxidation rate behaviour of the samples followed the parabolic rate characteristic for inward oxidation process driven by oxygen inward diffusion through the oxide scale with the maximum weight gain around of 60 g/m2. The oxidation resistance was strongly affected by the formation of the oxide protective layer on the surface. In so far, addition of zirconia particles has played no significant role to the oxidation behaviour.

1. Introduction

Oxide dispersion strengthening (ODS) of high chrome (Cr) steel is a promising way to get materials for high temperature application. Recently, ODS ferritic steels were developed as material of the 4th generation of high temperature nuclear power reactor due to their excellent strength and creep properties [1]. However, according to the environment of high temperature reactor application, the material should also be designed to address the problem of oxidation and irradiation damage. The dispersion of zirconia was used to replace the Yttria for synthesizing of ODS ferritic steel applied as nuclear reactor components because of the low cross section for neutron absorption of zirconium and the good corrosion resistance [2]. Previous work has concluded that addition of zirconia into a ferritic Fe-14Cr-5Al-3W-0.4Ti-0.25Y2O3 matrix could effectively lead in better performance of the ODS steels against high temperature and corrosion [3, 4]. Recently, some method for ODS steel manufacturing was also developed to improve the powder metallurgical route process which is limited by the amount of material and relatively high cost. The new method we used in this work for sintering of high temperature materials, Arc Plasma Sintering (APS), developed in PSTBM-BATAN has been successfully reduced time and energy consumption. This device was used for synthesizing many high temperature alloys such as Fe-Cr and the ferritic ODS steels with addition of nano Yttria and zirconia particles dispersed along the ally matrix to obtain the optimal properties on the high temperature material application [5]. In this work, oxide dispersion strengthened (ODS) ferritic steel with nominal

IOP Conf. Series: Materials Science and Engineering **176** (2017) 012047 doi:10.1088/1757-899X/176/1/012047

composition of Chromium as high as 12 wt.% dispersed with 0.5 and 1 wt.% ZrO2 was synthesized by the Arc Plasma Sintering (APS) method. The microstructure and the oxidation resistance were investigated to determine the oxidation characteristic concerning to the zirconia addition on the ODS ferritic steel.

2. Experimental

In this work the F-Cr ODS ferritic steel was developed with 12 wt.% Cr and 0.5 and 1 wt.% of the dispersion particles ZrO_2. The samples were prepared by mixing and milling all the materials (Fe, Cr and ZrO_2 powder) for 1, 2 and 3 hours. After milling the samples were pressed by isostatic compress machine with the compression load of 20 Ton to form sample coins. The samples were then consolidated by sintering in the APS for 4 minutes. The principal design of the APS is illustrated schematically in Fig. 1 [6]. The process was carried out in fully argon atmosphere in order to protect against oxidation. The sample coin was placed on a copper cup and exposed by the plasma with the power 300 Watt and variation of time.

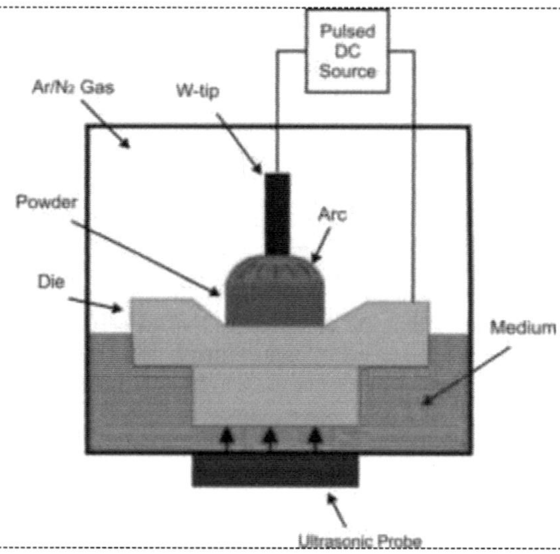

Figure 1. Schematic diagram of APS.

The morphology and composition of the alloys were studied by means of Scanning Electron Microscopy (SEM) coupled to Energy Dispersive Spectroscopy (EDX). The EDX spectrums illustrate the chemical composition of the samples. The mechanical properties of the consolidated samples were studied using the Vickers micro hardness (HV) measurement. The high temperature oxidation test was performed in the Magnetic Suspension Balance (MSB) workstation in air atmosphere at 700 °C to simulate the temperature in normal reactor operation. Detail information about MSB test is described elsewhere [7]. The oxidation time was carried out for 6 hours to get insight in to the early stage of oxide formation. The oxidation characteristic was extracted from the curve describing the relation of the weight gain to the oxidation time. The oxide scale and the effect to the mechanical properties were evaluated based on the microstructure and hardness measurements as described above.

3. Results and Discussion

3.1. Microstructure

The microstructure of ODS ferritic alloys dispersed by 0.5 and 1 wt.% were observed by SEM. The results are illustrated in Figure 2.

a. Fe-12Cr-0.5 ZrO_2

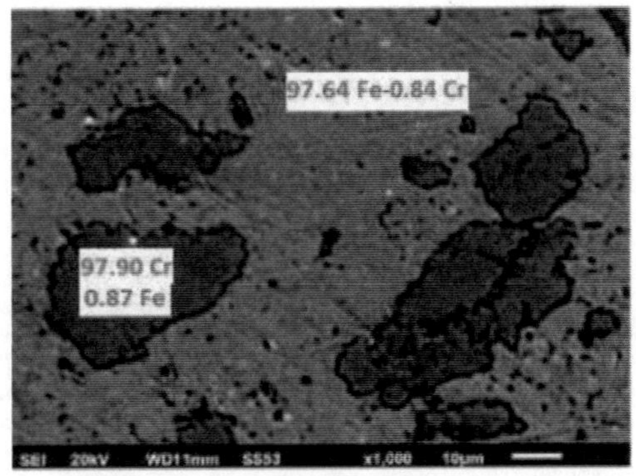

b. Fe-12Cr-1 ZrO_2

Figure 2. SEM images of the ODS steel milled 1 hour and sintered for 4 minutes.

The SEM images in Fig. 2 show the alloy microstructure consisting of bright matrix which dominated by Fe up to 98 wt.% and the dark second phases which is Cr-rich, about 98.8 wt.%. This observation can be concluded that the alloy formation begins by the formation of two single primary Fe and Cr phases. However, there are no significant differences between the sample with and without zirconia addition. Further, elemental transition from Fe to Cr phase interface is described by the results of EDX line scanning for the alloy with 0.5 % zirconia dispersion with milling time of 3 hours, shown in Figure 3 [8].

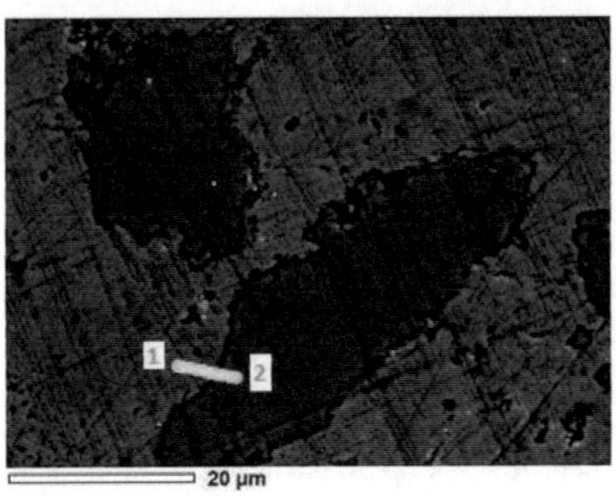

a. Fe-12Cr-0.5 ZrO_2

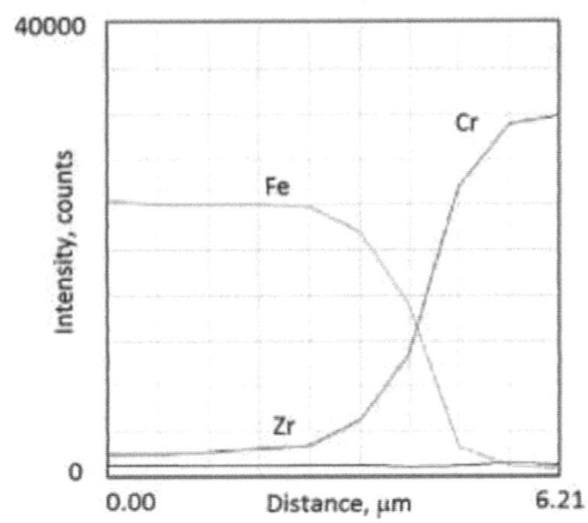

b. Fe-12Cr-1 ZrO_2

Figure 3. EDX line scan in the Fe-Cr phase interface

Based on the observation in Fig. 3, as expected the alloy formation is the result of inter-diffusion of both elements Fe and Cr in both direction forming a diffusion zone. However Cr moves faster into the new formed phase than Fe, as can be seen from the graphic showing concave structure. Both side of the phase exists. Detailed of microstructure and the alloying process was explained by the diffusion theory of Fe and Cr [6, 8]. The zirconia was finely dispersed in the matrix and in the grain boundary of both Fe and Cr phase as well. These development followed by the improvement of the mechanical properties that can be observed on the maximal hardness of 150 VHN after 4 minutes of sintering for sample of Fe-12Cr-0.5ZrO_2 with milling time for 3 hours.

3.2. Oxidation behavior

Figure 4 shows the oxidation characteristic curves of ODS steels with zirconia dispersion oxidized in air at 700°C for 6 hours.

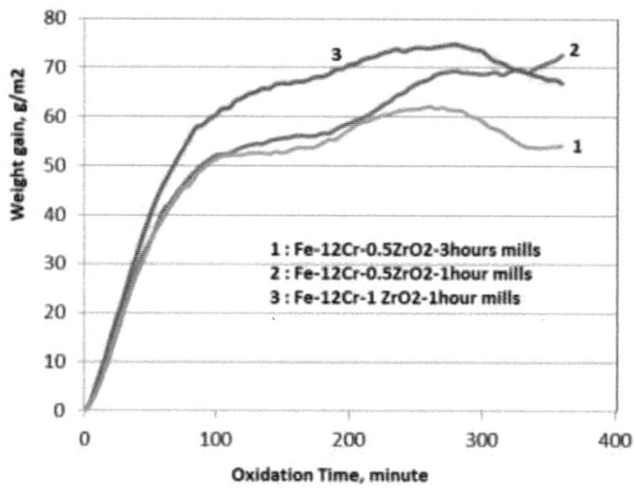

Figure 4. Oxidation curve of zirconia ODS Steel after MSB test.

The weight gains resulted from the oxidation in air are plotted as a function of the oxidation time. The early stage of oxidation during heating up to 150 minutes showed the oxidation rate that follows parabolic growth kinetics where the mass gain is proportional to the square root of time. This early stage was the most important stage that determine the total weight gain which elucidate the formation behaviour of the protective oxide layer. The characteristic of sample that follows the parabolic kinetic explains that the oxygen inward diffusion is the driven process during the formation of the oxide scale on the top of the sample surface. This parabolic characteristic stands in agreement with the well-known high temperature oxidation characteristic of Fe-Cr alloy [9]. The mechanism of high-temperature oxidation of Fe-Cr alloy is widely accepted to be the diffusion of oxygen anions through the iron oxide lattice [10]. The metal cations transport however is fully inhibited. The milling time seems to be not significantly influence the oxidation behavior, as obvious from the the oxidation curve. However, addition of zirconia particles 0.5 wt.% is likely to affect the weight gain. The weight gain of the samples with 0.5 wt.% ZrO_2 show significant decrease compared to sample containing 1 wt.% ZrO_2. This leads to the fact that the first two samples show higher oxidation resistance.

The result of the SEM analysis in Fig 5 shows the sample cross-sections revealing the oxide scales formed during oxidation test.

1

2

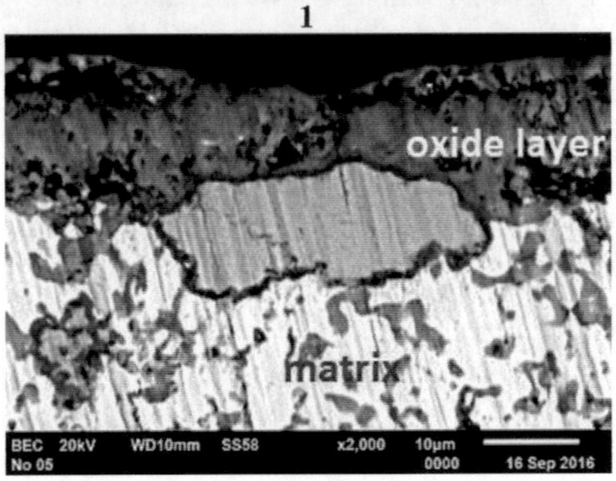

3

Figure 5. Oxidation curve of zirconia ODS Steel after MSB test.

1. Fe12Cr-0.5ZrO2-3h milling
2. Fe12Cr-0.5ZrO2-1h milling
3. Fe12Cr-1ZrO2-1h milling

From the EDX result the oxide layer is identified as stable Fe_2O_3.The thickness of the oxide layers of the samples around 15-20µm is believed to be a protective layer and explains the oxidation behaviour as shown in Fig. 3. Based on the layer thickness and oxidation behaviour of some Fe-Cr oxidation [9, 10], showed that the layer thickness and oxidation rate of the APS sintering product was good enough for high temperature corrosion behavior.

4. Conclusion

From this study, it can be concluded that APS experiment can describe the alloy formation sufficiently. The alloying of the ODS steel was inter diffusion process of Fe and Cr element that followed by the increasing of hardness. During the first stage of heating at 700°C, the oxidation behaviour of the ODS followed the parabolic growth kinetic. Addition of low concentration of zirconia in the alloy decreases the oxidation, whereas the addition of zirconia of higher concentration does not influence the oxidation rate. The effect may be caused by the influence of diffusion process due to the presence of nano oxide particle which is modify the oxide scale in the manner that decreases the oxygen diffusion channel. The oxide layer formed on the surface is a stable Cr_2O_3 and Fe_2O_3.

Acknowledgment

The authors thank to Ministry of Research, Technology and Higher Education (Kemenristekdikti) for financial support under the INSINas 2016 program, RT2016-0287.

Reference

[1] Klueh R L, J P R Shingledecker, R.W. Swindeman, D.T. Hoelzer 2005 *J. Nuclear Materials.*
[2] Toshinori Chuto, Fumihisa Nagase, Toyoshi Fuketa 2009 *Nuclear Engineering and Technology* **41** 2
[3] Nerea Garcia, Rodriguez 2015 *Journal of the American Ceramic Society*
[4] R Gao, Tao Zhang, Xu Ping Wang, C S Liu 2014 *Journal of Nuclear Materials*
[5] A. Dimyati 2016 *Proc. Asia- Pacific Microscopy Conference* **11**
[6] A. Dimyati 2015 *Technical Report Insentif Riset Sistem Inovasi Nasional (InSINas).*
[7] R. Salam, B. Bandriyana, A. Dimyati 2013 *Proc.Seminar IX SDM Teknologi Nuklir*
[8] Bandriyana, Agus Sujatno, Rohmad Salam, Arbi Dimyati 2016 *ASTECHNOVA (submitted)*
[9] A. Dimyati, H J Penkalla, P. Untoro, D Naumenko, W J Quadakkers, J Mayer 2003 *Metallkd.* **94**
[10] D S Khaerudin, M A Othman, S Mahzan, P Untoro, D Sebayang 2013 *Proc. Engineering* **23**

IOP Conf. Series: Materials Science and Engineering **176** (2017) 012048 doi:10.1088/1757-899X/176/1/012048

Molten Salt Synthesis and Structural Characterization of $BaTiO_3$ Nanocrystal Ceramics

S Ahda[1], S Misfadhila[2], P Parikin[1], T Y S P Putra[1]
[1]Center for Science & Technology of Advanced Materials, National Nuclear Energy Agency, Serpong, Indonesia
[2]Andalas University, Padang, Sumbar, Indonesia

E-mail: ahda@batan.go.id

Abstract. A new synthesis route to obtain high-purity barium titanate powder, $BaTiO_3$, using the molten salt method by reacting the raw materials ($BaCO_3$ and TiO_2) in an atmosphere of molten NaCl and KCl, has been developed. The synthesized $BaTiO_3$ ceramic particles have been successfully carried out at the sintering temperature 950°C for 4 hours. The Rietveld refinement of the XRD diffraction patterns was employed to characterize the structural information of the nanocrystalline $BaTiO_3$ ceramics. The lattice parameters (a=4.0043 Å, b=4.0308Å with space group P4mm) of tetragonal perovskite structure, as an indication of piezoelectric characteristics, have been successfully determined by the Rietveld refinement. While the crystallitte particle size and strains have been obtained for the values of 110.6 nm and 0.74 % respectively

1. Introduction

One of the ferroelectric materials is barium titanate ($BaTiO_3$) and has been a concern for several decades because of the attractive properties. Something interesting can be shown that it is chemically and mechanically very stable, and has also ferroelectric properties above room temperature. Therefore it can be easily prepared and applied in the form of a ceramic polycrystalline sample [1,2]. Material $BaTiO_3$ is also a strong candidate for the field induced piezoelectric transducers, due to their high polarization, high permittivity and large induced strain achievable [3]. Therefore to obtain good performance of materials, synthesis process which produces the perovskite crystal structure with a small particle size and uniform is needed.

The chemical formula of $BaTiO_3$, namely BT, is derived from an oxide of barium and titanium. It has five phases as a solid, listing from low temperature to high temperature: rhombohedral, orthorhombic, tetragonal, cubic and hexagonal crystal structure. Three of them show the piezoelectric effect, except cubic and hexagonal. At room temperature, BaTiO3 adopts a tetragonal perovskite type structure and high permittivity.It transforms to the cubic, paraelectric state at the Curie temperature,T_C, of approximates to 130 °C.

To improve the quality of materials (micro- and nano- $BaTiO_3$) has made various approaches to the process of synthesis which include a sol-gel [4], hydrothermal [5], high-temperature solid-state reaction [6,7], and the method of microwave synthesis method [8,9]. In order to synthesize nano-$BaTiO_3$, a low sintering temperature and a short isothermal holding (time) are both critical. In this study, we try to discuss the synthesis of the piezoelectric material with molten salt method which is

used to prepare $BaTiO_3$ fine powder. And so we will try to analyze the crystal structure of the synthesized product using the Rietveld refinement of X-ray diffraction (XRD) pattern in this study.
Common examples of the salt in the synthesis are molten salts of chloride and sulfate. In many experiments, the eutectic composition (0.5 NaCl - 0.5 KCl) of a mixture of salt is oftenly used and has a temperature lower than the temperature of liquid formation. While the melting point of NaCl and KCl are 801 °C and 770 °C, respectively, and for the eutectic composition is 650 °C [10,11]. At the heating temperature, the salt melts and the product particles form. The characteristics of the product powder are controlled by selecting the temperature and duration of the heating.

2. Experimental

Previously, the raw materials consisting of $BaCO_3$ (one mole) and TiO_2 (one mole) were mixed for $BaTiO_3$ and continued by grinding using a mortar for 4 hours. During grinding there were expected homogeneously mixed and particle size is getting smaller.The reaction mechanism in the molten salt method (MSS) is a little different from the solid-state reaction, due to the absence of the sample compaction, as long as the salt mixed in a reaction formation.

Carbonate and Oxides (as raw material, $BaCO_3$ and TiO_2) corresponding to a perovskite $BaTiO_3$ compound were mixed with salt of NaCl and KCl and then sintered at a temperature above the melting point (950 °C) of the salt. At this temperature, the oxides particles were rearranged and then diffused rapidly in a liquid state of the salt. With further sintering (during 4 hours), particles of the perovskite phase were formed through the nucleation and growth processes.

Schematic diagram of the molten salt method [7] is shown in figure 1. At the beginning a reaction batch was contained the reactants (raw materials) and salts. After the sintering process the existing salt was removed by washing with hot water repeatedly and then dried again. The chloride ions were detected by an $AgNO_3$ solution even after ten times of washing.

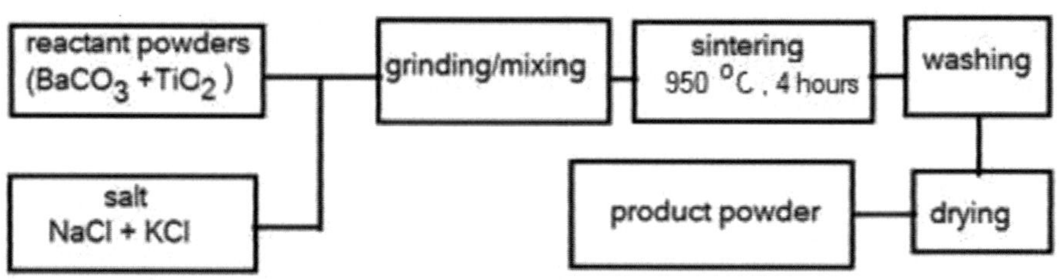

Figure 1. Preparation procedure in molten salt synthesis.

3. Result and Discussion

Regarding our experts in the preparation of ceramic powder, it was clear that the molten salt method that was used to obtain homogenous and ultrafine powder, whereas the conventional one (solid state reaction) needs high temperature of approximately (1000°C - 1400°C). Solid state reaction in this method was considered as an indirect method because the reaction was carried out in a molten salt directly between Ba^{+2}, Ti^{+4} and O^{-2} (diffuse) to obtain $BaTiO_3$ precursor. The synthesis products and materials used are clear by X-Ray Diffraction (XRD) as shown in Figure 2.

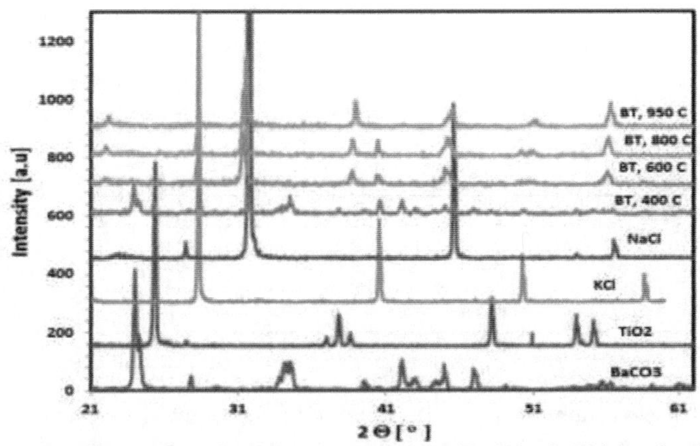

Figure 2. XRD pattern for the raw materials ($BaCO_3$ & TiO_2), the Salt (NaCl and KCl) and the sample prepared at the sintering temperature of 400, 600, 800, and 950 °C, respectively.

For the sample at sintering temperatureof 400, 600, 800 and 950 ° C (as the powder product), as shown at figure 2, the XRD patterns do not indicate the existence of raw material peaks of $BaCO_3$ and TiO_2 . However, samples at 400, 600 and 800 °C, that have not been washed with hot water, indicates the presence of NaCl and KCl peaks. And on the sample at 950 °C that has been washed does not show the peaks of the salt material. This means that the salt solution did not react with the sample. The resulting powder products show a distinct diffraction pattern to form new compounds. It explains that the synthesis process using molten salt method has been successfully carried out.

In the x-ray diffraction pattern for the sample at sintering temperature of 950 °C, we employed the Rietveld program to identify the crystal structure of the $BaTiO_3$ ceramic powder. The program also required and refined an approximate structural model for the actual structure. Firstly, we performed the Rietveld refinement based on the tetragonal perovskite crystal systems with space group of P4mm. The initial refinement was done by the zero-point shift, background and the unit-cell parameters.

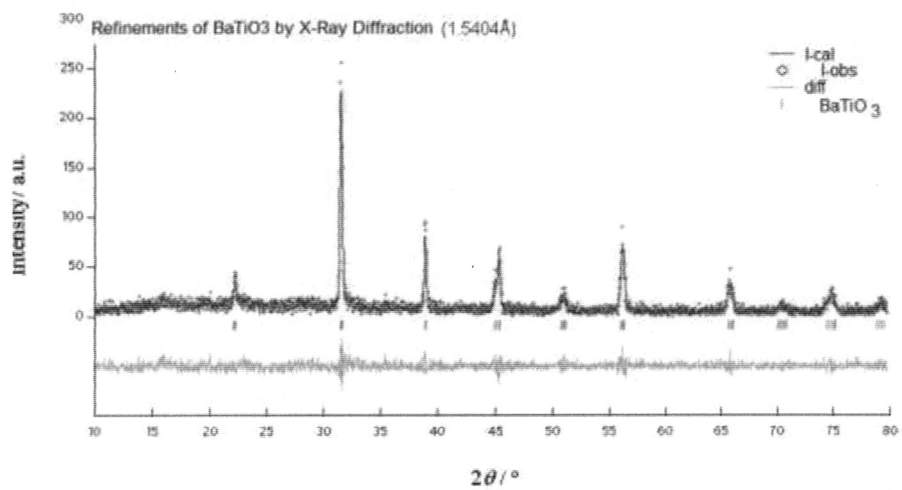

Figure 3. The structural refinement patterns of $BaTiO_3$ using X-ray powder diffraction.

The final results of this refinement were obtained the weighted R-factors of 33.63% (Rwp) and the goodness of fit indicator of 1.104 ($S=R_{wp}/R_e$). Figure 3 exhibits the Rietveld refinement (I_{cal}), experimental pattern (I_{obs}), differences ($diff = I_{obs} - I_{cal}$) and the peaks of diffraction planes.

Table 1. Structural parameters for $BaTiO_3$ obtained from the structural refinement using X-ray powder diffraction data at room temperature.

Symmetry and lattice parameter	Sites	Atomic coordinates		
Space group = P4mm		x	y	z
a = b = 4.0043(9) Å	Ba	0.5	0.5	0.5
c = 4.0308(12) Å	Ti	0	0	0
c/a = 1.0066	O(1)	0	0	0.475(69)
	O(2)	0.549(27)	0	0

So it could conclude that this approach may be suitable to determine the structural parameters for the nanocrystalline $BaTiO_3$ ceramics, such as lattice parameters, atomic coordinates, and isotropic thermal parameters (U_{iso}). Lattice parameters for the crystal system of perovskite tetragonal show that the value of a = 4.0043 Å and c = 4.0308 Å and has close to each other with the difference 0.66%.

The $BaTiO_3$ synthesis using different method (hydrothermal) by Song Wei Lu [12] has obtained cubic perovskite phase. It is possible to obtain a different phase, due to differences in lattice parameter for cubic (c/a = 1), while tetragonal (c/a~ 1) could close to cubic and also depends on the synthesis parameters performed. Cubic crystal system does not have piezoelectric properties. Diffraction pattern for standard cubic phase of $BaTiO_3$ [13], JCPDS No. 31-174, has a peak at diffraction angles (2□□ of about 31.5 and 45°, representing the plane of (110) and (200), respectively. Figure 4 shows the experimental $BaTiO_3$ peaks indicates the overlap of the two peaks (blue dot) and also in other words the refinemenet results can also show spliting (red line and dot) into two peaks. While in this experiment we can exhibite the presence of $BaTiO_3$ tetragonal phase, because there is the existence of two overlapping peaks at between 31 to 32° for planes of (101) and (110) and 44 to 47 ° for planes of (002) and (200), as shown in figure 4. In same way, the analysis of the peaks of the x-ray diffraction patterns to clarify the tetragonal phase was also done by some researchers [14,15].

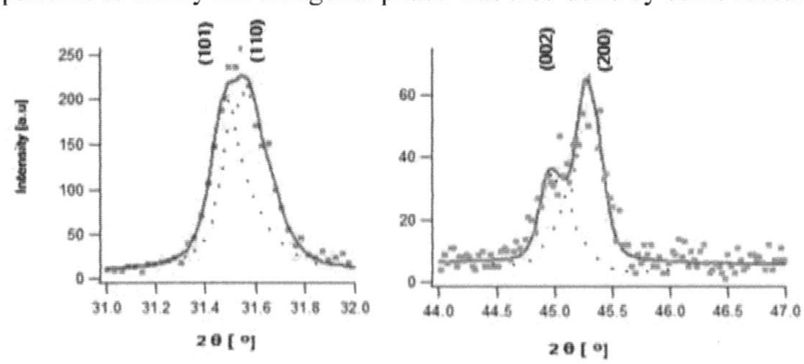

Figure 4. Peak broadening of BT nanocrystals (blue dot) and peak split of the tetragonal phase (red line-refinement out put and blue dot- experimeal data)

Polycrystalline aggregation formation during the synthesis process can also cause crystal defects. Deviations from perfect crystallinity can lead to broadening the diffraction peaks. Crystallite size and lattice strain are two main characteristics can be processed from the peak width analysis [16]. The size and strain can affect the peak broadening. By using **Williamson-Hall** analytical methods can estimate size-induced and strain as a function broadenings peak width and peak position 2θ [17,18]. The strain induced in powders due to crystal imperfection and distortion was calculated using the formula $\varepsilon = \beta/4\tan\theta$, where □ (full width at the half maximum, FWHM) was taken from the refinement results and

θ is the diffracted angle. With this approach, the Williamson_Hall plots can be drawn below in Fig. 5, and the examined formula [19,20] is given asfollows:

$$\beta \cos \theta = k\lambda/d + 4\ \varepsilon\{\sin \theta\} \quad (1)$$

where d is the crystallite particle size, (ε) elastic strain and k is Scherrer constant (somewhat arbitrary value that fall sin the range 0.87-1.0. We usually assume k = 1).

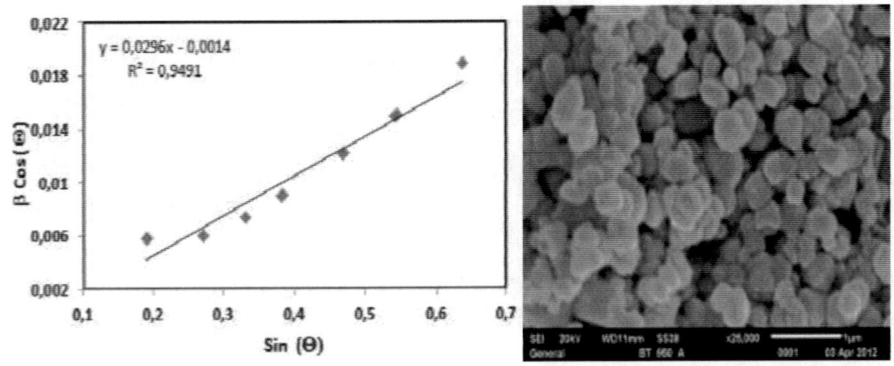

Figure 5. Williamson-Hall plot for molten salt method and the SEM (Scanning Electron Microscope) micrograph of a BT ceramic after sintering at 950°C/4hr

Crystallite particle size and strain can be calculated from the slope and y-intercept of the fitted line and obtained for d= 110,2 nm and ε=0,74 % respectively. The microstructure of a $BaTiO_3$ or a grain size sample can be shown in Fig 5. measured by SEM method. Analysis on SEM micrograph was done by measuring the grain size at horizontal with range of 116.7 to 500 nm. The crystallite size which was calculated by using the Wiliamson- Hall method and compared with the results estimated from SEM micrographs are able to show close to the same value range.

4. Conclusion.

Tetragonal phase of $BaTiO_3$ was achieved through molten salt method at the sintering temperature 950 °C within 2 hours in a solution of NaCl + KCl at mole ratio of 1: 1.

The presence of tetragonal perovskite system has been confirmed by the calculated lattice parameter using Rietveld refinement program and analyzed by the presence of overlapping peaks as its characteristic. Crystallite particle size of 110.2 nm can be estimated through the hall wiliamson- plot and close to as shown in micrograph SEM measurement, while the crystal strain of 0.74 % was obtained.

Acknowledgements

We would like to thank DR. Yetria Rilda, Mr.Engkir Sukirman and DR. Mardiyanto for scientific discussion in completing this study. We are gratefull to Mr. Gunawan (as Director of PSTBM-BATAN) and DR. Abu Khalid for suporting works, especially in the financial and administrative

Reference

[1] Jona, Shirane G 1993 *Ferroelectric crystals*, Dover Publications INC. New York.
[2] Burcu Ertuğ 2013 *American Journal of Engineering Research (AJER)* Volume-02 Issue-**08** pp-01-07.
[3] Sonia, Patel R K, Kumar P, Prakash C, Agrawal D K 2012 *Ceramics International* **38** p.1585-1589.
[4] Sharma N, Gaur A, Gaur U 2014 *Ceramics International* **40** **10** p.16441-16448.

[5] Cao Y, Zhu K, Wu Q, Gu Q, Qiu J 2014 *Materials Research Bulletin* **57** **9** p.162-169.
[6] Kambale K R, Kulkarnin A R, Venkataramani N 2014 *Ceramics International* **40** **1** p. 667-673.
[7] Syahfandi Ahda 2011(August) *Journal of Materials Science Indonesia* Vol. Materials for Sensor Special Edition page : 33 – 36.
[8] Fang C Y, Wang C, Polotai A V, Agrawal D K, Lanagan M T 2008 *Materials Letters* **62** **17-18** p. 2551-2553.
[9] Souza A E, Silva R A, Santosa G T A, Moreira M L, Volanti D P, Teixeira S R, Longo E 2010 *Chemical Physics Letters* **488** **1-3** p.54-56.
[10] Toshio Kimura 2011 *Molten Salt Synthesis of Ceramic Powders, Advances in Ceramics - Synthesis and Characterization*, Processing and Specific Applications, edited by Prof. Costas Sikalidis In Tech. (2011) p-75-100 ISBN 978-953-307-505-1.
[11] Syahfandi Ahda, Mardiyanto and Alimin Mahyudin 2013 *Journal of Materials Science Indonesia* Vol. **14** No. **4** p.284 – 289.
[12] Song W L, Burtrand I L, Wang Z L, William D S 2000 *Journal of Crystal Growth* **219** p.269-276.
[13] Vijayalaksmi R, Rajendran V 2010, *Digest Journal of Nanomaterials and Biostructures*, Vol. **5**, No 2, p. 511 – 517.
[14] Wu M, Xu R, Feng S H, Li L, Chen D and Luo Y J 1996*J. Mater. Sci.* **31** p.6201.
[15] Mehrnoosh Atashfaraz, Mojtaba Shariaty-Niassar, and Tadafumi Adschiri 2008 *Proceedings of the World Congress on Engineering and Computer Science 2008 (WCECS 2008*) October 22 - 24 San Francisco USA.
[16] Prabhu Y T, Rao K V, Kumar V S S and Kumari B S 2014 *World Journal of Nano Science and Engineering* **4** p.21-28.
[17] Suranarayana and Norton C M G 1998 *X-Ray Diffraction*: *A Practical Approach*. Springer, New York.
[18] Yendrapati Taraka Prabhu, Kalagadda Venkateswara Rao, Vemula Sesha Sai Kumar, Bandla Siva Kumari 2014 *World Journal of Nano Science and Engineering* 4 p.21-28.
[19] Mote VD, Purushotham Y and Dole BN 2012 *Journal of Theoretical and Applied Physics* **6**:6.
[20] Chrunika M J, Kosowicza S, Perkowskia P, Mrukiewicza M, Morawiaka P and Zasadab D 2013 *Acta Physica Polonica* A, Vol. **124.**

IOP Conf. Series: Materials Science and Engineering **176** (2017) 012049 doi:10.1088/1757-899X/176/1/012049

Synthesis and Characterization of Al doped ZnO (AZO) by Sol-gel Method

H Munawaroh[1], **S Wahyuningsih**[1], **A H Ramelan**[2],
[1]Inorganic Materials Research Group , Faculty of Mathematics and Natural Sciences, Sebelas Maret University, Ir. Sutami No. 36A Kentingan, Surakarta 57126, Central Java, Indonesia
[2]Electronic Materials Research Group, Faculty of Mathematics and Natural Sciences, Sebelas Maret University, Ir. Sutami No. 36A Kentingan, Surakarta 57126, Central Java, Indonesia

E-mail: sayekti@mipa.uns.ac.id

Abstract. Al doped ZnO (AZO) nanoparticles have been successfully synthesized by the simple sol-gel method. The starting materials of Al doped ZnO were $Zn(CH_3COO)_2{\cdot}2H_2O$ and $Al(OH)(CH_3COO)_2$. Preparation of AZO using polyethylene glycol as a surfactant. The solution of precursors was stirred at 60 °C for 2 hour in the conditions of Al contents are 0%, 2%, 3% and 4% (g/mL), respectivelly. In the last step reaction, gelation occurred from solution to sol gel. The sol gel then were dried at 60 °C following by annealing process for crystalization. By this simple sol gel method, the nanoparticles have been produced. The characterizations were conducted X-Ray Diffraction (XRD), Scanning Electron Microscopy (SEM), Fourrier Transform Infra-Red (FTIR) and X-Ray Fluorescence (XRF). XRD analysis reveals that all samples has crystallizes in polycrystalline nature and exhibit no other impurity phase. The variation of Al doped ZnO slighly affect the crystallinity and crystal size. Both crystallinity and crystal size decrease with increasing of Al content in AZO. Morphology of AZO shown the particle distribution more equitable with increased Al content. The synthesized AZO gaved shift peak absorption of asymetric and symetric vibrations of Zn-O-Zn around wavelengths of 680 cm^{-1} and 1630 cm^{-1} atributed of the uptake of the Al-O-Al bond instead Zn-O-Zn. XRF analysis shown that the increase ratio of Al entering into Zn influenced the Al dopant concentration.

1. Introduction

ZnO is a semiconductor which has direct band gap. The value of the energy band gap of ZnO monocrystalline is between 3.1-3.3 eV at room temperature and 3.44 eV at a temperature of 4 K and for polycrystalline ZnO films is between 3.28 to 3.30 eV [1]. ZnO exhibited some properties such as optical, acoustic, and attractive electrical hence it has a number of potential applications in the fields of electronics, optoelectronics and sensors. In addition, ZnO has several advantages such as stable chemical structure, non-toxic, and can be used as an additive in a variety of materials, as well as low cost material due to these materials have highly abundance in nature [2].

Due to it has large energy band gap, the ZnO has transparent to visible light (400-700 nm). Moreover, the wide bandgap ZnO only 5% efficiency of photocatalytic of sunlight. In order to be an affective, it is necessary to minimize the energy gap and increase the absorption of light. One this way is doping. Doping can be defined as the addition of impurities in the material hence modified of its electronic properties [3]. The material used as a doping material (Group III) such as Gallium (Ga), Fluorine (F), Indium (In), Manganese (Mn) and Aluminum (Al). Aluminum (Al) as a doping due to the seed has a morphology that increase the surface area coating and more efficient photocatalytic activity [4]. Selection of aluminum as a dopant also has advantages such as high conductivity, high transparent and low resistivity [5].

There are several methods to synthesis AZO such as sol-gel [6], spray pyrolysis [7], precipitation [8], and hydrothermal processes [9]. Among these methods, sol-gel method is in great demand due to it has thin and large layer, simple, and low cost [10]. Based on these issues, this study has been synthesis ZnO by sol-gel method. The study of temperature effect on crystalization ZnO as well as AZO, calcination treatments were done at 400 °C, 500 °C and 600 °C. In addition, to determine the effect of Al dopant like structural characteristics and physical properties, we investigated the synthesis of AZO by add the various Al dopand about 0%, 2%, 3% and 4% (w/v).

2. Experimental

2.1 Synthesis of Al doped ZnO nanoparticle

Al doped ZnO (AZO) samples were prepared by a simple sol-gel method. 0.5 M of zinc acetate dihydrate ($Zn(CH_3COO)_2 \cdot 2H_2O$, Aldrich) was dissolved in 30 ml of ethanol. Then, 10 ml of Polyethylene Glycol 1000 (PEG) was add into solution. The solution was stirred at 60 °C for 2 hour. To prepare the AZO nanoparticles, different amounts of Aluminum acetate ($Al(OH)(CH_3COO)_2$, Aldrich) were dissolved in a solution of zinc acetate dihydrate and polyethylene glycol to gain concentration at 0% (AZO 0%), 2% (AZO 2%), 3% (AZO 3%) and 4% (AZO 4%) (g/mL), respectively. Then, the solution was stirred at 60 °C for 1 hour till gel formation. After that, the solution was placed in an oven at 60 °C until the solvent evaporated. Finally, it was putted into an oxidizing furnace to perform heat treatment at 400 °C, 500 °C, and 600 °C. The final products were nanoparticles of Al doped ZnO (AZO).

2.2 Characterization

Crystal structures of the prepared AZO samples were examined by X-ray diffraction (XRD, Bruker D8 Advance) in the diffraction angle range 2θ = 10-80 °. The effect of doping concentration on crystallite size was also investigated and the crystallite size was calculated using the Scherrer's formula :

$$D\ (nm) = \frac{k\lambda}{B \cos\theta} \quad (1)$$

where D is the average of crystallite size, β is the full width at the half maximum of the diffraction peak, θ is the Bragg angle, λ is the wavelength of X-ray used and *k* is a constant.
The morphology of the synthesized AZO particles was visualized by Scanning Electron Microscopy (SEM). Fourrier Transform Infra-Red (FTIR) analyzed the groups of Al doped ZnO (AZO). XRF was also utilized to analyze the composition of various species in Al doped ZnO particles.

3. Result and Discussion

3.1 Effect of Annealing Temperature with Al doped ZnO

Annealing temperature can influence to rate of evaporation and crystallization process of the particles produced. The characterization results were shown in Fig. 1, which shows that the crystalline structure in accordance with the wurtzite hexagonal structure (JCPDS 80-0075). In Fig.1, it can be seen that the particles formed was Al doped ZnO and crystallinity of particle increase with increase annealing temperature. At high temperature, 600 °C, peak of ZnO seem sharp and clear, the particles have a high crystallinity. This is due to the higher temperature affect evaporation process and interparticle reaction more perfect, so can increased of crystalline formation [11]. These results are according with Zhang et al. (2013) [12] increase in temperature affects the formation of defects in ZnO crystals, such as substitution of Zn^{2+} atom by atom Al^{3+} is more easily and produce larger crystal size.

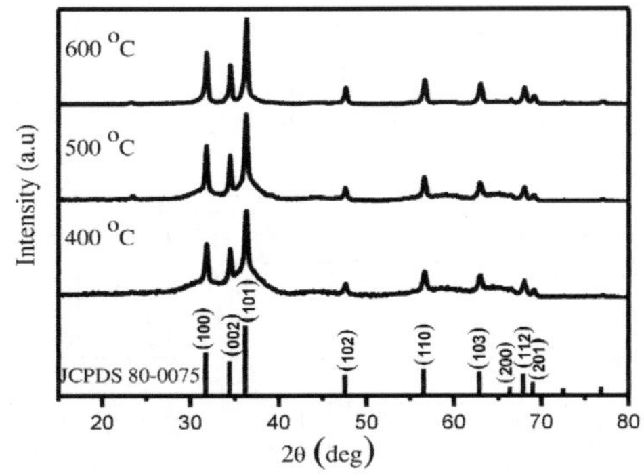

Figure 1. XRD patterns of 2% Al doped ZnO at various temperatures.

From these results, it was found that the optimum temperature to produce a large crystal size and fewer impurities at temperature 600 °C which shown in Table 1. This is because of the higher temperature, the higher particle nucleation rate and crystalite growth faster.

Table 1. Crystallite size of AZO 2% at various temperatures.

Temperatur (°C)	Crystallite size (nm)
400	1.290
500	12.659
600	20.247

3.2 Effect of Al doping

Fig. 2 shows the XRD pattern of Al doped ZnO with concentration addition of Al 0%, 2%, 3% and 4% w/v, respectively. Al dopant decreased the crystallinity described by broadening of peak width. In this study, the crystallinity of the highest AZO was obtained on the addition of Al dopants 2%. Increasing the concentration of the Al dopant will form stresses by difference size of ion Zn and Al growing and the formation of aluminium at the grain boundaries at a concentration of dopant higher [13].

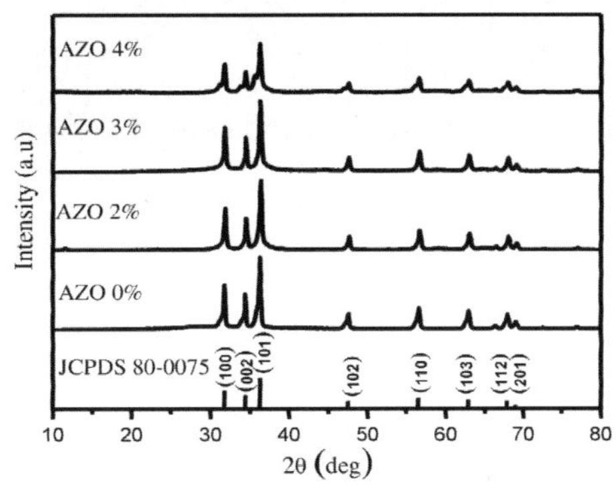

Figure 2. XRD patterns of Al doped ZnO nanoparticle.

All peaks were well matched with hexagonal structure of ZnO using standard data (JCPDS-80-0075). There is no other impurity peaks in XRD pattern which indicates the purity of nanoparticle formation as Al_2O_3. These result indicate that Al atoms into the structure of ZnO and ZnO is substituted in the lattice because of the size of atoms Al (0.54 Å) is smaller than that of Zn (0.74 Å). So that it can be seen that the Al doped ZnO substituting Zn atoms without changed the structure of the crystals formed. The crystallite size of AZO samples was also estimated from the XRD results by using Scherrer's equation and the results were listed in Table 2. The crystallite size of AZO particles decreased with increasing Al concentration and becomes 12-23 nm.

Table 2. Crystallite size of AZO particles.

Al content (%)	Crystallite size (nm)
0	28.651
2	20.247
3	20.215
4	10.779

From the results of XRD analysis, slit diffraction lower yield crystalline diameter size is smaller, so the degree of crystallinity is also low. This is due to the replacement of the position of Zn by Al during formation Al doped ZnO. Radius of Zn^{2+} ions is greater than Al^{3+} (r_{Al} = 0.054 nm and r_{Zn} = 0.074 nm), so it cause decrease the size of Al doped ZnO [15].

The study of morphology and structure were investigated using Scanning Electron Microscopy and shown in Fig. 3 for 2%, 3% and 4% Al doped ZnO. It was seem that agglomeration increases with increasing concentration of Al. The study of morphology and structure were investigated using Scanning Electron Microscopy and shown in Fig. 3. It was seen that agglomeration increases with increasing concentration of Al. Pure zinc oxide (AZO 0%) show that particle morphology broken, irregular and round shape was intact. With the addition of the dopant concentration of 2%, AZO particles ranging round and flat surface. The same result also occurs in the dopant concentration of 3% and 4%.

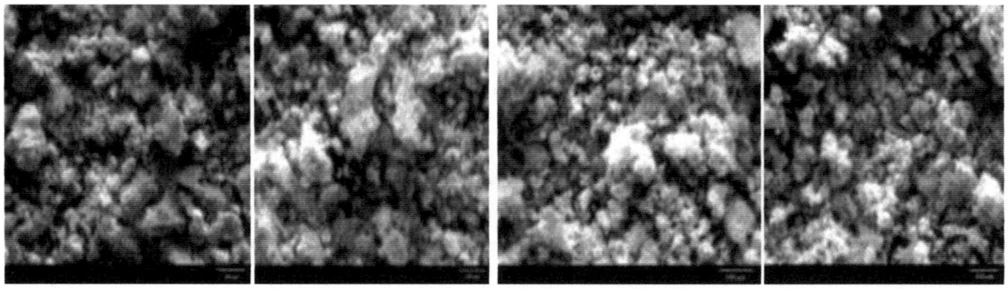

(a) (b) (c) (d)

Figure 3. SEM image of Al doped ZnO (a) 0% (b) 2% (c) 3% (d) 4%

Characterization of a finger vibration Al doped ZnO can be viewed using FTIR (Figure 4). Based on Fig. 4, AZO 0%, 2%, 3% and 4% gave the same weak absorption spectras at a wavenumber of 420 cm^{-1} atribute as a bending of Zn-O-Zn. The synthesized AZO gaved shift peak absorption of asymetric and symetric vibrations of Zn-O-Zn around wavelengths of 680 cm-1 and 1630 cm-1 atributed of the uptake of the Al-O-Al bond instead Zn-O-Zn. There are absorptions at the wavelength of 1360 cm^{-1} from an asymmetric stretching vibration Zn-O-Zn due to interference by Al atoms.

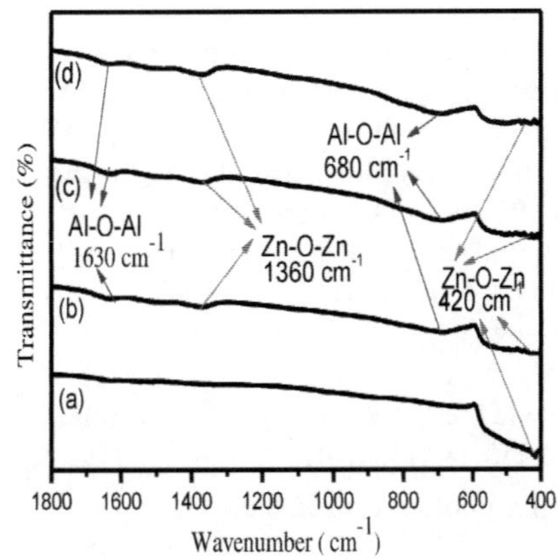

Figure 4. IR Spectra of Al doped ZnO (a) 0% (b) 2% (c) 3% (d) 4%

To check the chemical composition of the material were analyzed using X-Ray Fluorescence (XRF). Table 3. shows the results of the composition of materials Al doped ZnO. The Al doped ZnO contain elements of C, O, Zn, Al and Mg. Mg is fewer element in synthesized AZO nanoparticles. From the Table 2. shows that mass (%) of Zn decrease with increase Al doping concentration in AZO.

Table 3. XRF analysis result of Al doped ZnO (AZO).

Sample	**Mass (%)**						
	ZnO	**Al_2O_3**	**MgO**	**Zn**	**O**	**Al**	**Mg**
AZO 2 %	88.83	8.24	0	71.37	22.26	4.36	0
AZO 3%	85.39	10.53	1.09	68.60	23.09	5.57	0.66
AZO 4%	82.19	12.98	1.77	66.03	23.95	6.87	1.07

Furthermore, the calculation of ratio of Al/Zn in AZO 2%, AZO 3% and AZO 4% were shown in Fig. 5. the increasing Al dopand had beed added was capabled increase substitution of Zn atoms by Al. Replacement Zn by Al on AZO 2% have a ratio of Al /Zn = 0.061; on AZO 3% have a ratio of Al /Zn = 0.081 and on AZO 4% have a ratio of Al /Zn = 0.104.

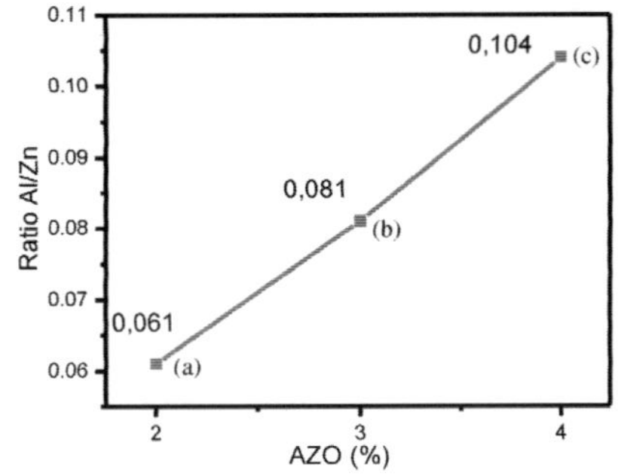

Figure 5. Influence of ratio Al addition in ratio of Al/Zn in AZO (a) 2%, (b) 3%, and (c) 4%

4. Conclusion

Al doped ZnO (AZO) were synthesized by sol-gel method. The structural and optical properties of AZO nanoparticle have been investigated. Annealing temperature influenced the crystalline quality of Al doped ZnO, where increasing of annealing temperature was followed by increase of crystallinity Al doped ZnO. The variation of Al content did not significantly influence the phase structure or the crystalline quality. SEM images of the samples show the agglomeration increases with increasing Al concentration. FTIR showed a new absorption at a wavelength of 680 cm^{-1} and 1630 cm^{-1} which is stretching vibration of symmetry and asymmetry of the Al-O-Al. XRF analysis shown that ratio of Al/Zn increase with increase Al doping concentration.

Acknowledgment

The authors would like to acknowledge the Ministry of Research, Technology and Higher Education (Kemenristek Dikti) REPUBLIK OF INDONESIA for financial assistance through the International research collaboration and scientific publications Project.

Reference

[1] Dengyuan S 2005 *Tesis Doktor at University of New South Wales.*
[2] Z H Astuti 2007 *Skripsi* Departemen Fisika ITB Bandung.
[3] J C fan, C C Ling and Z Xie 2011 *Prof.P.Preedep (Ed.)* ISBN: 978-953-307-276-0.
[4] Mahadik M A, Shinde S S, Hunge Y M, Mohite V S, Khumbhar S S, Moholkar A V, Rajpure K Y 2014 *UV Assisted photoelectrocatalytic oxidation of phthalic acid using spray deposited Al doped zinc oxide thin films* p 446-451.
[5] Gonzáles A E Jimenes 1998 *Journal of Crystal Growth* **192** p. 430-438.
[6] H Serier, M Gaudon and M Menetrier 2009 *Solid State Sci.* **11** 1192.
[7] A Verma, F Khan, D Kar, B C Chakravarty, S N Singh and M Husain 2010 *Thin Solid Films* **518** 2649.
[8] S F Du, Y J Tian, H D Liu, J Liu and Y F Chen 2006 *J. Am. Ceram. Soc* **89** 2440.
[9] W H Zhang, W D Zhang and J F Zhou 2010 *J. Mater. Sci.* **45** 209.
[10] Li Y and Meng J 2014 *Materials Letters* p. 260-262.
[11] Hou Q, Meng F and Sun 2013 *J. Nanoscale Research Letters* **8** p 144.
[12] Zhang Y, Liu L, Xing J, Yui L, Zhang J and Zhang Z 2013 *Materials Science in Semiconductor Processing* **16** p 1573–1579.
[13] Ayadi B, ElMir L, Alaya S, Djessas K, and Hernandez E 2007 *Nanotechnology* **18** p 445072-445706.
[14] Sahay P P and Nath R K 2008 *Sensors and Actuator B* **133** p 222-227.

Micromagnetic simulation of vortex-antivortex magnetization in permalloy nano particle

B Purnama[1], **S Muhammady and Suharyana**
Department of Physics, Faculty of Mathematics and Natural Sciences,Universitas Sebelas Maret, Jl. Ir. Sutami 36A Kentingan Surakarta 57126, Indonesia

E-mail: bpurnama@mipa.uns.ac.id

Abstract. A process of vortex-antivortex magnetization reversal in a Permalloy nano particle with uniform polarity of magnetization has been investigated numerically. Micromagnetic simulation is performed using the Landau–Lifshitz–Gilbert equation. A short field pulse is applied in a film plane anti parallel to magnetization direction. Sequences of simulation of reversals mechanism are evaluated for thickness of nano particle. As the results in the case of thickness of 20 nm thin layer, magnetization reversal realizes through a creation-annihilation of Neel-Bloch wall pair. Contrarily, reversal mechanism via a creation-annihilation process of vortex-antivortex pair occurs for thickness of 60 nm thin layer. By analyzing barrier energy of the sample, we find that a maximum barrier energy reaches a threshold value (e.g., $\sim 2.6\times10^6$ erg/cm^3 for Permalloy in this simulation).

1. Introduction

Nanometer-order size of magnetic devices is desired to be realized a high density hard disk drive (HDD), i.e. exceed to 10 Tbit/in^2 in the near future. For realizing, the most important effort up to now is how to maintain the magnetic property if a magnetic material scales down to nanometer-order without a significant thermal effect. As known well, the thermal effect causes the parts of magnetic property lost.

Generally, read-write mechanism of magnetic device employs magnetization reversal. Conventionally, magnetization reversal even occurs at infinite large medium such as continues films or bulk materials. There are two type of switching i.e. spontaneous magnetization via coherent rotation and magnetization reversal via domain nucleation following wall propagation to single domain configuration. Furthermore, configuration of domain wall should be Néel wall and Bloch wall. The magnetization reversals also may occur via vortex-antivortex formation [1-6]. In-depth information regarding the magnetization reversal mode will guide the design of nano devices with a minimum magnetic field require for reversal.

In this paper, a numerical investigation to study magnetization reversal on nanometer pattern Permalloy is carried-out using micromagnetic simulation by solved Landau-Lifshitz-Gilbert (LLG) equation. The discussion focus on the vortex-antivortex formation during magnetization reversal under magnetic field with nano second duration. In order to modify the vortex-antiovortex formation, thickness dependence of magnetic field required for switching is also investigated.

[1] To whom any correspondence should be addressed.

2. Experimental

Generally, magnetization under the influence of H_{eff} is given as Landau-Lifshitz-Gilbert (LLG) equation

$$\frac{dM^i}{dt} = -|\gamma| M^i \times H^i_{eff} + \frac{\alpha}{M_S} M^i \times \frac{dM^i}{dt} \tag{1}$$

where M_s is saturation magnetisation, $|\gamma|$ is gyromagnetic ratio ($|\gamma| = 1{,}76 \times 10^7$ $Oe^{-1}s^{-1}$ for free electron in metal) and α is Gilbert attentuation parameter. Magnetic energy consists of components, i.e. exchange energy, anisotropy energy, demagnetisation energy, and Zeeman energy. Without Zeeman energy, magnetic energy can be written as

$$E_{tot} = \int_V \left(A \sum_{i=x,y,z} (\nabla m_i)^2 + \varepsilon_a - \tfrac{1}{2} M \cdot H_d \right) dV \tag{2}$$

with $\boldsymbol{H}_d$ is demagnetisation vector, $\varepsilon_a = K[1 - m \cdot k]$, $m = M/M_S$ and k is unit vector directed to in-plane. In this simulation, 50 nm × 100 nm sized nanoparticel permalloy was subdivided into rectangular elements of 15 × 15. The nanosize permalloy is assumed here as uniaxial anisotropy structured in-plane nanoparticles with anisotropy constant K = 5000 erg/cm^3, exchange constant $A = 1{,}3 \times 10^{-6}$ erg/cm, saturation magnetisation $4\pi M_s = 1{,}0 \times 10^4$ gauss, $\alpha = 0{,}3$ and step time integraltion (dt) = 2,5 × 10^{-13} s [4-5].

3. Results and Discussion

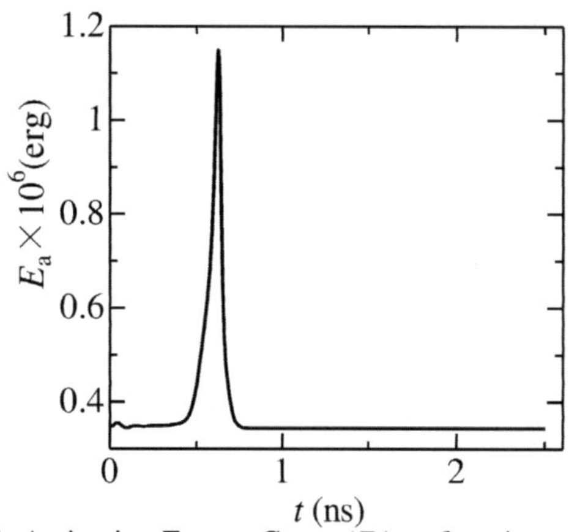

Figure 1. Activation Energy Curve (E_a) as function of time (t).

In this paper, energy which is needed to cause magnetisation reversal process is defined as activation energy (E_a) i.e. minimum energy that magnetic moment need to change from initial minimum condition into another minimum condition as shown at Figure 1. A left region of peak curve is initial condition and a right region of peak curve is other local minimum which obtains after the magnetization reversal. In this simulation, initial condition is set to be uniformly magnetization $M = 1$ when zero field is applied. Then, field with disagree orientation to initial magnetization is used to reverse the initial magnetization. So that magnetization gradually along to field direction. When the magnetization reduces to zero ($M = 0$) and starts to the reversed direction, the magnetization reversal occurs. The field is required for the magnetization reversal even define as switching field (H_{sw}).

IOP Conf. Series: Materials Science and Engineering **176** (2017) 012050 doi:10.1088/1757-899X/176/1/012050

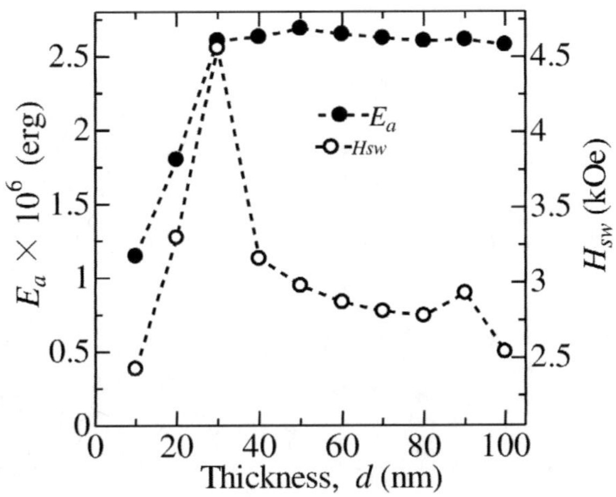

Figure 2. Thickness (d) dependence of the activation energy E_a and the switching field H_{sw}.

Figure 2 shows the thickness dependence of both the E_a and the H_{sw} under short pulse for 2.5 ns magnetic field of $H_{eff} = 5$ kOe. It is clear that the E_a goes up with the increase of the thickness, d and then becomes a constant at 2.6×10^6 erg after the critical thickness, d is obtained i.e. 30 nm. In this case, the maximum of the E_a is attained. Contrarily, the H_{sw} at the beginning linearly increases with the raise of the d, after $d = 30$ nm the remaining H_{sw} decreases with increasing the thickness. From the simulation results, it means that the addition mechanism for lowering the H_{sw} is necessary after the critical thickness is attained. When the d is equal 30 nm, the field required for switching H_{sw} is about 4.56 kOe. Moreover, the H_{sw} is only 2.54 kOe at $d = 100$ nm. Here, the lowering factor is about 44.3% (=((4.56-2.54)/4.56)×100%). The obtained result is well-suited to previous report by Zhang and Liu [7].

In more detailed, the gradually magnetization reversal on the micro-magnetic graph is analyzed as depicted at Figure 3. Three sequence of magnetization reversal for thickness d of 20 nm (after that call **X**), 30 nm (**Y**) and 60 nm (**Z**) is compared their type of nucleation mode. For X case, formation mode is most simple than for both Y and Z cases. Here, a very fast creation and domain wall propagation is appeared. And then the domain wall is quickly disappeared. There are 3 domains separated by Néel walls with typical S-state during in about 0.09 ns (90 ps). Contrast for the Y and the Z cases. A pairing vortex-antivortex is observed during the magnetization reversal. For the Y case, the first vortex-antivortex appears at $t = 1.11$ ns then makes an interaction with the other and finally disappeares as a single domain configuration at $t = 1.21$ ns. Duration time of the pairing vortex-antivortex is available about 0.1 ns (100 ps). For the Z case, very fast creation type of the vortex-antivortex is turned up at $t = 0.68$ ns and it makes interaction for $t = 0.13$ ns (130 ps) with other configuration to be a single domain formation via annihilation of domain wall. This means clearly that time for the first nucleation type of vortex-antivortex much effectively influences the assisted magnetization reversal. For the Y case, the time required for the first nucleation is much longer than the Z case because magnetic energy compensation of Bloch wall creation decreases with the increase of the thickness. From this fact, the field required for switching (H_{sw}) decreases with enlargement of the thickness (d) although the energy activation (E_a) is not so much different (see also the Fig. 2). From the above discussion, although the magnetic field require for switching H_{sw} is still ~ kOe orde, the critical thickness for a constant of the E_a is one point interesting for practical application. The high enough of the E_a will ensure thermal stability when sample down scale to nano meter orde without loss their magnetization at room temperature.

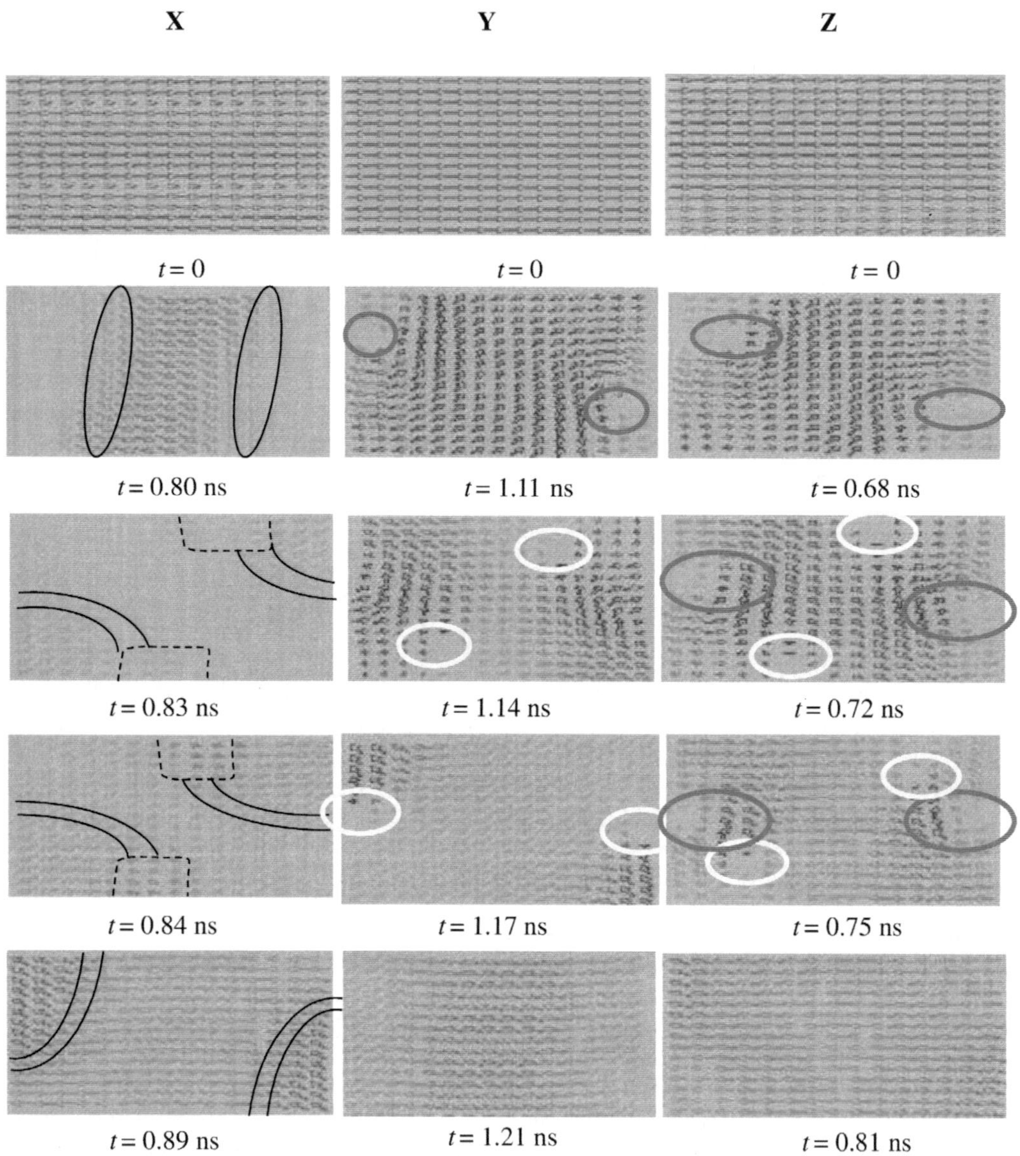

Figure 3. Micro-magnetic grapgh of magnetization reversal process in X (t = 20 nm), Y (t=30 nm) and Z (t = 60 nm). Yellow and red circles indicate vortex and anti-vortex.

4. Conclusion

We have studied a process of vortex-antivortex magnetization reversal in a Permalloy nano particle with uniform polarity of magnetization by Micromagnetic simulation. A short field pulse is applied in a film plane anti parallel to initial magnetization direction. Sequence of simulation of reversals mechanism is evaluated for thickness of nanometer order particle. In the case of thickness of 20 nm thin layer, magnetization reversal realized through a creation-annihilation of Neel-Bloch wall pair. Contrarily for the thickness of 60 nm thin layer, reversal mechanism via a creation-annihilation process of vortex-antivortex pair. By analyzing barrier energy of the sample we find that a maximum barrier energy reaches a threshold value (e.g., ~ 2.6×106 erg/cm^3 for Permalloy in this simulation).

References

[1] Yaying Z. 2003 *Magnetic Simulation on Advanced Recording Media*. Thesis. Department of Electrical and Computer Engineering, National University of Singapore.

[2] Schrefl T, Hrkac G, Goncharov A, Dean J, Bance S, Bashir MA. 2008 Finite Element/Boundary Element Simulation of Future Hard Disk Recording. *Applied Computing Conference*, Istanbul, Turkey, May 27-30.

[3] Jian-Gang, Z. 2003. New Weights for Hard Disk Drives. Materials Today July/August. Department of Electrical and Computer Engineering and Data Storage Systems Center, Carnegie Mellon University, Pittsburgh, PA

[4] Gupta R, Gupta M, and Gutberlet T. 2008 J. Phys., **71**, 1123-1127

[5] Ross CA. 2001 *Annu. Rev. Mater. Res.*, **31**, 203-235 (2001).

[6] Bolte MBW, Möller DPF, Meier GD, Thieme A. 2004 *Simulation of Micromagnetic Phenomena*. Proceedings 18th European Simulation Multiconference Graham Horton (c) SCS Europe.

[7] Zhang H and Liu Y. 2012 *J. of Nanosci. and nanotech.* Vol. **12**. 1063-1066

IOP Conf. Series: Materials Science and Engineering **176** (2017) 012051 doi:10.1088/1757-899X/176/1/012051

Development of Refined Natural Resin based Cashew Nut Shell Oil Liquid (CNSL) for Brake Pads Composite

S Wahyuningsih[1], A H Ramelan, P Rahmawati, B P N Tamtama, P P Sari, P L Sari, S Ichsan, Y R Kristiawan, F N Aini
Inorganic Material Research Group, Faculty of Mathematics and Natural Sciences, Sebelas Maret University, Ir. Sutami street No. 36 A, Surakarta, Central Java, 57125, Indonesia.

E-mail: sayekti@mipa.uns.ac.id

Abstract. Brake is one of the most important components in the vehicle. One type of brake that widely used is brake-based composites. One of the manufacture of composite material is resin. Cashew Nut Shell Liquid (CNSL) is a natural material which has chemical structure similar to synthetic phenol so it can be an alternative as a resin. Brake pads manufacture using CNSL as resin composites made to obtain the brake which is strong, wear-resistant, and environmentally friendly. The composite made using powder metallurgy techniques by mixing ingredients such as rubber, fibre glass, carbon, mineral sands and phenolic resin. Two formulas were composed by varying the resin and iron mineral sands in 5 grams. Composites were tested using Universal Testing Machine (UTM). The tensile strength result of those formulas are 600 N and 900 N and the elongations are 1.98 mm and 2.59 mm respectively. Formula 2 has a better tensile strength due to the addition of more resin is 15%. Since the better properties, formula 2 was derivated to 4 extended formulas and showed excellent pressure strength reached 20.000 N. It indicates that the addition of the resin can improve the mechanical properties of a composite.

1. Introduction

Brake pads is either of motor vehicles component that serves to slowing down or stop the vehicle, especially for a ground vehicle. The purpose of friction brakes is to decelerate a vehicle by transforming the kinetic energy of the vehicle to heat, via friction, and dissipating that heat to the surroundings. As a part of a commercial truck or automobile, brake materials have additional requirements, like corrosion resistance, light weight, long life, low noise, stable friction, low wear rate, and acceptable cost versus performance. In order to achieve the properties required of brakes, most brake materials are not composed of single element or compound, but rather are composites of many materials. More than 2000 different materials and their variants are now used in commercial brake components [1].

Development of the composite-based brake pads will reduce the damage of train wheels and sparks during braking continually, one way to do engineering materials using nature-based phenolic resin. Cashew Nut Shell Liquid (CNSL) is an oil composed from a complex phenolic compounds with

[1] sayekti@mipa.uns.ac.id

long carbon chain branched and unsaturated [2]. Utilization of CNSL and its derivatives product in various industries increasingly open, including pharmaceutical industry; insecticides; adhesives; varnishes and paints, brake pads and clutch plate vehicles, laminating resin, and epoxy resin [3]. The main components of CNSL composed by acidic compounds such as anacardic acid; cardanol; and cardol, which is a nature phenolic compound that has many advantages when it compared with synthetic from petroleum derivatives. CNSL is a renewable resource, where the availability of raw material can be guaranteed, different from synthetic phenolic of petroleum derivatives which the prices and existence depend on petroleum abundance. Indonesia had imported phenol in form phenol and phenol resin as much as 53,640 tonnes/year, so that the potential of CNSL in Cashew Nut skin can be explored properly, there will be foreign exchange savings due to a reduction in imports of phenol with their natural phenol substitution of CNSL. The CNSL compounds and their derivatives are environmental friendly products, so it chance to replace the phenolic resin as brake pads composite[4].

2. Experimental

2.1. Decarboxylation of CNSL

The extracted CNSL (100 g) was mixed with toluene (150 mL) in a round bottom flask and refluxed for 3 hours by using dean-stark apparatus and check the TLC for the absence of anacardic acid. Then the decarboxylated CNSL (50 g) and 200 mL methanol were placed in a 500 mL round bottom flask. Then 20 mL of 40% formaldehyde solution and 3.0 mL diethylenetriamine were added in this solution. This mixture was heated until boiling under reflux for 2 hours. After the solution was allowed to reach room temperature, a phase separation occurred, showing a slightly reddish upper solution, and a dark brown solidified lower phase. The upper phase was subsequently decanted, and treated with distilled water (40 mL) followed by petroleum ether. The petroleum ether layer was evaporated to dryness, yielding a reddish residue of cardanol (26 g).

2.2. Pellet preparation

Preparation of pellets were conducted by mixing all the ingredients named Formula 1 (F1) and Formula 2 (F2) shown in Table 1.

Table 1. Formulation of brake pads composites

Material	(F1)		(F2)	
	(%)	Gram	(%)	Gram
Resin	13	0.65	15	0.75
Elastomer	10	0.50	10	0.50
Fibre	13	0.65	13	0.65
$BaCO_3$	14	0.70	13	0.65
Al_2O_3	9	0.45	9	0.45
$CaCO_3$	14	0.70	13	0.65
$MgCO_3$	9	0.45	9	0.45
TiO_2	9	0.45	9	0.45
Carbon	9	0.45	9	0.45

After all the ingredients are mixed, and then pressed for 15 minutes at a temperature of 160 ºC use hot press, the resulting pellet was heated at 150 ºC for 30 minutes. Formula 2 was derivated to extended Formula 2 (Table 2.)

Table 2. Extended Formula 2 for brake pads composite preparation

Extended Formula 2					
Formula	**Concentration**				
Resin (CNSL/Phenolic)	0/100 (%)	5/95 (%)	10/90 (%)	15/85 (%)	20/80 (%)
	0/0.75	0.0375/0.712 5	0.065/0.675	0.097/0.6375	0.13/0.6
Fibre Glass	0.65	0.65	0.65	0.65	0.65
$BaCO_3$	0.65	0.65	0.65	0.65	0.65
Al_2O_3	0.45	0.45	0.45	0.45	0.45
$CaCO_3$	0.65	0.65	0.65	0.65	0.65
MgO	0.45	0.45	0.45	0.45	0.45
TiO_2	0.45	0.45	0.45	0.45	0.45
Carbon	0.45	0.45	0.45	0.45	0.45

2.3. Brake pad Composites Characterizations
The extracted cardanol was characteried by FTIR spectroscopy to determine the functional groups contained in the extract, especially hydroxyl groups that can be used to bind other compounds. While the Thermal properties of the cardanol was determined by simultant thermal analyzer. Tensile strength specimen was conducted by UTM M500 50CT to know how much resistance tensile strength of the specimen is and SEM was used to determine morphological appearance in micro scale.

3. Result and Discussion
Development the composite-based brake pads aim to reduce the damage of train wheels and sparks during braking continually. By the impruvitation of the compositions of brake pads as an engineering materials using nature-based phenolic resin, we hope the mechanical properties as well as chemical properties have been upgraded. Cashew Nut Shell Liquid (CNSL) is an oil composed from a complex phenolic compounds with long carbon chain branched and unsaturated [2].

CNSL extraction that's done using hot water, which is then dried to remove water from the extract of physical separation by centrifugation. Meanwhile, if we want a more pure fractions, it can be done by further distillation, or with the extraction treatment using an organic solvent. Based on the research results, the main structure CNSL is 90% anarcadic acid and 10% cardol, but in detail components in the CNSL separated will depend on the treatment path will be executed. The increase in the levels of cardanol done through extraction using semi-polar and non-polar solvent.

Raw CNSL has a high relative density because it has anarcadic acid as the major fraction, there is intermolecular attraction between the electronegative oxygen atom and the partially positive hydrogen atom of the phenol core as a result the molecules are closely packed together [5]. The structure of anacardic acid, cardol, cardanol and 2-methyl-cardol were founded in crude CNSL showed in Figure 1. CNSL compound was added as a resilience in the binder system and reduces brake noise.

IOP Conf. Series: Materials Science and Engineering **176** (2017) 012051 doi:10.1088/1757-899X/176/1/012051

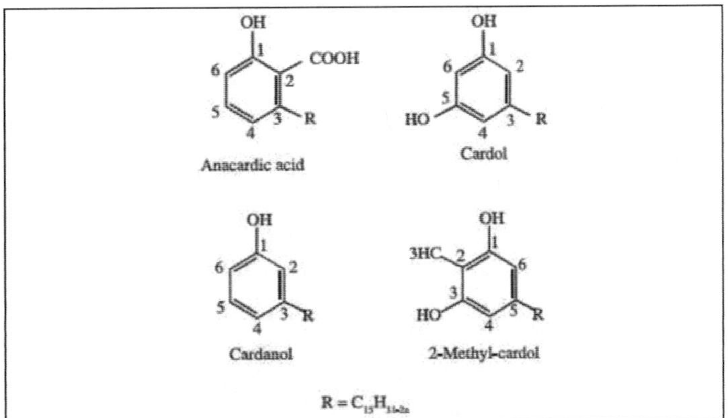

Figure 1. In Structure of anacardic acid,, cardol, cardanol and 2-methyl-cardol

FTIR analysis of technical CNSL and decarboxylated CNSL result were shown in Figure 2. Infrared spectra absorption were in the regions of 3424 cm^{-1}, 2700-2900 cm^{-1}, 1615 cm^{-1}, 1200-1600 cm^{-1}, and 500-700 cm^{-1} representing as -OH hydroxyl, C-H stretching, C-O, C=C aromatic stretching, and aromatic ring respectively. The presence of –OH hydroxyl has a role as a reactive group which will react with other components of composite material, so that CNSL can be used as an alternative of phenolic resin. But, raw CNSL still content impurities so that needed decarboxilation procedure to remove the impurities. The difference of two FTIR spectras indicate the lose of C-O from minor anacardic acid. CNSL based resin possess an outstanding resistance to the sulfiding action of mineral oils and high resistance to acids and alkalis [6]. In addition, high thermal stability also become one of requirement to be resin

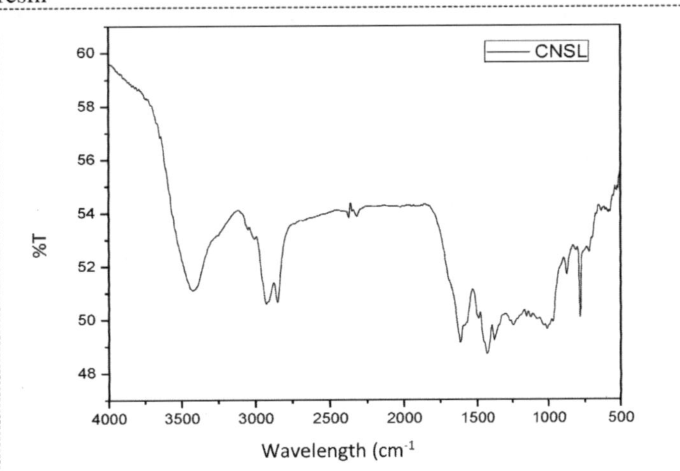

(a)

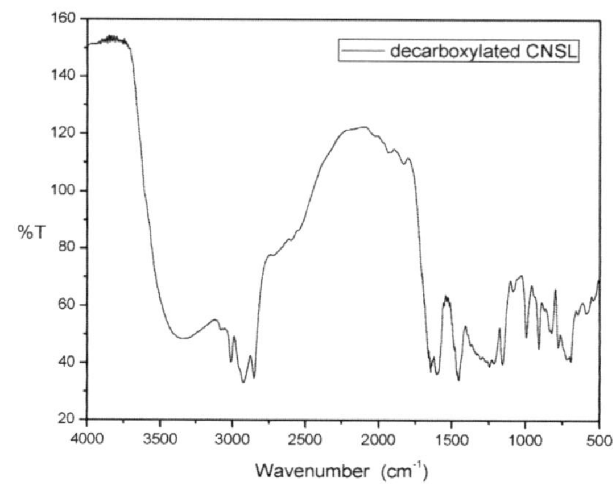

(b)

Figure 2. Characterization of raw CNSL (a) and Decarboxylated CNSL (b) using FTIR

Simulant Thermal Analysis (STA) for the determination of thermal resistance of materials. Based on thermal analysis using STA indicated that the CNSL has high stability in temperature under 400 °C, proved by the small decrease of its mass in the temperature range of 0 - 350 °C. Nevertheless, when the temperature up to 400 °C, the more mass loss because of the decomposition of the CNSL coumpound. The mass transformation was occured, so that the mss decreasing was appeared in the STA spectras (Figure 3). The quantity of mass loss has been retyped in Table 3.

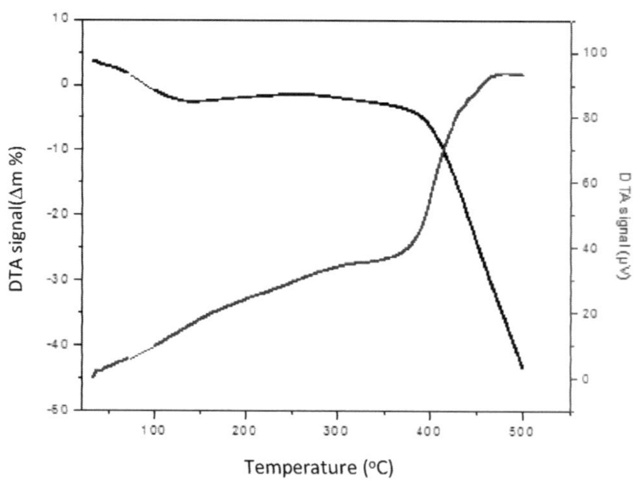

Figure 3. Simultant Thermal analysis of CNSL (– TGA Signal, – DTA Signal)

Table 3. The mass loss of CNSL during temperature treatment

Temperature (°C)	Mass (mg) %	Δ mass (%)
Room temperature		
34	-3.5	6.2
145	-2.7	
350	-3.0	40.3
500	-4.3	

3.1. Tensile strength

There are two specimens generated, those are specimen of Formula 1 (F1) and Formula 2 (F2) by the different amount of resin in each Formula. The tensile strength measurenment results are F1= 600 N and F2= 900 N, respectively. In addition, the elongation of F1=1.98 and F2=2.56. The higher mechanical properties results were affected by the amount of CNSL resin. Resin can bind to the filler materials such as Al_2O_3, Fe_2O_3, SiO_2, $MgCO_3$, $BaCO_3$ used in the composites preparation. The composition of each formula was showed in the Table 1. Specimen of F2 have more resin content that F1. The more resin, the more effective bonding occurs among the materials.

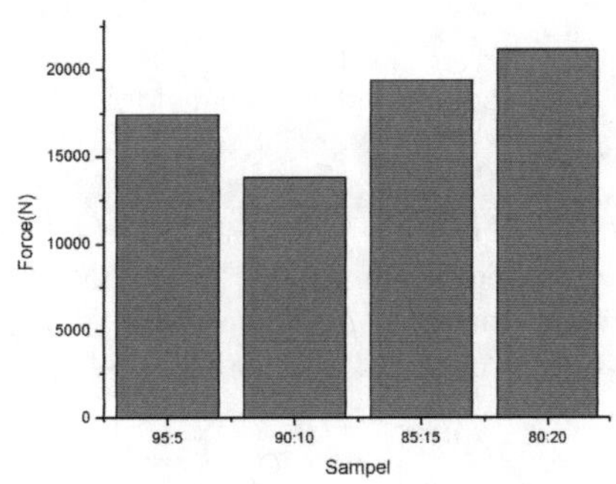

Figure 4. Pressure strenght test result of extended formula 2

Since formula 2 has better tensile strength result then it was derivated to 4 extra formulas. The pressure stregth of those formulas were tested using UTM machine and showed good results. The best pressure strength result was from formula with 20% CNSL resin content, reachs more than 20.000 N. The excellent pressure strength was supported by the hardness property of CNSL resin so that produce better mechanical property of composite based brake pads.

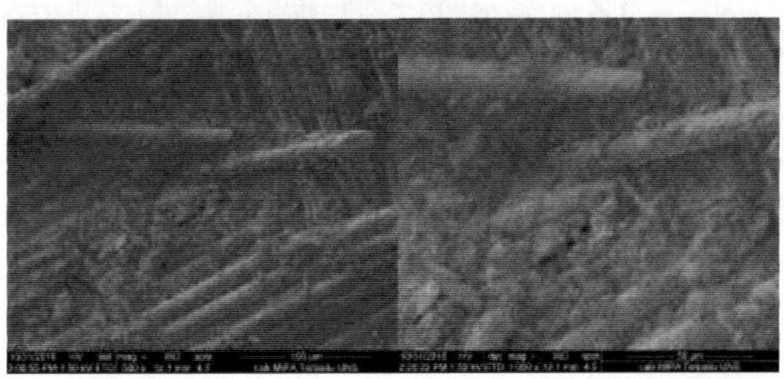

(a)

IOP Conf. Series: Materials Science and Engineering **176** (2017) 012051 doi:10.1088/1757-899X/176/1/012051

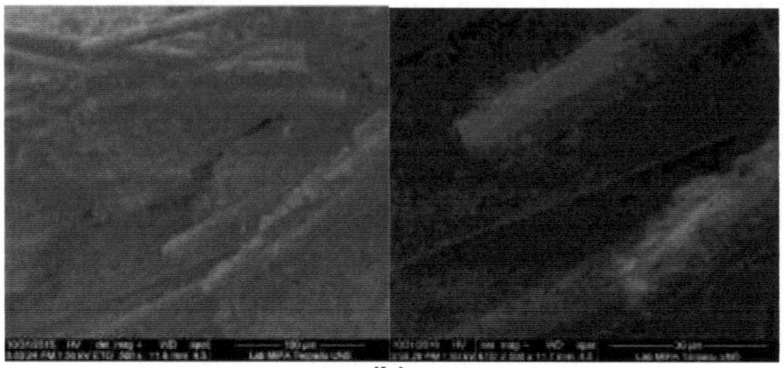

(b)

Figure 5. Scanning Eletron Microscopy analysis of brake pads composite microstructure of Formula 1 (a) and Formula 2 (b)

Scanning electron microscopy analysis showed in figure 4 for determining morphological appearance of composite shows excellent blend in microstructure scale of fibre glass and filler materials in each formula. It proves that the addition of CNSL resin improves physical bonding among all of composite components.

4. Conclusion

Cashew nut shell liquid (CNSL) have been used as a substituent of phenolic resin. The CNSL compound of cardanol, acid cardanol and cardol can act as a binder in the brake pads composite preparation due to the functional group in their compound. Infrared test showed the presence of hydroxyl groups belong Cardanol. The CNSL has high stability in temperature under 400 ºC, proved by the small decrease of its mass in the temperature range of 0 - 350 ºC. The addition of CNSL resin also improve mechanical properties of brake pads composites proven by higher tensile strenght, elongation and excellent pressure strength of formula 2, 900 N, 2,59 MPa and 20.000 N respectively.

Acknowledgement(s)

The researchers wish to express their gratitude to the Penelitian Unggulan UNS Program (PU UNS)) and Program Penelitian Teknologi untuk Industri (PPTI), Kemenristek dikti of Indonesia, for supporting this research and to the Integrated Mathematics and Natural Science Laboratory of Sebelas Maret University for supporting and providing the facilities for this research.

References

[1] Balau P J US 2001 *Departement of Energy* United State of America.
[2] Gandhi T S, Dholakiya B Z and Patel M L 2013 *Polish J. of Chem. Tech.* ***15(4)*** 24 — 27
[3] Kubo I, Ochi M, Vieira P C and Komatsu S 1993 *J. Agric. Food Chem.* **41** 1012-1015.
[4] Dallon J L 2004 *Approved non VOC Epoxy Protective Coating for Pipe and Storage Tank in Direct Contact with Potable Water Based On 100% Solid Fast Curing Phenalkamines.* Cardolite Corporation New Orleans USA.
[5] Njuku F W, Mwangi P M And Thiong'o G T 2014 *Int. J. Of Adv. Research* **2(3)** 928-941
[6] Knop A and Scheib W *J. Agric. Food Chem.* **41** 1012-1015

IOP Conf. Series: Materials Science and Engineering **176** (2017) 012052 doi:10.1088/1757-899X/176/1/012052

Preparation and Characterization of Cellulose and Nanocellulose from Agro-industrial Waste - Cassava Peel

S Widiarto[1,2], S D Yuwono[2], A Rochliadi[1] and I M Arcana[1]

[1]Inorganic and Physical Chemistry Research Division, Institut Teknologi Bandung, Jl. Ganesha 10, Bandung 40132-Indonesia

[2]Department of Chemistry, Lampung University, Jl. Prof Dr S Brodjonegoro 1, Bandar Lampung 35145-Indonesia

E-mail: sonny@fmipa.unila.ac.id, arcana@chem.itb.ac.id

Abstract. Cassava peel is an agro-industrial waste which is available in huge quantities in Lampung Province of Indonesia. This work was conducted to evaluate the potential of cassava peel as a source of cellulose and nanocellulose. Cellulose was extracted from cassava peel by using different chemical treatment, and the nanocellulose was prepared by hydrolysis with the use of sulfuric acid. The best methods of cellulose extraction from cassava peels are using alkali treatment followed by a bleaching process. The cellulose yield from this methods was 17.8% of dry base cassava peel, while the yield from nitric and sulfuric methods were about 10.78% and 10.32% of dry base cassava peel respectively. The hydrolysis was performed at the temperature of 50 °C for 2 hours. The intermediate reaction product obtained after each stage of the treatments was characterized. Fourier transform infrared spectroscopy showed the removal of non-cellulosic constituent. X-ray Diffraction (XRD) analysis revealed that the crystallinity of cellulose increased after hydrolysis. Morphological investigation was performed using Scanning Electron Microscopy (SEM). The size of particle was confirmed by Particle Size Analyzer (PSA) and Transmission Electron Microscopy (TEM).

1. Introduction

Recently, there is an increasing tendency towards more useful utilization of agro-industrial waste, such as sugarcane bagasse, cassava bagasse, and cassava peel, as raw materials for industrial use [1.2]. Cassava (Manihot esculenta Cranz) is the fifth most abundant starch crop produced in the world, the third most important food source for inhabitants in the tropical areas [3] and Indonesia is the third most cassava producing country in the world with around 24.5 million tonnes production in 2015 [4]. Cassava bagasse and cassava peel produced in large quantities by industrial tapioca starch. The bagasse is usually used as animal feed whereas cassava peel is very rarely used and wasted. The peels of cassava contribute to 15% of the tubular cassava root weight and the abundant production of this crop has resulted in the yield of a huge amount of peels. Cellulose content in cassava peel is quite promising at around 40-55% of dry base peel [5,6]. Therefore, the cassava peel has a potential as a source of cellulose.

Cellulose is one of the most abundant, renewable and biodegradable polymers [3]. Cellulose is a natural polymer which insoluble in water because of its long chain and high molecular weight (more than 500.000 Da). Cellulose in nanometers dimension has a range size of 10 nm to 350 nm. Nanocellulose are nanoparticles having outstanding mechanical properties can be used to enhance the mechanical properties of polymer nanocomposites [7].

Among the various methods for preparing nanocellulose, the acid hydrolysis is the most distinguished and broadly used [8]. This method is convenient and fast to generate nanocellulose with a preferable property. This process breaks the irregular and amorphous parts of cellulose, removing single and well-defined crystal. The important parameters that affect the properties of nanocellulose are the temperature, time of reaction, acid nature and ratio of the acid to cellulose. There are several stages in acid hydrolysis procedure i.e. strong acid hydrolysis of cellulose under controlled conditions; addition of water to stop hydrolysis process; centrifugation; dialysis with cellulosic membrane; ultrasonication and drying of the suspension [9,10].

The performance of nanocellulose as a reinforcing agent is dependent on the properties of cellulose and the properties of cellulose is depend mainly on source of original cellulose. Therefore, the development of nanocellulose from various source of cellulose is relevant. Thus, the purpose of this study was conducted to examine the potential of an agro-industrial waste i.e. cassava by isolating cellulose from its peel and preparing nanocellulose and then followed by determining its physicochemical and structural properties.

2. Experimental

2.1. Material

Cassava peel (CP) was collected from cassava crisp home industry located in Bandar Lampung, Lampung Province, Indonesia. The chemical used for cellulose isolation and nanocellulose preparation were: sodium hypochlorite, sodium hydroxide, sulfuric acid, nitric acid, sodium chlorite, sodium nitric, sodium sulfite, pure microcrystalline cellulose. All chemicals were pro analysis grade.

2.2. Methods

2.2.1. Isolation of cellulose. The isolation of cellulose from cassava peel was obtained by three different methods. Prior to cellulose isolation, cassava peel was separated from its outer skin, washed and milled into pulp. This pulp was then dried in sunlight for 2 days until reach a constant weight. *(Nitric methods)* 10 g dried pulp was dissolved in 200 mL of 3.5% HNO_3 and added 1.5 mg $NaNO_2$. Sample mixture is heated at 90 °C for 2 h with stirring. Then the mixture was filtered and the obtained residue was washed until neutral. The residue was refluxed with a mixture of 150 mL 2% NaOH and 2% $NaSO_3$ (1:1) solution for 2 hours. And then, the mixture was filtered and washed until the filtrate of obtained residue become neutral. The residue was refluxed use 2% $NaClO_2$ solution with a few drops of glacial acetic acid for 30 minutes. *(Sulphuric methods)* 10 g of dried pulp was dissolved in 200 mL of 0.5M H_2SO_4 by heating at 90 °C for 2 h with stirring. Subsequently the mixture is filtered and the obtained residue is washed until neutral. The next step is a bleaching process wherein the residue refluxed use 2% $NaClO_2$ solution plus a few drops of glacial acetic acid for 30 minutes. *(Alkali methods)* 10 g of dried pulp was refluxed with 200 mL of 4% sodium hydroxide for 2 hours with constant stirring at 90 °C. The mixture was bleached with 100 mL of 4% (w/v) sodium hypochlorite for 1 hour with constant stirring at 80 °C. The residue was washed with distilled water until a neutral pH, and it was dried at room temperature for 2-3 days.

2.2.2. Preparation of nanocellulose. Isolated cellulose from cassava peel was hydrolyzed with sulfuric acid with a ratio of cellulose to sulfuric acid 1:25. The hydrolysis of cellulose was carried out in 45% (w/w) sulfuric acid solution at 50 °C for 2 hours. The hydrolysis process was quenched by adding excess distilled water (250 mL) to the reaction mixture. The produced colloidal suspension was centrifuged at 14,000 rpm for 10 minutes at -4 °C. Then, it was dialyzed for 5 days to neutralize and eliminate the sulfate ions. The neutral colloidal suspension was sonicated for 30 minutes to homogenize the generated nanocellulose.

2.2.3. Cellulose and nanocellulose characterization. The purity of cellulose in isolated cellulose was determined by TAPPI 203 cm-99 methods. The morphology of untreated cassava peel, CP cellulose, and CP nanocellulose were characterized by SEM Zeiss MA 10. All samples were coated with gold prior to examination. The FTIR spectra of cellulose and nanocellulose were recorded on Fourier Transform Infrared Spectrometer Shimadzu Prestige 21. The samples were blended with KBr powder and then pressed to become thin pellets. The sample was measured in the wavelength range from 4000 cm^{-1} to 500 cm^{-1}. The particle size of nanocellulose was determined by Particle Size Analyzer (PSA) with a Beckman Coulter Delsa™ Nano. The nanocellulose was diluted with distilled water and inserted into the cuvette. The measurements were done at room temperature with the measurement range from 10 nm until 4000 nm. The morphology measurements of the nanocellulose colloids were carried out on a Transmission Electron Micrograph (TEM), JEOL JEM 1400. The X-ray diffraction data were collected using Bruker D8 Advance. The X-ray source is a 2.2 kW Cu anode long fine focus ceramic X-ray tube. The running conditions for the X-Ray tube are 40 kV and 40 mA.

3. Result and Discussion

3.1. Cellulose isolation

Three methods to isolate the cellulose from cassava peel have been tested where the result can be seen in Table 1. It is shown clearly that the alkali method is the best among the three methods tested.

Table 1. Yield percentage and cellulose purity from different methods

Methods	Yield, %	Cellulose, %
Nitric	10.78	92.83
sulphuric	10.32	90.87
Alkali	17.80	93.24

Fig. 1a shows the physical appearance of the CP raw material before any processing. After isolation of cellulose from the raw material, the result was a yellowish white cotton-like material (Fig. 1b). The purity of cellulose obtained by isolation of the cassava peel was fairly high with 90-94% of cellulose. The alkali method to produce cellulose is far more efficient than the two other methods. This is due to the minimum using of chemical in this method and at the other side; many small particles were discarded during washing step after strong acid hydrolysis at nitric and sulphuric methods.

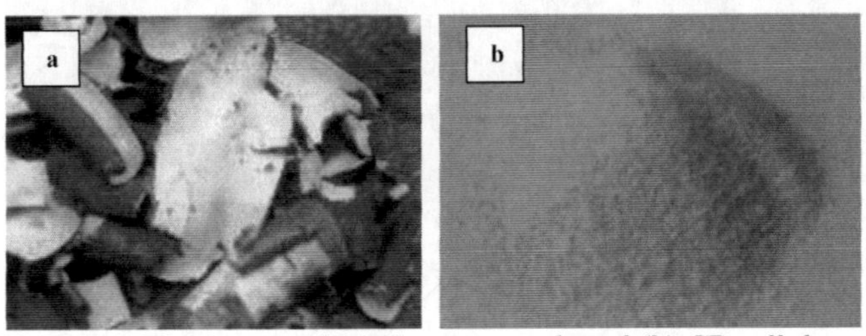

Figure 1. The image of (a) Cassava peel, and (b) CP cellulose,

3.2. FTIR Analysis

FTIR spectroscopy was used to confirm that the lignin and hemicellulose have been removed during cellulose isolation step through analysis of its functional groups. The FTIR spectra of untreated cassava peel and cassava peel cellulose are displayed in Fig. 2.

Based on FTIR spectrum, there are two peaks in untreated cassava peel which is not found in the spectrum of its cellulose. The peaks are 1502 cm^{-1}, and 1738 cm^{-1}. The peak at 1502 cm^{-1} is attributed to the C=C stretching vibration from the aromatic ring of lignin and the peak at around 1738 cm^{-1} is the C=O stretching vibration of carboxylic groups of hemicellulose and lignin [3,11,12].

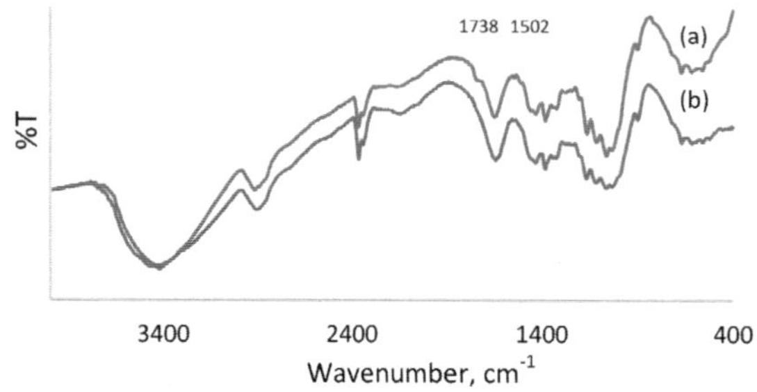

Figure 2. FTIR spectra untreated (a) cassava peel and (b) CP cellulose

3.3. *Preparation of nanocellulose*

Fig. 3 displays images of nanocellulose produced from CP isolated cellulose in the form of suspension colloid, gel and powder. The transparent gel was obtained after cold centrifugation at 14,000 rpm. The gel was freeze-dried for 24 hours to produce a white nanocellulose powder.

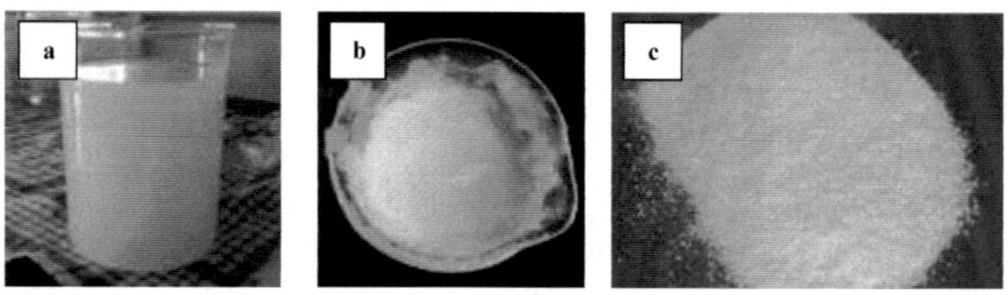

Figure 3. The image of nanocellulose in a form of (a) colloid, (b) gel and (c) powder

3.4. *Morphology study*

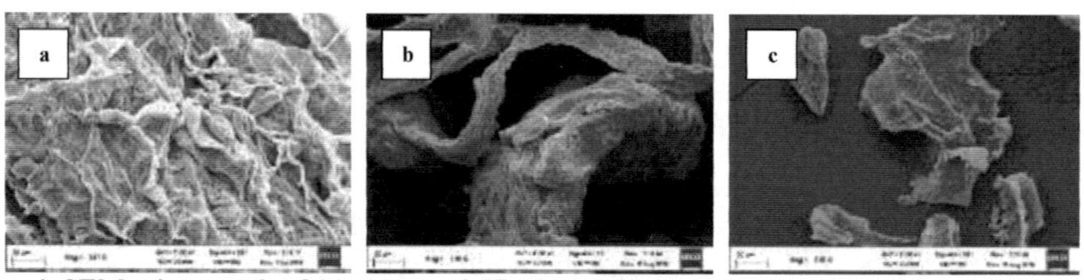

Figure 4. SEM micrograph of (a) untreated CP, (b) CP isolated cellulose and (c) CP nanocellulose

Scanning electron microscopy (SEM) was used to investigate the morphology of cassava peel, CP cellulose, and CP nanocellulose. The change of fiber surface after the isolation of cellulose was shown in Fig. 4 (a-b). It is shown that the granules of starch in untreated cassava peel (Fig. 4a) were disappeared after chemical treatment. Dried untreated cassava peel contain around 60-70% starch and 20-30% others impurities which were eliminated during cellulose isolation processes.

3.5. *Particle size distribution*

The size of nanoparticle was measured by Particle Size Analyzer with the mostly particle size around 150 nm (Fig. 5). However, the TEM images of nanocellulose confirm the size of particle was below 150 nm. The size varies from 100-300 nm in length and 3-8 nm in width (Fig. 6).

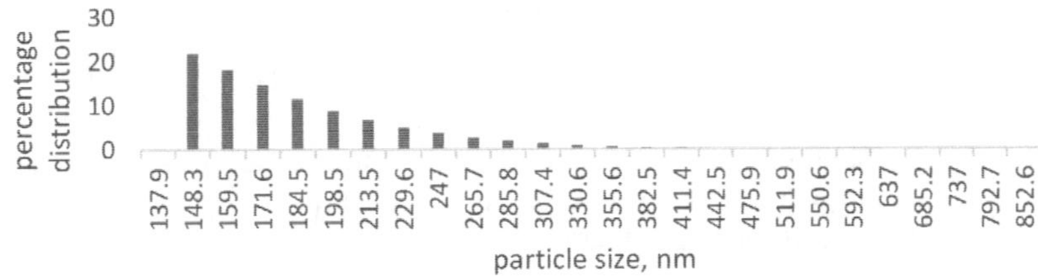

Figure 5. Particle size distribution of CP nanocellulose

3.6. Transmission Electron Microscopy (TEM)

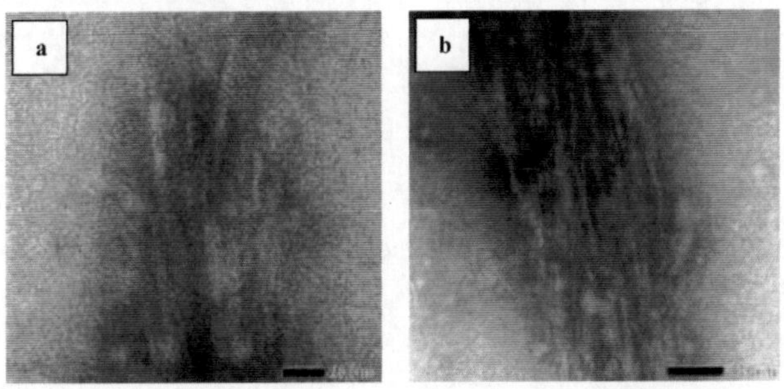

Figure 6. TEM images of CP nanocellulose in (a) 150,000 and (b) 80,000 times magnification

Preparation of nanocelulose from cassava peel was succesful as confirmed by obviously a needle-like (whiskers) nanoparticles [13,14]. Fig. 6a and 6b report TEM micrographs of the nanocellulose in 150,000 and 80,000 times magnification, respectively. The images show individual and aggregate nanocrystals. The appearance of aggregate nanoparticles is due to strong hydrogen bonding established between the whiskers [13]. It is also confirmed by the SEM image of the agglomerate particles (Fig. 4c).

3.7. Crystallinity

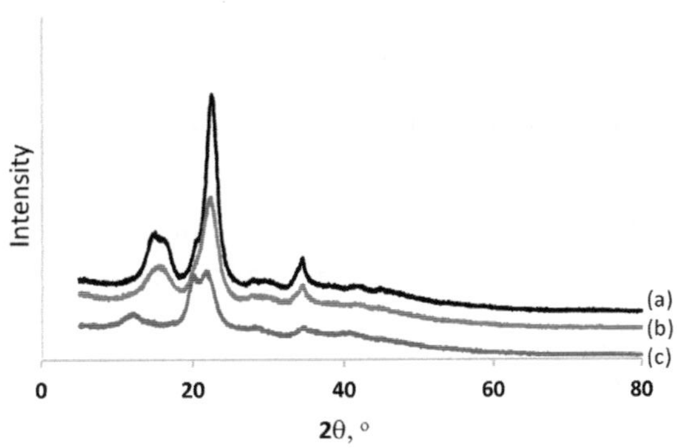

Figure 7. XRD of (a) pure MMC, (b) CP nanocellulose and (c) CP isolated cellulose

The crystallinity of isolated cellulose, nanocellulose and pure microcrystalline cellulose (MCC) were analyzed by X-ray diffraction (XRD). The XRD patterns show that there was a change in crystallinity type of cellulose before and after acid hydrolysis. Two major 2θ value of around 20° and 22° of CP isolated cellulose (Fig. 7c) are characteristic of cellulose II type crystalline weather the major intensity peak at around 22.6^o of CP nanocellulose (Fig. 7b) is related to cellulose I crystalline structure [15]. However, the XRD diffractogram of CP nanocellulose is very similar with pure MCC (Fig. 7a) and this is as evidence of the purity of the obtained nanocellulose.

Table 2 shows the crystallinity index and major XRD diffractogram peak of the samples. There was an increase of crystallinity index of CP isolated cellulose after acid hydrolysis due to the removal of remaining amorphous hemicellulose and lignin.

Table 2. Major peak and crystallinity index of CP isolated cellulose and nanocelulose

Samples	Major intensity (2θ), °	Cellulose type	Crystallinity index, %
CP isolated cellulose	20.2; 21.9; 34.7	II	50.5
CP nanocellulose	15.4; 22.3; 27.8; 34.4	I	51.2
Pure Microcrystalline cellulose	14.9; 16.0; 22.5; 28.0; 34.5	I	61.9

4. Conclusion

Cellulose can be isolated from the cassava peel excellently via alkali treatment followed by bleaching process. Nanocellulose was successfully obtained by sulfuric acid hydrolysis of the cassava peel isolated cellulose. According to TEM images, it was possible to produce a needle-like nanoparticle with sizes varying from 100-300 nm in length and 3-8 nm in width. Such a value is adequate for application CP nanocellulose as nanofillers in polymer matrices.

Acknowledgement

The authors would like to thank Indonesian Ministry of Research, Technology and Higher Education for scholarship and Hibah Bersaing Grant; UPT LTSIT of Lampung University for technical assistance during centrifugation, freeze-drying and SEM analysis.

References

[1] Wulandari W T, Rochliadi A and Arcana I M 2016 *IOP Conf. Ser.: Mater. Sci. Eng.* **107** *012045*
[2] Teixeira E M, Curvelo A A S, Correa A C, Marconcini J M, Glenn G M and Mattoso L H C 2012 *Ind. Crops Prod.* **37** 1 61-68
[3] Mandal A and Chakrabarty D 2011 *Carbohydr. Polym.* **86** 1291-1299
[4] FAO 2015 *Food Outlook Biannual Report on Global Food Market November 2015*
[5] Daud Z, Awang H, Kassim A S M, Hatta M Z M, and Aripin A M 2014 *Adv. Mat. Res.* **974** 384-388
[6] Versino F, Lopez O V and Garcia M A 2015 *Ind. Crops Prod.* **65** 79–89
[7] Martoïa F, Dumont P J, Orgéas L, Belgacem M N, and Putaux J L 2016 *RSC Adv.* **6** 47258
[8] Sheltami R M, Abdullah I, Ahmad I, Dufresne A and Kargarzadeh H 2012 *Carbohydr. Polym.* **88** 772-779
[9] Brinchi L, Cotana F, Fortunati E and Kenny J M 2013 *Carbohydr. Polym.* **94** 154-169
[10] Dong X M, Revol J F, and Gray D G 1998 *Cellulose* **5** 19-32
[11] Sain M and Panthapulakkal S 2006 *Ind. Crops Prod.* **23** 1-8
[12] Garside P and Wyeth P *Studies in Conservation* **48** 269-275
[13] Neto W P F, Silvérioa H A, Dantas N O, and Pasquini D 2013 *Ind. Crops Prod.* **42** 480–488
[14] Johar N, Ahmad I, and Dufresne A 2012 *Ind. Crops Prod.* **37** 93-99
[15] Zain N F M, Yusop S M, Ahmad I 2012 *J. Nutr. Food Sci.* **5** 334

Preparation of Silanized-Chitosan via Solution-Casting Method: Study on the Mechanical, Water Adsorption and Swelling Area Properties

O A Saputra[1], W A Lestari [1], M Nauqinida[1], W E Prasetyo[1], D S Handayani[1]
[1]Chemistry Department, Faculty of Mathematic and Natural Sciences, Sebelas Maret University Surakarta, Indonesia.

E-mail: oziadisaputra@student.uns.ac.id

Abstract. Silanizing of chitosan has been prepared using the solution-casting method. The silanizing agent is 3-glycydiloxypropyl trimethoxysilane (GPTMS). The compounding of chitosan/GPTMS was conducted for 1 hour at 80°C by varying GPTMS compound (5-20 wt %) and the final product is referred as Cs/GPTMS. The mechanical properties, including tensile strength, elongation and young modulus were tested according to ASTM D882-02. The addition of 15 % of GPTMS was able to improve the tensile strength value up to 55.66%. However, the elongation decreased to 64.89 %. Moreover, the water adsorption and swelling area properties were studied in this research to evaluate the material ability. Both of water adsorption and swelling area properties decreased by the presence of GPTMS in chitosan. The Cs/GPTMS materials to be promising candidate materials for plastic engineering due to their properties meet the requirement.

1. Introduction

Today, plastic technology has interestingly attention worldwide. Most important fields, including packaging, automotive, aerospace, house hold appliance, electronic device, and others are based on plastics [1-3]. Plastic as composites become strong hold in daily life in many application due their properties such as light weight, good mechanical properties, transparency, and low cost [4-6]. Polypropylene, polyethylene, polyvinyl cloride, and others based plastics are obtained from petroleum resources, becomes limited due to their availability in the nature decrease. Moreover, petroleum based plastics are non-biodegradable and most of them end up overburdening on landfill. This can lead to be a pollution and non-environmental friendly [7-8]. Environmental and sustainable issue challenged research and technology to replaced petroleum based plastic with the biodegradable material, like bio-plastic. Lately, the shift toward sustainable production systems has focused attention on bioplastics. Bioplastics are types of plastics that are made from plant sources and/or can be decomposed by microbial activity. It has more environmental friendly properties and their using same as conventional plastics [8].

Production and export of shrimp in Indonesia continues to increase and cause of increasing of waste shrimp shells. Waste of shrimp heads are reaching 35-50% from the total weight of the shrimp

[1] Chemistry Department, Faculty of Mathematic and Natural Sciences, Sebelas Maret University Surakarta, Indonesia.

and have chitin content of 60-70%. Chitin can be processed into chitosan with a yield of 15-20%. Chitosan has biodegradable, renewable and non-toxic properties [9-10]. Thus, chitosan can be one of the promising alternative candidates for bioplastic fabrication due their environmentally friendly and biodegradation properties. In recent years, significant attention has been given to a new class of bioplastic materials represented by chitosan bioplastic as the promising alternative to conventional plastics due to their superior properties. However the main drawback of chitosan films with their combined properties has been restricted due to their inherent water sensitivity and relatively low stiffness and strength, especially in moist environments [11]. Numerous studies have focused on improving the physical properties of biopolymer based films by decreasing the hydrophilicity and improving the mechanical properties. Modification of chitosan to improve the mechanical properties has been carried out by several researchers, such as the addition of filler compound [12], fiber reinforcement [13], crosslinked [14] and other. Beside its poor mechanical properties, chitosan also has a high water uptake. In several applications, such as bio-plastic packaging, this water absorption capability must be reduce without eliminating the biodegradation properties. Consequently, chitosan modification is necessary in order to improve the adhesion between fibre and matrix which yield composites with poor properties. Moreover, for the mechanical properties improvement and decrease the water adsorption as bioplastics which compare to virgin chitosan [26-29].

Silane coupling agent, 3-glycydiloxypropyl trimethoxysilane (GPTMS) have methoxysilane group that able to form a siloxane network [15]. The presences of this network are able to act as cross-linking in the polymer chain and improve the mechanical properties. Saputra et. al. [16] has been reported that the addition of hybrid filler GPTMS against polypropylene/empty fruit bunch fiber (PP/EFB) bio-composites was able to increase the tensile strength and Young's modulus but decrease the elongation while providing the degradable properties of the plastic. It also investigated that the presence of GPTMS can reduce the water uptake of PP/EFB bio-composites. So in this study will be investigated the effect of GPTMS on mechanical properties and water absorption of chitosan membrane as an environmentally friendly bioplastics and behave like conventional plastic.

2. Experimental

3. 1. Materials

Materials which used for the study includes chitosan (Cs) was supplied from PT BIOTECH SURINDO, Indonesia with medical grade, Acetic acid (CH_3COOH) and sodium hydroxide (NaOH) were analytical purity grade and purchase from Merck. 3-glycydiloxypropyl trimethoxysilane (GPTMS) was obtained from Sigma Aldrich. Acetic acid used in this study was prepared in 1.5% of concentration. Sodium hydroxide also prepared in 1 M of concentration.

3. 2. Fabrication of Bio-Plastic Chitosan/GPTMS

About 2 gr of chitosan was dissolved in 1.5% acetic acid (100 mL) using three neck bottom flask. The solution process was performed in room temperature about 1 hour. The addition of acetic acid was slowly until gel-solution formed. Then, temperature was raised to 70 ^{0}C under stirring condition. Amount of GPTMS (0, 5, 10, 15 and 20 %wt) was added and the mixture was stirred for 1 hour. The mixture then poured in stainless steel template and dried at 50 ^{0}C for 24 hours. The bio-plastic Cs/GPTMS washed with 1 M NaOH for acetic acid removal, then washed again several times with distilled water. The bio-plastic dried at room temperature for one to two days. The product was characterized using FTIR, also for raw material (chitosan).

2.3. Mechanical and Swelling Test of Cs/GPTMS Bio-plastic

Mechanical properties of bio-plastic were tested with according to ASTM D882-02 standard method. The mechanical properties which measured in this study are tensile strength (MPa), elongation (%) and Young's Modulus (MPa). The applying cross head speed is 10 mm/min and the thickness of the sample is 4 mm.

IOP Conf. Series: Materials Science and Engineering **176** (2017) 012053 doi:10.1088/1757-899X/176/1/012053

Samples with initial weight (Wo) such as chitosan, Cs/GPTMS (0-20%) immersed in a container of water with a variety of immersion time 0, 24, 48 and 72 hours. Sample weight after soaking was recorded as the final weight (Wi), then determined the percentage of water absorption capacity (% WA) by equation 1.

$$\%WA = \frac{(Wi - Wo)}{Wo} \times 100\% \quad \text{.....(1)}$$

3. Result and Discussion

3.1. Chemical Interaction of Cs/GPTMS Bio-plastic

Manufacturing of Cs/GPTMS bio-plastic has been successfully performed via casting solution method. Chemically, chitosan and GPTMS were able to perform chemical interactions. Since, both chitosan and GPTMS had an active group as their active site for chemically reaction. While, chitosan has hydroxyl and amine group. In other hand, GPTMS has unstable oxirene ring as it active group [15]. The chemical interaction possibility between chitosan and GPTMS was showed in Fig 1, where the opening bond of the oxirene group cause the lone pair electrons of the hydroxy group in chitosan was possible to attract with the carbon atom partial positive of oxirene group. Another possibility may also occur, where the amine group of chitosan can undergo chemically interactions with GPTMS, due nitogen atom in the amine group has a lone-pair electrons that can be donated to a carbon atom partial positive in the GPTMS compound.

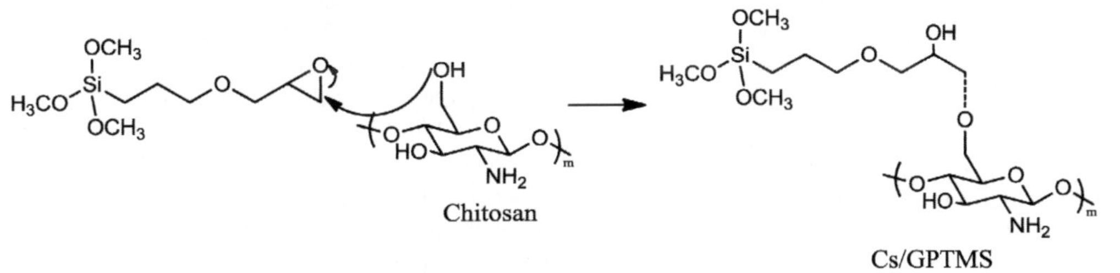

Figure 1. Possible chemical interaction between Chitosan and GPTMS

To proving the existence of chemical interaction between chitosan and GPTMS can be analyzed using FTIR spectroscopy. Fig 2 shows a shift of the $-NH_2$ typical peak from 3080 cm^{-1} and 2886 cm^{-1} to 3395 cm^{-1} and 3141 cm^{-1}, as well as the loss of absorption at 3581 cm^{-1} which is the characteristic band of hydroxy group (OH), was indicate that chitosan was chemically interaction with the other compound, in this case referred as GPTMS. This chemical interaction formed the ether linkage, which is evidenced by the appearance of absorption at 1020 cm^{-1} which characteristic of the ether absorption peak. In addition, a strong absorption at 1087 cm^{-1} which is the typical peak of the Si-O-Si group also shifted to 1050 cm^{-1} in the FTIR spectrum of Cs/GPTMS. Even more interesting is the disappearance of the medium absorption at 910 cm^{-1} which is the characteristic absorption oxirene group, thus strengthening the allegations of chemical interactions between hydroxyl groups of chitosan and oxirene group on GPTMS. Therefore, the loss of the hydroxyl absorption of chitosan and oxirene on GPTMS proved that both these groups interact chemically in accordance with the illustration in Fig 1.

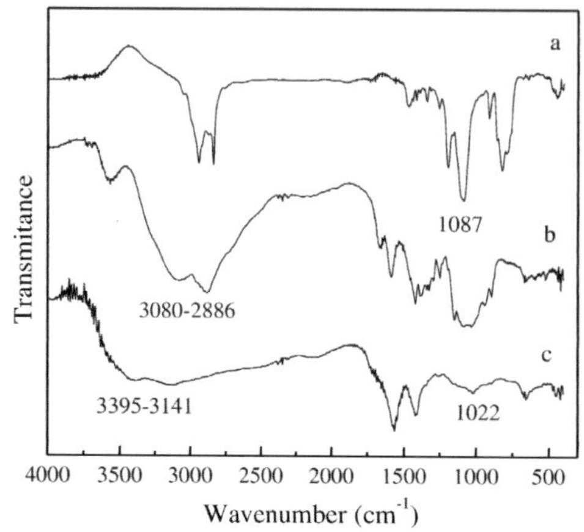

Figure 2. FTIR spectrum of (a) GPTMS, (b) chitosan and (c) Cs/GPTMS

3.2. Mechanical Properties of Cs/GPTMS Bio-plastic

The mechanical properties such as tensile strength (TS) of the Cs/GPTMS bio-plastics can be seen in Fig. 3. Chitosan as the polymer matrix has a tensile strength of 24.5 MPa. The addition of the GPTMS with the lowest (5%) and highest (20%) concentration was turned out to have an impact on the decreasing of tensile strength which compare to virgin chitosan by 14.48% and 6.79%, respectively. It was possible due at low concentrations, distribution of GPTMS was uneven, thus impact on the decreasing TS value. Whereas at high concentrations (20%), the GPTMS was gathering at a certain point and caused this point become brittle. So, in the tensile strength testing, the C4 formula has lower tensile strength than chitosan, which is caused by the ease of sample to fracture. However, the C2 (GPTMS 10%) and C3 (GPTMS 15%) formulas was led to the increase of the TS value, which is 9.95% and 55.66%, respectively. The increasing of the TS value was possible by homogenous distribution of the GPTMS on chitosan matrices. In addition, GPTMS also can formed a crosslink network, thereby increase the mechanical strength of the Cs/GPTMS bio-plastic [14,15]. The significantly increasing of the TS value was obtained in the C3 formula (GPTMS 15%). This allows, the addition of GPTMS can form a good siloxane network, so it has a high tensile strength due to strong chemical interaction between the matrix of chitosan by the presence of a cross-link network of the silane groups [16]. Fig. 4 showed the possibility of siloxane network which acts as crosslinker in the Cs/GPTMS bio-plastics.

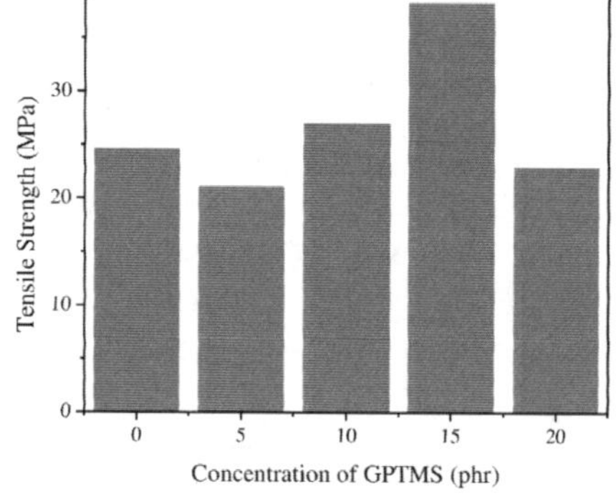

Figure 3. Tensile strength of composites

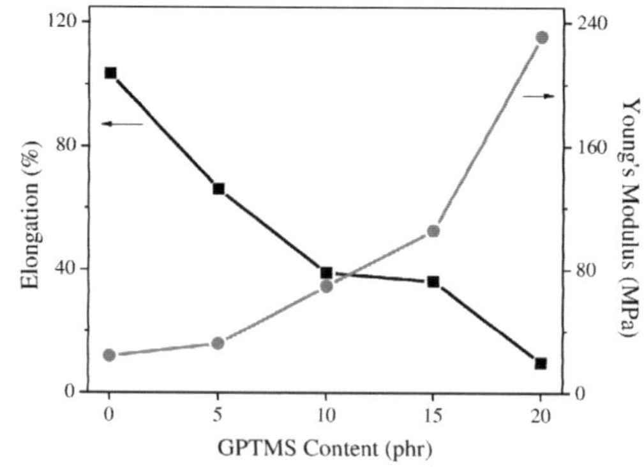

Figure 4. Elongation and YM of composites.

The existence of the siloxane network as a crosslinker turned out to have an impact on the stiffness of the Cs/GPTMS bio-plastics. It can be seen in Fig. 4., that the increasing of GPTMS content in the chitosan matrix reduced significantly of chitosan elongation. Decreasing of elongation or elasticity properties of this material prove that GPTMS able to act as a crosslinker agent causing decreased elasticity of the chitosan due to the decreasing of polymer chain flexibility. The addition of 20% GPTMS was cause to lowering of the elongation percentage by 90.42%. This is because, not only to forming crosslinking, the uneven distribution of GPTMS at high concentration may also conducted at a certain point that can decrease the elasticity of the chitosan.

Stiffness of the Cs/GPTMS bio-plastics increased with increasing of GPTMS content. Addition of the excess amount GPTMS was given the brittle nature of the chitosan matrix, wherein the Young's modulus value of the C4 formula (GPTMS 20%) increased by 873.32% or 231.2 MPa. While the GPTMS 15% (C3) composition was increase to 343.28% or 105.3 Mpa. The increasing of YM value on C3 is not as significant in the C4, it is possible at C3 flexibility of the polymer chain of the chitosan was greater than C4, where the presence of the silane as crosslinker group at C3 is less than the C4 so that the level of rigor on C4 higher than C3.

3. 3. Water Absorption Properties

It is very significant to study in detail the water absorption behavior in order to measure not only the consequences that the water absorbed may have, but also how this water uptake can be minimized in some way. The water absorption testing was intended to determine of bio-degradation properties of the Cs/GPTMS bio-plastics which seen from their ability to absorb water. Water absorption testing on this research was carried out for 4 days, where every 24 hours, the weight after immersion was recorded. The weight increasing of the bio-plastics after soaking is demonstrate of easiness to degrade due high surface polarity. The water adsorption of various bioplastics is described in Fig. 5. Chitosan (C0) has high water adsorption percentage which is 88.9% after saturated condition. This is due to chitosan is hydrophilic polymer, thus the water is easy to adsorbed. However, after modification with 15% GPTMS, the water adsorption decreased to 555.0% at immersing time 4 days (Water adsorption C3= 13.57%). The addition of GPTMS which high concentration was able to decreased the water adsorption. This is due to, the presence of siloxane network in chitosan surface can act as water barrier due their hydrophobic properties.

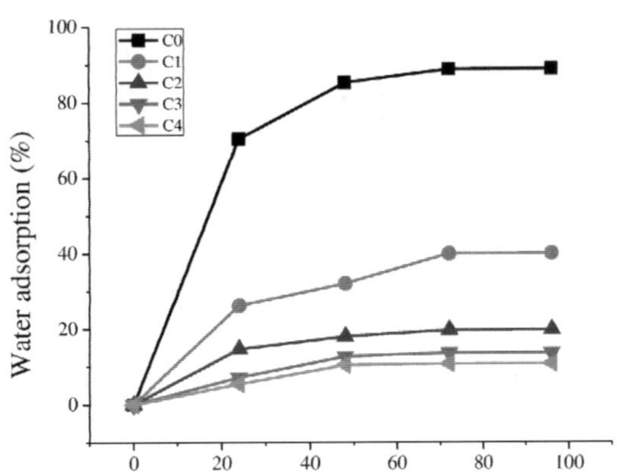

Figure 5. Water adsorption of various bioplastic at different contact time

Chitosan has degradable properties (indicated by the ability of water absorption), non-toxic and environmentally friendly [17,18]. Chitosan also classified as a biocompatible material [19]. Therefore, after water adsorption testing, chitosan without the addition of GPTMS has very high water absorption ability. However, the higher addition of the GPTMS content was able to reduce water absorption ability of chitosan, but not to eliminate the water absorption ability of the material. Although the percentage of water absorption of bio-plastics chitosan filled with GPTMS was decreased, but the Cs/GPTMS still be categorized as bio-degradable material due to have water absorption ability though not significant as chitosan. The low water absorption capabilities of Cs/GPTMS bio-plastics due to the presence of crosslinked between the chitosan, causing the material more dense and the water was difficult to get in on to chitosan matrix.

4. Conclusion

The addition of the GPTMS on chitosan matrices was successfully achieved due both active groups of the raw materials can perform chemically interactions as evidenced by losing of typical absorbance band of the each functional group in Chitosan and GPTMS. The addition of GPTMS has affected on the mechanical properties of the chitosan, while GPTMS with 15% content was significantly improve the tensile strength value, by 55.66% compared to the virgin chitosan. Moreover, the addition of GPTMS also increased the rigidity of the virgin chitosan by the presence of siloxane network as cross-linker. The existence of this cross-linker also decreased the water absorption properties or bio-degradation ability. Nonetheless, Cs/GPTMS bio-plastics still categorized as bio-degradable material because it still has the water absorption ability but not significant like virgin chitosan.

Acknowledgement

Authors would like to acknowledge to Indonesian Directoral General of Higher Education for financial support from Program Kreatifitas Mahasiswa (PKM) 2015.

References

[1] Jeyanthi S, Rani J J 2012 *Int. J. Physc. Sci.* **7** 5765
[2] Hufenbach W, Bohm R, Thieme M, Winkler A, Mader E, Rausch J, Schade M 2011 *Mater. Des*, **32** 1468
[3] Tall S, Albertsson A C, Karlsson S 2001 *Polym. Adv. Technol* **12** 279
[4] Garces J M, Moll D J, Bicerano J, Fibiger R, Mcleod D G 2000 *Adv. Mater* **12** 1835
[5] Friedrich K, Almajid A A 2013 *Appl. Compos. Mater* **20** 107
[6] Xiao K Q, Zhang L C, Zarudi I 2007 *Compos. Sci. Technol* **67** 177
[7] Suharty N S, Ismail H, Dihardjo K, Nizam M, Firdaus M 2014 *Adv. Mater. Res* **950** 18
[8] Storz H, Vorlop K D 2013 *App. Agric. Forestr. Res* **63** 321
[9] Pal P, Banat F 2014 *J. Nat. Gas Sci Eng* **18** 227
[10] Chao A C, 2008 *J. Membrane Sci* **311** 306
[11] Prashanth K V, Harish K L, Shamalab T R, Tharanathan R N 2005 Int. Biodeter. Biodegr **56** 115
[12] Spinks G M, Shin S R, Wallace G G, Whitten P G, Sun I Kim S I, Kim S 2006 *J Sensor. Actuat. B-Chem* **115** 678
[13] Albanna M Z, Bou-Akl T H, Walters III H L, Matthew H W T 2012 *J. Mech. Behav. Biomed* **5** 171
[14] Lopez C R, Bodmeier R 1997 *J. Control. Release* **44** 215
[15] Ren L, Tsuru K, Hayakawa S, Osaka A 2002 *Biomater* **23** 4765
[16] Saputra O A, Prasetyo W E, Pramono E, Handayani D S, Kurnia, Pujiasih S, Arifagama I 2015 *Prosiding Seminar Nasional Kimia* 181
[17] Kim J H, Jegal J, Lee K H J 2003 *Membrane Sci* **213** 273
[18] Kumar M N V R, Muzzarelli R A A, Muzzarelli C, Sashiwa H, Domb A 2004 *J. Chem. Rev.* **104** 6017
[19] Jameela S R, Jayakrishnan A 1995 *Biomater* **16** 769

Effect of cassava starch-based edible coating incorporated with lemongrass essential oil on the quality of papaya MJ9

D Praseptiangga, R Utami, L U Khasanah, I P Evirananda, and Kawiji
Department of Food Science and Technology, Sebelas Maret University (UNS), Jl. Ir. Sutami 36 A Kentingan 57126, Surakarta, Indonesia

E-mail : praseptiangga_de@yahoo.com

Abstract. Edible films and coatings have emerged as an alternative packaging in food applications and have received much attention due to their advantages. The incorporation of essential oils in film matrices to give antimicrobial properties had been observed recently, and could be used as promising preservation technology. In this study, cassava starch-based edible coating incorporated with lemongrass essential oil (1%) was applied by spraying and dipping methods to preserve papaya MJ9 during storage at room temperature. The quality of papaya MJ9 was analyzed based on its physicochemical and microbiological properties. The addition of lemongrass essential oil (1%) significantly inhibited the microbial growth on papaya MJ9 by reducing the value of total yeast and mold as compared to the control. This study also showed that for parameters of weight loss, total soluble solid, vitamin C, and total titratable acid, papaya MJ9 with cassava starch-based edible coating incorporated with lemongrass essential oil (1%) had the lower values than control, however, they had the higher value than control on firmness parameter. These results indicate that cassava starch-based edible coating incorporated with lemongrass essential oil (1%) can be used as an alternative preservation for papaya MJ9.

1. Introduction

Tropical fruits have been received much attention on international trade in recent years, because of consumer preferences for experiencing different, exotic, and diversified products [1]. Papaya is one tropical fruits which occupies a prominent place among the most important tropical fruits [2]. Indonesia as one of tropical countries has been producing papaya 840,000 tons in 2014, while Central Java province has been producing 105,624 tons in 2014 [3]. Papaya MJ9 is one of papaya varieties that cultivated in Indonesia especially by farmers' groups in Boyolali, Central Java province, Indonesia. The importance of papaya from the economy perspectives is shown by its substantial production. However, papaya is a perishable fruits and postharvest diseases caused by microorganisms, particularly fungi, reduce the shelf-life and market value of papaya [2].

Among the available preservative treatments to control fungi decay, edible coatings has been remarked to be used to prolong the shelf-life and prevent the quality loss of papaya. Edible coatings is a method to maintain the original appearance of products, inhibit the release of gas, water vapor, and avoid directly contact with oxygen by providing a thin layer that is safe to be eaten and environmentally friendly [4] [5]. The materials used for edible coating could be proteins, lipids, and polysaccharides that can be utilized individually or as mixed composite blends [6] [7]. Several studies have shown that edible coatings made of different coating materials could maintain quality and prolong the shelf-life of fruit products such as fresh-cut pineapple [8] and apples [9]. Moreover, cassava starch is one of potential edible coatings materials due to its properties, including isotropic, odorless, tasteless, colorless, non-toxic, biologically degradable, have good flexibility, and low water permeability [10].

Edible coatings incorporated with essential oils also have been used in food preservation since addition of some active substances have been useful to improve edible coatings properties and enhance the protective purposes [11]. Cinnamon oil incorporated on chitosan coating improved the storage quality of frozen rainbow trout [12]. Essential oils of *Alpinia purpurata* and *Kaempferia rotunda* enrichment on cassava starch-based edible coating could maintain refrigerated patin fillets for several days [11] [13] [14]. However, the addition of lemongrass essential oil on cassava starch-based edible coatings have limited found to extend papaya fruits shelf-life. Thus, this study aimed at evaluating the effect of cassava starch-based edible coating incorporated with 1% lemongrass essential oil on the quality of papaya MJ9.

2. Experimental

Papaya MJ9 were obtained from farmers' groups in Mojosongo, Boyolali (Central Java province, Indonesia) and the characteristics of them were as follows, weight range of 950-1250 g, clean, with no diseases and bruises, and the homogeneity of maturity level that were shown by orange stripe on the surface of papaya. Essential oil of lemongrass was produced by water-vapor distillation for 4 hours [15]. Prior to distillated, lemongrass (purchased from local market) was sliced (10 mm) and air dried. Cassava starch "Rose Brand" was obtained from local market and other supporting materials, including chemicals for analysis were supplied by local distributor and were analytical grade.

The preparation of cassava starch-based edible coating solution followed previous procedure [14]. Briefly, coating formula were 5 g cassava starch, 150 ml of distilled water, and 2 ml glycerol. 1% (v/v) lemongrass essential oil and 0.1 % (v/v) tween 80 were mixed after the last heating of the solution. Edible coating solution was applied to papaya by two methods (dipping and spraying methods). For the dipping method, coating solution was put into boxes, furthermore, papaya MJ9 was dipped into a solution of edible coating for 2 minutes, and drained, while for the spraying method, a solution of edible coating (± 40 ml) was sprayed on papaya MJ9 with a pressure of 3-4 bar by sprayer gun and compressor [16]. Samples were then dried with a hair dryer to dry the coating layer and kept at room temperature for examining the quality parameters of papaya during storage. Samples were analyzed microbiologically and chemically, including weight loss analysis [9], firmness analysis [9], pH analysis [17], vitamin C analysis [18], Total Soluble Solids analysis (TSS) [9], Total Titratable Acid analysis (TTA) [18], and Total Molds and Yeasts analysis [19] at 0, 3, 6, 9, and 12 days of storage at room temperature. All samples and all analysis were made in duplicates. A completely randomized design (CRD) with one factor was used in this study. There were five treatment variations, F1 (control, without coating), F2 (cassava starch-based edible coating, dipping method), F3 (cassava starch-based edible coating, spraying method), F4 (cassava starch-based edible coating + 1% lemongrass oil, dipping method), and F5 (cassava starch-based edible coating + 1% lemongrass oil, spraying method). The data were then statistically analyzed by one way analysis of variance (ANOVA) at 0.05 significance level and differences in the mean values were determined with Duncan's (DMRT) test ($p < 0.05$) to determine the differences among concentration treatment and differences among storage time at room temperature. SPSS Statistics 16 program was used for statistical analysis.

3. Results and discussion

3.1. Weight loss

Weight loss of all formulas were increased significantly ($p <0.05$) in line with the storage time at room temperature, however edible coating treatment could inhibit the occurrence of weight loss on comparison to the control (Table 1). Polysaccharides-based edible coating acts as a permeable membrane that is selective for the exchange of O_2 and CO_2 gases, therefore can reduce the level of respiration in fruits and vegetables [20]. Lemongrass essential oil acted as hydrophobic component in the layer edible coating, so that the weight loss of papaya MJ9 on the treatment of F4 and F5 are smaller (p <0.05) than the treatment of F2 and F3, and control.

Table 1. Weight loss of papaya MJ9 during storage (%)

Formula	Storage (Days)				
	0	3	6	9	12
F1	0.00^{Aa}±0.000	4.51^{Ab}±0.132	14.78^{Cc}±0.078	25.69^{Dd}±0.049	31.48^{Ce}±0.651
F2	0.00^{Aa}±0.000	3.44^{Bb}±0.417	12.21^{ABc}±0.431	18.04^{BCd}±0.240	24.72^{Be}±0.813
F3	0.00^{Aa}±0.000	4.27^{Bb}±0.011	13.22^{Bc}±0.205	19.14^{Cd}±0.368	25.28^{Be}±2.164
F4	0.00^{Aa}±0.000	4.23^{Bb}±0.176	11.16^{Ac}±0.545	14.28^{Ad}±0.665	18.08^{Ae}±0.205
F5	0.00^{Aa}±0.000	4.23^{Bb}±0.099	11.32^{Ac}±00.587	16.64^{Bd}±0.997	20.31^{Ae}±0.057

$^{A\text{-}E}$ Mean with different letter within each column are significantly different (P < 0.05)
$^{a\text{-}e}$ Mean with different letter within each line are significantly different (P < 0.05)

3.2. Firmness

Table 2 shows that during the 12 days of storage at room temperature, the firmness of papaya decreased significantly (p<0.05), however coatings treatment can maintain the firmness of papaya MJ9 compared to control, as reported in the study of strawberry [21]. Other study showed that fruit without treatment (or control) was less able to withstand the metabolic processes occur during storage, thus the reform process protopectin into soluble pectin would become faster, [22] this means that edible coating can prevent softening of the fruit [23].

It has been reported that the firmness of papaya treated with gum arabic coating incorporated with 0.05% lemongrass essential oil could maintain the firmness better value compared to the control and treatment of gum arabic coating alone (p <0.05) [19]. For firmness parameter, coating methods (dipping and spraying) did not show any significant difference.

Table 2. Firmness of papaya MJ9 during storage (N)

Formula	Storage (Days)				
	0	3	6	9	12
F1	38.97^{Ac}±1.598	31.68^{Ab}±3.274	27.62^{Aab}±3.387	24.84^{Aa}±0.297	21.74^{Aa}±0.169
F2	38.87^{Ac}±2.079	31.72^{Ab}±2.679	30.69^{Ab}±1.689	25.29^{ABa}±0.283	22.99^{Ba}±0.289
F3	37.63^{Ab}±4.441	35.33^{Ab}±2.199	28.76^{Aa}±2.136	25.23^{ABa}±0.085	23.27^{Ba}±0.021
F4	38.41^{Ac}±0.806	33.18^{Ab}±2.956	29.15^{Aab}±2.758	27.87^{Ba}±1.329	25.52^{Ca}±0.028
F5	37.57^{Ad}±0.177	33.95^{Acd}±3.253	30.08^{Abc}±1.039	26.98^{ABab}±2.029	25.02^{Ca}±0.417

$^{A\text{-}E}$ Mean with different letter within each column are significantly different (P < 0.05)
$^{a\text{-}e}$ Mean with different letter within each line are significantly different (P < 0.05)

3.3. Total soluble solid (TSS)

Table 3 shows an increasing rate of TSS during storage at room temperature that is caused by the hydrolysis of starch into simple sugars [24]. Coating treatment either with or without the incorporation of 1% lemongrass oil had a significant effect (p <0.05) on TSS of papaya after 9 days of storage compared with control. It is in line with the other study about papaya treated with gum arabic coating [19]. The results also showed that the incorporation of 1% lemongrass oils had a significant effect on the TSS value of papaya after 12 days of storage, whereas dipping or spraying methods did not show any significant difference for TSS parameter.

Table 3. Total soluble solid (TSS) of papaya MJ9 during storage (°Brix)

Formula	Storage (Days)				
	0	3	6	9	12
F1	$6.88^{Aa}\pm0.459$	$8.40^{Ab}\pm0.424$	$9.60^{Ac}\pm0.283$	$10.38^{Bcd}\pm0.035$	$11.05^{Cd}\pm0.071$
F2	$7.00^{Aa}\pm0.884$	$8.30^{Ab}\pm0.707$	$9.43^{Abc}\pm0.601$	$10.00^{Acd}\pm0.000$	$10.78^{Bd}\pm0.035$
F3	$6.60^{Aa}\pm0.424$	$8.10^{Ab}\pm0.141$	$9.70^{Ac}\pm0.141$	$10.10^{Ac}\pm0.141$	$10.73^{Bd}\pm0.035$
F4	$6.83^{Aa}\pm0.530$	$8.55^{Ab}\pm0.106$	$9.13^{Ab}\pm0.177$	$10.05^{Ac}\pm0.071$	$10.45^{Ac}\pm0.141$
F5	$7.05^{Aa}\pm0.212$	$8.00^{Ab}\pm0.141$	$9.70^{Ac}\pm0.141$	$10.10^{Ad}\pm0.000$	$10.55^{Ae}\pm0.141$

$^{A\text{-}E}$ Mean with different letter within each column are significantly different ($P < 0.05$)
$^{a\text{-}e}$ Mean with different letter within each line are significantly different ($P < 0.05$)

3.4. pH

Table 4 shows the decreasing rate of pH value in all treatments ($p < 0.05$) in line with the storage time at room temperature. The decreasing rate of pH value is possibly because of the production of organic acids, such as malic acid and citric acid during storage as a result of biochemical reactions [25]. Increased levels of total acid is accompanied by a decrease in pH during storage indicating that the organic acid in releasing hydrogen ions contributed on lowering the pH during storage period [26]. Starting on the storage days of 6th, F1 treatment showed the smallest value of pH although it did not show any significant differences to the coating treatment. Further, on the 12th days of storage, coating treatment with the incorporation of 1% lemongrass essential oil could inhibit a decrease in the pH value, while dipping or spraying methods did not show any significant difference for pH parameter.

Table 4. pH of papaya MJ9 during storage

Formula	Storage (Days)				
	0	3	6	9	12
F1	$7.08^{Ac}\pm0.035$	$6.63^{Ac}\pm0.035$	$5.70^{Ab}\pm0.424$	$5.08^{Aa}\pm0.106$	$4.90^{Aa}\pm0.000$
F2	$7.08^{Ad}\pm0.248$	$6.58^{Ac}\pm0.106$	$6.15^{ABb}\pm0.071$	$5.70^{Ba}\pm0.000$	$5.48^{ABa}\pm0.035$
F3	$6.70^{Bd}\pm0.000$	$6.53^{Acd}\pm0.248$	$6.20^{ABc}\pm0.000$	$5.70^{Bb}\pm0.283$	$5.20^{ABa}\pm0.071$
F4	$6.90^{ABd}\pm0.071$	$6.60^{Ac}\pm0.141$	$6.13^{ABb}\pm0.035$	$6.03^{Bab}\pm0.035$	$5.90^{Ba}\pm0.000$
F5	$6.85^{ABb}\pm0.141$	$6.45^{Aab}\pm0.141$	$6.25^{Bab}\pm0.071$	$5.95^{Ba}\pm0.071$	$5.75^{Ba}\pm0.064$

$^{A\text{-}E}$ Mean with different letter within each column are significantly different ($P < 0.05$)
$^{a\text{-}e}$ Mean with different letter within each line are significantly different ($P < 0.05$)

3.5. *Vitamin C*

Table 5 shows that cassava starch-based edible coating treatments (with or without incorporation of 1% lemongrass oil) could inhibit the increase in value of vitamin C in papaya. A papaya ripening is characterized by impairment of firmness, increased dissolved solids and the amount of vitamin C [27]. Edible coatings treatment on the surface of the papaya acts as a barrier to water vapor transmission, so that the respiration process is inhibited and the increasing value of vitamin C is also inhibited. Dipping or spraying methods of coatings did not show any significant difference for this parameter.

Table 5. Vitamin C of papaya MJ9 during storage (%)

Formula	Day				
	0	3	6	9	12
F1	$0.055^{Aa}\pm0.0031$	$0.084^{Bb}\pm0.0000$	$0.097^{Bb}\pm0.0125$	$0.121^{Cc}\pm0.0093$	$0.184^{Bd}\pm0.0078$
F2	$0.053^{Aa}\pm0.0062$	$0.073^{ABab}\pm0.0031$	$0.086^{ABb}\pm0.0093$	$0.097^{ABb}\pm0.0031$	$0.123^{Ac}\pm0.0187$
F3	$0.046^{Aa}\pm0.0000$	$0.059^{Aab}\pm0.0031$	$0.067^{Aab}\pm0.0078$	$0.079^{Ab}\pm0.0062$	$0.125^{Ac}\pm0.0156$
F4	$0.046^{Aa}\pm0.0031$	$0.064^{Aa}\pm0.0093$	$0.086^{ABb}\pm0.0031$	$0.092^{ABb}\pm0.0000$	$0.133^{Ac}\pm0.0140$
F5	$0.055^{Aa}\pm0.0031$	$0.062^{Aa}\pm0.0062$	$0.086^{ABb}\pm0.0093$	$0.101^{Bb}\pm0.0125$	$0.143^{Ac}\pm0.0093$

$^{A\text{-}E}$ Mean with different letter within each column are significantly different ($P < 0.05$)
$^{a\text{-}e}$ Mean with different letter within each line are significantly different ($P < 0.05$)

3.6. Total titratable acid (TTA)

Based on Table 6, the coating treatments could inhibit the increasing value of TTA on papaya during storage at room temperature, while dipping or spraying methods of coatings did not show any significant difference for this parameter. It is in line with other studies that reported an increasing value of TTA in control was greater ($p < 0.05$) than papaya treated with coating of gum arabic 5% and 10% [19]. Other study showed that pectin-based edible coating can maintain the integrity of the cell wall thereby slowing the release of acid and inhibit the increase TTA [26].

Table 6. Total titratable acid (TTA) of papaya MJ9 during storage (%)

Formula	Storage (Days)				
	0	3	6	9	12
F1	$0.29^{ABa}\pm0.057$	$0.25^{Aa}\pm0.000$	$0.49^{Cb}\pm0.000$	$0.53^{Bb}\pm0.000$	$0.70^{Bc}\pm0.057$
F2	$0.21^{Aa}\pm0.064$	$0.25^{Aa}\pm0.000$	$0.35^{ABb}\pm0.028$	$0.37^{Ab}\pm0.000$	$0.49^{Ac}\pm0.000$
F3	$0.21^{Aa}\pm0.064$	$0.25^{Aa}\pm0.000$	$0.29^{Aa}\pm0.057$	$0.33^{Aa}\pm0.000$	$0.54^{Ab}\pm0.064$
F4	$0.37^{Ba}\pm0.057$	$0.37^{Ba}\pm0.057$	$0.41^{BCa}\pm0.000$	$0.49^{Bab}\pm0.057$	$0.56^{Ab}\pm0.354$
F5	$0.25^{ABa}\pm0.000$	$0.25^{Aa}\pm0.000$	$0.29^{Aab}\pm0.057$	$0.33^{Ab}\pm0.000$	$0.49^{Ac}\pm0.000$

$^{A-E}$ Mean with different letter within each column are significantly different ($P < 0.05$)
$^{a-e}$ Mean with different letter within each line are significantly different ($P < 0.05$)

3.7. Yeast and mold

As seen in Table 7, coating treatments without the incorporation of 1% lemongrass oil can suppress the growth of molds and yeasts up to 0,455 log CFU/g after 12 days of storage at room temperature, while the coating treatment with the incorporation of 1% lemongrass oil can suppress the growth of molds and yeasts up to 1.48 log CFU/g. Dipping or spraying methods of coatings did not show any significant difference for this parameter. Other study showed that total mold and yeast of fresh-cut pineapple treated with coating incorporated with 1% peppermint oil (1.11 log CFU/g) was smaller than control and coating without incorporation of essential oil [28].

Table 7. Total yeast and mold of papaya MJ9 during storage (log CFU/g)

Formula	Storage (Days)				
	0	3	6	9	12
F1	$1.35^{Aa}\pm0.071$	$2.76^{Bb}\pm0.078$	$3.77^{Dc}\pm0.078$	$4.24^{Cd}\pm0.120$	$4.99^{Ce}\pm0.169$
F2	$1.24^{Aa}\pm0.085$	$2.64^{Bb}\pm0.219$	$3.26^{Cc}\pm0.057$	$3.89^{Bd}\pm0.177$	$4.44^{Be}\pm0.078$
F3	$1.15^{Aa}\pm0.212$	$2.59^{Bb}\pm0.262$	$2.96^{Bb}\pm0.021$	$3.63^{Bc}\pm0.134$	$4.34^{Bd}\pm0.219$
F4	$1.24^{Aa}\pm0.085$	$2.10^{Ab}\pm0.141$	$2.57^{Ac}\pm0.028$	$3.02^{Ad}\pm0.092$	$3.40^{Ae}\pm0.014$
F5	$1.15^{Aa}\pm0.212$	$2.42^{ABb}\pm0.021$	$2.72^{Ac}\pm0.106$	$3.29^{Ad}\pm0.064$	$3.73^{Ae}\pm0.049$

$^{A-E}$ Mean with different letter within each column are significantly different ($P < 0.05$)
$^{a-e}$ Mean with different letter within each line are significantly different ($P < 0.05$)

4. Conclusion

The incorporation of lemongrass essential oil (1%) on cassava starch-based edible coating significantly inhibited the microbial growth on papaya MJ9 by reducing the value of total yeast and mold, weight loss, total soluble solid, vitamin C, and total titratable acid as compared to the control, however, they had the higher value than control on firmness parameter, suggesting that cassava starch-based edible coating incorporated with lemongrass essential oil (1%) could be used as an alternative preservation for papaya MJ9.

Acknowledgements

This work was financially supported by research project of PUPT Ministry of Research, Technology, and Higher Education of Republic of Indonesia No: 041/SP2H/LT/DRPM/II/2016, Indonesia.

References

[1] Nunes M C N, Emond J P, Brecht J K 2006 *Postharvest Biology and Technology* **41**(3) 328-340
[2] Espitia P J P, Soares N F F, Botti L C M, Melo N R, Pereira O L, Silva W A 2012 *Campinas* **15**(4) 307-316
[3] Taufik Y 2015. Statistik Produksi Hortikultura Tahun 2014, Direktorat Jendral Hortikultura, Kementerian Pertanian, Jakarta
[4] Trisnawati E, Dewi A, Abdullah S 2013 *Jurnal Teknik Kimia* **19**(2) 17-26
[5] Garnida Y 2006 *Infomatek* **8**(4) 207 – 222
[6] Karbowiak T, Debeaufort F, Champion D, Voilley A 2006 *Journal of Colloid and Interface Science* **294** 400-410
[7] Praseptiangga D, Fatmala N, Manuhara G J, Utami R, Khasanah L U 2016 *AIP Conf. Proc.* **1746** 020036
[8] Azarakhsh N, Osman A, Ghazali H M, Tan C P, Adzahan N M 2014 *Postharvest Biology and Technology* **88** 1-7
[9] Jo W S, Song H Y, Song N B, Lee J H, Min S C, Song K B 2014 *LWT - Food Science and Technology* **55** 490-497
[10] Maran J P, Sivakumar V, Sridhar R, Thirugnanasambandham K 2013 *Carbohydrate Polymers* **92** 1335-1347
[11] Utami R, Praseptiangga D, Nurhartadi E, Manuhara G J, Khasanah L U, Azzahra F A, Indrayati F 2016 *Jurnal Teknologi (Sciences and Engineering)* **78**(4-2) 193-198
[12] Ojagh S M, Rezaei M, Razavi S H 2014 *Journal of Aquatic Food Product* **23** 146-154
[13] Utami R, Kawiji, Nurhartadi E, Kurniasih M, Indianto D 2013 *Agritech* **33**(4) 399-406
[14] Utami R, Kawiji, Nurhartadi E, Putra A Y T, Setiawan A 2014 *International Food Research Journal* **21**(1) 413-419
[15] Kawiji, Khasanah L U, Christian A P 2010 *Jurnal Teknologi Hasil Pertanian* **1**(1)
[16] Masruroh H, Fauzi A F, Dwi A S, Paramita V 2013 *Prosiding SNST Ke-4* Fakultas Teknik, Universitas Wahid Hasyim, Semarang
[17] Apriyantono A, Fardias D, Puspitasari N L, Sedamawati, Budiyanto S 1989 *Analisa Pangan Petunjuk Laboratorium* IPB Press, Bogor
[18] Syafutri M I, Pratama F, Saputra D 2006 *Jurnal Teknologi dan Industri Pangan* **17**(1) 1-11
[19] Addai Z R, Abdullah A, Mutalib S A, Musa K H 2013 *International Journal of Chemtech Research* **5**(6) 2854-2862
[20] Winarti C, Miskiyah, Widaningrum 2012 *J. Litbang Pert.* **31**(3) 85-93
[21] Ribeiro C, Vicente A A, Teixeira J A, Miranda C 2007 *Postharvest Biology and Technology* **44** 63-70
[22] Purwoko B S, Fitradesi P 2000 *Buletin Agron.* **28**(2) 66-72
[23] Raghav P K, Agarwal N, Saini M 2016 *International Journal of Scientific Research And Modern Education* **1**(I) 188
[24] Arifiya N, Purwanto Y A, Budiastra I W 2015 *Jurnal Keteknikan Pertanian* **3**(1) 41-48
[25] Vega W R C, Piotrowicz I B B, Prentice C, Borges C D 2014 *International Food Research Journal* **21**(5) 2017-2023
[26] Scalon S D P Q, Oshiro A M, Dresch D M 2012 *Rev. Bras. Frutic. Jaboticabal-SP* **34**(4) 1022-1029
[27] Hernandez Y, Lobo M G, Gonzalez M 2006 *Food Chemistry* **96** 654-664
[28] Bitencourt R G, Possas A M M, Camilloto G P, Cruz R S, Otoni C G, Soares N F F 2014 *Ciência Rural, Santa Maria* **44**(6) 1119-1125

IOP Conf. Series: Materials Science and Engineering **176** (2017) 012055 doi:10.1088/1757-899X/176/1/012055

Preparation of activated carbon from waste plastics polyethylene terephthalate as adsorbent in natural gas storage

Yuliusman[1], Nasruddin[2], A Sanal[1], A Bernama[1], F Haris[1], I T Ramadhan[1]
[1]Chemical Engineering Department, University of Indonesia, Kampus Baru UI Depok, Indonesia
[2]Mechanical Engineering Department, University of Indonesia, Kampus Baru UI Depok, Indonesia

E-mail: usman@che.ui.ac.id; yuliusman@yahoo.com

Abstract. The main problem is the process of natural gas storage and distribution, because in normal conditions of natural gas in the gas phase causes the storage capacity be small and efficient to use. The technology is commonly used Compressed Natural Gas (CNG) and Liquefied Natural Gas (LNG). The weakness of this technology safety level is low because the requirement for high-pressure CNG (250 bar) and LNG requires a low temperature (-161°C). It takes innovation in the storage of natural gas using the technology ANG (Adsorbed Natural Gas) with activated carbon as an adsorbent, causing natural gas can be stored in a low pressure of about 34.5. In this research, preparation of activated carbon using waste plastic polyethylene terephthalate (PET). PET plastic waste is a good raw material for making activated carbon because of its availability and the price is a lot cheaper. Besides plastic PET has the appropriate characteristics as activated carbon raw material required for the storage of natural gas because the material is hard and has a high carbon content of about 62.5% wt. The process of making activated carbon done is carbonized at a temperature of 400 ° C and physical activation using CO_2 gas at a temperature of 975 ° C. The parameters varied in the activation process is the flow rate of carbon dioxide and activation time. The results obtained in the carbonization process yield of 21.47%, while the yield on the activation process by 62%. At the optimum process conditions, the CO_2 flow rate of 200 ml/min and the activation time of 240 minutes, the value % burn off amounted to 86.69% and a surface area of 1591.72 m^2/g.

1. Introduction

Natural gas has the potential of reducing the gasoline consumption of Indonesia due to Indonesia's high reserve amounting to 103.3 trillion cubic feet.One of the problems in the use of natural gas is in the process of storage and distribution, because in normal conditions (pressure 1 atm) natural gas is in its gas phase so the storage capacity will be small henceinefficient to use. Currently, the technology that is generally used is Compressed Natural Gas (CNG) and Liquefied Natural Gas (LNG). However, because CNG requires a high pressure of 250 bar and LNG requires a low temperature (-161°C), it become harmful in their uses. Adsorbed Natural Gas(ANG) technology uses activated carbon as an adsorbent, causing natural gas to be able to be stored at low pressure of about 34.47 bar with large storage capacity [6].

Activated carbon in ANG is made from plastic waste PET (polyethylene terephthalate) due to its abundant availability in Indonesia. The amount of plastic waste PET in Indonesia reached 26,500 tons per day, as stated by the Ministry of Environment of Indonesia. Aside from that, activated carbon made from PET has the greatest surface area compared to activated carbon from waste PVC (polyvinyl chloride) and tires [3]. This study, therefore, aims to make activated carbon from plastic waste PET by physical activation using a variation of the flow rate of CO_2 and to obtain the optimal activation time in terms of creating activated carbon with high surface area.

2. Experimental

2.1. Activated Carbon Preparation

PET plastic waste is cleaned of impurities and then dried. PET plastic waste is crushed and screened to obtain the size of about 1-3 mm. This study has fixed parameters such as temperature and carbonation time, activation temperature and the rate of N_2 gas. After tools and materials preparation then we do the carbonization process to increase the carbon content of the PET plastic waste by releasing flammable substances such as CO_2, CH_4, H_2. Carbonization is carried out at 400°C with the gradual heating rate of 10 ° C / min, for 240 minutes. Then, the experiment proceeded with physical activation using CO_2 gas, at a temperature of 975°C for 60 minutes.

2.2. Characterization of Activated Carbon

Characterization of activated carbon was ensued by iodine number method with the concept of measuring the strength of adsorption of activated carbon in a solution of iodine. Iodine that is not absorbed in the activated carbon will be titrated using Sodium Thiosulphate ($Na_2S_2O_3$) until the solution becomes a pale-yellow color, then the solution is provided with an indicator which will change the color back to black and redo the titration process until the color becomes transparent. The less Sodium Thiosulphate is used, the greater the power of activated carbon adsorption. The equation for calculating the iodine number is as follows.

$$Adsorbed\ Iodine, \frac{mg}{g} = \frac{10 - \frac{V1xN}{0.1}}{W} \times 12.69 \times \frac{V2}{10} \tag{1}$$

where

V1	Use of sodium thiosulphate (ml)
V2	Volume of Iodine Solution (ml)
N	Normality of sodium thiosulphate
12,69	The amount of Iodine needed for a 1 mL sodium thiosulphate 0.1 N
W	Sample weight (gram)

To change the iodine number into surface area a linear regression equation is used that refers to ASTM D-4607-94. The equation is as follows

$$Iod\ Number = 0.6366\, x\ surface\ area + 174.34 \tag{2}$$

In addition to iodine number, this study also used the characterization of SEM + EDX to see the morphology and composition of activated carbon.

3. Results and Discussion

3.1. Activated Carbon Preparation

3.1.1. Carbonization Process

Carbonization process aims to vaporize the compounds that are volatile which will create pores due to the loss of such compounds. Volatile compounds which are in PET plastic can be decomposed at a temperature of 480°C comprising of water content that evaporates on 100°C- 150°C. Terephthalic Acid

(TPA) contained from PET are sublimate so that the TPA molecule will form CO_2, CO, and CH_4 gases [5]. This carbonization process lasted for 210 minutes that resulted in a smoke cloud. This happens because many volatile substances contained in plastic PET evaporated during carbonization. The results of the carbonization process produce % gain on average by 21.47%. These results correspond to the value that is a 17.4% fixed carbon value. This result is said to be good enough for the carbonization process because it can be said that almost all volatile compounds in PET has been lost.

3.1.2. Physical Activation Process

Activated carbon activation process is aimed at the formation of the macro pore structure and microporous activated carbon. In this study, physical activation is done by using carbon dioxide gas (CO_2), which will open the pores in activated carbon. Carbon dioxide gas serves as the activating gas where the gas will react with carbon so as to open the pores which will increase surface area. Aside from that, carbon dioxide gas also acts to prevent the entry of oxygen gas into the activator. A decrease in activated carbon formation can be viewed along with increased gas flow rate; it is in line with the principle of Le Chatelier's where an additional amount of reactant, in this case CO_2, will add to the result of the product and the bigger the amount of gas that reacts with carbon, the bigger the number of pores that will form. After getting a flow rate that produces the largest surface area, the activation process was repeated with variations of activation time using the optimal flow rate that has been obtained previously. It can be seen in Table 1, bigger reduction in the mass of activated carbon corresponds with increased activation time. This indicates that there had been a reaction and the formation of pores in activated carbon. However, the reduction of the resulting mass is less than the reduction in the process of CO_2 gas flow rate variation. This indicates that the reaction and the formation of pores in activated carbon are already approaching their optimum conditions.

Table 1. Impact of time to process yield

Variable	Mass (g)		
Activation Time (minute)	Carbon (g)	Activation Carbon (g)	%Yield
120	10	6.46	64.6
150	10	6.2	62
240	10	6.2	62

3.2. Characterization of Activated Carbon

3.2.1. Burn Off

Burn Off is the value derived from the reduction of the initial mass of raw material to the final mass of activated carbon. Burn off value can be assumed as the formation of pore (volume) in the activated carbon. Calculation of burn off was done by using the following equation.

$$burnoff = \frac{initialmass - finalmass}{initialmass} x\ 100\% \tag{3}$$

The burn off value is in accordance with the number of volatile matter and ash content residing on PET plastic waste with a slight reduction due to small particles of activated carbon that got swept by the flowing gas. The resulting calculation implies that the value of % burn off at the optimum operating conditions, namely the CO_2 flow rate of 200 ml/min and the activation time of 240 minutes, is 86.69%; this corresponds to the amount of volatile matter content and ash from plastic waste PET which is 82.6 % and less than 0.1% [4]. Data from the burn off process: the initial mass of raw materials and the mass of active carbon after activation can be seen in the following table.

Table 2. Impact of time and gas flow rate of CO_2 to process' yield

Gas Flow Rate (ml/minute)	Activation Time (minute)	Raw Material (g)	Activated Carbon (g)	% Yield
100	60	46,57	7,04	15,12
150	60	46,57	6,9	14,82
200	60	46,57	6,37	13,68
200	120	46,57	6,46	13,87
200	180	46,57	6,2	13,31
200	240	46,57	6,2	13,31

3.2.2. Iodine Number Testing

The surface area of activated carbon as an adsorbent is one of the most important parameter. Activated carbon adsorbent can be regarded as a good adsorbent if it has a high surface area. This is because the surface area of the adsorbent is one of the main factors affecting the adsorption process. For the measurement of the surface area of the activated carbon samples, iodine number method and BET method was used. Iodine number testing aims to determine the surface area of activated carbon through absorption of activated carbon to the solution of iodine. The values of iodine number obtained equates to the milligrams of iodine absorption capability per unit gram of activated carbon (mg/g). Iodine number obtained can be converted into the surface area using Equation 2.

Based on the graph in Figure 1, the surface area after physical activation is higher compared with the results of carbonization. The increase in surface area between the results obtained using physical activation with the results of carbonization with optimal flow rate of 200 ml/min is 799.7 mg/g so that it can be concluded that physical activation can magnify the width of the pores so as to increase the surface area. In regards to the variation of flow rate of CO_2, an increase in surface area corresponded from the increase of flow rate of CO_2 that indicates the high amount of CO_2 that reacted and opened the pores. The optimum flow rate is the flow rate with the highest surface area that is at a flow rate of 200 ml/min.

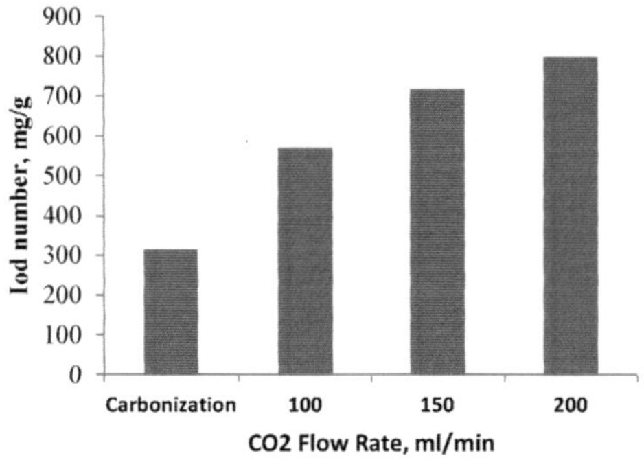

Figure 1. Effect of gas flow rate of CO_2 to surface area

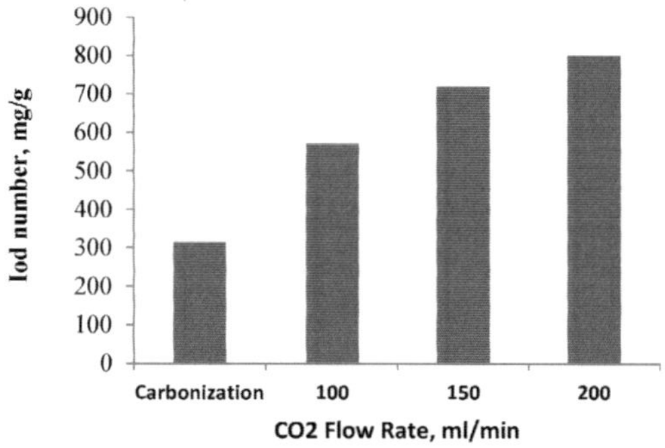

Figure 2. Surface area in different activation time

Based on Figure 2, an increase of activation time corresponding to an increase of surface area proves that the formation of pores still happens up to the highest increase of activation time. A comparison of the activation time of 120 minutes with activation time of 60 minutes' shows that there's a significant increase in surface area. In contrast, there is no significant increase of surface area if a comparison was to be made between 180 minutes of activation time with 240 minutes of activation times; this signals

IOP Conf. Series: Materials Science and Engineering **176** (2017) 012055 doi:10.1088/1757-899X/176/1/012055

that the activation time is nearing its optimum value hence it is concluded that the optimum value of activation time to create activated carbon with the highest surface area is 240 minutes. From the results, it can be seen that the optimal value of operating conditions in terms of creating activated carbon with the main parameter of surface area is at temperature of 975°C, carbon dioxide gas flow rate of 200 ml/min, and activation time of 240 minutes that created a surface area of 1591,72 m^2/g.

3.2.3. SEM (Scanning Electron Microscopy) Characterization

SEM characterization of the activated carbon aims to look at the structural morphology of activated carbon as well as see the pores formed in the activated carbon. Characterization is done on carbon and carbonization results from the variation of the activation time to see the difference between the results of carbonization and physical activation. The results of the characterization of the samples can be seen in Figure 3a, 3b, 3c and 3d, with the magnification of FE-SEM is 5000 X.

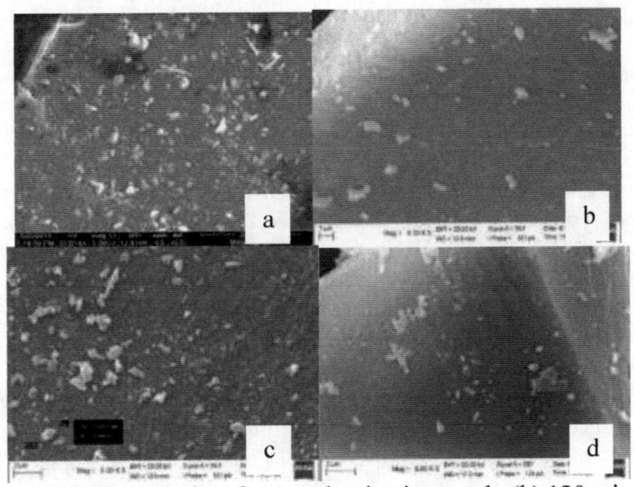

Figure 3. FE-SEM characterization results of; (a) carbonization result, (b) 120 minutes activation time, (c) 180 minutes activation time, (d) 240 minutes activation time

It can be seen in the results of FE-SEM characterization of the activated carbon samples that they do not show a lot of macro pores and are dominated by micro pores. This is due to the physical activation done using CO_2 gas in which the CO_2 gas maintained the condition of activated carbon by making it avoid O_2 gas that can cause burning and turn the activated charcoal into ash. CO_2 also reacts with carbon to form pores in activated carbon and increase the surface area of the activated carbon.

It can be seen that on activated carbon with an activation time of 120 minutes there's little to no impurities that can be seen on the carbon from carbonization which proves that the sample has reacted with CO_2 and open up the pores on the sample and also the higher heat in the activation process aids in removing impurities that are still trapped in the carbon sample. This happens because at a higher temperature, combustion process occurs better than at low temperatures. In the activated carbon with an activation time of 180 minutes and 240 minutes, it can be seen that there is a change on the surface of activated carbon in the sense of how it is more rugged than the activated carbon with a time of 120 minutes. This indicates that it has formed pores in activated carbon.

3.2.4. EDX (Energy-Dispersive X-Ray Spectroscopy) Characterization

EDX characterization aims to determine the composition of activated carbon made in this study. Figure 4 shows that the resulting activated carbon holds elements other than carbon such as oxygen, sulfur, iron, and copper. The carbon content of raw materials is 63% whereas the activated carbon with optimal activation time reached 83.2%, which means an increase of 20.2%. This indicates that the volatile matter has been greatly reduced, causing the formation of pores that provide increased surface area on the activated carbon.

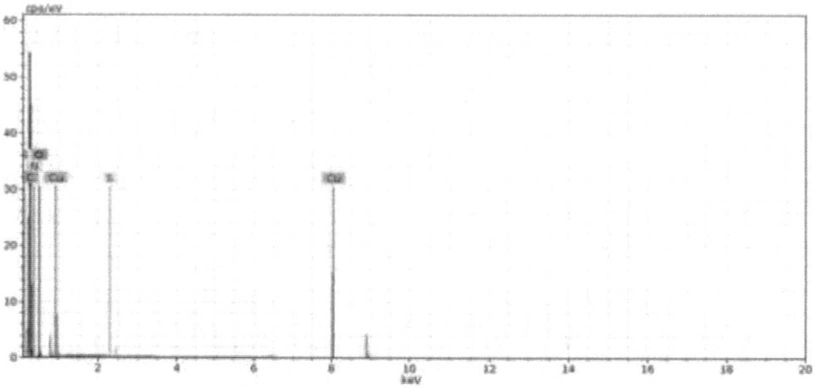

Figure 4. EDAX result of activated carbon with 240 minutes activation time

The physical activation treatment with CO_2 can significantly increase the carbon composition, which is about 7.4% when compared with activated carbon with the activation time of 240 minutes. This indicates that the reaction between carbon dioxide and carbon opened the pores on the sample. It can be seen in the results of EDX characterization which found that the carbon content in the activated carbon that is made has a corresponding value where, according to Duong (2008), the elements contained in the activated carbon equal to 85-95% of carbon and other elements are hydrogen, nitrogen, sulfur and oxygen. The iron contained in the activated carbon is a result of the physical activation container tube peeling off and going into the activated carbon, while the copper contained in the activated carbon is caused by the EDX testing sample container containing copper.

4. Conclusion

It can be concluded from the study entitled Preparation of Activated Carbon from Waste Plastics Polyethylene Terephthalate as Adsorbent Natural Gas Storage that:
The carbonization process obtained a yield of 21.47% and the physical activation process obtained a yield of 62%. Overall yield of 13.3% was obtained.Increasing the flow rate of CO_2 can increase the surface area of activated carbon produced. The optimum flow rate of CO_2 obtained at 200 ml/min.
Increasing activation time can increase the surface area of activated carbon produced. The maximum activation time is 240 minutes.The highest surface area was obtained when the gas flow rate of $CO_2$is 200 ml/min and activation time is 240 minutes, which is1591,72 m^2/g.Aside from carbon, other elements that resided within the activated carbon is oxygen, iron, and copper.

References

[1] Do D D 1998 *Adsorption Analysis: Equilibria and Kinetics* (London: Imperial College Press)
[2] Esfandiari A, Kaghazchi T and Soleimani M 2012 *J. Taiwan. Inst. Che. Eng.* **43** 631–637
[3] Lian F, Xing B and Zhu L 2011 *J. Colloid. Interface. Sci.***360** 725–730
[4] Melendi-Espina S, Alvarez R, Diez M A and Casal M D 2015 *Fuel. Proces. Technol.* **13** 351-358
[5] Rachmawati Q and Herumurti W 2015 *Jurnal. Teknik.ITS.***4**
[6] Samid D D 2011 *Optimasi Karakteristik Karbon Aktif Pada Sistem Adsorbed Natural Gas Dalam Kondisi Dinamis* (Depok, West Java: Universitas Indonesia)
[7] Yuliusman, Widodo W P, Yulianto S N and Rany A 2014 *The 9th Joint Conf. Che.* 130-134
[8] Yuliusman, Widodo W P, Yulianto S N 2015 *Int. J. Technol.* **6** 492-503

Institute of Physics
Dirac House, Temple Back
Bristol BS1 6BE UK

ISSN: 1757-8981
ISBN 978-1-5108-3744-7

CURRAN ASSOCIATES INC.
proceedings.com